TABLE 3b. Cumulative Distribution of S

Cumulative probability (shaded area) $\Phi(z) = Pr(Z \leq z)$ for positive z

z	0.00	0.01	0.02	0.03	0.04	0.05	0.06	0.07	0.08	0.09
0.0	0.50000	0.50399	0.50798	0.51197	0.51595	0.51994	0.52392	0.52790	0.53188	0.53586
0.1	0.53983	0.54380	0.54776	0.55172	0.55567	0.55962	0.56356	0.56749	0.57142	0.57535
0.2	0.57926	0.58317	0.58706	0.59095	0.59483	0.59871	0.60257	0.60642	0.61026	0.61409
0.3	0.61791	0.62172	0.62552	0.62930	0.63307	0.63683	0.64058	0.64431	0.64803	0.65173
0.4	0.65542	0.65910	0.66276	0.66640	0.67003	0.67364	0.67724	0.68082	0.68439	0.68793
0.5	0.69146	0.69497	0.69847	0.70194	0.70540	0.70884	0.71226	0.71566	0.71904	0.72240
0.6	0.72575	0.72907	0.73237	0.73565	0.73891	0.74215	0.74537	0.74857	0.75175	0.75490
0.7	0.75804	0.76115	0.76424	0.76730	0.77035	0.77337	0.77637	0.77935	0.78230	0.78524
0.8	0.78814	0.79103	0.79389	0.79673	0.79955	0.80234	0.80511	0.80785	0.81057	0.81327
0.9	0.81594	0.81859	0.82121	0.82381	0.82639	0.82894	0.83147	0.83398	0.83646	0.83891
1.0	0.84134	0.84375	0.84614	0.84849	0.85083	0.85314	0.85543	0.85769	0.85993	0.86214
1.1	0.86433	0.86650	0.86864	0.87076	0.87286	0.87493	0.87698	0.87900	0.88100	0.88298
1.2	0.88493	0.88686	0.88877	0.89065	0.89251	0.89435	0.89617	0.89796	0.89973	0.90147
1.3	0.90320	0.90490	0.90658	0.90824	0.90988	0.91149	0.91308	0.91466	0.91621	0.91774
1.4	0.91924	0.92073	0.92220	0.92364	0.92507	0.92647	0.92785	0.92922	0.93056	0.93189
1.5	0.93319	0.93448	0.93574	0.93699	0.93822	0.93943	0.94062	0.94179	0.94295	0.94408
1.6	0.94520	0.94630	0.94738	0.94845	0.94950	0.95053	0.95154	0.95254	0.95352	0.95449
1.7	0.95543	0.95637	0.95728	0.95818	0.95907	0.95994	0.96080	0.96164	0.96246	0.96327
1.8	0.96407	0.96485	0.96562	0.96638	0.96712	0.96784	0.96856	0.96926	0.96995	0.97062
1.9	0.97128	0.97193	0.97257	0.97320	0.97381	0.97441	0.97500	0.97558	0.97615	0.97670
2.0	0.97725	0.97778	0.97831	0.97882	0.97932	0.97982	0.98030	0.98077	0.98124	0.98169
2.1	0.98214	0.98257	0.98300	0.98341	0.98382	0.98422	0.98461	0.98500	0.98537	0.98574
2.2	0.98610	0.98645	0.98679	0.98713	0.98745	0.98778	0.98809	0.98840	0.98870	0.98899
2.3	0.98928	0.98956	0.98983	0.99010	0.99036	0.99061	0.99086	0.99111	0.99134	0.99158
2.4	0.99180	0.99202	0.99224	0.99245	0.99266	0.99286	0.99305	0.99324	0.99343	0.99361
2.5	0.99379	0.99396	0.99413	0.99430	0.99446	0.99461	0.99477	0.99492	0.99506	0.99520
2.6	0.99534	0.99547	0.99560	0.99573	0.99585	0.99598	0.99609	0.99621	0.99632	0.99643
2.7	0.99653	0.99664	0.99674	0.99683	0.99693	0.99702	0.99711	0.99720	0.99728	0.99736
2.8	0.99744	0.99752	0.99760	0.99767	0.99774	0.99781	0.99788	0.99795	0.99801	0.99807
2.9	0.99813	0.99819	0.99825	0.99831	0.99836	0.99841	0.99846	0.99851	0.99856	0.99861
3.0	0.99865	0.99869	0.99874	0.99878	0.99882	0.99886	0.99889	0.99893	0.99896	0.99900
3.1	0.99903	0.99906	0.99910	0.99913	0.99916	0.99918	0.99921	0.99924	0.99926	0.99929
3.2	0.99931	0.99934	0.99936	0.99938	0.99940	0.99942	0.99944	0.99946	0.99948	0.99950
3.3	0.99952	0.99953	0.99955	0.99957	0.99958	0.99960	0.99961	0.99962	0.99964	0.99965
3.4	0.99966	0.99968	0.99969	0.99970	0.99971	0.99972	0.99973	0.99974	0.99975	0.99976
3.5	0.99977	0.99978	0.99978	0.99979	0.99980	0.99981	0.99981	0.99982	0.99983	0.99983
3.6	0.99984	0.99985	0.99985	0.99986	0.99986	0.99987	0.99987	0.99988	0.99988	0.99989
3.7	0.99989	0.99990	0.99990	0.99990	0.99991	0.99991	0.99992	0.99992	0.99992	0.99992
3.8	0.99993	0.99993	0.99993	0.99994	0.99994	0.99994	0.99994	0.99995	0.99995	0.99995
3.9	0.99995	0.99995	0.99996	0.99996	0.99996	0.99996	0.99996	0.99996	0.99997	0.99997
4.0	0.99997	0.99997	0.99997	0.99997	0.99997	0.99997	0.99998	0.99998	0.99998	0.99998

DUXBURY

Statistics *for the* Sciences

Martin Buntinas
Gerald M. Funk

Loyola University Chicago

BROOKS/COLE

Australia • Canada • Mexico • Singapore • Spain • United Kingdom • United States

THOMSON
BROOKS/COLE

Statistics Editor: *Carolyn Crockett*
Assistant Editor: *Ann Day*
Editorial Assistants: *Julie Bliss,*
    Rhonda Letts
Technology Project Manager: *Burke Taft*
Marketing Manager: *Joseph Rogove*
Marketing Assistant: *Jessica Perry*
Advertising Project Manager:
    Tami Strang

Project Manager, Editorial Production:
    Hal Humphrey
Print/Media Buyer: *Jessica Reed*
Permissions Editor: *Sommy Ko*
Production Service: *Summerlight Creative*
Copy Editor: *Steven M. Summerlight*
Cover Image: *Todd Daman*
Compositor: *Martin Buntinas*
    and Gerald M. Funk

COPYRIGHT © 2005 Brooks/Cole, a division of Thomson Learning, Inc. Thomson Learning™ is a trademark used herein under license.

ALL RIGHTS RESERVED. No part of this work covered by the copyright hereon may be reproduced or used in any form or by any means—graphic, electronic, or mechanical, including but not limited to photocopying, recording, taping, Web distribution, information networks, or information storage and retrieval systems—without the written permission of the publisher.

Printed in the United States of America

1 2 3 4 5 6 7   07 06 05 04 03

For more information about our products,
contact us at:
**Thomson Learning
Academic Resource Center
1-800-423-0563**

For permission to use material from this text,
contact us by:
Phone: 1-800-730-2214
Fax: 1-800-730-2215
Web: http://www.thomsonrights.com

ISBN 0-534-38774-8

**Brooks/Cole—Thomson Learning
10 Davis Drive
Belmont, CA 94002
USA**

Asia
Thomson Learning
5 Shenton Way #01-01
UIC Building
Singapore 068808

Australia/New Zealand
Thomson Learning
102 Dodds Street
Southbank, Victoria 3006
Australia

Canada
Nelson
1120 Birchmount Road
Toronto, Ontario M1K 5G4
Canada

Europe/Middle East/Africa
Thomson Learning
High Holborn House
50/51 Bedford Row
London, WC1R 4LR
United Kingdom

Latin America
Thomson Learning
Seneca, 53
Colonia Polanco
11560 Mexico D.F.
Mexico

Spain/Portugal
Paraninfo
Calle Magallanes, 25
28015 Madrid
Spain

Duxbury Titles of Related Interest

Daniel, *Applied Nonparametric Statistics,* 2nd ed.

Derr, *Statistical Consulting: A Guide to Effective Communication*

Durrett, *Probability: Theory and Examples,* 2nd ed.

Graybill, *Theory and Application of the Linear Model*

Johnson, *Applied Multivariate Methods for Data Analysts*

Kuehl, *Design of Experiments: Statistical Principles of Research Design and Analysis,* 2nd ed.

Larsen, Marx, & Cooil, *Statistics for Applied Problem Solving and Decision Making*

Lohr, *Sampling: Design and Analysis*

Lunneborg, *Data Analysis by Resampling: Concepts and Applications*

Minh, *Applied Probability Models*

Minitab, Inc., *MINITAB^{TM} Student Version 12 for Windows*

Myers, *Classical and Modern Regression with Applications,* 2nd ed.

Newton & Harvill, *StatConcepts: A Visual Tour of Statistical Ideas*

Ramsey & Schafer, *The Statistical Sleuth,* 2nd ed.

SAS Institute Inc., *JMP-IN: Statistical Discovery Software*

Savage, INSIGHT: *Business Analysis Software for Microsoft® Excel*

Scheaffer, Mendenhall, & Ott, *Elementary Survey Sampling,* 5th ed.

Shapiro, *Modeling the Supply Chain*

Winston, *Simulation Modeling Using @RISK*

To order copies, contact your local bookstore or call 1-800-354-9706. For more information, contact Duxbury Press at 10 Davis Drive, Belmont, CA 94002, or go to www.duxbury.com.

Contents

PREFACE xi

1 WHAT IS STATISTICS?   1
    1.1   SAMPLING AND ESTIMATION 1
    1.2   SIMULATION 3
    1.3   ERROR ANALYSIS 5
    1.4   PRECISION VERSUS CONFIDENCE 9
    1.5   HOW HYPOTHESES ARE TESTED 12
    1.6   PREVIEW 13
            REVIEW EXERCISES 15

2 HOW TO DESCRIBE AND SUMMARIZE DATA   19
    2.1   VARIABLES AND DATA SETS 20
    2.2   CATEGORICAL DATA 22
    2.3   ORDINAL DATA 25
    2.4   RATIO DATA 26
    2.5   FREQUENCY TABLES AND HISTOGRAMS 27
    2.6   GROUPED DATA AND STURGE'S RULE 31
    2.7   STEM-AND-LEAF PLOT 35
    2.8   FIVE-NUMBER SUMMARY 37
    2.9   BOX PLOT 40

2.10	THE MEAN	42
2.11	VARIANCE AND STANDARD DEVIATION	45
2.12	OGIVES AND QUANTILES	52
2.13	EXPLORATORY DATA ANALYSIS	56
2.14	FORMULAS FOR HISTOGRAMS (optional)	64
	REVIEW EXERCISES	67

3 PROBABILITY — 75

- 3.1 OVERVIEW . . . 75
- 3.2 DEFINITIONS . . . 78
- 3.3 PROBABILITIES OF EVENTS . . . 83
  - 3.3.1 THEORETICAL PROBABILITY . . . 83
  - 3.3.2 EMPIRICAL PROBABILITY . . . 86
  - 3.3.3 MONTE CARLO METHOD (optional) . . . 88
  - 3.3.4 SUBJECTIVE PROBABILITY . . . 89
- 3.4 RULES OF PROBABILITY . . . 90
  - 3.4.1 ADDITION RULES . . . 92
  - 3.4.2 INDEPENDENCE AND CONDITIONAL PROBABILITY . . . 97
- 3.5 TREE DIAGRAMS . . . 105
- 3.6 BAYES' METHOD . . . 107
- REVIEW EXERCISES . . . 114

4 DISCRETE RANDOM VARIABLES — 121

- 4.1 INTRODUCTION . . . 121
- 4.2 BASIC PROPERTIES . . . 123
- 4.3 PROBABILITY HISTOGRAMS . . . 128
- 4.4 EXPECTED VALUE OR MEAN . . . 132
- 4.5 FUNCTIONS OF RANDOM VARIABLES . . . 133

4.6	VARIANCE AND STANDARD DEVIATION	135
4.7	DISCRETE UNIFORM RANDOM VARIABLES	138
4.8	LAW OF LARGE NUMBERS (optional)	140
	REVIEW EXERCISES	141

5 RANDOM VARIABLES FOR SUCCESS/FAILURE EXPERIMENTS 143

5.1	BERNOULLI RANDOM VARIABLES	143
5.2	BINOMIAL RANDOM VARIABLES	147
5.3	HYPERGEOMETRIC RANDOM VARIABLES	155
	REVIEW EXERCISES	160

6 INTRODUCTION TO HYPOTHESIS TESTING 163

6.1	OVERVIEW	164
6.2	TWO TYPES OF ERROR	164
6.3	THE SIGN TEST	170
6.4	BINOMIAL EXACT TEST	176
6.5	TRIVIAL EFFECT CAN BE SIGNIFICANT	185
6.6	FISHER'S EXACT TEST (optional)	186
	REVIEW EXERCISES	191

7 CONTINUOUS RANDOM VARIABLES 195

7.1	BASIC PROPERTIES	195
7.2	PERCENTILES AND MODES	202
7.3	EXPECTED VALUE OR MEAN	205
7.4	FUNCTIONS OF RANDOM VARIABLES	207

7.5 VARIANCE AND STANDARD DEVIATION 208
7.6 CHEBYSHEV'S INEQUALITY 212
    REVIEW EXERCISES 215

8 NORMAL RANDOM VARIABLES 219

8.1 INTRODUCTION 219
8.2 NORMAL APPROXIMATION OF BINOMIAL 227
8.3 CONTINUITY CORRECTION 230
8.4 CENTRAL LIMIT THEOREM 233
8.5 PROCESSES THAT FOLLOW THE NORMAL CURVE . . 234
    REVIEW EXERCISES 236

9 WAITING TIME RANDOM VARIABLES 241

9.1 GEOMETRIC RANDOM VARIABLES 241
9.2 EXPONENTIAL RANDOM VARIABLES 246
9.3 POISSON RANDOM VARIABLES 251
9.4 POISSON APPROXIMATION OF BINOMIAL 255
9.5 POISSON AS INVERSE EXPONENTIAL 259
    REVIEW EXERCISES 261

10 TWO OR MORE RANDOM VARIABLES 267

10.1 JOINT RANDOM VARIABLES AND DENSITIES 267
10.2 INDEPENDENCE OF RANDOM VARIABLES 275
10.3 EXPECTATION, COVARIANCE, AND CORRELATION . 278
10.4 LINEAR COMBINATION OF RANDOM VARIABLES . . 284
     REVIEW EXERCISES 288

11 SAMPLING EXPERIMENTS AND THE LAW OF AVERAGES — 291

- 11.1 POPULATIONS AND PARAMETERS 292
- 11.2 SAMPLES AND STATISTICS 296
- 11.3 LAW OF AVERAGES FOR THE SAMPLE COUNT 299
- 11.4 LAW OF AVERAGES FOR THE SAMPLE SUM 305
- 11.5 LAW OF AVERAGES FOR THE SAMPLE PROPORTION 312
- 11.6 LAW OF AVERAGES FOR THE SAMPLE MEAN 316
- 11.7 THE z STATISTIC 320
- 11.8 THE T STATISTIC 321
- 11.9 ESTIMATORS OF PARAMETERS; ACCURACY AND PRECISION 323
- REVIEW EXERCISES 327

12 THE z AND t TESTS OF HYPOTHESES — 333

- 12.1 THE z TEST 333
- 12.2 TWO-SIDED z TEST 339
- 12.3 BOOTSTRAPPING AND THE t TEST 345
- 12.4 WHICH IS THE NULL HYPOTHESIS? 347
- REVIEW EXERCISES 349

13 ESTIMATION WITH CONFIDENCE — 351

- 13.1 DIFFERENCE BETWEEN CONFIDENCE AND PROBABILITY 351
- 13.2 TWO-SIDED CONFIDENCE INTERVALS 352
- 13.3 ONE-SIDED CONFIDENCE INTERVALS 356
- 13.4 BOOTSTRAPPING AND THE t CURVES 358
- 13.5 MARGIN OF ERROR AND SAMPLE SIZE 367
- 13.6 INTERVAL ESTIMATE OF PROPORTION 369

13.7	SMALL SAMPLE INTERVAL ESTIMATES OF PROPORTIONS: BINOMIAL EXACT INTERVALS	374
	REVIEW EXERCISES	380

14 TWO-SAMPLE INFERENCE — 385

14.1	MATCHED PAIR SAMPLES	385
14.2	INDEPENDENT SAMPLES	389
14.3	WELCH'S FORMULA	393
14.4	INDEPENDENT SAMPLES WITH EQUAL VARIANCES	396
	REVIEW EXERCISES	402

15 CORRELATION AND REGRESSION — 407

15.1	INTRODUCTION	407
15.2	SCATTER PLOTS	408
15.3	THE CORRELATION COEFFICIENT	411
15.4	FITTING A SCATTER PLOT BY EYE	418
15.5	THE REGRESSION LINE	423
15.6	ESTIMATION WITH REGRESSION	428
15.7	THE REGRESSION PARADOX	431
15.8	TESTING FOR CORRELATION	434
15.9	CORRELATION IS NOT CAUSATION	439
	REVIEW EXERCISES	441

16 INFERENCE WITH CATEGORICAL DATA — 443

16.1	INTRODUCTION	443
16.2	COMMENTS ON THE DEFINITION OF χ^2	447
16.3	TESTING GOODNESS OF FIT	450
16.4	CONTINGENCY TABLE TESTS	457

16.5	ONE-SIDED CHI-SQUARE TESTS FOR THE **2** × **2** CONTINGENCY TABLE	464
16.6	COMPOUND HYPOTHESES	467
	REVIEW EXERCISES	469

17 RESAMPLING METHODS   471

17.1	OVERVIEW	471
17.2	PARAMETRIC BOOTSTRAPPING	473
17.3	NONPARAMETERIC BOOTSTRAPPING	479
17.4	PERMUTATION TESTS	488
17.5	COMPUTER-ASSISTED PARAMETRIC RESAMPLING	491
17.6	COMPUTER-ASSISTED NONPARAMETRIC RESAMPLING	495
	REVIEW EXERCISES	501

STATISTICAL TABLES   503

ANSWERS TO SELECTED EXERCISES   515

INDEX   528

PREFACE

This book is an introduction to statistics for students who have completed at least one quarter or semester of calculus. This includes students of mathematics; statistics; computer science; the physical, biological, and social sciences; economics; and management science.

Statistics is the body of knowledge that deals with inductive reasoning. In our everyday life, we use "common sense" to deal with situations where we have to act without having all of the facts needed to make decisions with complete certainty. Statistics is "common sense" raised to the level of a formal study.

A decision on whether a vaccine is effective, or whether an electrical switch will last the life of an appliance, or whether a new teaching method improves reading skills of children cannot be based on wishful thinking or haphazard judgments. Ideally, sources of variability are identified, a controlled experiment is performed, and the plausibility of various conclusions are measured and assessed before a decision is made.

Yet no discipline is misused and misunderstood as much as statistics. We see evidence of this every day in the media and even in scientific papers. Nowadays, running data through statistical software is easy. **But understanding the purposes and the limitations of statistical procedures as well as the correct interpretation of the results is essential to any meaningful application.**

Thus, an understanding of statistical principles is essential for nearly all science and mathematics programs. Elementary statistics courses taught to nonscience majors do not integrate the student's knowledge of calculus and would not satisfy this need. Generally, the only other option is a course in mathematical statistics, which typically consists of one semester of probability followed by one semester of statistics. To include this in the curriculum of a typical science major, could only be done at the expense of some other requirement and would thus distort the balance in the program. Consequently, most mathematics and science programs cannot include statistics in the required part of their curricula.

In teaching our course, we struggled for some time with picking topics from various two-term textbooks. Neither we nor our students found this satisfactory. We decided that a new kind of textbook was needed that was specifically designed for a calculus-based one-semester or one-quarter statistics course. This is the result. The book has been extensively tested by several instructors over several years. Most of the book can be covered in a one-term course intended for science and mathematics students.

KEY FEATURES

Inference before midterm. The traditional format of books at this level is to develop the necessary mathematical probability tools before introducing statistical inference. Unfortunately, postponing this statistical inference to the end of the term may have negative consequences in that often what is learned last is first forgotten. The approach of this book is to introduce statistical inference concepts early and often. These ideas are reinforced by repeatedly reworking old ground at ever-increasing levels of awareness.

Statistical inference is introduced in the first chapter. Then, because students have been introduced to error analysis, precision, and confidence, it is possible to discuss standard error as early as Chapter 2. This allows students the time to get comfortable working with standard errors and related concepts well before the formal study of sampling experiments in Chapters 11.

Immediately after binomial and hypergeometric probability models have been studied in Chapter 5, the formal vocabulary and structure of statistical hypothesis tests are introduced in Chapter 6 in the context of success–failure experiments. We start with single-sample experiments, extend to paired samples, and then to two-sample experiments. Tests such as the **sign test, binomial exact test**, and **Fisher's exact test** are developed here. We think that this approach is the best one for science students for the following reasons:

1. The road from the Introduction to Inference is as short as possible. Students see real statistics using real data without having to endure lengthy and unmotivated theory.

2. Procedures for these tests are intuitive and easily understood. Students thus find the resulting test decisions convincing.

3. Statistical inference at this stage is based only on the elementary binomial and hypergeometric probability models. The Central Limit Theorem is not needed until later (Chapter 8).

4. All of this can be completed well before midterm. In our approach, the more commonly used z and t tests are introduced in Chapter 12. By this time, many of the basics of inference are already familiar to the students. This approach also allows for a more complete discussion of confidence interval methods in Chapter 13 than is found in competing texts. For example, we discuss small-sample confidence intervals for proportions and explicitly discuss the coverage curves associated with interval estimation procedures.

5. Finally, in Chapter 17 we revisit inference through the methods of **resampling**.

And all of this can be accomplished in a single term.

Instructors who prefer to use the traditional approach of completing the study of probability models and large sample distributions before beginning the study of formal statistical inference can easily accomplish this by postponing Chapter 6 until after the Central Limit Theorems, normal approximations (Chapter 8), and sampling distributions of statistics (Chapter 11) have been completed.

Students who have completed a "short" or reform calculus course are prepared for the calculus used in the book. Multivariate calculus and linear algebra are not required. For example, Chapter 10, which covers two or more random variables, can be understood without a background in multivariate calculus; the discussion is restricted to discrete random variables with the explanation that all results apply to the continuous case as well. The statistical ideas covered in this book are not too different from those in introductory texts for students who are not mathematically prepared. We do not believe that it is necessary to have a calculus background to understand these statistical ideas. Rather, we believe that students who have a calculus background should have the opportunity to see statistics in this setting. Calculus is used to build standard models. We give our readers the chance to hone their skills by exploring simple probability models used in statistical inference.

Examples are targeted for a science curriculum. The order of presentation in this book is generally to start each new topic with examples that lead to a discussion of new concepts, definitions, and specific results. These are then followed by a more formal treatment of the topic. There are over 150 worked-out examples and over 700 exercises in this book. Almost half of the exercises have answers in the back of the book, and a *Student's Solution Manual* is available. A CD-ROM of all solutions is available to instructors. We try to structure them on science themes. Many are based on laboratory measurements and error analysis. Bayes' Theorem is explained in context of diagnosis of illness, forecasting weather, and predicting the economy. Fisher's exact test and the binomial exact test are presented in the context of public health experiments and drug trials. Tests of hypotheses involve computer programs, physics, chemistry, archaeology, and psychology.

Sections that involve worked-out examples are generally followed by section exercises. At the end of each chapter are review exercises. Nearly half of the exercises have answers in the back of the book. Furthermore, solutions to selected exercises are available on the website

http://www.stat.luc.edu/StatisticsForTheSciences

Exercises are marked by symbols as follows:

-     indicates that there is an answer in the back of the book.

★   indicates a thought-provoking question.

C   involves programming a graphing calculator or computer.

The separator

$$* * * * * * * * *$$

is followed by exercises that go beyond the basics.

Probability theory and theories of inference have been kept to a minimum. Our goal is to develop statistical intuition and statistical common sense. We have omitted topics such as moment-generating functions and likelihood functions because they would seriously reduce the time one can spend on statistical inference and exploratory data analysis. We explore relatively few probability models so that sufficient time can be spent in their development.

Technology is used, but the presentation is independent of specific software or technology. It is assumed that students have access to a graphing calculator such as the TI-83 Plus[1], statistical software such as Minitab[2], or a spreadsheet such as Microsoft Excel[3]. In Chapter 17 on resampling methods, Maple[4] is also used. However, the treatment in the book is independent of the technology used. Subsections entitled **Notes on Technology** are included next to exercise sets throughout the text to give students instructions on how to use Minitab, Excel, and the TI-83 Plus graphing calculator. This frees the instructor from having to lecture on which buttons to press.

---

[1] TI-83 Plus is a registered trademark of Texas Instruments Incorporated.
[2] Minitab is a registered trademark of Minitab Incorporated.
[3] Microsoft Excel is a registered trademark of Microsoft Corporation.
[4] Maple is a registered trademark of Waterloo Maple Incorporated.

Graphing-calculator technology or computer software is used to simulate models and to explore the relationships between data and models.

Computer simulations and computer-drawn graphs are used to illustrate sampling distributions. Emphasis is on data analysis with careful attention paid to the assumptions behind statistical inference. Some of the data sets we use are available to the students online at **StatLib**, hosted by the Statistics Department at Carnegie Mellon University, and in particular at **The Data and Story Library, DASL**. Other data sets are available online through various federal government agencies. We also use the data in Paul Meier's essay *The Biggest Public Health Experiment Ever: The 1954 Field Trial of the Salk Poliomyelitis Vaccine*, which is available on the Web at http://www.stat.luc.edu, and other data drawn from various journal and newspaper articles.

This book is a gateway to further courses in statistics. We hope that this text will motivate some students to take additional courses in statistics. Students who complete a course from this textbook should be ready for most courses in applied statistics, such as regression analysis, design and analysis of statistical experiments, stochastic processes, and data analysis. Furthermore, this course will complement research methods courses in various disciplines.

The syllabus is flexible. There is flexibility in the syllabus of a course using this textbook. Most of the book can be covered in one semester.

- Some sections can be skipped and others emphasized at the instructor's preference. Usually, the last section of each chapter can be covered or skipped at the option of the instructor. Some interior sections are also marked as optional.

- Chapter 6 introduces testing of hypotheses immediately after the study of binomial and hypergeometric random variables. For those who prefer a more traditional order, these tests—including the sign, binomial exact, and Fisher's exact tests—can be presented at the same time as the z and t tests in Chapter 12. Also, Chapter 16 (Inference with Categorical Data) can precede Chapter 15 (Correlation and Regression).

- Chapter 9 (**Waiting Time Random Variables**) on probability models involving Poisson and exponential random variables can be postponed to the end of the term.

- Time spent on Chapter 10 **Two or More Random Variables** can be reduced by discussing only the boxed results of Section 10.4 along with some hand waving.

Acknowledgments. A special thanks to E. N. Barron, Anthony Giaquinto, Steven L. Jordan, Richard Maher, Henry Park, Hans-Juergen Petersen, and Alan Saleski as well as their students for classroom testing earlier versions of this text and for providing numerous suggestions. We also wish to thank the following reviewers for their numerous insights and suggestions: Linda Brant Collins, University of Texas at San Antonio; Jack Tubbs, University of Arkansas; Guang-Hwa (Andy) Chang, Youngstown State University; and R. B. Campbell, University of Northern Iowa. Finally, many thanks to Carolyn Crockett, Kipp Blackburn, Joseph Rogove, S. M. Summerlight, Ann Day, and others on the editorial staff of Duxbury Press for their encouragement and support during the writing of this book.

Chapter 1

WHAT IS STATISTICS?

1.1 SAMPLING AND ESTIMATION

1.2 SIMULATION

1.3 ERROR ANALYSIS

1.4 PRECISION VERSUS CONFIDENCE

1.5 HOW HYPOTHESES ARE TESTED

1.6 PREVIEW

Commonly, the word *statistics* means the arranging of data into charts, tables, and graphs along with the computations of various descriptive numbers about the data. This is a part of statistics, called **descriptive statistics**, but it is not the most important part. The most important part is concerned with reasoning in an environment where one does not know, or cannot know, all of the facts needed to reach conclusions with complete certainty. One deals with judgments and decisions in situations of incomplete information. In this chapter, we will give an overview of statistics along with an outline of the various topics in this book.

1.1 SAMPLING AND ESTIMATION

Let's begin with an example.

Example 1.1   *Harris Poll.*   Louis Harris and Associates[1] conducts polls on various topics, either face-to-face, by telephone, or by the Internet. In one survey

---

[1] www.harrisinteractive.com

on health trends of adult Americans conducted in 1991, surveyors contacted 1256 randomly selected adults by phone and asked them questions about diet, stress management, seat belt use, and so on. One of the questions asked was "Do you try hard to avoid too much fat in your diet?" Harris reported that 57% of the people responded Yes to this question. The article stated that the margin of error of the study was plus or minus 3%.

This is an example of an inference made from incomplete information. The group under study in this survey is the collection of adult Americans, which consists of more than 200 million people. This is called the **population**. If every individual of this group were to be queried, the survey would be called a **census**. Yet of the millions in the population, the Harris survey examined only 1256 people. Such a subset of the population is called a **sample**.

Once every ten years, the U.S. Bureau of the Census conducts a survey of the entire U.S. population. The year 2000 census cost the government billions of dollars. For the purposes of following health trends, it's not practical to conduct a census. It would be too expensive, too time-consuming, and too intrusive into people's lives. We shall see as we progress through this text that, if done carefully, 1256 people are sufficient to make reasonable estimates of the opinion of all adult Americans. Samuel Johnson was aware that there is useful information in a sample. He said that you don't have to eat the whole ox to know that the meat is tough.

The people or things in a population are called **units**. If the units are people, they are sometimes called **subjects**. A characteristic of a unit (such as a person's weight, eye color, or the response to a Harris Poll question) is called a **variable**. If a variable has only two possible values (such as a response to a Yes or No question, or a person's sex) it is called a **dichotomous variable**. If a variable assigns a number to each individual (such as a person's age, family size, or weight), it is called a **quantitative variable**.

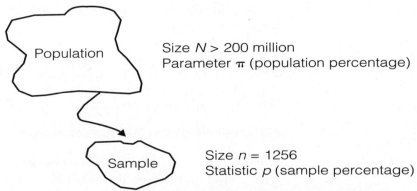

Figure 1.1 Parameter and statistic for a dichotomous variable

A number derived from a *sample* is called a **statistic**, whereas a number derived from the *population* is called a **parameter**. Parameters are usually denoted by Greek letters, such as π, for population percentage of a dichotomous variable, or μ, for population mean of a quantitative variable. For the Harris study, the **sample percentage** $p = 57\%$ is a statistic. It is not the (unknown) **population percentage** π, which is the percentage that we would obtain if it were possible to ask the same question of the entire population.

Inferences we make about a population based on facts derived from a sample are uncertain. The statistic p is not the same as the parameter π. In fact, if the study had been repeated, even if it had been done at about the same time and in the same way, it most likely would have produced a different value of p, whereas π would still be the same. The Harris study acknowledges this variability by mentioning a margin of error of $\pm 3\%$.

How can the researchers say that the margin of error is plus or minus 3% when such a small sample of all adult Americans was contacted? This is one of the questions that we will deal with in this book.

1.2 SIMULATION

Consider a box containing chips or cards, each of which is numbered either 0 or 1. We want to take a random sample from this box in order to estimate the percentage of the cards that are numbered with a 1. The population in this case is the box of cards, which we will call the **population box**. The percentage of cards in the box that are numbered with a 1 is the parameter π. In the Harris study, the parameter π is unknown. Here, however, in order to see how samples behave, we will make our model with a known percentage of cards numbered with a 1—say, $\pi = 60\%$. At the same time, we will estimate π, pretending that we do not know its value, by examining 25 cards in the box.

We take a **simple random sample with replacement** of 25 cards from the box as follows. Mix the box of cards, choose one at random, record it, replace it, and then repeat the procedure until we have recorded the numbers on 25 cards. Although survey samples are not generally drawn *with replacement*, our simulation simplifies the analysis because the box remains unchanged between draws; so, after examining each card, the chance of drawing a card numbered 1 on the following draw is the same as it was for the previous draw—in this case a 60% chance. Let's say that, after drawing the 25 cards this way, we obtain the following results, recorded in 5 rows of 5 numbers:

$$\begin{array}{ccccc}
0 & 1 & 1 & 1 & 1 \\
1 & 0 & 1 & 1 & 0 \\
1 & 0 & 1 & 0 & 1 \\
0 & 0 & 0 & 0 & 1 \\
1 & 0 & 1 & 0 & 1
\end{array}$$

Based on this random sample of 25 draws, we want to guess the percentage of 1's in the box. There are 14 cards numbered 1 in the sample. This gives us a sample percentage of $p = 14/25 = 0.56 = 56\%$. If this is all of the information we have about the population box, and we want to estimate the percentage of 1's in the box, our best guess would be 56%. Notice that this sample value $p = 56\%$ is 4 percentage points below the true population value $\pi = 60\%$. We say that the **random sampling error** (or simply **random error**) is -4%.

Equivalently, instead of using a box of cards, we can simulate this experiment by generating random numbers using a programmable calculator, such as the **TI-83 Plus**, a computer program, such as **Excel** or **Minitab**, or a table of random digits such as Table 1 on page 503. In later chapters, we will give instructions for using technology in sections marked **Notes on Technology**, but for now we will use a table of random digits to demonstrate the procedure. For convenience, the random digits in Table 1 on page 503 are grouped into fives. Each of the 45 rows holds 10 groups of fives. The entire table holds $45 \times 10 \times 5 = 2250$ random digits. To guard against using the same random numbers in every simulation, we should start in a random row and column—say, row 16, column 6. We can obtain a random starting point by tossing a paper clip or coin on the page. The first 25 digits in row 16, starting with column 6, are

99109   14827   24949   16210   95105

Because we want a 60% chance of drawing a 1, we assign the value 1 to the six random digits $0, 1, 2, 3, 4, 5$ and assign the value 0 to the four digits $6, 7, 8, 9$; this results in the following values:

00110   11010   11010   10111   01111

In this simulation, we ended up with 16 1's, which results in the statistic $p = 16/25 = 0.64 = 64\%$. The random sampling error is $+4\%$.

Continuing where we left off in row 16, the next 25 digits go into row 17 and result in

00100   00101   01000   11111   11110

This time we have 13 1's to obtain a statistic of $p = 52\%$ and a random sampling error of -8%.

One more time:

11101   11110   11110   11101   01111

gives us 20 1's with $p = 80\%$, and a random error of $+20\%$.

The random errors were -4% from the box of cards and $+4\%, -8\%,$ and $+20\%$ from the table of random numbers.

1.3 ERROR ANALYSIS

An **experiment** is a procedure that results in a measurement or observation. The Harris Poll is an experiment that resulted in the measurement (statistic) of 57%. An experiment whose outcome depends upon chance is called a **random experiment**. On repetition of such an experiment, one will typically obtain a different measurement or observation. So, if the Harris Poll were to be repeated, the new statistic would very likely differ slightly from 57%. Each repetition is called an **execution** or **trial** of the experiment.

The four simulations above are trials of a random experiment that resulted in four different percentages. The random sampling errors of the four simulations average out to

$$\textbf{AV:} \quad \frac{-4\% + 4\% - 8\% + 20\%}{4} = +3\%$$

Note that the cancellation of the positive and negative random errors results in a small average. Actually, with more trials, the average of the random sampling errors tends to zero.

So in order to measure a "typical size" of a random sampling error, we have to ignore the signs. We *could* just take the **mean of the absolute values (MA)** of the random sampling errors. For the four random sampling errors above, the MA turns out to be

$$\textbf{MA:} \quad \frac{|-4\%| + |+4\%| + |-8\%| + |+20\%|}{4} = 9\%$$

The MA is difficult to deal with theoretically because the absolute value function is not differentiable at 0. So in statistics, and error analysis in general, the **root mean square (RMS)** of the random sampling errors is generally used. For the four random sampling errors above, the RMS is

$$\textbf{RMS:} \quad \sqrt{\frac{(-4\%)^2 + (+4\%)^2 + (-8\%)^2 + (+20\%)^2}{4}} = \sqrt{124}\,\% = 11.14\%$$

The RMS is a more conservative measure of the typical size of the random sampling errors in the sense that MA $\leq$ RMS (see Review Exercise 1.21 at the end of this chapter).

For a given experiment, the RMS of *all* possible random sampling errors is called the **standard error (SE)**. We will study the standard error more formally later in the book. For example, whenever we use a random sample of size n and its percentages p to estimate the population percentage π of a dichotomous population, we will show that

$$\text{SE}_p = \sqrt{\frac{\pi(1-\pi)}{n}} \leq \frac{1}{2\sqrt{n}} \qquad (1.1)$$

which for $n = 25$ comes out to $\text{SE} = 0.097979\ldots \leq 0.10$. Notice that the estimate $\text{SE}_p \leq 10\%$ is reasonably close to the RMS $= 11.14\%$ that we obtained in our four simulations. As the number of simulations increases, the RMS will tend to the SE. A proof of the inequality in Formula 1.1 is an exercise in calculus and is left to the reader as Review Exercise 1.13. This inequality is useful because, in a situation such as the Harris Poll, the value of the parameter π is generally unknown.

Notice from Formula 1.1 that, other things being equal, as the size n of the random sample goes up, the standard error goes down in proportion to the square root of the sample size. So for samples four times as large, the standard error will be $1/\sqrt{4} = 1/2$ as large. This means that, for samples of size 100, the standard error will be no more than approximately $\frac{1}{2} \times 10\% = 5\%$. If we combine our four samples in our simulations of size 25, we notice that we have a total of $14 + 16 + 13 + 20 = 63$ cards numbered 1, which gives us $p = 63\%$, and thus a random error of $+3\%$. Another sample of 100 would likely give us a different random sampling error.

Although the standard error will be studied in more detail in later chapters, at this point we want to observe an interesting property. The standard errors of independent trials of a random experiment have a Pythagorean relationship, just as do right triangles in geometry. If A is a measurement of a parameter α with a standard error SE_A, and B is an independent measurement of a parameter β with a standard error SE_B, then $A + B$ is a measurement of $\alpha + \beta$ with a standard error $\sqrt{\text{SE}_A^2 + \text{SE}_B^2}$. In error analysis, it is convenient to write

$$A = \alpha \pm \text{SE}_A$$

$$B = \beta \pm \text{SE}_B$$

and

$$A + B = \alpha + \beta \pm \sqrt{\text{SE}_A^2 + \text{SE}_B^2}$$

Note that the measurements add, but not the SE's. This property results in one of the important parts of the law of averages.

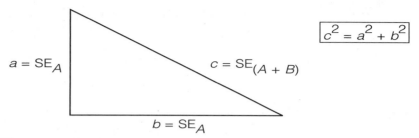

Figure 1.2 Pythagorean property of measurement error

To illustrate this relationship, let's say that a person commutes to work by car. She uses the car's trip odometer to measure the distance traveled. After many trips to work, she observes an average reading of 24.3 miles. Due to changes in traffic, weather, and other conditions, each trip results in a slightly different odometer reading, from which she estimates that the SE is approximately 0.4 miles. If A represents the distance measured on any given trip, we can write

$$A = 24.3 \pm 0.4 \text{ miles}$$

This does not mean that *every* trip varies by 0.4 miles; rather, it means that for such trips, odometer readings are approximately 24.3 miles and vary by about 0.4 miles, on the average.

She also often travels to a cottage on weekends. The distance to the cottage is greater than the commute, and the odometer readings are more variable because the trips involve stops for fuel and fast food. Let's say that an odometer reading B for a trip to the cottage has the formula

$$B = 67.2 \pm 1.1 \text{ miles}$$

A trip from work followed by a trip to the cottage will have a total odometer reading C, which can be calculated by adding the readings A and B. The Pythagorean relationship results in

$$\begin{aligned} A + B &= 24.3 + 67.2 \pm \sqrt{(0.4)^2 + (1.1)^2} \text{ miles} \\ C &= 91.5 \pm 1.2 \text{ miles} \end{aligned}$$

EXERCISES FOR SECTION 1.3

- **1.1** A medium apple has an average of 80 calories. However, the number of calories A varies with the size and type of apple. Let's say that the standard error is $SE_A = 10$ calories. Similarly, the number of calories B in a bran muffin is 400 calories with a standard error of $SE_B = 40$ calories. Let C be the total number of calories in a lunch consisting of a medium apple and a bran muffin.

    a. Find the estimated total number of calories of C along with the standard error SE_C.

    b. Do the same for a lunch that also includes a cup of yogurt, which has 220 calories, give or take 15 calories or so.

1.2 For a biology lab experiment, you have to randomly select four mice. The mice to be chosen come from a population bred to weigh 30 grams, give or take 5 grams or so.

The total weight T of the four mice will be _____ grams, give or take _____ grams or so. Use the Pythagorean property of standard errors.

1.3 Repeat the simulation described in this section by taking another four samples of size $n = 25$ from a population box with $\pi = 0.60$. Use either Table 1 of random numbers[2] or a calculator with a RAND function.[3]

    a. What average value of p do you get?

    b. What average value of the random sampling errors do you get?

    c. What MA of the random sampling errors do you get?

    d. What RMS of the random sampling errors do you get?

C 1.4 Use a computer program or a programmable calculator to simulate 100 samples of size $n = 25$ from the above population box with $\pi = 0.60$. The values of p will vary from simulation to simulation.

    a. What average value of p do you get?

    b. What average value of the random sampling errors do you get?

    c. What MA of the random sampling errors do you get?

    d. What RMS of the random sampling errors do you get?

Students doing this experiment will have varying results, but the RMS should be close to that predicted by Formula 1.1 on page 6 above.

C 1.5 Continuing with Exercise 1.4, do the same for samples of size $n = 100$.

C 1.6 Continuing with Exercise 1.4, do the same for samples of size $n = 1256$.

---

[2] Obtain a starting point by tossing a coin or paper clip on the table on page 503.

[3] The RAND function on a graphing calculator has to be initialized on its first use. You can use your Social Security number as a seed. Initializing is like choosing a random starting point on the table. Do this only once, otherwise the calculator will produce the same sequence of numbers every time.

1.4 PRECISION VERSUS CONFIDENCE

The "margin of error" of 3% mentioned in the Harris survey article is not the standard error. It should be more precisely stated as the margin of error for 95% confidence; that is, approximately 95% of the time, the random error will be within this "margin of error." In the news media, the 95% condition is generally assumed but not always explicitly mentioned. We will show in later chapters, that, for large random samples as in the Harris survey, this 95% margin of error is approximately 1.96 times (roughly *twice*) the standard error. Using the inequality in Formula 1.1 on page 6, we note that for $n = 1256$, we obtain $\text{SE}_p \leq \frac{1}{2\sqrt{1256}} = 0.0141 = 1.41\%$, which is close to half of the researchers' reported 3%. Because they obtained a statistic of $p = 57\%$, they state that they are 95% confident that $p = 57\%$ is within 3 percentage points of the true, but unknown, parameter π.

The margin of error is a measure of the **precision** of the estimate. However, it is inversely related to the level of **confidence** in this estimate. For a given sample size, if we want to *increase* confidence, we have to *decrease* the precision, and vice versa. For example, we can be 100% confident that π is somewhere between 0% and 100%. Although our confidence is high, the precision is so poor, it makes the statement worthless. On the other hand, we will show later that a margin of error of ±*three* standard errors, which gives us a precision of $\pm 3 \frac{1}{2\sqrt{1256}} = \pm 0.0423 = \pm 4.23\%$, will give us a statement with 99.7% confidence. And to get a statement with 99% confidence we need a margin of error of approximately ±2.6 standard errors, which is approximately ±3.6%. Using a margin of error of 1 standard error will result in a statement with only 68% confidence. At this point, do not worry about the specific percentages and levels of confidence. Details and derivations will come in later chapters.

> *In statistics, for a given data set, precision is inversely related to the level of confidence. However, by increasing the sample size both precision and level of confidence can be improved.*

This Harris Poll example shows how an estimate of an unknown parameter is made by examining a random sample. This is part of the study of **inferential statistics**. Our simulation, on the other hand, involves a population box whose content is known. Examining the different samples that are possible and likely from a population with a known parameter is part of **sampling theory**. An understanding of the behavior of samples from a known population is a prerequisite of inferential statistics.

Note on estimation errors

It is important to realize that random sampling error is not the only source of error in a survey.[4] In Chapter 2 on page 22, we comment on a news brief by Susan Hill, Senior Program Analyst at the National Science Foundation Division of Science Resource Studies. This news brief includes data on the number and types of degrees offered by postsecondary schools in the United States. The target population consists of all 2-year and 4-year institutions of higher learning in the United States. This survey is actually a census, because all institutions accredited to award degrees are queried in the initial mailing (close to 6000 in all). Follow-ups are conducted by means of mail and telephone. One source of error is **nonresponse**. Nonresponse rates for institutions ranged from 4% to 15%. The study attempted to correct for this by the imputation of missing values using prior-year returns, whenever these were available.

This report also discusses nonresponse rates for racial and ethnic data when bachelor's degree recipients were queried. The report estimates that for the year 1991, racial and ethnic nonresponse rates ranged from a low of 2.4% of the white, non-Hispanic recipients to a high of 14.9% of the Asian or Pacific Islander recipients. Nonresponse is a problem encountered whenever a survey or census is attempted. As you see from this example, nonresponse rates are not uniform across all segments of a population, so that nonresponse could affect some groups more than others, **biasing** the results.

Measurement error can also bias a survey. In this study, degrees are grouped in various ways (science and engineering degrees, chemistry, computer science, etc.). Institutions in different regions may use groupings of degrees (**taxonomies**) that differ from those of the survey. They may classify double majors or dual B.S. and M.S. degrees differently. The institutional representatives who fill out the survey forms may find the survey's taxonomy difficult and confusing. This could cause measurement errors.

Another source of bias is **response bias**, which occurs when a respondent gives an incorrect response. The respondent may be influenced by the phrasing of the question, or may not recall something correctly, or may simply be lying.

> *Whereas random sampling error can be reduced by increasing the size of the sample, sampling bias cannot. Sampling bias can only be reduced by changing the method of collecting the data.*

---

[4] For example, the National Science Foundation publication NSF 95-318, *Guide to NSF Science and Engineering Resources Data*, describes several ongoing surveys designed to obtain information about science education and funding in the United States. This report includes a frank discussion of potential errors in these studies.

EXERCISES FOR SECTION 1.4

- **1.7** As part of a project for a political science class, Tina and Karen plan to take a random phone survey similar to the Harris Poll, except that they want to estimate the public sentiment on a bill before the U.S. Congress.

    a. Use Formula 1.1 on page 6 to estimate the size needed for their sample if they want results with a standard error of less than 5%.

    b. What size should the sample be if they want the 95% margin of error to be less than 5%?

    c. What size for the 99.7% margin of error to be less than 5%?

1.8 In another Harris Poll of 1144 adult Americans, 306 people felt that the U.S. Constitution should be amended to have presidential elections decided by popular vote rather than by the electoral college. Find the statistic p and estimate the standard error SE_p.

- **1.9** As part of a 1990 study on the causes of asthma, the parents of 939 seven-year-old children in five German cities were interviewed.[5] Of these children, 57 had had a doctor's diagnosis of asthma at some time in their lives. Find the statistic p and estimate the standard error of this statistic.

1.10 *The New York Times* reported on a poll conducted October 18-21, 2000. This random phone survey found that among 1010 registered voters, 45% responded Yes to the question "Does the presidential candidate George W. Bush have the ability to deal wisely with an international crisis?" *The New York Times* explained, "In theory, in 19 cases out of 20 the results based on such a sample will differ by no more than three percentage points in either direction from what would have been obtained by seeking out all American adults." Fill in the following, choosing from these options:

population, parameter, sample, statistic, unknown, π, p, 45%, 3%, 1.5%

The symbol for the parameter is _____. The value of the parameter is _____. The collection of 1010 registered voters is called the _____. The symbol for the statistic is _____. The value of the statistic is _____. The standard error is approximately _____. The 95% margin of error is approximately _____.

---

[5]Susanne Lau et al., *Early exposure to house-dust mite and cat allergens and development of childhood asthma: A cohort study*, The Lancet **356** (2000), pp. 1392–97.

1.5 HOW HYPOTHESES ARE TESTED

Let's consider an example along with two hypotheses.

Example 1.2 *Salk vaccine trials.* According to an article by Paul Meier[6], over a million children participated in a trial in 1954 of the Salk vaccine to see whether it would protect children against polio. In one part of the study, as summarized in Table 1.1 below, 401,974 children were injected with either the Salk vaccine or a salt solution placebo. The injections of the vaccine and placebo were assigned to the children at random. Furthermore, the trial was **double-blind**; that is, neither the children nor the diagnosing physicians were aware of who had been given vaccine or the salt solution.

Table 1.1 Summary of randomized double-blind Salk vaccine trials

Group	Number of subjects	Fatal polio	Paralytic & fatal polio	Nonparalytic polio	Total cases
Vaccine	200,745	0	33	24	57
Placebo	201,229	4	115	27	142
Totals	401,974	4	148	51	199

Does Salk vaccine prevent death from polio?

None of the children who received the Salk vaccine died of polio whereas four of the children in the placebo group died of polio. This seems to be evidence for the effectiveness of the vaccine, but how strong is the evidence? Polio was not a common disease, and the incidence of it would vary from year to year. Before the vaccine was developed, it was conceivable to have only four deaths in a group of approximately 400,000, which is a rate of 1 per 100,000.

For the sake of argument, let us suppose that the Salk vaccine was completely ineffective in preventing death from polio (this is called the **null hypothesis**); that is, suppose the vaccine prevented no deaths from polio, so that the four unfortunate children who died of polio just happened to fall into the placebo group by chance.

Assuming the null hypothesis, the vaccine had no effect and the four children would have died with or without the vaccine. The four children fell into the placebo group by chance, as if by the toss of a fair coin—say, heads for vaccine and tails for placebo. What is the chance that, for all four of the children who died, the coin came up tails? It is certainly not impossible for a coin to come up tails four times in a sequence. The chance of tails on each toss of a fair coin is 50%, so the chance of four tails in sequence is 50% of

---

[6]Paul Meier, *The Biggest Public Health Experiment Ever: The 1954 Field Trial of the Salk Poliomyelitis Vaccine.* This article appears in Judith Tanur, et al. (editors), *Statistics a Guide to the Unknown*, third edition, Duxbury Press, 1989.

50% of 50% of 50%, or $(0.50)^4 = 0.0625 = 6.25\%$. Such a low percentage is evidence for the effectiveness of the Salk vaccine in preventing death, but the evidence is not overwhelming. It would *not* be considered to be proof beyond a reasonable doubt.

This does not show that the vaccine is ineffective in preventing death from polio. It just means that there is insufficient evidence to conclude that the vaccine prevents death from polio.

> *Absence of evidence is not evidence of absence.*[7]

Does it prevent paralytic polio?

Of the children who were injected with the Salk vaccine, 33 were later diagnosed with paralytic polio compared with 115 of the children in the placebo group. Using a similar analysis, we presume the null hypothesis that the vaccine was completely ineffective against paralytic polio; that is, the $33 + 115 = 148$ would have gotten paralytic polio with or without the vaccine. This is as if 148 tosses of a fair coin come up 33 heads and 115 tails. In 148 tosses, you expect approximately half of them to be heads; that is, 74 should be heads or so. But what is the chance that 148 tosses result in as few as 33 heads? Although it is not impossible to get so few heads, we shall see later when we study **probability theory** that the probability is less than 1 in 255 billion. Given this evidence, retaining the null hypothesis would be outrageous. We have no reasonable choice but to reject the null hypothesis and conclude that the Salk vaccine is effective against paralytic polio.

1.6 PREVIEW

Here are some of the issues that we will consider in this book:

- How do you take a sample? What is a random sample? In the Harris study, how do you account for people who cannot be reached by phone, do not answer the phone, refuse to participate, misunderstand the question, lie, and so on? These are all parts of the topic of the **design of experiments**. Although no chapter of this text is dedicated exclusively to this topic, the issues involved are pointed out throughout the book.

---

[7]Douglas G. Altman, J. Martin Bland, *Statistical Notes: Absence of evidence is not evidence of absence*, British Journal of Medicine **311** (1995), p. 485. This article can be found on the Web at http://bmj.com/cgi/content/full/311/7003/485.

- Once you have a sample, which consists of a collection of data, you want to organize and summarize this data. This can be done by the use of tables, graphs, and numbers (statistics) such as the mean, median, range, and standard deviation. This topic, **descriptive statistics**, is discussed in Chapter 2.

- **Probability theory**, studied in Chapter 3, is the theoretical tool of statistics. We saw in Example 1.2 how probability theory is used in the testing of hypotheses. There were no deaths in the treatment group, yet the evidence is not convincing that the Salk vaccine prevented any deaths. On the other hand, 22% of the 148 cases of paralytic polio fell into the treatment group, which is overwhelming evidence of the effectiveness of the vaccine in preventing paralytic polio.

- In Chapters 4, 5, 7, 8, and 9, **probability models** are developed. We show that many of these models can be described in terms of standard population box models.[8] By translating story problems into probability box models, we can eliminate many of the extraneous details that complicate the analysis of case studies. We then develop procedures for analyzing the population models, whose solutions can be translated back into the story problems.

- In Chapter 11, **sampling theory** is developed. It is essential to know the behavior of random samples from known populations before we do the inverse of making inferences about unknown populations by examining random samples.

- In Chapters 6, 12, and 14, we will consider the statistical rules of evidence and how data can be used in **statistical testing of hypotheses**. We examine samples from unknown populations and, with the help of sampling theory, determine the plausibility of various hypotheses about the populations.

- Chapters 13 and 14 deal with the examination of data from unknown populations in order to make **estimates of parameters** of populations. We consider precision of estimates and the level of confidence of estimates.

- In Chapter 15, we consider the **statistical relationship** of variables. In algebra, an equation such as $y = x^2$ describes a functional (or deterministic) relationship between x and y. Statistical relationships are not

---

[8]Population box models are, in turn, related to Polya Urn models as described in Marek Fisz, *Probability Theory and Mathematical Statistics*, third edition, John Wiley & Sons, Inc., 1963.

functional. For example, given a person's height x at age 14 and the same person's height y at age 21, the variables x and y are statistically (or stochastically) related, but not functionally related. **Correlation** measures the strength of the relationship of two statistical variables. Generally, tall 14-year-olds become tall 21-year-olds; and short 14-year-olds become short 21-year-olds, but you could not write a deterministic equation between x and y. We say that the two variables are positively correlated. Similarly, there is a negative correlation between the weight of a car and its fuel economy. We can use the strength of a relationship to predict the value of one variable from another. For example, if you had to predict the height of a 21-year-old person without knowing any other facts, your best guess is the average height for all 21-year-olds, plus or minus a standard error. However, if you know the person's height at age 14, along with other facts such as the correlation coefficient, the prediction can be greatly improved. The prediction will not be perfect, because the relationship is statistical. But the prediction will have more precision than one made without the knowledge of the height at age 14. The prediction of one variable from others is known as **regression analysis**.

- In Chapter 16, we continue with statistical inference by comparing multiple dichotomous populations. We can test whether the historical data of a state lottery fit a model of equal probability for each of the possible outcomes. Or we can test whether any of a number of treatments makes a difference in the outcome of an illness.

- Finally, Chapter 17 covers resampling methods where we discuss parametric and nonparametric bootstrapping. This chapter is computer intensive and deals with inferences from small data sets.

Advice to the reader   At this point, it is a good idea to **review calculus**. It is not used heavily in the first week or so of the course but, beginning with Chapter 3, you will need to know the basic facts of differentiation and integration.

REVIEW EXERCISES FOR CHAPTER 1

- **1.11** Cyberchondriacs are people who go online to search for information about health, medical care, or particular diseases. In a nationwide Harris Poll of 1001 adults surveyed between May 26 and June 10, 2000, 56% said that they were online from home, office, school, library, or another location. Also, 86% of these 56% said that they have looked for health information online. This means that 48% of all American adults have looked for health information online. Based on the U.S. census bureau estimate of 204 million American adults, this amounts to 98 million people. In this situation, the number 56% is which of the following:

(a) a population (b) a sample (c) a parameter (d) a statistic

The number 98 million is an estimate of which of the above?

1.12 *The New York Times*/CBS News Poll uses a computer to randomly select phone numbers within each of the 42,000 residential exchanges in the country. This way the pollsters have access to both listed and unlisted numbers. In one survey, they called 1279 numbers. Of these, 352 were unlisted. This is not surprising because 29% of all numbers are unlisted. Select one option:

(a) 352 and 29 are both parameters;
(b) 352 and 29 are both statistics;
(c) 352 is a parameter and 29 is a statistic;
(d) 352 is a statistic and 29 is a parameter;
(e) none of the above.

1.13 (Proof) Recall from calculus the techniques of finding maximum values of functions.

  a. Show that the maximum of the function $f(x) = x(1-x)$ occurs when $x = \frac{1}{2}$.
  b. Use this to prove the inequality in Formula 1.1

$$\sqrt{\frac{\pi(1-\pi)}{n}} \leq \frac{1}{2\sqrt{n}}$$

1.14 Tossing a fair coin n times and counting heads is like n draws from a population box model consisting of cards numbered with 0's and 1's, and with a population percentage of $\pi = 0.50 = 50\%$. Suppose I toss a fair coin 100 times and compute the sample percentage p of heads. Use Formula 1.1 on page 6 to compute the standard error for the statistic p. Now, considering 400 tosses, what happens to SE_p?

- **1.15** Darius submits a computer-programming assignment. He claims that the program, when run on the busy university network, will have a run time to completion of 20 seconds, with a standard error of 5 seconds.

  a. Assuming this to be true, and if his instructor were to test this program by running it twice, the total run time should be _____ seconds, give or take _____ seconds, on the average.

b. If the instructor runs Darius's program four times, the total run time should be _____, give or take _____, on the average.

1.16 Continue with the previous Exercise 1.15.

    a. If the instructor were to execute the program a total of 100 times, the total run time should be _____, give or take _____, on the average.

    b. Afterward, the instructor takes an average of the 100 runs. If the student's estimates are correct, the instructor should obtain an *average* run time of _____ seconds, give or take _____ seconds or so.

• **1.17** Consider walnuts packed in "one-pound" bags. Although the nominal weights of the bags are one pound, it's hard to put exactly one pound in each bag. As it turns out, the bags vary in weight by a standard error of approximately 1 ounce.

    a. Use the Pythagorean property of standard errors to estimate the packaging error for the total weight of two packages.

    b. What about the packaging error in a box of 16 bags?

★ **1.18** If the children who took part in the placebo control study of the Salk vaccine trials were assigned to the vaccine and placebo groups by chance, why weren't the two groups of the same size?

★ • **1.19** Why do you think so many children were needed for the Salk vaccine trials?

★ **1.20** Give an example of a study where a double-blind experiment is not possible.

    * * * * * * * *

1.21 (Proof) Show that for two measurements, with random errors a and b, we have MA $\leq$ RMS; that is, show that the following is true for all $a \geq 0$, $b \geq 0$:

$$\frac{a+b}{2} \leq \sqrt{\frac{a^2+b^2}{2}}$$

Chapter 2

HOW TO DESCRIBE AND SUMMARIZE DATA

2.1 VARIABLES AND DATA SETS

2.2 CATEGORICAL DATA

2.3 ORDINAL DATA

2.4 RATIO DATA

2.5 FREQUENCY TABLES AND HISTOGRAMS

2.6 GROUPED DATA AND STURGE'S RULE

2.7 STEM-AND-LEAF PLOT

2.8 FIVE-NUMBER SUMMARY

2.9 BOX PLOT

2.10 THE MEAN

2.11 VARIANCE AND STANDARD DEVIATION

2.12 OGIVES AND QUANTILES

2.13 EXPLORATORY DATA ANALYSIS

2.14 FORMULAS FOR HISTOGRAMS (optional)

2.1 VARIABLES AND DATA SETS

A characteristic of individuals in a population is called a **variable**. The individuals may be people, in which case they are called **subjects**; or they may be animals or things, in which case they are often called **cases**. A variable can take different values for different individuals. Examples of variables for people are age, hair color, sex, height, weight, blood type, annual salary, and years of education. The actual values of a variable are called **observations.** If the observations are numbers, they are often called **measurements**. A collection of observations or measurements of a variable forms a **data set**. Such a data set may be obtained by taking a random sample from a population as was discussed in Chapter 1.

In this chapter, we will learn how to organize, display, and summarize a data set. **However, we have to keep in mind that the goal of statistics is to use the information provided by a data set to study the population from which it came. To study the data set itself is not the goal.**

Although it is important that the description of data must give an undistorted and unbiased representation, we must be mindful that an obsession with details can create clutter that hinders a clear view of the population.

Before we can describe a data set, we need to think a little about its source. We need to study the process that generated the data. Here is a simple example.

Example 2.1   **What is the length of the coastline of Florida?** Of Michigan? Which is longer? Before you run that by your research librarian, think a little. There are many things that we just assume can be measured without paying any attention to how it is measured. Benoit Mandelbrot has pointed out that, if coastlines were measured in great enough detail, then their lengths are essentially infinite. Specifically, he said:

> The question I raised in 1967 is "how long is the coast of Britain?" The answer is... It depends on the size of the instrument used to measure length... As measurement becomes more refined, the measurement length will increase. Thus, all coastlines are of infinite length in a certain sense. But of course some are more infinite than others.[1]

---

[1] This quotation is from an interview with Anthony Barcellos as reported in Donald J. Albers and G. L. Alexanderson (editors), *Mathematical People Profiles and Interviews*, Birkhäuser, 1985.

Included with the interview is a fake map of the coast of Britain that looks realistic but is actually an artificial fractal image of a coastline. Fractals are used in the computer modeling of irregular patterns in nature. Because each enlargement of a fractal image appears to have the same level of detail as the original, it is easy to imagine its length to be infinite.

Clearly, when thinking of a coastline, we usually choose to imagine a smooth curve that passes through points on the coastline that have been surveyed. The length of this curve will of course be finite. We tend to forget that the amount of smoothing matters a lot. Our idea of the length of the coastline depends on the smoothing. Before we can compare smoothed coastline measurements of Florida and Michigan, we need to find out if the same method for measuring coastlines was used. If not, we will not know whether the difference in numerical values reflects a difference in coastline lengths or a difference in measuring techniques.

Watch out when comparing unemployment rates for different countries, or crime statistics from different time periods, or other situations where data are collected by different organizations using slightly different definitions or methods of measuring. Just because the numbers reside in the same database under similar headings need not mean that they have much else in common.

Now suppose we have assembled a set of measurements that we believe are comparable. How do we describe a data set? There are various ways to categorize data. The type of data affects the way that we can describe it. Suppose that you asked the other students in your class to fill out a survey asking for their name, age, gender, blood type, whether or not they lived in a university dorm, and number of semesters of calculus completed. Assuming your classmates were good-natured enough to respond to your little survey, how would you go about describing the distribution of ages? Blood types? Semesters of calculus? And so on. In trying to describe a data set, a basic task is to make the data set more easily and effectively understood. Data sets do not exist in isolation, and it is often helpful to make comparisons with other data.

The written description is often accompanied with tables and graphs. These tables and displays should show the data without distorting it. They should bring coherency to large data sets and encourage comparison of different parts of the data set.

The most basic form of data description is sorting the data by some attribute and displaying the data in tabular form or as bar graphs. When deciding on a technique to describe a data set, ask the obvious questions: What do we want the data to explain? What are we trying to show? Are we looking at the data for the first time? If so, we want to check the data for internal consistency. We should look for reporting errors. We may want to look for relationships between two or more variables. This is all in the domain of exploratory data analysis. Decisions will have to be made about how the data should be represented. The way we go about summarizing data depends on the kind of data it is.

2.2 CATEGORICAL DATA

Some variables—such as gender, religion, race, occupation, and blood type—are **categorical**; that is, they assign which of several categories a person or thing is in. For such data, one can prepare tables, bar graphs, or pie charts. A variable such as height, weight, age, or years of education that assigns a number to a person or thing is called a **quantitative** variable.

Example 2.2   *Data brief.*   Below is part of a data brief entitled *Trends in Bachelor's Degrees Awarded to Racial/Ethnic Minorities in Science and Engineering.*[2] It includes the following:

- **Direct reference to data**—"In 1994, there were similar numbers of black bachelor's degree recipients in Science and Engineering (26,289) and Asian recipients (26,420)."

- **Comparison with other data**—"Minority students still comprised 28 percent of the "college-age" population (18–24 years old) in 1994, whereas baccalaureates earned by underrepresented minority students accounted for only 12 percent of the total Science and Engineering degrees in 1994."

---

[2]The data brief summarizes information from the NSF report *Science and Engineering Degrees, by Race/Ethnicity of Recipients: 1987–1994*, written by Susan T. Hill. In this report, which is available at **www.nsf.gov/sbe/srs/nsf96329**, she describes data based on the National Center for Educational Statistics Completions Survey.

- **Comparison of different parts of the same data set**—See the table below.

S&E bachelor's degree recipients, by broad field: percentage distribution, 1994			
Race or ethnicity	Engineering	Natural sciences	Social sciences
Underrepresented minorities	12	29	59
Whites	15	31	54
Asians	25	39	36

- **Comparison over time**—See bar graph in Figure 2.1 below.

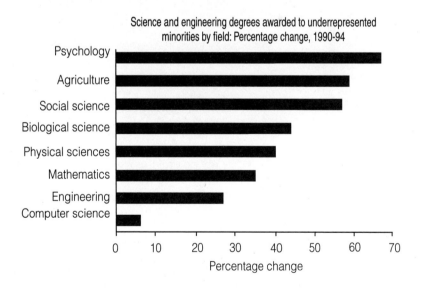

Figure 2.1 Comparison over time

Things to notice

- The author chose to use percentages instead of raw numbers when comparing the Science and Engineering degrees for racial and ethnic groupings.

- She used a bar chart showing percentage change to describe differences between the results of the 1990 and 1994 surveys.

- To make a point about growth or change, it is often more effective to use percentages, or some other easily understood transformation of the data (e.g., adjusting dollar amounts for inflation), rather than report raw numbers.

- The data presentation should be sorted so that comparisons the writer wants the reader to make are in close proximity. Notice that the bars in the last chart have been sorted from largest to smallest percentage change. It is easy to see that from 1990 to 1994 the percentage changes in baccalaureates ranged from an increase of 67% in psychology to an increase of only 6% in computer science.

Warning

Because this is a brief, not a full report, much useful information has been omitted. For example:

- We are not told how the various classifications were defined.

- We are not told what constitutes the underrepresented minority group, nor whether race and ethnicity are self-reported.

- We are not told how the survey was conducted or what schools were selected to be included in the survey.

- We do not have the information to judge the scope or reliability of the National Center for Educational Statistics Completions Survey whose results are being described.

EXERCISES FOR SECTION 2.2

2.1 Consider the following blood types of 40 children:

$$
\begin{array}{cccccccccc}
B & A & O & O & O & A & B & O & A & B \\
AB & O & O & B & O & A & O & A & AB & A \\
AB & A & A & A & O & AB & O & O & A & O \\
O & O & O & B & O & AB & O & A & B & B.
\end{array}
$$

Make a table of percentages and a bar graph to compare this data with the frequencies of the blood types of the population of the United States as given below:

Blood type	O	A	B	AB
Percentage	45.4	39.5	11.1	4.0

2.3 ORDINAL DATA

For some categorical variables, the categories can be ranked in some meaningful order—for example, from smallest to largest. Such a variable is said to be **ordinal**. Variables that are qualitative measurements, such as "poor, fair, good, excellent," or "cold, cool, warm, hot," or "strongly disagree, disagree, neither agree nor disagree, agree, strongly agree," are ordinal. Review Exercise 2.38 on page 67 is such an example.

Categorical data that are not ordinal are called **nominal**. Blood type, as considered in Exercise 2.1, is nominal. We could record whether someone has Type O blood, but we could not reasonably rank blood types from least to most. Of course, we could assign a numerical code to each blood type and even compute the average of these coded values. But that average would not describe anything meaningful.

Example 2.3 **Class grades.** Consider the following grades of 20 students in a statistics class:

$$
\begin{array}{cccccccccc}
D & B & D & B & F & C & A & B & B & D \\
A & B & B & C+ & C & C+ & B & A & A & B+
\end{array}
$$

Because we can order grades from worst to best, this is ordinal data. Moving along the data set, we can make a tally of grades as follows:

Grade	Tally	Frequency f	Relative frequency $f\% = f/n$
A	////	4	0.20
B+	/	1	0.05
B	///// //	7	0.35
C+	//	2	0.10
C	//	2	0.10
D+		0	0.00
D	///	3	0.15
F	/	1	0.05
Totals:		$n = 20$ ✓	1.00 ✓

This permits us to put the grades in order:

$$
\begin{array}{cccccccccc}
A & A & A & A & B+ & B & B & B & B & B \\
B & B & C+ & C+ & C & C & D & D & D & F
\end{array}
$$

The **median**, which divides the *ordered* list into two equal parts, is halfway between the 10th and 11th ordered grade, namely, the grade of B. The **mode**, which is the most frequent grade, is also the grade B.

2.4 RATIO DATA

Quantitative data for which it is meaningful to form quotients is called **ratio data**. A person's age, which is the time elapsed between birth and the present, is a ratio variable. It is meaningful to talk about half of a person's age. On the other hand, it is not meaningful to consider double the calendar date July 4, 1776, so calendar date is not a ratio variable. Other examples of ratio measurements are height, weight, salary, and number of semesters of calculus a person has taken. The distinction between ordinal and ratio variables can be blurry. A data set of grades, as in Example 2.3 above, is ordinal. But when points are assigned to grades, the data set becomes ratio. By assigning 4 points to the grade A, 3.5 points to B+, 3 points to B, and so on, we can compute the class grade point average (GPA) to be GPA = 2.775.

Ratio variables can be **discrete** or **continuous**. The possible measurement of discrete data can be put in one-to-one correspondence with a subset of integers.

For discrete variables, there are gaps between possible values. A variable that can have only integer values will be discrete.

2.5 FREQUENCY TABLES AND HISTOGRAMS   27

> *Continuous data can be described by points in an interval of the real line. There are no gaps between two possible values of a continuous variable.*

The number of semesters of calculus that a person has completed is a discrete variable. Height and weight are continuous variables. In principle, a discrete variable can be measured without error. On the other hand, for a continuous variable, measurement error is unavoidable.

Example 2.4 **Defective items.** A production line inspector in a manufacturing plant records the number of defective items produced each hour of an eight-hour shift:

$$4 \quad 2 \quad 4 \quad 5 \quad 10 \quad 5 \quad 3 \quad 6$$

Example 2.5 **Men's heights.** Below we have the heights of 33 men, measured in inches:

72.0	69.9	66.8	66.0	70.3	67.8	70.1	65.2	70.4	69.8	67.5
64.9	75.3	69.6	66.0	68.9	68.3	73.9	71.8	72.6	65.9	65.1
69.6	69.9	70.2	66.3	70.1	71.0	69.2	69.8	66.0	71.0	70.7

The defective items data set is clearly discrete. The *concept* of height is a continuous variable. A man's height can, in theory, take any real number in some reasonable interval. However, notice that the *measurements* above are all rounded to one decimal place. Although the concept is continuous, the set of measurements is discrete. If the concept is continuous, it's common practice to treat the data as continuous, even though measurements have been rounded. The data set has repeated values because the measured heights are rounded to one decimal place, but the actual heights of the men are all different.

2.5 FREQUENCY TABLES AND HISTOGRAMS

We consider discrete data first. The tally and **frequency table** of the defective items example looks like this:

28   CHAPTER 2.   HOW TO DESCRIBE AND SUMMARIZE DATA

Tally and frequency table

Number of items	Tally	Frequency f	Relative frequency $f\% = f/n$
2	/	1	$1/8 = 0.125$
3	/	1	$1/8 = 0.125$
4	//	2	$2/8 = 0.250$
5	//	2	$2/8 = 0.250$
6	/	1	$1/8 = 0.125$
7		0	$0/8 = 0.000$
8		0	$0/8 = 0.000$
9		0	$0/8 = 0.000$
10	/	1	$1/8 = 0.125$
Totals:		$n = 8$ ✓	1.000 ✓

The median is 4.5, which is halfway between the 4th and the 5th ordered observation. The most frequent observations are 4 and 5 defective items. Thus, the data set has two modes, 4 and 5.

A **histogram** is a graphical display based on the frequency table.

Areas of rectangles

> *Unlike a bar graph for categorical data, where the heights of rectangles are proportional to the frequencies, in a histogram the **areas** of the rectangles are proportional to the frequencies.*

In the graph of Figure 2.2, because the widths of the rectangles are all equal (to 1), the heights of the rectangles are the relative frequencies (or we could have made them the frequencies). The next section deals with histograms in which the class widths are not all equal.

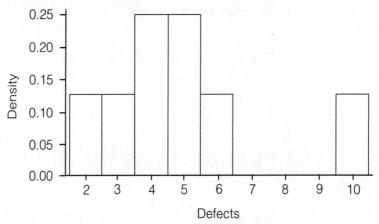

Figure 2.2   Histogram of the number of defective items

2.5 FREQUENCY TABLES AND HISTOGRAMS

Although there are no observations of 7, 8, and 9 defective items, these values are *not* left out of the horizontal scale. Those three rectangles are of zero height, however. Nor is the horizontal scale foreshortened there.

Uniform scale

> *The horizontal scale is uniform.*
> *The histogram fits the horizontal scale,*
> *not the other way around.*

If it is not practical to show the origin on either the horizontal or vertical scale, a zigzag is sometimes drawn.

No gaps

> *Note that, unlike the bar graphs for categorical data, there are no gaps between the rectangles. For discrete data, the left and right boundaries of the rectangles on the horizontal scale occur halfway between possible values.*

In this case, the centers of the rectangles are at the possible data values 2, 3, 4, ..., 10, and the **boundaries** of the rectangles occur at the impossible values of 1.5, 2.5, 3.5, 4.5, ..., 10.5.

The data set is clearly not symmetric about the median. Rather, it is skewed toward larger values (**skewed to the right**) because the right tail of the histogram extends much further from the median than the left tail. A skew to the right makes sense for these data. One cannot observe fewer than 0 defective items, yet during a bad hour one can have a large number of defective items.

Frequency tables and histograms for continuous data sets are made differently. For a continuous variable, it makes no sense to talk about the frequency of a specific value. The data set of men's heights in Example 2.5 has repeated values only because the measured heights are rounded to one decimal place, but the actual heights of the men are all different. For continuous data sets, all frequency tables must necessarily group data into **intervals** or **classes**. The frequency table below has 6 classes of equal width $w = 2$ (inches). The first class has $f = 4$ observations. Because the class width is 2, the first class has **absolute density** of $d = f/2 = 4/2 = 2$ observations per unit (inch), or a **relative density** of $d\% = d/n = 2/33 = 0.061 = 6.1\%$ of the observations per unit.

Frequency table of men's heights with 6 classes

Class	Tally	x	f	$d = f/w$	$d\% = d/n$
64–66	////	65	4	2.0	0.061
66–68	///// //	67	7	3.5	0.106
68–70	///// ////	69	9	4.5	0.136
70–72	///// ////	71	9	4.5	0.136
72–74	///	73	3	1.5	0.045
74–76	/	75	1	0.5	0.015

$$n = \Sigma f = 33$$
(Each class contains its left endpoint but not its right one.)

With discrete variables, class boundaries occur halfway between possible values of the variable. With continuous variables, this cannot be done because all values are possible. Some convention must be made for placing boundary values. The most common convention, which we have followed in the table above, is to make the class intervals closed on the left and open on the right, such as [64,66), [66,68), and [68,70). Thus, the measurements that occur at boundaries are always put into the classes to the right. This may create a slight bias to the right. A less common convention that avoids this bias is to alternate and put the first observed boundary measurement into the class to the right, the second into the class to the left, the third into the class to the right, and so on.

> *The midpoint of each class, denoted by x in the above table, is called the* **class mark**.

The **class mark** is the representative value of the class used in numerical computations, and each histogram rectangle is centered on its class mark (see Figure 2.3).

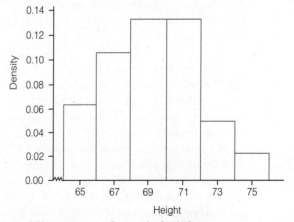

Figure 2.3 Histogram of men's heights

Another graphical display that depicts the same information as the histogram is the **frequency polygon**. Plot points at the midpoints of the tops of the rectangles of the histogram. Also adjoin classes of 0 frequency at both the left and right ends, and plot points at their class marks on the horizontal axis. Then connect the plotted points. The purpose of the new classes at the left and right ends is to anchor the frequency polygon to the horizontal axis. For the men's heights example, the frequency polygon joins the points (63,0), (65,4), (67,7), (69,9), (71,9), (73,3), (75,1), (77,0) (see Figure 2.4).

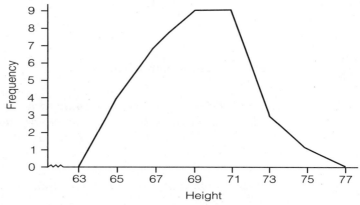

Figure 2.4 Frequency polygon of men's heights

The distinction between discrete and continuous variables leads to different methods of descriptive statistics. However, the distinction can be blurry. The variable of a person's age, which is the time between birth and the present, is continuous. Age can be expressed in years, days, hours, minutes, seconds, and even fractions of seconds. In everyday usage, however, age is generally rounded down to a discrete whole number of years. On the other hand, annual salary, which can be measured exactly to the penny, is clearly discrete. However, the unit of a penny is so small in comparison to the range of annual salaries that annual salary is generally treated as a continuous variable, with clean class boundaries such as \$0, \$10,000, and \$20,000. Were we to treat annual salary as a discrete variable, the boundaries would be at awkward impossible values such as $-\$0.005, \$999.995, \$19,999.995$, etc, with class midpoints \$4,999.995, \$14,999.995, \$24,999.995, etc.

2.6 GROUPED DATA AND STURGE'S RULE

The frequency table of the defective items data in Example 2.4 has $k = 9$ rows (classes) but only $n = 8$ observations. There are too many classes, which results in a histogram without form. In order to better summarize the data set, we will group the data into classes.

To decide on the number of classes, one can use **Sturge's Rule**, which states that for a data set of size n, the number of classes k should be approximately $\log_2 n$ rounded up to the next whole number. In our case, $n = 8$, so $\log_2 8 = 3$, which means that k should be approximately 4.

Sturge's Rule

$$2^{k-1} \approx n$$

Sturge's Rule is only a guide.

There are many factors particular to a data set that often makes it more convenient to choose fewer or more classes than given by Sturge's Rule. However, if we depart too far from the rule, we can get a histogram that loses too much information (it has too few classes) or one that is so cluttered with detail that the histogram loses shape (it has too many classes).

For the men's height example, Sturge's Rule for $n = 33$ suggests a frequency table with 6 classes; and we used 6 classes of equal width 2. For the defective items example, we choose $k = 4$ classes below. The classes are not of equal width.

Grouped frequency table

Class	Frequency f	Relative frequency $f\% = f/n$
0–2	1	0.125
3–4	3	0.375
5–6	3	0.375
7–11	1	0.125
Totals:	$n = 8$ ✓	1.000 ✓

Notice that the first class contains three possible values of 0, 1, and 2. We say that the first class has a width 3. The second class has width 2 because it contains two possible values of 3 and 4. The third class, also of width 2, contains possible values of 5 and 6. And the fourth class, with width 5, contains possible values 7, 8, 9, 10, and 11. If we took a sample from another day of production, we would obtain other values, possibly values not obtained this day. Because we want the table and histogram to be representative of daily results, we extend the lower limit of the first class to the possible value 0. The fourth class goes to an arbitrary upper limit of 11. If, on another day, a further class were needed, it could be the class 12–16 of width 5, or perhaps the class 12–21 of width 10.

> *If a data set has long tails, it's often a good idea to make the first and last classes wider than the interior ones. Remember that our goal is to use the data set to make a representation of the population from which it came. This is often achieved by reducing the amount of detail in a histogram.*

The total of the frequency column is the size of the data set n. The total of the relative frequency column is always 1; it is sometimes convenient to use this as a check of one's calculations.

> *Special care must be taken if the classes are of unequal width. For histograms, the area (not the height) of each rectangle is proportional to the data frequency in that class.*

The unit of measure for the vertical axis of a histogram is thus either **absolute density** $d = f/w$ or **relative density** $d\% = d/n = f\%/w$.

Working copy of the frequency table for the defective items example

Class number	Class	Class width w	Frequency f	Density $d = f/w$	Relative frequency $f\% = f/n$	Relative density $d\% = f\%/w$
1	0–2	3	1	$1/3 = 0.333$	$1/8 = 0.125$	0.0417
2	3–4	2	3	$3/2 = 1.500$	$3/8 = 0.375$	0.1875
3	5–6	2	3	$3/2 = 1.500$	$3/8 = 0.375$	0.1875
4	7–11	5	1	$1/5 = 0.200$	$1/8 = 0.125$	0.0250
			$n = \Sigma f = 8$		$\Sigma f\% = 1$	

Because these are discrete data, the possible values of $0, 1, 2, 3, \ldots, 11$ are interior to the classes. The boundaries of the rectangles are at the impossible values of $-0.5, 2.5, 4.5, 6.5,$ and 11.5 (see Figure 2.5).

34   CHAPTER 2.   HOW TO DESCRIBE AND SUMMARIZE DATA

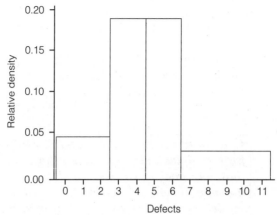

Figure 2.5 Histogram of data with unequal class widths

Because there is one observation in the first class, and the width of the first class is 3 units, there are 1/3 observations per unit in the first class. This is the density of the first class, which is the height of the first rectangle on the absolute density scale.

In general, for the absolute density scale,

$$\text{Height of Rectangle} = \text{Density} = \frac{\text{Frequency}}{\text{Class Width}}$$

For example, the height of the 2nd class is the density $3/2 = 1.50$, and the height of the 4th class is the density $1/5 = 0.20$.

If the relative density scale $d\%$ is used instead of the density scale d, as is in the graph above, then the vertical scale is changed by a factor of n.

Thus, for the relative density scale,

$$\text{Height of Rectangle} = \text{Relative Density} = \frac{\text{Relative Frequency}}{\text{Class Width}}$$

This changes the vertical scale, but the graph remains the same. **Actually a histogram need not display a vertical scale at all.** In comparing histograms of data sets of different sizes, it's important to use the relative density scale because it keeps the total area of each histogram constant.

Important features of the histogram

- Areas of rectangles, not heights of rectangles, represent the class frequencies.

- There are no gaps between adjacent rectangles; that is, adjacent classes have boundaries in common.

- The horizontal scale is uniform. Classes of various widths should have rectangles with proportional widths.

- A histogram need not have a vertical scale. However, if it does, it must be based on the density (or relative density) scale.

EXERCISES FOR SECTION 2.6

2.2 Sketch frequency polygons of the defective items of Example 2.4.

a. First use the table of the ungrouped data given on page 28.

b. Then use the grouped version of the same data set given in the table on page 32.

2.3 Twenty children in a kindergarten class were asked to count the number of people in their families. These are the data they reported:

$$
\begin{array}{cccccccccc}
2 & 8 & 3 & 4 & 4 & 6 & 6 & 5 & 3 & 4 \\
4 & 3 & 5 & 5 & 4 & 3 & 7 & 4 & 5 & 6
\end{array}
$$

Group the data set into a frequency table according to Sturge's Rule and sketch a histogram.

2.4 Below we have the birth weights of 16 babies, measured in pounds:

$$
\begin{array}{cccccccc}
9.2 & 6.0 & 4.2 & 8.7 & 3.3 & 7.1 & 8.0 & 6.2 \\
7.4 & 7.1 & 3.1 & 10.0 & 10.5 & 6.8 & 7.1 & 7.2
\end{array}
$$

Group the data set into a frequency table according to Sturge's Rule, using classes of equal width. Sketch a histogram and the corresponding frequency polygon.

2.7 STEM-AND-LEAF PLOT

A type of display that makes it easy to read and describe data is a **stem-and-leaf plot** or **stem plot**. It is a pattern for displaying ordered data sets of small to moderate size quickly and without losing information. It is especially convenient for displaying numerical data sets of data consisting of two digits, but it can be used for the men's height data of Example 2.5 with three digits:

```
72.0  69.9  66.8  66.0  70.3  67.8  70.1  65.2  70.4  69.8  67.5
64.9  75.3  69.6  66.0  68.9  68.3  73.9  71.8  72.6  65.9  65.1
69.6  69.9  70.2  66.3  70.1  71.0  69.2  69.8  66.0  71.0  70.7
```

For each observation, the decimal point is ignored, and the number is split into the **stem number**, consisting of the leading digits, and the **leaf number**, consisting of the trailing digits. Usually the leaf number has only one digit. The first man's height of 72.0 inches has stem number 72 and leaf number 0.

First display Start with a vertical line and write the stem numbers on the left in increasing order. Running through the data, in the order observed, write the leaves to the right of the vertical line next to the appropriate stem: Leaf 0 is "hung" on the stem 72, then leaf 9 is "hung" on stem 69, then leaf 8 is "hung" on stem 66, then leaf 0 is "hung" on the same stem 66, and so on, to obtain the stem plot below.

```
64 | 9
65 | 2 9 1
66 | 8 0 0 3 0
67 | 8 5
68 | 9 3
69 | 9 8 6 6 9 2 9
70 | 3 1 4 2 1 7
71 | 8 0 0
72 | 0 6
73 | 9
74 |
75 | 3
```

Ordered leaves Next arrange the leaves in order:

```
64 | 9
65 | 1 2 9
66 | 0 0 0 3 8
67 | 5 8
68 | 3 9
69 | 2 6 6 8 8 9 9
70 | 1 1 2 3 4 7
71 | 0 0 8
72 | 0 6
73 | 9
74 |
75 | 3
```

From the stem plot, it is easy to write the data in a sorted list.

Sorted Men's heights

64.9	65.1	65.2	65.9	66.0	66.0	66.0	66.3	66.8	67.5	67.8
68.3	68.9	69.2	69.6	69.6	69.8	69.8	69.9	69.9	70.1	70.1
70.2	70.3	70.4	70.7	71.0	71.0	71.8	72.0	72.6	73.9	75.3

For the first display, do not worry about arranging the leaves in order. In particular, do *not* put the data in order before constructing the stem plot; that is doing the job backward, and it's prone to errors.

The stem plot has the shape of a histogram laid on its side.

Example 2.6   **Defective items.**   If there are too many leaves on the stems or if there are only a few stems, then the leaf values can be split into two (or more) substems. For example, the defective items data set

$$4 \quad 2 \quad 4 \quad 5 \quad 10 \quad 5 \quad 3 \quad 6$$

can be displayed in a stem plot with the 0–9 and 10–19 stems split into substems 0–4, 5–9, and 10–14, 15–19, as follows:

```
0 | 2 3 4 4
0 | 5 5 6
1 | 0
1 |
```

2.8 FIVE-NUMBER SUMMARY

The eight values on the above stem plot of the defective items data set can be written in ranked order:

Rank	1	2	3	4	5	6	7	8
Value	2	3	4	4	5	5	6	10

The values range from the **minimum** value $min = 2$ to the **maximum** value $max = 10$. These extreme values can be used as the starting point of a **five-number summary** of a data set. The average rank is $(1 + 2 + \cdots + 8)/8 = 9/2 = 4.5$. For a list containing an even number of values, there is no value in the exact center, so the center of this sorted list is halfway between the 4th value of 4 and the 5th value of 5. This is the **median** value.

The median is now used to split the original data into two equal-sized parts as follows.

The lower-ranking values: { 2, 3, 4, 4 } and

the higher-ranking values: { 5, 5, 6, 10 }

> *If there are an odd number of measurements, include the median in both parts.*

The median of the lower-ranking values is called the **first quartile**; $Q_1 = \frac{3+4}{2} = 3.5$. Similarly, the median of the upper ranking values is called the **third quartile**; $Q_3 = \frac{5+6}{2} = 5.5$. Sometimes the symbol Q_2 is used for the median.

> *The* **five-number summary** *of a data set consists of the smallest and largest ranking measurements along with the median and the quartiles, written in ascending order:*
>
> min Q_1 med Q_3 max

For the defective items data, the five-number summary is

2 3.5 4.5 5.5 10

For the men's height data, the five-number summary is

64.9 66.8 69.8 70.4 75.3

2.8 FIVE-NUMBER SUMMARY

We should point out that other computational procedures are commonly used to find the quartiles. For example, the Texas Instruments programmable calculator TI-83 Plus and the Minitab statistical software find slightly *different* values for the quartiles than our method.[3] In Section 2.12, we show a procedure for computing all percentiles. This procedure leads to values for the quartiles that differ slightly from the above.

The five-number summary can be used to describe various data features. We can describe the center of a data set by its median. We can describe two measures of the spread of the data. The **range** is the difference between the largest and smallest value:

$$\text{range} = \text{max} - \text{min}$$

For the men's height data, $range = 75.3 - 64.9 = 10.4$ (inches).

The **interquartile range (IQR)** is the difference between the third and first quartile. This is the range of the middle 50% of the data.

$$\text{IQR} = Q_3 - Q_1$$

For the men's height data, $IQR = 71.0 - 66.8 = 4.2$ (inches).

EXERCISES FOR SECTION 2.8

2.5 Make a stem plot and find the five-number summary for the family sizes given in Exercise 2.3 on page 35:

```
2  8  3  4  4  6  6  5  3  4
4  3  5  5  4  3  7  4  5  6
```

2.6 Make a stem plot and find the five-number summary for the birth weights given in Exercise 2.4 on page 35:

```
9.2  6.0  4.2   8.7   3.3   7.1  8.0  6.2
7.4  7.1  3.1  10.0  10.5   6.8  7.1  7.2
```

---

[3] Our definitions of quartiles and the box plot are consistent with those in John Tukey's groundbreaking book *Exploratory Data Analysis*, Addison-Wesley, 1977.

2.9 BOX PLOT

A **box plot** (or **box-and-whisker plot**) in its simplest form is just a graphical display of the five-number summary. A rectangle with width extending from the first to the third quartile and arbitrary height is drawn. A vertical line segment at the median is drawn within this "box." Then horizontal line segments (called **whiskers**) are extended from the midpoint of the sides to the *min* and *max*. The width of the box is equal to the interquartile range *IQR*.

Figure 2.6 shows a box plot of the men's height data.

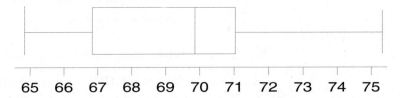

Figure 2.6 Box plot of men's heights

Often the box plot is modified to explicitly show outlying values. The whiskers are drawn only to the most extreme values within 1.5 box widths of the sides of box. Any data values between 1.5 and 3.0 box widths from the sides of the box are explicitly labeled with one type of symbol (indicating possible **outliers**). Those that are more than 3.0 box widths from the sides are labeled with another symbol (indicating probable outliers; these are sometimes called **far out points**).

Figure 2.7 shows a box plot of the defective items data; the symbol ∗ shows the outlying value 10.

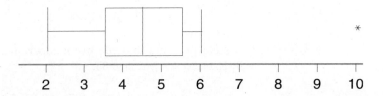

Figure 2.7 Modified box plot of number of defective items

EXERCISES FOR SECTION 2.9

- **2.7** Make a box plot for the family sizes given in Exercise 2.3 on page 35:

  2 8 3 4 4 6 6 5 3 4
  4 3 5 5 4 3 7 4 5 6

2.8 Make a box plot for the birth weights given in Exercise 2.4 on page 35:

  9.2 6.0 4.2 8.7 3.3 7.1 8.0 6.2
  7.4 7.1 3.1 10.0 10.5 6.8 7.1 7.2

NOTES ON TECHNOLOGY

Getting Started and Graphs

Getting Started

These **Notes on Technology** that appear throughout the textbook cannot replace the manual and help button for your calculator or software. Learn the basics of starting and exiting the technology from your manual or, if you are using a computer lab, from a computer technician. Become familiar with working with the technology and interpreting the output.

Minitab: While in Minitab, at any point you can obtain online help by clicking on the **Help** button and going down to **Search for help on** choice.

Excel: You may first have to add **Analysis ToolPak**. You can check by clicking **Tools** to see whether **Data Analysis** is an option. If not, then install **Analysis ToolPak** by selecting **Add-Ins** under the **Tools** menu, then selecting **Analysis ToolPak**, and then clicking **OK**. While in Excel, at any point you can obtain online help by clicking on the **F1** button.

TI-83 Plus: Many programmable calculators, such as the TI-89 and TI-92, do not have the statistical capabilities of the TI-83 Plus, but these statistical programs can be downloaded from the manufacturer's Web site. Learn how to make and store data in lists. To store the Defective Items Example 2.4 data, you can enter {4,2,4,5,10,5,3,6} on your display (including the brackets and commas) and then hitting **STO**→ followed by **L1**, or you can use the instructions below. A list of available functions can be found by hitting the **CATALOG** button but this will not show you how to use them. For this reason, it's important to keep the manual that came with your calculator handy.

Graphs Below are instructions for graphing the Defective Items Example 2.4.

Minitab: Type in the numbers 4 2 4 5 10 5 3 6 into column **C1** (you could give the column a name like **Defects** if you'd like). You can get a histogram of the data by clicking on **Graph**, then on **Histogram**, typing **C1** under **X** (or double clicking on **C1** while the cursor is under **X**), and hitting the **OK** button. You should get a histogram similar to the one shown in Figure 2.2. It is difficult to control the data grouping and the class widths. In most situations, you can accept the default settings. Stem plots may be made using **Graph→Stem-and-Leaf** and box plots can also be made using **Graph→Boxplot**.

Excel: Type in the numbers 4 2 4 5 10 5 3 6 into column **A**. As an option, in column **B**, you may enter the upper limit of the classes 2 3 4 5 6 7 8 9 10. Select **Tools→ Data Analysis** and select **Histogram**. Type the **Input Range**, in this case **A1:A8**, and then type the **Bin Range**, in this case **B1:B9**. Check off the **Chart Output** option and hit **OK**. You will get a bar graph. To remove the gaps between the rectangles, right click you mouse while in one of the rectangles. Then click **Format Data Series→Options**, change **Gap Width** to 0, and hit **OK**. The Excel histogram is not standard in several ways. Each class contains its right endpoint but not its left one, the horizontal scale is shifted to the left by one class width, and, if you are using classes of unequal widths, the horizontal scale is not uniform. Excel does not make stem plots and box plots without downloading additional Ad-Ins.

TI-83 Plus: Press **STAT**, select **Edit**, and hit **ENTER**. Enter the data 4 2 4 5 10 5 3 6 in the L1 column. Hit the **STAT PLOTS** button (this is **2nd→Y=**), select **Plot1**, and hit **ENTER**. Then select **On** and the picture of a histogram for **Type**. Then click **ZOOM→ZoomStat**. If you wish to control the class width, hit **WINDOW** and set the variable X_{scl}; then hit **GRAPH** to see the histogram. Similarly, you can make box plots and frequency polygons by making other selections under **STAT PLOTS→Plot1→Type**.

2.10 THE MEAN

The median, mode, and **(arithmetic) mean** are commonly used measures of the center of a data set. The **mean** is just the average of all the values in a data set. If we label the data set x and the elements of a data set $x_1, x_2, x_3, \ldots, x_n$, then the formula for the mean, computed from the raw data, is

$$\textit{Sample mean:} \quad \bar{x} = \sum_{i=1}^{n} x_i / n$$

For a finite population of size N, the mean of the entire population is denoted by

$$\textbf{\textit{Population mean:}} \quad \mu = \sum_{i=1}^{N} x_i / N$$

For the defective items example

$$4 \quad 2 \quad 4 \quad 5 \quad 10 \quad 5 \quad 3 \quad 6$$

the mean is

$$\bar{x} = \frac{4+2+4+5+10+5+3+6}{8} = \frac{39}{8} = 4.875$$

The calculation of the mean is an arithmetic procedure, which is quite easy for calculators and computers. However, the mean is not necessarily a possible value of the data set. In this example, it is clear that one cannot observe 4.875 defective items.

It is often necessary to compute a mean from a grouped frequency table. Sometimes, the data set is too large for a computation from the raw data. At other times, the raw data are unavailable.

Working grouped frequency table

Class	Mark x	frequency f	Relative frequency f/n	xf	xf/n
0–2	1.0	1	0.125	1.0	0.1250
3–4	3.5	3	0.375	10.5	1.3125
5–6	5.5	3	0.375	16.5	2.0625
7–11	9.0	1	0.125	9.0	1.1250
Sums:		$n=8$	1.000	37.0	4.6250

For this grouping, we are treating the data set as if it had one observation of 1.0, three observations of 3.5, three of 5.5, and one of 9.0.

The **(grouped) mean** is the sum of the column labeled xf/n:

$$\textbf{\textit{Grouped Mean:}} \quad \bar{x} = \sum x(f/n) \qquad (2.1)$$

44   CHAPTER 2.   HOW TO DESCRIBE AND SUMMARIZE DATA

For humans, it is generally easier to divide the sum $\sum xf = 37$ by the number of observations $n = \sum f = 8$:

$$\text{Grouped Mean: } \bar{x} = \frac{\sum xf}{n} = \frac{37}{8} = 4.625$$

In this case, the grouped mean came within $|4.875 - 4.625| = 0.25$ of the raw mean. In general, the grouped mean and the raw mean are close, and never farther than a half of the maximum class width from each other.

Specific features of the median, mean, and mode

- The median divides the histogram into two parts of equal area.
- The mean is at the vertical line that would be the balance point of the histogram constructed as a lamina from a material of homogeneous density.
- If a histogram is close to being symmetric, then the mean and median will be close to each other.
- If the histogram is skewed to the right (left), then the mean is to the right (left) of the median.
- The median is not very sensitive to the choice of interval widths or the exact placement of an upper bound for an open-ended interval.
- The mean is not too sensitive to choice of interval widths for interior intervals, but it is sensitive to choice of end intervals.
- The mode is the midpoint of the class with the highest density. It is sensitive to both choice of interval widths and end intervals.
- Furthermore, a data set may have more than one mode.

EXERCISES FOR SECTION 2.10

- **2.9** Find the grouped mean of the following continuous data:

Class	Frequency
$[-1, 1)$	1
$[1, 3)$	2
$[3, 4)$	4
$[4, 8)$	1

2.10 Find the grouped mean of the following discrete data:

Class	Frequency
0–4	1
5–7	4
8	3
9–11	2
12–14	2

• **2.11** A survey of a small town results in the following table of family sizes:

Family size	Frequency
2	15
3	12
4	15
5–9	18

    a. Sketch the histogram.
    b. Find the mean.
    c. Suppose we assign the size 7 to the 18 families in the last class. Then there are 252 people in the 60 families in this town. If you asked each of the 252 people to state the size of his or her family, what mean would you get? Explain why this is larger than the mean family size?

2.12 Suppose you polled all students at your college and asked them to list the sizes of their classes. In view of the previous Exercise 2.11, explain why most students would be in classes larger than average.

2.11 VARIANCE AND STANDARD DEVIATION

The following two data sets have the same size and the same median, mode, and mean:

| Data set 1 | 48 | 49 | 50 | 50 | 51 | 52 |
| Data set 2 | 0 | 10 | 50 | 50 | 90 | 100 |

    They are very different in that the values of the second data set are more variable. In order to describe a data set well, we need to measure not only the center of the data but also its variability.

    A simple measure of the variation of data is the range. It is the difference between the largest value and the smallest value in the data set. The range is the simplest measure of variation to calculate because it involves only the two most extreme data values; in the defective items example, we have *range* =

$10 - 2 = 8$. But this is also its disadvantage. If only a single value of a data set is very large, the range will make the data set appear to be very dispersed. The IQR, which measures the range of the middle half of the data, is a more robust measure in the sense that it is not so sensitive to a few extreme values.

A good measure of variability should take into account the spread of *all* of the values of the data set. One measure that does this is the **mean absolute deviation,** abbreviated **MAD**. For each element x_i of a data set $\{x_1, x_2, \ldots, x_n\}$, the **deviation from the median** is $x_i - m$ where m denotes the median. The MAD computed from the raw data is the average of the absolute values of the deviations from the median.

$$\textit{Mean absolute deviation: } \mathrm{MAD} = \frac{1}{n}\sum_{i=1}^{n} |x_i - m|$$

For the raw data of the defective items example, we have

$$\mathrm{MAD} = \frac{|2-4.5| + |3-4.5| + 2|4-4.5| + 2|5-4.5| + |6-4.5| + |10-4.5|}{8} = \frac{16}{8} = 2$$

The MAD seems like a reasonable way of measuring the variability of a data set because, unlike the range, it takes into account the deviations of *all* of the observations. For our example, it says that on the average the observations deviate from the median by 2 units. However, the MAD is difficult to deal with theoretically because the formula uses the absolute value function, which is not differentiable everywhere.

The **standard deviation** is the most commonly used measure of the variability of a data set. It is the RMS of the deviations from the mean; that is, it is the square **R**oot of the **M**ean of the **S**quared deviations from the mean.

$$\textit{Sample standard deviation: } s = \sqrt{\frac{1}{n-1}\sum_{i=1}^{n}(x_i - \overline{x})^2}$$

The average is taken with respect to $n-1$ **degrees of freedom** whenever there are n elements of the data set. As a partial explanation for division by $n-1$, note that the sum of all of the deviations from the mean $\sum(x_i - \overline{x})$ is always zero; so as soon as we know $n-1$ deviations, the last one can be found. Because there are only $n-1$ freely varying deviations, the standard deviation is computed by taking an average with respect to $n-1$ independent deviations.

The standard deviation of the entire population is denoted by σ and will be discussed in more detail in later chapters. For a finite population of size N, the **population standard deviation** is defined as follows:

2.11 VARIANCE AND STANDARD DEVIATION

> **Population standard deviation:** $\sigma = \sqrt{\frac{1}{N}\sum_{i=1}^{N}(x_i - \mu)^2}$

The effect of using the population mean μ instead of the sample mean $\bar{x}$ results in division by N instead of $n-1$.

Calculators generally give you a choice between division by n or division by $n-1$. Choose $n-1$ when computing s. When dividing by $n-1$, the standard deviation is sometimes denoted by s^+.

The **variance** s^2 is the square of the standard deviation. Similarly, σ^2 is the population variance.

Unbiased estimator Division by $n-1$ in the computation of s makes the sample variance s^2 an **unbiased estimator** of the population variance σ^2. This means that for different samples from a population, the values of s^2 tend to cluster about the population variance σ^2. If we were to divide by n in the computation of s, then the sample variances would tend to underestimate the population variance σ^2.

Example 2.7 *Defective items, raw data.* For the raw data of the defective items example,

$$4 \quad 2 \quad 4 \quad 5 \quad 10 \quad 5 \quad 3 \quad 6$$

recall that the mean was computed to be $\bar{x} = 4.875$. Thus, the variance is

$$s^2 = \frac{(2-4.875)^2 + (3-4.875)^2 + 2(4-4.875)^2 + 2(5-4.875)^2 + (6-4.875)^2 + (10-4.875)^2}{8-1}$$

$$= 5.839$$

Thus, the standard deviation is the square root

$$s = \sqrt{5.839} = 2.416$$

Example 2.8 *Defective items, grouped data.* Recall, for the grouped data of the defective items example, the grouped mean was computed to be $\bar{x} = 4.625$. We compute the variance from the grouped data as follows.

Working table for grouped frequency

Class	Mark x_j	Frequency f_j	Deviation $x_j - \bar{x}$	$f_j(x_j - \bar{x})^2$
0–2	1.0	1	−3.625	13.1406
3–4	3.5	3	−1.125	3.7969
5–6	5.5	3	0.875	2.2969
7–11	9.0	1	4.375	19.1406
Sums:		$n = 8$		38.3750

In the last column, be careful to square the deviations before multiplying by f, and not the other way around.

$$\text{Grouped Sample Variance: } s^2 = \frac{\sum f_j(x_j - \bar{x})^2}{n - 1} \quad (2.2)$$

Then we take the square root to obtain the (grouped) standard deviation:

$$s = \sqrt{\frac{38.375}{7}} = \sqrt{5.482} = 2.341$$

Because the concept of the standard deviation is not immediately understandable, we offer some rules. The standard deviation measures how far away elements of the data set are from the mean. It is sort of an average of deviations from the mean, or more precisely, the RMS of the deviations. Clearly, if all the numbers in a data set are the same, then the standard deviation will be zero. Larger values for standard deviation indicate more variability. Most of the data will be within a few standard deviations of the mean.

Random error rule

If an element x of a data set is chosen at random and you have to guess its value, your best guess is the mean $\bar{x}$. The difference between your guess and the true value of the element is called the random error. If you guess the mean, then the absolute value of your random error $|x - \bar{x}|$ will be roughly the standard deviation s or so.

2.11 VARIANCE AND STANDARD DEVIATION

To further understand the standard deviation, here are some **common** or **empirical rules**.

68% rule
> *Approximately 68% of the measurements in a data set are within 1 standard deviation of the mean; that is, about 68% of the measurements lie in the interval* $[\bar{x} - s, \bar{x} + s]$.

95% rule
> *Approximately 95% of the measurements in a data set are within 2 standard deviation of the mean; that is, in the interval* $[\bar{x} - 2s, \bar{x} + 2s]$.

99.7% rule
> *Approximately 99.7% (that is, virtually all) of the measurements in a data set are within 3 standard deviation of the mean.*

The empirical rules are based on the "normal bell-shaped" curve, so they are not exact for all data sets. However, these rules are good approximations for surprisingly many data sets.[4] The 99.7% rule could be interpreted as

$$range \approx 6s$$

However, this is generally not a good approximation of the range.

For the raw data of the defective items example, 75% of the measurements lie in the interval $[\bar{x} - s, \bar{x} + s] = [2.459, 7.291]$, and 90% of the measurements lie in the interval $[\bar{x} - 2s, \bar{x} + 2s] = [0.042, 9.708]$, and 100% of the measurements lie in the interval $[\bar{x} - 3s, \bar{x} + 3s] = [-2.374, 12.124]$. For this size data, this is almost as close as one can get to the common rules. However, $6s = 14.499$ is a substantial overestimate of $range = 10 - 2 = 8$.

Standard error of $\bar{x}$

For a given sample, the mean $\bar{x}$ is a measure of the center of the sample and the standard deviation s is a measure of the variability of the sample. The **standard error** of the sample mean $\bar{x}$, written $\sigma_{\bar{x}}$, is a measure of the accuracy of the sample mean $\bar{x}$ as an estimate of the population mean μ. Just as the standard deviation s measures the variability of the observations

---

[4]We will later prove **Chebyshev's Inequality** in Section 7.6 which states that, for each $k \geq 1$, the fraction of the data that is more than k standard deviations from the mean is always at most $1/k^2$. For example, for $k = 2$, Chebyshev's Inequality says that *at most* $1/4 = 25\%$ of the values of a data set are outside of the interval $[\bar{x} - 2s, \bar{x} + 2s]$; that is, *more than* 75% of the measurements fall inside the interval. The corresponding common rule says that for many "normal" data sets, it's around 95%.

in a sample, the standard error of $\bar{x}$ measures the variability of $\bar{x}$ between samples.

As a sample size n gets larger, the sample mean $\bar{x}$ approaches the population mean μ and the sample standard deviation s approaches the population standard deviation σ. However, at the same time, the standard error $\sigma_{\bar{x}}$ approaches zero because $\bar{x}$ becomes a better estimate of μ. The formula for the standard error of $\bar{x}$ is given by

$$\textbf{Standard Error of } \bar{\textbf{x}}: \quad \sigma_{\bar{x}} = \frac{\sigma}{\sqrt{n}}$$

If the population standard deviation σ is unknown, then we can estimate it using the sample standard deviation s. This gives us the **estimated standard error** of $\bar{x}$.

$$\textbf{Estimated Standard Error of } \bar{\textbf{x}}: \quad \text{SE}_{\bar{x}} = \frac{s}{\sqrt{n}}$$

The terms *standard error* and *estimated standard error* are commonly not distinguished. However, *standard deviation* (SD) and *standard error* (SE) are very different; it is extremely important they not be confused.

Example 2.9 **Defective items.** In the defective items example, we computed from the grouped data, the standard deviation $s = 2.341$. This makes $\text{SE}_{\bar{x}} = \frac{s}{\sqrt{n}} = \frac{2.341}{\sqrt{8}} = 0.8278$.

This means that, if another random hour was sampled, we expect to observe approximately $\bar{x} = 4.625$ defective items, give or take $s = 2.341$ or so.

However, if we were to take more random samples of size $n = 8$, the sample means would vary. Each sample mean would differ from the population mean μ by approximately $\text{SE}_{\bar{x}} = 0.8278$ or so.

> The **SD** measures the typical random error for a single observation. The **SE** measures the typical random error for the mean of multiple observations.

EXERCISES FOR SECTION 2.11

- **2.13** Find the mean and standard deviation of the grade point averages of 10 students:

    2.4 2.2 2.4 1.8 3.5 1.8 1.2 2.4 1.9 3.4

2.14 Find the standard deviations of the following data sets:
   a. −1, 1
   b. −1, 0, 1
   c. −2, 0, 2
   d. −1, 0, 0, 1
   e. −1, −1, 0, 0, 1, 1
   f. 9, 10, 11

- **2.15** Find the standard deviation and the estimated standard error of $\bar{x}$ for the grouped data of Exercise 2.9 on page 44.

2.16 Find the standard deviation and the estimated standard error of $\bar{x}$ for the grouped data of Exercise 2.10 on page 45.

- **2.17** Find $\bar{x}$ and s for the raw data of family sizes given in Exercise 2.3 on page 35.

2.18 Consider the raw data of men's heights of Example 2.5 on page 27.
   a. Find the mean and standard deviation of the raw data
   b. Compare the raw data to the three common rules.
   c. If one of the men were to be randomly selected and you had to estimate his height, your best guess would be _____; your estimate would differ from his actual height by approximately _____ or so.
   d. Compute the estimated standard error.
   e. The population mean μ is approximately _____ give or take _____ or so.

- **2.19** Consider the birth weights of 16 babies as given in Exercise 2.4 on page 35.
   a. Find $\bar{x}$, s and $SE_{\bar{x}}$.
   b. If another baby from the same population were to be weighed, your best estimate for its weight would be about _____. Your estimate would differ from its actual weight by about _____ or so.
   c. The population mean μ differs from $\bar{x}$ by about _____ or so.

2.20 The men's heights in Example 2.5 have been grouped below into 7 classes of width 2, starting with interval [63, 65):

52   CHAPTER 2.   HOW TO DESCRIBE AND SUMMARIZE DATA

Frequency table with 7 classes

Class	Tally	x	f
63–65	/	64	1
65–67	///// ///	66	8
67–69	////	68	4
69–71	///// ///// //	70	12
71–73	///// /	72	6
73–75	/	74	1
75–77	/	76	1
			$\sum f = 33$

a. Sketch the histogram. Note that in spite of Sturge's Rule, this histogram has a less defined form than the histogram 6 classes in Figure 2.3 on page 30. This shows that, by phasing the classes differently, the mode changes considerably. In particular, one histogram has a sharp peak and the other is bimodal. **Sharp peaks and flat tops are often the result of phasing of the classes; they are not necessarily the result of inherent properties of the raw data.**

b. Find the mean and standard deviation of this grouped data and then compare them with those obtained for the raw data in Exercise 2.18.

- **2.21** Find the mean and standard deviation of the grouped data of men's heights in Example 2.5 as given in the table on page 30, and compare them with the mean and standard deviation obtained for the raw data in Exercise 2.18.

* * * * * * * * *

2.22 (Proof) For a data set $x_1, x_2, x_3, \ldots, x_n$ define the function

$$f(x) = \sqrt{\frac{(x_1 - x)^2 + (x_2 - x)^2 + (x_3 - x)^2 + \cdots + (x_n - x)^2}{n - 1}}$$

This function defines an RMS deviation from x. Use the second derivative test from calculus to show that the minimum value of this function occurs at the mean of the data set $x = \bar{x}$. This result shows that if you have to guess the value of a randomly chosen element of a data set, your best guess is the mean $\bar{x}$.

2.12   OGIVES AND QUANTILES

Let $x_1, x_2, x_3, \ldots, x_n$ denote a data set of n numbers. The **(empirical) cumulative distribution function** is defined by the equation

$$F(x) = \frac{\text{Number of } x_i's \leq x}{n}$$

2.12 OGIVES AND QUANTILES

Recall the defective items Example 2.4 and the table on page 28:

Number of items	Tally	Frequency f	Relative frequency $f\% = f/n$
2	/	1	$1/8 = 0.125$
3	/	1	$1/8 = 0.125$
4	//	2	$2/8 = 0.250$
5	//	2	$2/8 = 0.250$
6	/	1	$1/8 = 0.125$
7		0	$0/8 = 0.000$
8		0	$0/8 = 0.000$
9		0	$0/8 = 0.000$
10	/	1	$1/8 = 0.125$
Totals:		$n = 8$ ✓	1.000 ✓

Figure 2.8 shows the cumulative distribution function for this table.

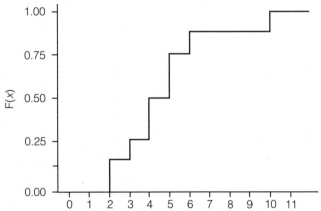

Figure 2.8 Cumulative distribution function for the number of defective items of the preceding table

An **ogive**[5] is a cumulative relative frequency graph. If the data set

$$x_1, x_2, x_3, \ldots, x_n$$

is in the interval $[a, b]$, and it is grouped into classes

$$a = a_0 < a_1 < a_2 < \cdots < a_k = b$$

then the ogive associated with this partition is the broken-line graph formed by connecting the points

$$(a_0, 0), (a_1, F(a_1)), (a_2, F(a_2)), \ldots, (a_{k-1}, F(a_{k-1})), (a_k, 1)$$

---
[5] Pronounced $\bar{o}'j\bar{\imath}v$

54   CHAPTER 2.   HOW TO DESCRIBE AND SUMMARIZE DATA

The ogive function $H(x)$ consists of this broken line, extended to all real numbers by defining

$$H(x) = 0 \text{ if } x < a \text{ and } H(x) = 1 \text{ if } x > b$$

> *For an ogive, points are plotted over the class boundaries, whereas for a frequency polygon, points are plotted over the class marks.*

Example 2.10 ***Defective items.*** Let's consider the defective items example again, where the data are grouped into 4 classes with endpoints $a = a_0 = -0.5, a_1 = 2.5, a_2 = 4.5, a_3 = 6.5$, and $b = a_4 = 11.5$ as in the table on page 32.

Class	Frequency f	Relative frequency $f\% = f/n$
0–2	1	0.125
3–4	3	0.375
5–6	3	0.375
7–11	1	0.125
Totals:	$n = 8$ ✓	1.000 ✓

This table results in the following histogram as shown in Figure 2.5.

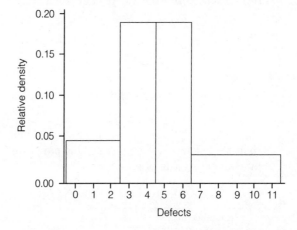

2.12 OGIVES AND QUANTILES

Figure 2.9 shows the ogive for this table and histogram.

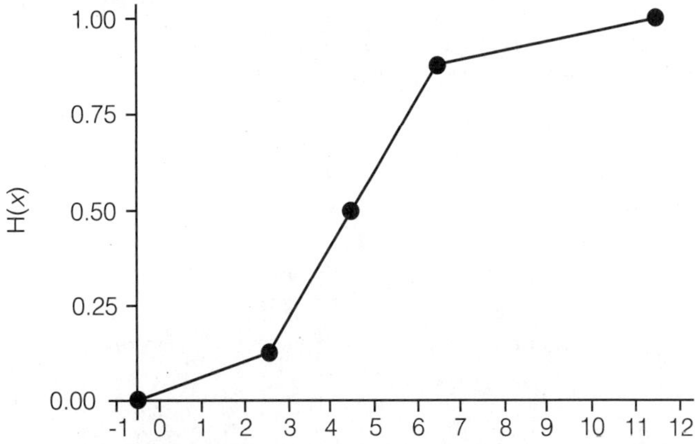

Figure 2.9 Ogive for the number of defective items

Notice that this ogive is plotted over the *class boundaries* $-0.5, 2.5, 4.5, 6.5,$ and 11.5; not over the class marks.

***

How are quantiles computed?

The qth **quantile** (or 100q **percentile**) x_q is supposed to be a value for which the fraction of the data no greater than x_q is at least q and the fraction of data no less than x_q is at least $1 - q$. The qth quantile x_q can be found graphically from the ogive by solving

$$q = H(x_q)$$

In particular, the 80th quantile $x_{.80}$ for the defective items example is the solution of

$$0.80 = H(x_{.80})$$

According to Figure 2.9 above, the solution can be seen to be approximately

$$x_{.80} = 6.1$$

EXERCISES FOR SECTION 2.12

- **2.23** For the defective items example above, find the 80th quantile analytically using the fact that $H(4.5) = 0.50$ (note that $4.5 = x_{.50}$ is the median), $H(6.5) = 0.875$, and the fact that the function H is a straight line between 4.5 and 6.5. Similarly, find the 30th percentile $x_{.30}$.

2.24 Sketch an ogive from the table in Exercise 2.20 on page 51 and compute the 50th and 90th percentiles.

- **2.25** Sketch an ogive for the family sizes given in Exercise 2.3 on page 35 and compute the 25th percentile. Compare with the value of Q_1 computed from the stem plot:

$$\begin{array}{cccccccccc} 2 & 8 & 3 & 4 & 4 & 6 & 6 & 5 & 3 & 4 \\ 4 & 3 & 5 & 5 & 4 & 3 & 7 & 4 & 5 & 6 \end{array}$$

2.26 Sketch an ogive for the birth weights given in Exercise 2.4 on page 35, using the five classes $[2,4)$, $[4,6)$, $[6,8)$, $[8,10)$, and $[10,12)$.
Also compute the 50th and 90th percentiles:

$$\begin{array}{cccccccc} 9.2 & 6.0 & 4.2 & 8.7 & 3.3 & 7.1 & 8.0 & 6.2 \\ 7.4 & 7.1 & 3.1 & 10.0 & 10.5 & 6.8 & 7.1 & 7.2 \end{array}$$

2.13 EXPLORATORY DATA ANALYSIS

Example 2.11   ***Speed of light.***   Simon Newcomb measured the time required for light to travel from his laboratory on the Potomac River to a mirror at the base of the Washington Monument and back, a total distance of about 7400 meters. These measurements were used to estimate the speed of light. Theory has it that the speed of light in a vacuum (or any homogeneous material) is constant, so we would anticipate that the speed of light through the atmosphere would be essentially constant as well, although atmospheric changes will have some effect. The data set[6] is listed below.

Speed of Light Data

28	22	36	26	28	28	26	24	32	30	27
24	33	21	36	32	31	25	24	25	28	36
27	32	34	30	25	26	26	25	−44	23	21
30	33	29	27	29	28	22	26	27	16	31
29	36	32	28	40	19	37	23	32	29	−2
24	25	27	24	16	29	20	28	27	39	23

By itself, this data set is not very helpful. The numbers are supposed to measure passage time of light, but the units of measure are not given. This is

---

[6]From S. M. Stigler, *Do robust estimators work with real data?*, Annals of Statistics **5** (1977), pp. 1055–78. These data can be downloaded from the *Data with Stories* archive maintained at Carnegie Mellon University; the URL is http://lib.stat.cmu.edu/DASL/

2.13 EXPLORATORY DATA ANALYSIS

an example of *coded data*. Data coding is used to make it easier to read. To uncode the data, we must add 24800 and then divide by 10^9. The uncoded measurements in the original units of seconds and looks like this:

0.000024828	0.000024822	0.000024836	0.000024826	0.000024828	0.000024828
0.000024826	0.000024824	0.000024832	0.000024830	0.000024827	0.000024824
0.000024833	0.000024821	0.000024836	0.000024832	0.000024831	0.000024825
0.000024824	0.000024825	0.000024828	0.000024836	0.000024827	0.000024832
0.000024834	0.000024830	0.000024825	0.000024826	0.000024826	0.000024825
0.000024756	0.000024823	0.000024821	0.000024830	0.000024833	0.000024829
0.000024827	0.000024829	0.000024828	0.000024822	0.000024826	0.000024827
0.000024816	0.000024831	0.000024829	0.000024836	0.000024832	0.000024828
0.000024840	0.000024819	0.000024837	0.000024823	0.000024832	0.000024829
0.000024798	0.000024824	0.000024825	0.000024827	0.000024824	0.000024816
0.000024829	0.000024820	0.000024828	0.000024827	0.000024839	0.000024823

The leading digits of the original data are all the same. As you can see, this repetition makes it difficult to focus on the variability that does exist.

Stem plot Arranging the coded speed of light data into a stem plot will also make it easier to read and describe.

```
   1   -4 | 4
   1   -3 |
   1   -3 |
   1   -2 |
   1   -2 |
   1   -1 |
   1   -1 |
   1   -0 |
   2   -0 | 2
   2    0 |
   2    0 |
   2    1 |
   5    1 | 669
  18    2 | 0112233344444
 (28)   2 | 5555566666777777888888899999
  20    3 | 0001122222334
   7    3 | 666679
   1    4 | 0
```

In the first column, we have also included the **depth** of the stem. The depth is the accumulation of the number of leaves of the corresponding or more extreme stems. However, the depth of the stem in the middle is not recorded. Instead, the frequency (28), which is the number of leaves in the middle stem, is recorded.

For Newcomb's data, we cannot help but notice a large gap between the two smallest measurements. There is also a gap between the second and third smallest observations. The data set is **symmetric** if data values at the same depth are roughly the same distance from the median. Looking at the stem-and-leaf plot for Newcomb's data, it is easy to see that the data are not symmetric. Rather, they are skewed toward smaller values (skewed to the left) because the smaller values extend much further from the median than do the larger values. Stem plots are also used to detect unusual round-off patterns such as most of the trailing digits being 0 or 5, or all trailing digits being even.

Five-number summary

The median $med = 27$ is halfway between the 33rd and 34th measurement of the stem plot, the first quartile is the 17th value $Q_1 = 24$, and the third quartile is the 50th value (or the 17th from the top) $Q_3 = 31$.

The five-number summary is thus

$$-44 \qquad 24 \qquad 27 \qquad 31 \qquad 40.$$

We compute $range = 40 - (-44) = 84$ and $IQR = 31 - 24 = 7$.

Box plot

Figure 2.10 shows a box plot of Newcomb's data using the software package Minitab. The o's indicate "far out" values. Obviously, Newcomb's estimate of the speed of light will be greatly influenced by these two values.

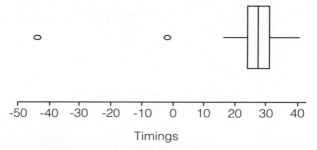

Figure 2.10 Box plot of Newcomb's coded speed of light data (66 observations)

Frequency table

Stem plots become unwieldy when the data set has thousands of measurements instead of dozens. However, some of the same information contained in a stem plot can be summarized in a frequency table. The variable consisting of Newcomb's timings of light is continuous. A continuous data frequency table for Newcomb's data looks like this:

Interval	Frequency	Cumulative frequency
[−45, −40)	1	1
[−40, −35)	0	1
[−35, −30)	0	1
[−30, −25)	0	1
[−25, −20)	0	1
[−20, −15)	0	1
[−15, −10)	0	1
[−10, −5)	0	1
[−05, −0)	1	2
[0, +05)	0	2
[5, +10)	0	2
[+10, +15)	0	2
[+15, +20)	3	5
[+20, +25)	13	18
[+25, +30)	28	46
[+30, +35)	13	59
[+35, +40)	6	65
[+40, +45)	1	66

This table includes all of the information of the stem plot, but it does not include the exact data values contained in the leaves. Some information is lost in the frequency table. The classes above are all 5 units wide.

Because so many of the classes have 0 frequency, we obtain a more parsimonious description of the data by combining classes into wider ones. Sturge's Rule suggests 8 classes, but a grouping into 9 classes is more convenient:

Interval	Frequency	Cumulative frequency
[−50, −30)	1	1
[−30, −10)	0	1
[−10, +10)	1	2
[+10, +20)	3	5
[+20, +25)	13	18
[+25, +30)	28	46
[+30, +35)	13	59
[+35, +40)	6	65
[+40, +45)	1	66

Histogram The histogram in Figure 2.11 was constructed using Minitab. Spreadsheets such as Microsoft Excel also create bar charts; however, the default options usually do not work well for histograms. This histogram would have looked better if the vertical axis had been shifted to the left or omitted.

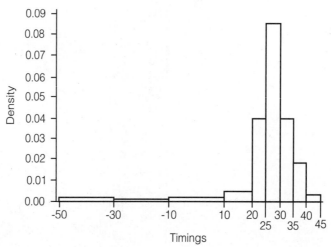

Figure 2.11 Histogram of Newcomb's coded speed of light data (66 observations)

What to look for when viewing a histogram Look for intervals where the data's density is high; intervals where the density is zero; patterns of increase and decline of the density; peaks and valleys; sharp changes in density; symmetry.

Newcomb's coded data has a median of 27, mode of 27.5, and mean of 26.21. The standard deviation of Newcomb's coded data is $s = 10.75$.

The value -44 is more than 6.5 standard deviations below the mean. If the data are clustered about the mean, then it is highly unusual to have data values more than 5 standard deviations from the mean. Notice the high concentration of data in the 25–30 interval, the left skew, and the gap between -30 and -10.

Newcomb wanted his estimate of the data to be consistent with the bulk of the data. He based his estimate on the mean of 65 measurements, leaving out the measurement -44.[7]

Leaving out the value -44, the five-number summary becomes

$$-2 \qquad 24 \qquad 27 \qquad 31 \qquad 40$$

---

[7] Ibid. Stigler's article looks at modern methods of dealing with seemingly aberrant values (or extreme outliers).

Also, the mean becomes 27.292 and the standard deviation becomes 6.249. Notice that leaving out the value −44 changes neither the median nor the IQR, but it does increase the mean and shrinks the standard deviation.

Example 2.12   ***Describing U.S. census data.***   U.S. census data are readily available in electronic form.[8] We are interested in comparing family incomes in four areas in metropolitan Chicago: Rogers Park and Water Tower within the city limits and Maywood and Wilmette in the suburbs. The respective postal zip codes of these areas are 60626, 60611, 60153, and 60091.

Tables of family income in these four zip codes, based on the 1990 U.S. census, are reproduced in the following pages.[9] For easy comparison, each of the corresponding figures (Figures 2.12 through 2.15) was drawn to the same scale, and tick marks were included on the axis for improved readability. The classes correspond to those in the U.S. census tables with the following exception. The "over \$150,000 family income" class was arbitrarily truncated to range from \$150,000 to \$300,000. For example, in the zip code area around Water Tower, 1468 of the 4555 or 32.2% of these family incomes were \$150,000 or more. So the family income density per \$1000 of this class is $0.322/(300 - 150) = 0.002147$. However, in the zip code area in Rogers Park, 102 of 11243, or 0.907%, of these family incomes were \$150,000 or more. So the family income density per \$1000 is $0.00907/(300 - 150) = 0.000060$.

When comparing the histograms for the four regions, a first impression is that the family income distribution around Rogers Park and Maywood are quite distinct from those around Water Tower and Wilmette. There was a much larger spread in reported family incomes in the areas around Water Tower and Wilmette, while most family incomes were concentrated in a more narrow range of values in the areas around Rogers Park and Maywood. This is indicated in the histograms for these areas by the relatively large densities.

The zip code areas around Water Tower and Wilmette had a high proportion of families with reported incomes in excess of \$150,000 in 1989. There were also some family incomes in the \$0 to \$5000 interval. The overall impression is that these family income distributions are heterogeneous and are skewed toward higher values. The zip code areas around Rogers Park and Maywood had only a small proportion of family incomes in excess of \$100,000; their income distributions are more homogeneous than the other two, but they still appear skewed toward higher incomes.

---

[8] For online information about data availability and form, visit the census bureau's website at www.census.gov.

[9] Histograms were generated using Minitab 10.5. The newer versions of Minitab allow for unequal interval lengths, but you must remember to choose the Density option or else you will get a bar graph with heights proportional to frequencies, which is not consistent with good statistical practice.

Cook County: zip=60611
(Water Tower area in Chicago)
FAMILY INCOME IN 1989 ($)

Range	Count
Less than 5,000	79
5,000 to 9,999	103
10,000 to 12,499	20
12,500 to 14,999	14
15,000 to 17,499	26
17,500 to 19,999	38
20,000 to 22,499	52
22,500 to 24,999	39
25,000 to 27,499	78
27,500 to 29,999	54
30,000 to 32,499	74
32,500 to 34,999	29
35,000 to 37,499	101
37,500 to 39,999	36
40,000 to 42,499	78
42,500 to 44,999	63
45,000 to 47,499	140
47,500 to 49,999	123
50,000 to 54,999	250
55,000 to 59,999	147
60,000 to 74,999	366
75,000 to 99,999	483
100,000 to 124,999	434
125,000 to 149,999	260
150,000 or more	1468

Cook County: zip=60091
(Wilmette, IL)
FAMILY INCOME IN 1989 ($)

Range	Count
Less than 5,000	46
5,000 to 9,999	53
10,000 to 12,499	23
12,500 to 14,999	33
15,000 to 17,499	48
17,500 to 19,999	22
20,000 to 22,499	80
22,500 to 24,999	101
25,000 to 27,499	122
27,500 to 29,999	121
30,000 to 32,499	123
32,500 to 34,999	130
35,000 to 37,499	109
37,500 to 39,999	116
40,000 to 42,499	199
42,500 to 44,999	120
45,000 to 47,499	182
47,500 to 49,999	106
50,000 to 54,999	414
55,000 to 59,999	318
60,000 to 74,999	1039
75,000 to 99,999	1236
100,000 to 124,999	881
125,000 to 149,999	565
150,000 or more	1592

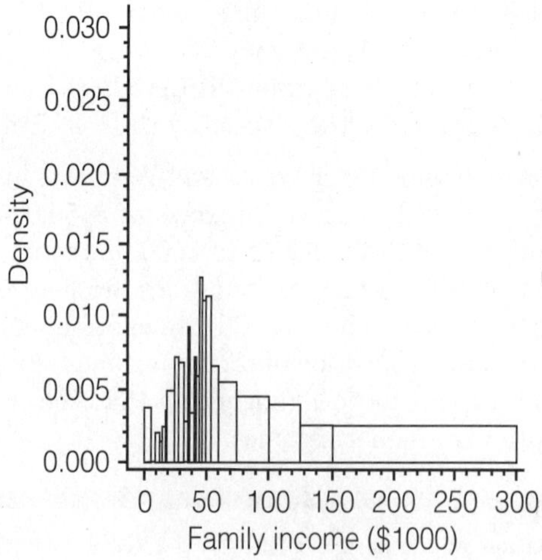

Figure 2.12 Family income zip 60611

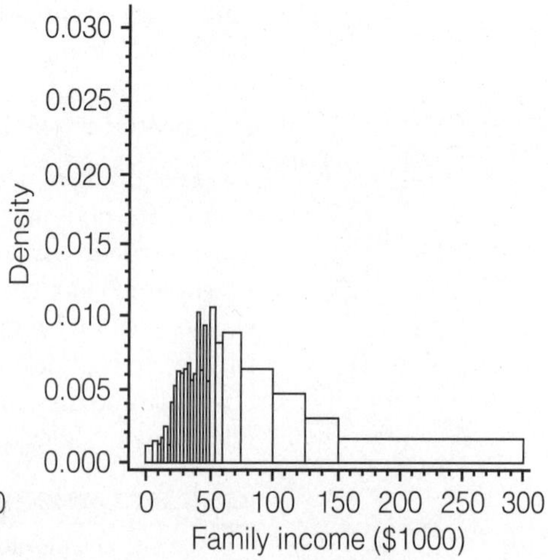

Figure 2.13 Family income zip 60091

2.13 EXPLORATORY DATA ANALYSIS

Cook County: zip=60153
(Maywood, IL)
FAMILY INCOME IN 1989 ($)

Range	Count
Less than 5,000	248
5,000 to 9,999	435
10,000 to 12,499	198
12,500 to 14,999	275
15,000 to 17,499	464
17,500 to 19,999	310
20,000 to 22,499	344
22,500 to 24,999	412
25,000 to 27,499	386
27,500 to 29,999	319
30,000 to 32,499	539
32,500 to 34,999	292
35,000 to 37,499	427
37,500 to 39,999	254
40,000 to 42,499	544
42,500 to 44,999	363
45,000 to 47,499	374
47,500 to 49,999	306
50,000 to 54,999	459
55,000 to 59,999	444
60,000 to 74,999	779
75,000 to 99,999	586
100,000 to 124,999	232
125,000 to 149,999	20
150,000 or more	52

Cook County: zip=60626
(Rogers Park area in Chicago)
FAMILY INCOME IN 1989 ($)

Range	Count
Less than 5,000	758
5,000 to 9,999	731
10,000 to 12,499	745
12,500 to 14,999	548
15,000 to 17,499	676
17,500 to 19,999	503
20,000 to 22,499	679
22,500 to 24,999	579
25,000 to 27,499	507
27,500 to 29,999	473
30,000 to 32,499	584
32,500 to 34,999	364
35,000 to 37,499	456
37,500 to 39,999	329
40,000 to 42,499	365
42,500 to 44,999	272
45,000 to 47,499	273
47,500 to 49,999	250
50,000 to 54,999	527
55,000 to 59,999	322
60,000 to 74,999	587
75,000 to 99,999	463
100,000 to 124,999	80
125,000 to 149,999	70
150,000 or more	102

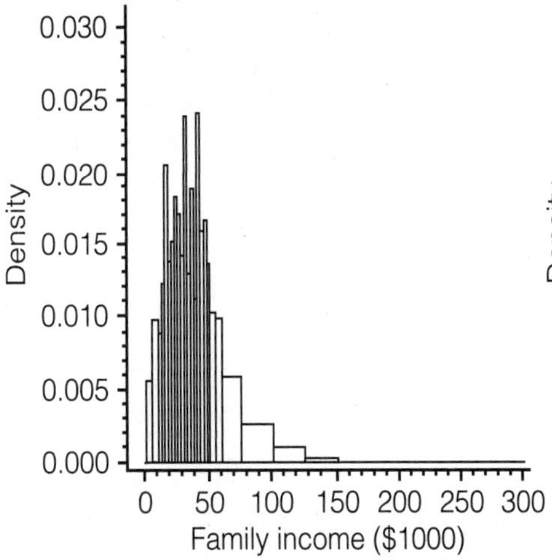

Figure 2.14 Family income zip 60153

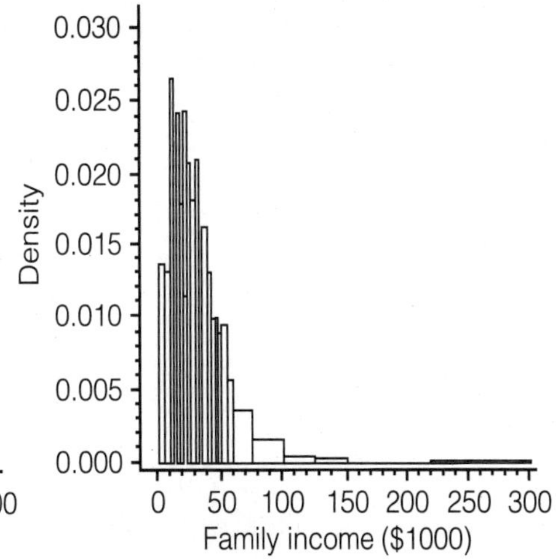

Figure 2.15 Family income zip 60626

EXERCISES FOR SECTION 2.13

★ • **2.27** Calculate the mean family income around the Water Tower area in Chicago. Use the table for zip 60611 with $a = a_0 = 0, a_1 = 5000, a_2 = 10000, \ldots, a_{24} = 150000, a_{25} = 300000 = b$. Show that this mean family income is \$118,193. Notice that b, the upper bound for family incomes, was chosen somewhat arbitrarily. Show how to change b so that the computed mean family income around Water Tower would be \$125,000. Change b again so that mean family income around Water Tower would be \$150,000.

2.28 Half the area under the histogram is supposed to be on each side of the median. Explain how the choice of b affects the computation of the median.

• **2.29** Use the table on page 62 to calculate the median family income around the Water Tower area of Chicago. The median is smaller than the mean because these family incomes are skewed toward larger values.

2.30 Use the table on page 62 to find the five-number summary for family income around the Water Tower area of Chicago. Sketch a box plot.

2.31 Use the table on page 62 to calculate the median family income for Wilmette.

★ **2.32** Download Stephen M. Stigler's paper *Do robust estimators work with real data?* The Annals of Statistics, **5**:4 (1977), page 1075 from the *Data with Stories* archive http://lib.stat.cmu.edu/DASL. Describe A. A. Michelson's measurements of the speed of light in air. The 100 measurements are grouped by trials. Construct side-by-side box plots. What do you see? Compute summary statistics. Describe between-trial variation and within-trial variation. Construct and describe the histogram for all 100 measurements.

2.14 FORMULAS FOR HISTOGRAMS (optional)

Consider a data set $x_1, x_2, x_3, \ldots, x_n$ where all values are in an interval $[a, b]$. Group the data into k classes with end points

$$a = a_0 < a_1 < a_2 < \cdots < a_k = b$$

Let the frequency of the data in each class $[a_{j-1}, a_j)$ be denoted by f_j and define the width of each class by $w_j = a_j - a_{j-1}$.

2.14 FORMULAS FOR HISTOGRAMS (optional)

For each class $[a_{j-1}, a_j)$, let the indicator function $I_j(x)$ be the function that has the value 1 on the class $[a_{j-1}, a_j)$ and zero elsewhere:

$$I_j(x) = \begin{cases} 1 & : \quad \text{if } a_{j-1} \leq x < a_j \\ 0 & : \quad \text{otherwise} \end{cases}$$

Recall that the height of each rectangle in the histogram is the class density. Let the histogram function $h(x)$ consist of the tops of the rectangles defining the histogram, which can be written as

$$h(x) = \sum_{j=1}^{k} I_j(x) \frac{f_j}{nw_j}$$

The ogive function can be obtained by

$$H(x) = \int_a^x h(t)\,dt$$

Note that $\int_a^b h(x)\,dx$ is the area under the histogram, which is the sum of the areas of the rectangles. This sum must be equal to 1.

The modal class is that subinterval that contains the maximum of $h(x)$. (It is not helpful to differentiate $h(x)$ to find the maximum because $h(x)$ is constant on each class.)

The median of the histogram is that solution med of the equation

$$\int_a^{med} h(x)\,dx = \int_{med}^b h(x)\,dx$$

More generally, the qth quantile is the solution x_q of the equations

$$\int_a^{x_q} h(x)\,dx = q \quad \text{and} \quad \int_{x_q}^b h(x)\,dx = 1 - q \qquad (2.3)$$

The (grouped) mean of the histogram is

$$\bar{x} = \int_{-\infty}^{\infty} xh(x)\,dx = \int_a^b xh(x)\,dx \qquad (2.4)$$

Note that within each class $h(x)$ is constant, so

$$\int_{a_{j-1}}^{a_j} xh(x)\,dx = \frac{f_j}{nw_j} \int_{a_{j-1}}^{a_j} x\,dx = \frac{a_j + a_{j-1}}{2} \frac{f_j}{n} = m_j \frac{f_j}{n}$$

where $m_j = \frac{a_j + a_{j-1}}{2}$ denote the class marks. Thus, the mean can be expressed as the weighed average of the class marks:

$$\bar{x} = \sum_{j=1}^{k} m_j \frac{f_j}{n} = \frac{a_0 f_1 + a_1(f_1 + f_2) + \cdots + a_{k-1}(f_{k-1} + f_k) + a_k f_k}{2n} \quad (2.5)$$

which agrees with Equation (2.6) on page 43. The last expression shows that the mean of a histogram is also the weighted average of the endpoints $a_0, a_1, \ldots a_k$ of the classes. You can make the mean as large or as small as you want by manipulating your choice of the boundaries of the subintervals, so be careful when you choose endpoints a and b.

EXERCISES FOR SECTION 2.14

- **2.33** (Proof) Compute the total area of the rectangles formed by the histogram function $h(x)$.

- **2.34** (Proof) Show that, when the derivative of the ogive curve exists, it is the same as the histogram; that is, show $H'(x) = h(x)$ whenever $x \neq a_j, j = 0, 1, 2, \ldots, k$.

- **2.35** (Proof) Show that, when the derivative of the cumulative distribution function exists, it is zero; that is, show $F'(x) = 0$

- **2.36** Compute the quartiles for Newcomb's data (Example 2.11) using Equations 2.3 on page 65 with $a = a_0 = -45$, $a_i = i - 45$ for $1 \leq i \leq 85$ and $a_{86} = b = 41$. Notice that the method of computing quartiles in the five-number summary (often called **hinges**), as shown on page 58, is slightly different.

- **2.37** Compute the mean for Newcomb's data using Equation (refgroupedmean2Eqn above with a_i defined as in Exercise 2.36 above, for $i = 0, 1, 2, \ldots, 86$.

NOTES ON TECHNOLOGY

Summary Statistics

Minitab: Enter your data in column **C1**. Then select **Stat→Basic Statistics→Display Descriptive Statistics**. Enter **C1** in the **Variables** box and hit **OK**. This will display most of the statistics we have discussed, as well as the trimmed mean TrMean, where the highest 5% and lowest 5% of the data are ignored.

Excel: Recall from the previous **Notes on Technology** on page 41 that you may have to install **Analysis ToolPak** by selecting **Add-Ins** under **Tools**. Enter your data in column **A**. Click **Tools→Data Analysis**. Choose **Descriptive Statistics** and click **OK**. Then enter the input range (in the form **A1:A11** if the data are in cells A1 through A11), choose **Summary statistics**, and click **OK**.

TI-83 Plus: Enter your data in list L1. Then choose **STAT→CALC→1-Var Stats** and hit **ENTER** twice. Use the down arrow key to display the five-number summary statistics.

REVIEW EXERCISES FOR CHAPTER 2

2.38 Physician H. W. Darrell devised a method of ranking facial wrinkles on a scale ranging from 1 (no wrinkles) to 6 (severely wrinkled). He considered 152 photographs of women in their forties, of whom 77 were smokers and 75 were nonsmokers. Make a table and a bar graph to compare the two groups.

Score	1	2	3	4	5	6
Smokers	14	30	17	11	4	1
Nonsmokers	21	27	22	5	0	0

• **2.39** Below we have the daily sulfur oxide emissions of an industrial plant for 15 days, measured in pounds:

$$
\begin{array}{ccccc}
21.0 & 18.9 & 11.4 & 17.5 & 18.5 \\
23.3 & 9.9 & 13.2 & 17.7 & 24.5 \\
13.9 & 13.9 & 19.4 & 13.0 & 19.7
\end{array}
$$

a. Group the data into a frequency table of equal widths starting with class [7.5,12.5).

Find the following statistics from the table:

b. mode **c.** mean **d.** standard deviation

e. Find the five-number summary.

f. Find the IQR.

2.40 The number of books checked out at a public library by a sample of eight primary school children were recorded:

$$6 \quad 4 \quad 5 \quad 7 \quad 5 \quad 5 \quad 6 \quad 10$$

For this data, sketch and find the following:

a. histogram   **b.** mean   **c.** mode (modes)

d. median   **e.** variance   **f.** standard deviation

g. range   **h.** stem plot   **i.** box plot

- **2.41** Below are the heights of 33 women, measured in inches:

69.1	64.9	68.2	67.0	67.6	65.5
66.2	66.2	68.0	64.9	65.8	65.9
67.8	67.6	66.8	66.8	66.7	64.9
66.8	65.9	64.5	70.2	65.0	66.0
65.5	72.2	64.3	68.7	64.8	67.8
71.0	67.8	69.2			

  a. Group the data into a frequency table of equal widths starting with [64,66).

  Find the following statistics from the table:

  b. mode   **c.** mean   **d.** standard deviation

  e. Find the five-number summary.   **f.** Find the IQR.

2.42 Below are the birth lengths of 20 babies born in the Kaiser Foundation Hospital, Oakland, California, measured in inches.[10]

19.3	20.8	22.0	16.5	19.5	21.5	19.5	18.0	20.0	20.0
22.0	19.5	21.5	20.0	21.5	19.3	20.0	21.0	20.8	21.5

  a. Why do you think that there are so many recorded measurements that end in .0, .3, .5, and .8?

  b. Group the data into a frequency table.

  Find the following statistics from the table:

  c. mode   **d.** mean   **e.** standard deviation

  f. Find the five-number summary.   **g.** Find the IQR.

---

[10] These data were randomly selected from the records of the *Child Health and Development Study* conducted under the supervision of Prof. Jacob Yerushalmy, School of Public Health, University of California, Berkeley.

REVIEW EXERCISES FOR CHAPTER 2

- **2.43** Consider the following biweekly salaries in dollars of 10 part-time workers:

    225.36 187.49 188.95 200.00 155.45 182.35 86.50 115.00 263.25 126.88

    a. Group the data into a table of equal widths starting with class [0.00, 50.00).

    b. Sketch a histogram for the table.

    c. Find the mean and standard deviation of the grouped data from the table.

    d. Find the mean and standard deviation of the ungrouped data.

    e. Find the mode of the grouped data.

2.44 Consider the following grade point averages for 10 students:

   2.4 2.2 2.4 1.8 3.5 1.8 1.2 2.4 1.9 3.4

   a. Group the data into a table with classes 0–0.9, 1–1.9, 2–2.9, and so on.

   b. Sketch a histogram for the table.

   c. Find the mean and standard deviation of the grouped data and compare to the values found from the raw data in Exercise 2.13.

   d. Find the mode of the grouped data.

   e. Make a stem plot.

   f. Sketch a box plot.

- **2.45** The histogram below was drawn without its horizontal scale. Its standard deviation is about 20. Below it are four scales. One of these scales belongs to the histogram. Which one and why?

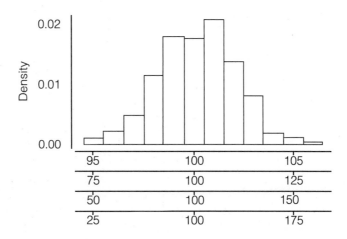

2.46 We present Prussian army data investigated by L. Bortkiewicz.[11] He considered the number y of soldiers in each cavalry corps who died each year from a kick by a horse. He looked at 10 cavalry corps over a period of 20 years, which constitutes a sample of $n = 10 \times 20 = 200$ observations.

Prussian army data

Number of deaths y	Frequency f
0	109
1	65
2	22
3	3
4	1

a. Compute the relative frequencies. b. Sketch a frequency polygon.
c. Compute the mean.   d. Compute the standard deviation.
e. Sketch an ogive.   f. Compute the 80th and 95th percentiles.

- **2.47** Consider the data on the number of library books checked out as given in Exercise 2.40 on page 68. Suppose the last number had been recorded in error as 100 books, instead of 10. Find the resultant mean, median, standard deviation, and IQR. Which of these statistics are resistant to errors and which ones are sensitive to errors?

2.48 A Geiger counter registers a click whenever an alpha particle[12] is detected. The table below gives results of a famous experiment by Rutherford and Geiger. They observed the number of alpha particles emitted by radioactive iodine in 2,608 periods of 7.5 seconds each.

Geiger Counter Data

Number of clicks	Frequency
0	57
1	203
2	383
3	525
4	532
5	408
6	273
7	139
8	45
9	27
10	16
Total	2608

---

[11] L. Bortkiewicz, *Das Gesetz der kleinen Zahlen*, Teubner, Leipzig, 1898. We return to this example in Review Exercise 9.44.

[12] An alpha particle is a helium nucleus emitted by a radioactive substance.

a. Compute the relative frequencies.

b. Sketch a frequency polygon.

c. Compute the mean.

d. Compute the standard deviation.

- **2.49** Below is an ogive for men's heights in Example 2.5 grouped as in the table on page 30. The vertical scale is in terms of the cumulative frequency. Change the vertical scale to cumulative percent and find the 60th percentile.

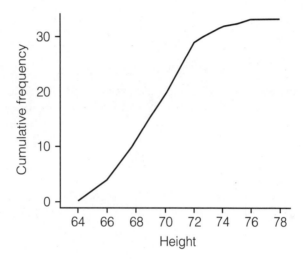

2.50 Compare and contrast the box plots below for the heights of the 33 men in Example 2.5 and of the 33 women below.

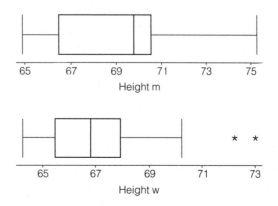

2.51 Find the five-number summary for family income in Wilmette, using the table on page 62. Sketch the box plot and compare it to the box plot for the Water Tower area of Chicago as done in Exercise 2.30.

2.52 Find the five-number summary for family income in Maywood, using the table on page 63.

2.53 Find the five-number summary for family income in the Rogers Park area of Chicago using the table on page 63. Sketch the box plot and compare it to the box plot for Maywood, as done in the previous Exercise 2.52

* * * * * * * * *

2.54 Consider the following table.

Marital Status and Living Arrangements of Adults 18 Years Old and Over
March 1995   (Numbers in thousands)

Characteristics of adults 18 yrs and over	All ages	18 to 24 years	25 to 34 years	AGE 35 to 44 years	45 to 64 years	65 to 74 years	75 and over
MARITAL STATUS							
Males	92,008	12,545	20,589	20,972	24,900	8,097	4,908
never-married	24,628	10,726	8,019	3,631	1,708	342	201
Females	99,588	12,613	20,800	21,363	26,550	10,117	8,147
never-married	19,312	9,289	5,540	2,286	1,426	408	360
LIVING ARRANGEMENTS							
Males	92,008	12,545	20,589	20,972	24,900	8,097	4,908
living alone	10,160	626	2,221	2,265	2,793	1,134	1,121
Females	99,588	12,613	20,800	21,363	26,550	10,117	8,147
living alone	14,640	591	1,448	1,400	3,594	3,247	4,360

    a. Construct histograms to compare the age distributions of adult males and adult females.

    b. Compare and contrast the age distributions of never-married males with never-married females.

    c. Compare the percentage of never-married males by age.

    d. Compare the percentage of never-married females by age.

★   e. Write a news brief describing the age and gender distribution of the 24.8 million adults living alone in the United States.

2.55 The data shown in the stem plot below are for 254 lottery payoffs in millions of dollars (for example, 3 : 5 refers to a payoff of 3.5 million dollars). The stem plot is severely rounded. For example, the largest payoff was 8,695,000 dollars, and the smallest was 830,000 dollars. We have $n = 254$, median = 2.7025, mean = 2.9036, $Q_1 = 1.94$, $Q_3 = 3.65$, standard deviation = 2.9034.

```
0 : 8
1 : 00001112223333333333344444
1 : 555555666666777777788888888999999999999
2 : 000000011111111111122222223333333444444444
2 : 5555566666666677777888999999999999999
3 : 000000001111112222333333333444
3 : 55555555666667777777888888899999999
4 : 0122234
4 : 55555678888889
5 : 111111134
5 : 555667
6 : 44
6 : 7
7 :
7 : 6
8 :
8 : 7
```

a. Construct a box plot for these data.

b. Make a histogram. Use the histogram to approximate the proportion of payoffs that are between 3 and 4 million dollars. Compare this approximation with the one obtained from the stem plot.

c. After examining the histogram, stem plot, and your box plot, describe the data set. Is it skewed or symmetric? If it is skewed, in which direction is the skew? Are there outlying values? Are there gaps? Are there intervals where the data density is high? Are there sharp changes in density? What patterns of increase and decline in density do you notice? Are there any (other) unusual features?

2.56 (Proof) Show that the formula

$$s^2 = \frac{\sum_{i=1}^{n}(x_i - \bar{x})^2}{n-1}$$

is the algebraically the same as

$$s^2 = \frac{\sum_{i=1}^{n} x_i^2 - \frac{(\sum_{i=1}^{n} x_i)^2}{n}}{n-1}$$

Chapter 3

PROBABILITY

3.1 OVERVIEW

3.2 DEFINITIONS

3.3 PROBABILITIES OF EVENTS

    3.3.1 THEORETICAL PROBABILITY

    3.3.2 EMPIRICAL PROBABILITY

    3.3.3 MONTE CARLO METHOD

    3.3.4 SUBJECTIVE PROBABILITY

3.4 RULES OF PROBABILITY

    3.4.1 ADDITION RULES

    3.4.2 INDEPENDENCE AND CONDITIONAL PROBABILITY

3.5 TREE DIAGRAMS

3.6 BAYES' METHOD

3.1 OVERVIEW

Recall that in statistics our goal is to make inferences about a population using information obtained from a sample of the population. In the preceding chapter, we studied methods of organizing and summarizing information obtained from a sample. Conclusions that we draw about a population based on a random sample from the population involve probability. In this chapter, we learn the basic definitions and rules of probability theory.

Probability theory has its origins in the analysis of games of chance dating back to the 1600s. Gamblers realized that probability will not tell you whether or not some event will happen but, for repeatable experiments, it will tell you about what percentage of times the event will happen. As such, it is a guide to actions and strategies.

We all use probability in an informal way every day; it's the basis of what we call common sense. We do not *know* that we will be hit by a car if we cross a busy street without watching, but common sense tells us not to do so because we believe that the *chance is high* of being hit. Similarly, we decide on the fastest route home, not because we *know* it will be quickest every time, but because, based on our experience of traffic flow, we judge it *likely* to be quickest. Although these are trivial examples that do not need rigorous analysis, many serious decisions in science require the use of formal rules of probability.

For an event that is not repeatable, probability is a measure of the belief that the event is true or will happen. Suppose a surgeon tells you that there is a 20% chance that a lump on your body is cancerous. The surgeon means that about 20% of the lumps of this type have turned out to be cancerous in the past. For you, however, this is not a repeatable experiment; you have only one body. The value 20% will not tell you whether your lump is cancerous or not. Nevertheless you can use this fact in weighing your options.

Probability theory and statistics both deal with uncertainty. Yet there is a fundamental difference between probability and statistics. Probability deals with *deductive* reasoning, whereas statistics deals with *inductive* reasoning. Probability assumes that the contents of a population (the group under study) are known. Statistics uses facts observed in samples to make conclusions about an unknown population.

Example 3.1 **Genetics.** The following example from genetics illustrates the difference between probability and statistics. The cells in a malignant tumor form a **monoclonal** population if they are all descendants of a single ancestor. Otherwise, they form a **polyclonal** population. There is evidence to suggest that most malignant tumors are monoclonal.[1]

In mammals, including humans, females have two X chromosomes in all their cells but males have only one. Female mammals inactivate one of their X chromosomes. The inactivation occurs

---

[1] Harold Varmus and Robert A. Weinberg, *Genes and the Biology of Cancer*, Scientific American Library, 1993.

after a female embryo reaches many cell divisions. This X chromosome inactivation *takes place randomly*; some cells have their maternal X chromosome suppressed, and others have their paternal X chromosome suppressed. However, once this happens, all the descendants of these cells follow the same inactivation pattern as their ancestor cell. So in the tissues of adult females, both inactivation patterns are usually observed. However, all the cells within a malignant tumor have the same X chromosome inactivated. This is taken as evidence that each malignant tumor forms a monoclonal population.

Probability problem

Suppose there are many female laboratory rats, each with a malignant lung tumor. Suppose both types of X chromosome suppressions are present in cells drawn from nonmalignant lung tissue, but only one type is present in the malignant lung tumor tissue. If malignant tumors are polyclonal populations, what is the probability that for each rat all the malignant cells follow the same X chromosome inactivation pattern? Developing mathematical models to answer this question is the realm of probability.

The part of probability theory that deals with drawing some cells from many is called **sampling theory**. Questions such as the following are questions of sampling theory. If twenty cells were to be selected at random from nonmalignant lung tissue, what is the chance that the sample has no maternal X chromosome suppressed? Or only five? Or more than twice as many cells with one kind of X chromosome suppression than the other?

Statistical problem

Suppose many female laboratory rats, each with a malignant lung tumor, are observed. Suppose both types of X chromosome suppressions are observed in cell samples drawn from nonmalignant lung tissue but only one type is present in samples of the malignant lung tumor tissue. Which hypothesis is better supported by this evidence? The hypothesis that malignant tumors are polyclonal or the hypothesis that malignant lung tumors are monoclonal? How strong is the evidence? Statistical inference techniques attempt to provide answers to questions like these.

The questions of statistics we leave to later chapters. Here we start with a systematic study of probability.

3.2 DEFINITIONS

An **experiment** is a procedure by which an observation or measurement is obtained. An execution of an experiment is called a **trial**. The observation or measurement obtained is called the **outcome** of the trial. A **random experiment** is one where the outcomes depend upon chance.

As we saw in Newcomb's measurements of the speed of light of Example 2.11, each trial resulted in a different number. The variability was not due to bad instruments or bad techniques. In Newcomb's day, the standard meter had been defined by international agreement as the distance between two fine lines on a bar of platinum-iridium alloy kept in Paris, France. The speed of light in a vacuum was a physical constant that was measured (with some variability) in the meter-kilogram-second system.[2]

The National Institute of Standards and Technology (formerly, the National Bureau of Standards), using today's best equipment, would get different outcomes on each trial, were it to measure the time it takes light to traverse the distance between a point at the base of the Washington Monument and a point in the Smithsonian Institution.

In fact, most scientific experiments are conducted under conditions that are repeatable, but not perfectly repeatable. There is an element of randomness in physical measurements. They are *random experiments*.

Hereafter, any experiments we discuss are assumed to be random experiments. The simplest types of random experiments are those that involve tossing coins, rolling dice, and drawing cards. We will spend some time on these types of problems but only to the extent that they will help us understand the process of observing samples from a population.

The set of all possible outcomes of an experiment is called the **outcome set**. It is also called the **sample space**, but one should not confuse the term with a sample of a population.

Example 3.2   **Toss a coin.** Let's toss a coin and observe whether it comes up heads or tails. This is called a **dichotomous experiment** because there are only two possible outcomes. The outcome set is $S = \{H, T\}$, where H stands for heads and T for tails.

---

[2] Several researchers' estimates of the speed of light and their reported errors are shown in Figure 1 of the paper W. J. Youden, *Enduring Values*, Technometrics **14** (1972), Number 1, p. 4. Since 1983, the meter has been defined as the length of the path traveled by light in a vacuum during a time interval of $1/299{,}792{,}458$ of a second. Thus, the modern definition of the length of a meter makes the speed of light in a vacuum a constant (exactly 299,792,458 meters per second).

Example 3.3   ***Count heads.***   Change the experiment to the following: Toss a coin and count the number of heads. There are only two possibilities: no head or one head. The process generating the observations is the same as in Example 3.2, but the way the outcomes are recorded has changed. Here the outcomes are measurements instead of observations, and the outcome set becomes $S = \{0, 1\}$.

Example 3.4   ***Three tosses.***   Now let's toss the coin three times. We obtain the outcome set

$$S = \{\text{HHH, HHT, HTH, HTT, THH, THT, TTH, TTT}\}$$

where, for example, HTT denotes that heads will appear on the first toss followed by two tails, THT denotes alternating outcomes with the first one tails, followed by heads, and then tails.

Example 3.5   ***Count heads.***   Again change the experiment. Toss the coin three times and count the number of heads. To each of the eight outcomes of Example 3.4, we assign a number 0, 1, 2, or 3, depending on the number of H's. Here

$$S = \{0, 1, 2, 3\}$$

In Example 3.4, the outcome set contained eight possible outcomes; whereas in Example 3.5 there are four.

Example 3.6   ***Colored cards.***   A box contains six colored cards: 3 red, 2 white, and 1 blue. Select a card at random and observe the color. The outcome set is below:

$$S = \{\text{red, white, blue}\}$$

Example 3.7   ***Colored cards.***   Suppose the above box contains a different mix of red, white, and blue cards—say, 12 red, 6 white, and 1 blue. This is a different experiment from the one in Example 3.6, but the outcome set is the same.

Definition

> An event *is something that may or may not happen in an experiment.*

Example 3.8 **Roll a die.** Let's roll a standard six-sided die and count the number of dots on the top face. The outcome set is below:

$$S = \{1, 2, 3, 4, 5, 6\}$$

Here are some events for this experiment:

E: "observe an even number"

F: "observe a number larger than 2"

G: "do not observe the number 5"

H: "do not observe an odd number"

With each event, one can associate a subset of the outcome set consisting of the favorable outcomes. The event E is associated with the favorable outcomes $2, 4$, and 6; the event F is associated with the favorable outcomes $3, 4, 5$, and 6; the event G is associated with $1, 2, 3, 4$, and 6; and H is associated with $2, 4$, and 6.

> *Each event is determined by its favorable outcomes, not by the words used to describe it.*

Thus, every event can be identified with the subset of favorable outcomes. Without ambiguity, we can refer to E by either the phrase "observe an even number" or the subset $\{2, 4, 6\}$ of S. The events E: "observe an even number" and H: "do not observe an odd number" have the same favorable outcomes; thus, E and H are the same event, and we write $E = H = \{2, 4, 6\}$.

One can combine events to create new ones. For example,

E *and* F: "observe an even number which is greater than 2"

can be written

$$\text{"}E \text{ and } F\text{"} = E \cap F = \{4, 6\}$$

Similarly, we can write

$$\text{``}E \text{ or } F\text{''} = E \cup F = \{2, 3, 4, 5, 6\}$$

and

$$\text{``not } E\text{''} = E' = \{1, 3, 5\}$$

Note that the event "E or not E" is $E \cup E' = \{1, 2, 3, 4, 5, 6\} = S$. This is called the **certain event**.

Also, "E and not G" $= E \cap G' = \{\ \} = \emptyset$. This is called the **impossible event**. There is *only one* impossible event. Hence, "E and not E" is the same event as $E \cap G'$, which is the same as the event "observe fifteen dots."

Because E and *not* G cannot happen in the same execution of an experiment—that is, $E \cap G'$ is the impossible event—we say that E and *not* G are **incompatible** (or **mutually exclusive** or **disjoint**) **events**.

Example 3.9   *Waiting for a head.*   Toss a coin until heads appears. Then count the number of tosses. Any number of tosses is possible; this is an example of an experiment with an infinite outcome set. You can obtain heads on the first toss; or one tail followed by heads; or tails, tails, then heads; and so on. The outcome set is $S = \{1, 2, 3, 4, \cdots\}$.

Example 3.10   *Continuous outcome set.*   A paramedic team is responsible for answering calls in a certain region of southwest Michigan including Interstate Highway 94 between New Buffalo (mile marker 3) and Union Pier (mile marker 6). This section is a three-mile stretch of highway that is unremarkable, so that no part of the highway is more accident prone than any other. Suppose that a call comes in that an accident occurred on this section of the highway. Let X be this point measured in miles from the beginning of the highway. This means that X is a random number between 3 and 6. Identifying the exact location of an accident is a random experiment that can be modeled as follows. Imagine tossing a point onto the closed interval $[3, 6]$ of the real line, as if it were a pinball. Let the point bounce around until it comes to rest at a random spot X. This is a random experiment whose outcome set is the continuum of real numbers from 3 to 6.

Example 3.11   *Geiger counter.*   A Geiger counter registers a click whenever an alpha particle is detected. In the famous experiment described in Review Exercise 2.48 (on page 70), Rutherford and Geiger counted the number of alpha particles emitted by radioactive iodine in a 7.5-second interval. The experiment has an outcome set $S = \{0, 1, 2, \ldots\}$. Although the outcome set is presumably finite, it's not clear what the upper bound should be. Rutherford and Geiger observed no count larger than 10.

Example 3.12   ***Waiting for a click.*** Instead of counting the number of clicks of a Geiger counter in a time period, consider the time between consecutive clicks in seconds. This is a random experiment whose outcome can be any positive real number. The outcome set is the infinite interval $(0, \infty)$. Note that it is not possible to observe an outcome of 0 seconds because two alpha particle emissions occurring at the same time produce a single click.

A random experiment is **discrete** if there is only a finite or countably infinite number of outcomes. This means that it is possible to represent the possible outcomes by an exhaustive list $\{e_1, e_2, e_3, \ldots\}$ that either terminates or is infinite. Examples 3.1 through 3.9 are all discrete; all are finite except Example 3.9. Example 3.10 is not discrete because there is a continuum of possible outcomes; it is a fundamental result of Georg Cantor that this set cannot be formed into an exhaustive list. Similarly, Example 3.12 is not discrete.

How many distinct events can you describe for each of these experiments? Because events are determined by the subset of favorable outcomes, there are as many events as there are subsets of the outcome set. Obviously, for any infinite outcome set, such as in Examples 3.9–3.12, there are infinitely many events. However, for finite outcome sets, there is only a finite number of distinct events.

Example 3.13   ***Toss a coin.*** There are four events for the experiment of tossing a coin once.

$$\begin{aligned} \text{The impossible event:} \quad & \emptyset & = & \ \{\ \} \\ \text{Observe heads:} \quad & H & = & \ \{\text{Heads}\} \\ \text{Observe tails:} \quad & T & = & \ \{\text{Tails}\} \\ \text{The certain event:} \quad & S & = & \ \{\text{Heads, Tails}\} \end{aligned}$$

Example 3.14   ***Roll a die.*** In the rolling-a-die experiment of Example 3.8 on page 80, there are 2^6 possible events because a set of 6 elements has 2^6 subsets (the number 6 corresponds to the six elements and the number 2 corresponds to the two possibilities of each element either belonging or not belonging to the subset). There are[3]:

$$\binom{6}{0} = \tfrac{6!}{0!6!} = 1 \text{ event containing no outcomes}$$

$$\binom{6}{1} = \tfrac{6!}{1!5!} = 6 \text{ events with one outcome}$$

$$\binom{6}{2} = \tfrac{6!}{2!4!} = 15 \text{ events with two outcomes}$$

$$\binom{6}{3} = \tfrac{6!}{3!3!} = 20 \text{ events with three outcomes}$$

$$\binom{6}{4} = \tfrac{6!}{4!2!} = 15 \text{ events with four outcomes}$$

$$\binom{6}{5} = \tfrac{6!}{5!1!} = 6 \text{ events with five outcomes}$$

$$\binom{6}{6} = \tfrac{6!}{6!0!} = 1 \text{ event with six outcomes}$$

For observing three tosses of a coin as in Example 3.4, there are eight possible outcomes and 2^8 distinct events. For counting the number of heads in three tosses of a coin as in Example 3.5, there are 2^4 distinct events because S contains 4 possible outcomes.

3.3 PROBABILITIES OF EVENTS

The **probability** or **chance** of an event E is a measure of the likelihood of the event occurring. This measure is a number between 0 and 1 (or between 0% and 100%) written $Pr(E)$. There are three major ways of obtaining such a measure: theoretically, empirically, and subjectively.

3.3.1 THEORETICAL PROBABILITY

Theoretical probability appeals to **symmetry** in an experiment. This symmetry makes it reasonable to assign equal probability to certain events. Some-

---

[3]The symbol $\binom{n}{k}$, read "n take k" or "n choose k" is discussed in more detail in Chapter 5. Another symbol often used for this same quantity is $_nC_k$.

84   CHAPTER 3.   PROBABILITY

times you have to find the right way of looking at an experiment to reveal equally likely outcomes.

Example 3.15   **Roll a die.**   In rolling a die, assume that the die is fair, and hence all possible outcomes are equally likely. The probability of each possible outcome is $\frac{1}{6}$. Hence:

$$Pr(\text{"observe exactly 1 dot"}) = \frac{1}{6}$$
$$Pr(\text{"observe exactly 2 dots"}) = \frac{1}{6}$$
$$Pr(\text{"observe exactly 3 dots"}) = \frac{1}{6}$$
$$Pr(\text{"observe exactly 4 dots"}) = \frac{1}{6}$$
$$Pr(\text{"observe exactly 5 dots"}) = \frac{1}{6}$$
$$Pr(\text{"observe exactly 6 dots"}) = \frac{1}{6}$$

If all possible outcomes of an experiment are equally likely, then

$$\Pr(E) = \frac{\text{number of favorable outcomes}}{\text{number of possible outcomes}}$$
$$= \frac{\text{count of E}}{\text{count of S}}$$

This rule can immediately be used to compute the probabilities of all events for Example 3.15. For example,

$$Pr(\text{"observe less than 3 dots"}) = 2/6$$

$$Pr(\text{"do not observe 4 dots"}) = 5/6$$

The same rule applies to Examples 3.2, 3.3, and 3.4. However, to compute the probabilities for Examples 3.5, 3.6, 3.7, 3.9, and 3.10, we have to look at each of the experiments in a different way.

Example 3.16   **Roll a pair of dice.**   Roll a pair of dice and count the number of dots on the top faces. There are 11 possible outcomes, ranging from 2 dots to

12 dots but they are not equally likely. Now imagine that the dice can be distinguished— say, one of them is red and the other is green. The colors have no effect on the outcomes of the experiment but they permit us to break up the experiment in a way that has equally likely outcomes. We can list all of them in the following matrix of 36 pairs of numbers, where the first denotes the number of dots on the red die and the second number denotes the number of dots on the green die.

$$
\begin{array}{c|cccccc}
 & \multicolumn{6}{c}{\text{Green}} \\
 & 1 & 2 & 3 & 4 & 5 & 6 \\
\hline
1 & 11 & 12 & 13 & 14 & 15 & 16 \\
2 & 21 & 22 & 23 & 24 & 25 & 26 \\
\text{Red} \quad 3 & 31 & 32 & 33 & 34 & 35 & 36 \\
4 & 41 & 42 & 43 & 44 & 45 & 46 \\
5 & 51 & 52 & 53 & 54 & 55 & 56 \\
6 & 61 & 62 & 63 & 64 & 65 & 66 \\
\end{array}
$$

We can use this symmetry to compute the probabilities of each of the possible outcomes:

$$
\begin{aligned}
Pr(\text{``2 dots on top faces''}) &= Pr(\{11\}) = \frac{1}{36} \\
Pr(\text{``3 dots on top faces''}) &= Pr(\{12, 21\}) = \frac{2}{36} \\
Pr(\text{``4 dots on top faces''}) &= Pr(\{13, 22, 31\}) = \frac{3}{36} \\
Pr(\text{``5 dots on top faces''}) &= Pr(\{14, 23, 32, 41\}) = \frac{4}{36} \\
Pr(\text{``6 dots on top faces''}) &= Pr(\{15, 24, 33, 42, 51\}) = \frac{5}{36} \\
Pr(\text{``7 dots on top faces''}) &= Pr(\{16, 25, 34, 43, 52, 61\}) = \frac{6}{36} \\
Pr(\text{``8 dots on top faces''}) &= Pr(\{26, 35, 44, 53, 62\}) = \frac{5}{36} \\
Pr(\text{``9 dots on top faces''}) &= Pr(\{36, 45, 54, 63\}) = \frac{4}{36} \\
Pr(\text{``10 dots on top faces''}) &= Pr(\{46, 55, 64\}) = \frac{3}{36} \\
Pr(\text{``11 dots on top faces''}) &= Pr(\{56, 65\}) = \frac{2}{36} \\
Pr(\text{``12 dots on top faces''}) &= Pr(\{66\}) = \frac{1}{36} \\
\end{aligned}
$$

Example 3.17 **Count heads.** The experiment of tossing a fair coin three times and observing the number of heads as in Example 3.5 has four possible outcomes

86   CHAPTER 3.   PROBABILITY

that are not equally likely. However, using Example 3.4, which has equally likely outcomes, one can obtain the following probabilities:

$$\begin{aligned}
Pr(\text{``0 heads''}) &= Pr(\{\text{TTT}\}) = 1/8 \\
Pr(\text{``1 head''}) &= Pr(\{\text{HTT,THT,TTH}\}) = 3/8 \\
Pr(\text{``2 heads''}) &= Pr(\{\text{HHT,HTH,THH}\}) = 3/8 \\
Pr(\text{``3 heads''}) &= Pr(\{\text{HHH}\}) = 1/8
\end{aligned}$$

Example 3.18   *Box of numbered cards.*   Consider a box containing three cards marked with the numbers $0, 0$, and 1. The experiment consists of randomly drawing a card from this box and observing the number. The outcome set is $S = \{0, 1\}$. Becasue the possible outcomes are not equally likely, one cannot compute the probabilities by using the counting rule above. Instead, one pretends that the two cards marked 0 can be distinguished. For example, imagine that one of them is colored green, the other red. The colors result in three equally likely outcomes, but they have no effect on the chance of drawing a 0 or a 1. One can thus see that the chance of drawing 1 is 1/3 and the chance of 0 is 2/3.

Example 3.19   *Colored cards.*   Similarly, if a box contains three red cards, two white ones, and one blue one, as in Example 3.6 on page 79, then $S = \{\text{red, white, blue}\}$ and

$$\begin{aligned}
Pr(\text{``red''}) &= 3/6 \\
Pr(\text{``white''}) &= 2/6 \\
Pr(\text{``blue''}) &= 1/6
\end{aligned}$$

3.3.2   EMPIRICAL PROBABILITY

Often experiments are too complicated to break up into symmetrical parts. One can then use empirical evidence to obtain an **empirical probability** .

Example 3.20   *Illinois Lottery.*   One of the games in the Illinois Lottery is called Pick Four. Machines are designed to each randomly draw one ball out of ten balls numbered 0 through 9. Twice a day, except Sunday afternoons, four such machines choose a winning sequence. For example, on Friday, December 27, 1996, the winning sequences were 0 6 5 9 at midday and 2 2 9 2 in the evening. Appealing to symmetry, the theoretical probability that a machine chooses any one digit is $\frac{1}{10} = 0.10$, but some players contend that there may be small deviations from the theoretical probability. For example, there may be deviations due to differences in the weight of the painted numbers on the balls. Using historical data for the first 2684 draws in the year 1996, which we obtained from www.illinoislottery.com, we calculate the following empirical probabilities:

$$Pr(0) = \frac{265}{2684} = 0.09873323$$
$$Pr(1) = \frac{255}{2684} = 0.09500745$$
$$Pr(2) = \frac{261}{2684} = 0.09724292$$
$$Pr(3) = \frac{296}{2684} = 0.11028316$$
$$Pr(4) = \frac{263}{2684} = 0.09798808$$
$$Pr(5) = \frac{262}{2684} = 0.09761550$$
$$Pr(6) = \frac{249}{2684} = 0.09277198$$
$$Pr(7) = \frac{290}{2684} = 0.10804769$$
$$Pr(8) = \frac{284}{2684} = 0.10581222$$
$$Pr(9) = \frac{259}{2684} = 0.09649776$$

In Chapter 16, we will return to this example and consider whether these empirical probabilities differ significantly from the theoretical probabilities of 0.10. In other words, are the differences between the empirical values and the theoretical values about what you would expect from randomness? Or are the probabilities of the various numbers really different from the theoretical ones?

Example 3.21 ***Predicting weather.*** Historical data of atmospheric conditions can be used to predict the weather. If a meteorologist says that the chance of rain tomorrow is 30%, it means that, based on the history of this region, under similar atmospheric conditions it has rained 30% of the time.[4]

---

[4] There is an article by Robert G. Miller of the National Weather Service called *Very Short Range Weather Forecasting Using Automated Observations* in the book Judith Tanur, et al. (editors), *Statistics a Guide to the Unknown*, third edition, Duxbury Press, 1989. This article describes how automatically gathered weather data at airports are used to make predictions, on visibility, for example, 10 minutes hence.

3.3.3 MONTE CARLO METHOD (optional)

The Monte Carlo method, named after the city in the Monaco principality, uses computer simulations to find numerical solutions for various mathematical problems. It was developed during World War II to empirically solve complicated probabilities needed in the development of the atomic bomb. For example, we can use the Monte Carlo Method to find areas of complicated regions in a plane. Consider a region A contained in the rectangle $R = [a,b] \times [c,d] = \{(x,y) \mid a < x < b, c < y < d\}$. The Monte Carlo method consists of choosing random pairs of numbers (x,y) in R and observing whether or not they fall in the region A. The fraction of times the points fall inside the region A is approximately the area of A divided by the area of the rectangle R, $(b-a) \times (d-c)$. The same idea can be used to find volumes of solids and to evaluate multiple integrals. The Monte Carlo method is generally better than other numerical methods in evaluating multiple integrals of more than four variables.

Example 3.22   *Estimating Euclid's number π.*   You can write a simple computer program and use the Monte Carlo method to make a good approximation of the number π. Let A be the first quadrant of a circle of radius 1 centered at the origin. According to the formula πr^2 for the area of a circle, the area of A is $\pi/4$. Use a random number generator to find random points (x,y) with $0 \leq x < 1$, and $0 \leq y < 1$. Let n be the number of random points, and let k be the number of times the random point falls inside A (check whether $x^2 + y^2 < 1$). The number k/n is approximately $\pi/4$.

We tried this on a simple calculator with just $n = 25$ points and obtained $k = 22$. This makes our estimate of π equal to $4 \cdot \frac{k}{n} = 4 \cdot \frac{22}{25} = 3.52$. One can easily repeat this for n in the thousands to obtain excellent approximations of π.

NOTES ON TECHNOLOGY

Random Numbers

Minitab: To generate 100 random numbers between 0.0 and 1.0, select **Calc→Random Data→Uniform**. Fill in the dialog boxes to generate **100** rows of data, store in column(s) **C1**, and enter the lower limit **0.0** and upper limits **1.0**. If you want random integers, then select **Calc→Random Data→Integer** instead.

Excel: Type **=RAND()** in cell A1. Copy cell A1 to cells A2 through A100. This can be done by clicking on cell A1, clicking **Ctrl-C**, highlighting cells A2 through A100, and then hitting **ENTER**. The random numbers between 0.0 and 1.0 will appear in the first 100 rows of column A. For integer data, use the function =**Randbetween**. You can find a list of functions such as Rand and Randbetween by clicking on the icon **f$_x$**.

TI-83 Plus: Press **MATH**→**PRB**→**rand** and hit **ENTER**. Complete the entry **rand(100)** and then press **STO**→ followed by **L1**. You should see $rand(100) \to L_1$ on the display. Hit **ENTER**. This will place 100 random numbers between 0.0 and 1.0 in list L1. For random integers between 0 and 9, use the function **randint** instead of **rand** to display $randint(0, 9, 100) \to L_1$

3.3.4 SUBJECTIVE PROBABILITY

A geologist may state that there is a 25% chance of oil at a certain location underground. This is a **subjective probability**. If the expert is reliable, then approximately 25% of locations with such a rating will yield oil. Similarly, a financial analyst may predict at certain levels of confidence whether the value of specific stocks will rise or fall. Even though such predictions are based on subjective considerations (which are sometimes just educated intuitions), many decisions important to investors, business, and governments are based on subjective probabilities.

Suppose some expert says that based on meteorite evidence there is a 40% chance of life existing or having existed on Mars. Life on Mars is not a repeatable experiment. The 40% is a measurement of the expert's belief it is so. In theory, it should mean that for every 100 predictions made by this expert, each at a 40% level, you would expect about 40 of them to be true and the rest false. Of course, in practice it all depends on the expert.

Intuition can be misleading. For example, subjective probabilities about rare events are often biased by news media reports. In one study,[5] college students, among others, judged the frequency of 41 causes of death. After being told that the annual death toll from motor vehicle accidents was 50,000, they were asked to estimate the frequency of death from the other 40. These researchers found that dramatic and sensational causes of death tended to be overestimated and unspectacular causes tended to be underestimated. As

---

[5] Paul Slovic, Baruch Fischhoff, and Sarah Lichtenstein, *Understanding Perceived Risk*, which appears in the book Daniel Kahneman, Paul Slovic and Amos Tversky (editors), *Facts in Judgment Under Uncertainty: Heuristics and Biases*, Cambridge University Press, 1982. The authors do not describe what sampling procedures they used, so it is inappropriate to assume that other college students would have responded similarly.

examples, they point out that on average homicides were judged to be about as frequent as strokes but public health statistics indicate that strokes are roughly 11 times more frequent. The frequencies of death from tornadoes, pregnancy, and botulism were all greatly overestimated as well. Severely underestimated causes of death included asthma, diabetes, and tuberculosis. The authors noticed that many of the most severely underestimated causes of death are either unspectacular, claim victims one at a time, or are common in nonfatal form.

Probability evaluation is especially difficult when it involves events that are rare. Unfortunately, there are times when subjective probabilities are all that one has to work with. For example, insurance companies have to estimate the risks to nuclear power plants for events that have never happened, such as serious earthquakes, floods, direct-hit plane crashes, and sabotage.

EXERCISES FOR SECTION 3.3

C • **3.1** Use a graphing calculator or computer program to repeat the above Monte Carlo estimate of Euclid's number π, using at least 100 points.

C **3.2** Use a graphing calculator or computer program to make a Monte Carlo estimate of the area $\int_0^1 x^2 \, dx$. Use at least 100 points.

3.4 RULES OF PROBABILITY

Regardless of which interpretation of probability one uses, the theory assumes certain basic rules or axioms. The rules and axioms that we give are not intended to be a minimum set; it is possible to prove some of them from others:

For any event **E**, *we have* $0 \leq \Pr(\mathbf{E}) \leq 1$

Impossible event: $\Pr(\emptyset) = 0$

There is a difference between an event having probability zero and an event being impossible. If one randomly chooses a point in the closed interval $[0, 1]$, the probability of getting a point in a subinterval I is the length of I. The probability of getting a single point—say, the point $1/2$—is zero, because the length of a point is zero. Yet this is not impossible. In the accident on the road Example 3.10, each accident must occur at some point, but at each point the probability is zero. If $Pr(E) = 0$, we say that the event E is **almost impossible** or of **measure zero**.

$$\textit{Certain event: } \mathbf{Pr(S) = 1}$$

Just as there is a difference between an event having probability zero and being impossible, there is a difference between an event having probability one and the event being certain. In Example 3.9 (tossing a fair coin until a head appears), the event "number of tosses is finite" has probability 1, but it is not certain. If $Pr(E) = 1$, we say that the event E is **almost certain** or **sure**.

For a given event E, the **opposite event** (also called the **complement event**) is the event that E does not happen. It is sometimes written E' or "not E."

$$\textit{Opposites: } \mathbf{Pr(\textit{not } E) = 1 - Pr(E)}$$

The complement of an almost impossible event is almost certain.

Example 3.23 *Dots on a pair of fair dice.* Roll a pair of fair dice and find the probability that the numbers on the top faces are different. The opposite event is the event that the two faces show the same number of dots. Notice that, of the 36 possible outcomes listed in Example 3.16, there are six ways of getting the same number on both rolls: $\{11, 22, 33, 44, 55, 66\}$.

$$Pr(\text{different}) = 1 - Pr(\text{same}) = 1 - 6/36 = 30/36 = 5/6$$

Example 3.24 **Count the aces.** Consider the experiment of rolling a pair of fair dice and then counting the aces. We use the rule of opposites to compute the following probabilities:

$$Pr(\text{not all aces}) = 1 - Pr(\text{all aces}) = 1 - \frac{1}{36} = \frac{35}{36}$$

and

$$Pr(\text{at least one ace}) = 1 - Pr(\text{no aces}) = 1 - \frac{25}{36} = \frac{11}{36}$$

Inclusion: **If $E \subset F$, then $Pr(E) \leq Pr(F)$**

3.4.1 ADDITION RULES

Example 3.25 **Draw a card.** Suppose we draw a card from a well-shuffled standard 52-card deck[6] and observe the card drawn. For those who are unfamiliar with such a deck, there are four suits: ♡ (hearts), ♢ (diamonds), ♣ (clubs), and ♠ (spades). Each of the suits has 13 cards labeled A (ace), $2, 3, 4, 5, 6, 7, 8, 9, 10, J$ (jack), Q (queen), and K (king).

The entire outcome set can be represented as follows.

$$S = \left\{ \begin{array}{cccccccccccccc} \heartsuit A & \heartsuit 2 & \heartsuit 3 & \heartsuit 4 & \heartsuit 5 & \heartsuit 6 & \heartsuit 7 & \heartsuit 8 & \heartsuit 9 & \heartsuit 10 & \heartsuit J & \heartsuit Q & \heartsuit K \\ \diamondsuit A & \diamondsuit 2 & \diamondsuit 3 & \diamondsuit 4 & \diamondsuit 5 & \diamondsuit 6 & \diamondsuit 7 & \diamondsuit 8 & \diamondsuit 9 & \diamondsuit 10 & \diamondsuit J & \diamondsuit Q & \diamondsuit K \\ \clubsuit A & \clubsuit 2 & \clubsuit 3 & \clubsuit 4 & \clubsuit 5 & \clubsuit 6 & \clubsuit 7 & \clubsuit 8 & \clubsuit 9 & \clubsuit 10 & \clubsuit J & \clubsuit Q & \clubsuit K \\ \spadesuit A & \spadesuit 2 & \spadesuit 3 & \spadesuit 4 & \spadesuit 5 & \spadesuit 6 & \spadesuit 7 & \spadesuit 8 & \spadesuit 9 & \spadesuit 10 & \spadesuit J & \spadesuit Q & \spadesuit K \end{array} \right\}$$

Consider the event Spade = "draw a spade" =

$$\{\spadesuit A, \spadesuit 2, \spadesuit 3, \spadesuit 4, \spadesuit 5, \spadesuit 6, \spadesuit 7, \spadesuit 8, \spadesuit 9, \spadesuit 10, \spadesuit J, \spadesuit Q, \spadesuit K\}$$

Because all possible outcomes are equally likely,

$$Pr(\text{Spade}) = 13/52$$

---

[6] The online Chance Web page www.geom.umn.edu/docs/education/chance/ has an example showing that shuffling a new deck in the usual manner a few times is not enough to make every sequence of cards equally likely.

Similarly, the event

$$\text{Ace} = \text{"draw an ace"} = \{\heartsuit A, \diamondsuit A, \clubsuit A, \spadesuit A\}$$

has probability
$$Pr(\text{Ace}) = 4/52$$

What about the event "Spade or Ace"? Let's count the number of favorable outcomes. There are 13 spades and 4 aces, but one outcome $\spadesuit A$ has been counted twice, so it must be subtracted to get a total of $13 + 4 - 1 = 16$ favorable outcomes. Thus,

$$Pr(\text{Spade or Ace}) = 16/52 \quad \text{or}$$

$$Pr(\text{Spade or Ace}) = Pr(\text{Spade}) + Pr(\text{Ace}) - Pr(\text{Spade and Ace})$$

In general, we have the basic addition rule, which we write in set notation:

Basic Addition Rule

$$\boxed{\mathbf{Pr(E_1 \cup E_2) = Pr(E_1) + Pr(E_2) - Pr(E_1 \cap E_2)}} \qquad (3.1)$$

We can see this from the **Venn diagram** in Figure 3.1.

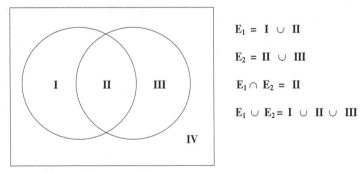

Figure 3.1 Venn diagram for two events

Similarly, guided by the Venn diagram in Figure 3.2, we can write a formula for three events:

$$\begin{aligned}
Pr(E_1 \cup E_2 \cup E_3) &= Pr(E_1) + Pr(E_2) + Pr(E_3) \\
&\quad - Pr(E_1 \cap E_2) - Pr(E_1 \cap E_3) - Pr(E_2 \cap E_3) \\
&\quad + Pr(E_1 \cap E_2 \cap E_3)
\end{aligned} \qquad (3.2)$$

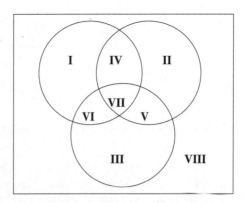

Figure 3.2 Venn diagram for three events

Example 3.26   *Reading news.*   Consider a particular group of students and choose a student at random. Let T be the event that the student has seen the latest issue of *Time* magazine, N the event that the student has seen the latest issue of *Newsweek*, and W the event that the student has seen the latest issue of *U.S. News & World Report*. Suppose we have the following probabilities:

$$Pr(T) = 0.23, \ Pr(N) = 0.25, \ Pr(W) = 0.30$$
$$Pr(T \cap N) = 0.10, \ Pr(T \cap W) = 0.15, \ Pr(N \cap W) = 0.18$$
$$Pr(T \cap N \cap W) = 0.05$$

These probabilities correspond to the percentage of students in each of the sets. To compute the percentage of students who have seen at least one of these news magazines, one can use Equation (3.2) above:

$$Pr(T \cup N \cup W) = 0.23 + 0.25 + 0.30 - 0.10 - 0.15 - 0.18 + 0.05 = 0.40 = 40\%$$

What percentage of the students did not see any of these news magazines? The event of having seen none is the opposite of the event of having seen at least one. Using the rule of opposites, we obtain the following probability:

$$\begin{aligned} Pr(\text{"student has seen none"}) &= 1 - Pr(\text{"student has seen at least one"}) \\ &= 1 - Pr(T \cup N \cup W) \\ &= 1 - 0.40 = 0.60 \end{aligned}$$

Alternatively, one can compute the same probabilities from the Venn diagram above by assigning regions I, II, and III to T, N, and W, respectively. Then the region VII has the probability $Pr(T \cap N \cap W) = 0.05$, the region V has probability $Pr(N \cap W) - 0.05 = 0.18 - 0.05 = 0.13$, and so on, until finally region VIII ends up with what is left over, namely, $1 - 0.40 = 0.60$.

For four events—E_1, E_2, E_3, and E_4—a Venn diagram will lead to the following formula with 15 terms:

$$Pr(E_1 \cup E_2 \cup E_3 \cup E_4) = Pr(E_1) + Pr(E_2) + Pr(E_3) + Pr(E_4)$$
$$- Pr(E_1 \cap E_2) - Pr(E_1 \cap E_3) - Pr(E_1 \cap E_4) - Pr(E_2 \cap E_3) - Pr(E_2 \cap E_4) - Pr(E_3 \cap E_4)$$
$$+ Pr(E_1 \cap E_2 \cap E_3) + Pr(E_1 \cap E_2 \cap E_4) + Pr(E_1 \cap E_3 \cap E_4) + Pr(E_2 \cap E_3 \cap E_4)$$
$$- Pr(E_1 \cap E_2 \cap E_3 \cap E_4) \tag{3.3}$$

For five events, the corresponding formula has 31 terms. As you can see, as the number of events goes up, the number of terms in the addition formula increases rapidly.

There is one important special case that is an exception. Recall that two events E_1 and E_2 are disjoint if "E_1 and E_2" is the impossible event. A collection of events is said to be **pairwise disjoint** if, for all $i \neq j$, E_i and E_j are disjoint ($E_i \cap E_j = \emptyset$). A Venn diagram for three disjoint events is shown in Figure 3.3.

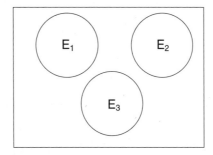

Figure 3.3 Venn diagram for three disjoint events

For pairwise disjoint events, the probabilities of all the intersections in Equations 3.1, 3.2, and 3.3 are zeros. Thus, we obtain the following *Addition Rule*.

Addition Rule

> If $E_1, E_2, E_3, \ldots$ *are pairwise disjoint, then*
>
> $$\Pr(E_1 \cup E_2 \cup E_3 \cup \cdots) = \Pr(E_1) + \Pr(E_2) + \Pr(E_3) + \cdots.$$
>
> *This rule holds for both a finite and an infinite number of events.*

One has to be careful in checking that events are pairwise disjoint. The pairwise disjoint condition is quite strict.

Example 3.27   **Toss three coins.**   In tossing three coins, consider the following three events:

$E_1 =$ "at least two heads" $= \{$HHH, HHT, HTH, THH$\}$

$E_2 =$ "first toss tails" $= \{$THH, THT, TTH, TTT$\}$

$E_3 =$ "last toss tails" $= \{$HHT, HTT, THT, TTT$\}$

Note that $E_1 \cap E_2 \cap E_3 = \emptyset$; that is, the three events are *disjoint*, but they are not *pairwise disjoint* because

$$E_1 \cap E_2 = \{\text{THH}\}$$
$$E_1 \cap E_3 = \{\text{HHT}\} \text{ and}$$
$$E_2 \cap E_3 = \{\text{THT, TTT}\}$$

> *CAUTION:* **Disjoint** *is not the same as* **pairwise disjoint.**

Example 3.28   **Rolls of a fair die.**   A fair die is rolled four times. On each roll, the probability of an ace is 1/6, but the events are not pairwise disjoint. So to find the probability of at least one ace in four rolls, we cannot simply add the probabilities $1/6 + 1/6 + 1/6 + 1/6$. Because the experiment can come out $6^4 = 1296$ ways, counting the favorable outcomes is not practical. In the next section (e.g., Example 3.31 on page 103), we will show how to calculate this probability by using the rule of opposites and then applying a multiplication rule.

3.4.2 INDEPENDENCE AND CONDITIONAL PROBABILITY

You might think that half of all married people are male, and the other half female. However, let's look at the U.S. census bureau data entitled *Marital Status and Living Arrangements of Adults 18 Years Old and Over (March 1995)*, which has some surprises.[7] The following table summarizes these data (numbers are in thousands):

Marital status	Males	Females	Total
Married	57,730	58,931	116,661
Unmarried	34,277	40,658	74,935
Total	92,007	99,589	191,596

How many married couples are there in the United States? A more detailed breakdown of the marital status data is given in the following table.

Marital status	Males	Females	Total
Married, spouse present	54,934	54,905	109,839
Married, spouse absent	2,796	4,026	6,822
Never married	24,628	19,312	43,940
Widowed	2,282	11,080	13,362
Divorced	7,367	10,266	17,633
Total	92,007	99,589	191,596

Converting to percentages, the table looks like this:

---

[7] The U.S. Bureau of the Census Web page is www.census.gov. The data collected by the census bureau are provided by household members. We do not know why the reported data are inconsistent with the obvious (one husband and one wife per marriage). We can only speculate as to why some men would deny being married or married more than once, why some women would claim a marriage that does not exist, or why men and women have different ideas as to what constitutes a marriage. Further investigation is needed to determine whether these unusual data values were caused by an important phenomena or if this is just an artifact of the data-collection method.

Marital status	Males (%)	Females (%)	Total (%)
Married, spouse present	28.7	28.7	57.3
Married, spouse absent	1.5	2.1	3.6
Never married	12.9	10.1	22.9
Widowed	1.2	5.8	7.0
Divorced	3.9	5.4	9.2
Total	48.0	52.0	100.0

Notice that 9.2% of all adults in the United States are divorced, and 52.0% of all adults are female. It's a common mistake to conclude that 52.0% of 9.2% of all people are divorced females. That calculation leads to the value 4.8%, yet the table shows that actually 5.4% of all adults in the United States are divorced females.

This effect is even greater for the category of people never married. Choose an adult (age 18 years and over) from the population at random. For each event E, $Pr(E)$, is the proportion of adults in the population of type E. In particular,

$$Pr(\text{never married}) = \frac{43940}{191596} = 0.2293$$

If we restrict our outcome set to females, the event of being never-married is called a conditional event, written "never married | female" and read "never married given female." Although, for the entire adult U.S. population, the chance that an adult has never been married is 0.23, looking only at the females, the chance that she has never been married is only

$$Pr(\text{never married} \mid \text{female}) = \frac{19312}{99589} = 0.1939$$

Because the probabilities are different, we say that the events of "never married" and "female" are **dependent events**.

Definition

> *Events* **E** *and* **F** *are said to be* **independent** *if*
> **Pr(E | F) = Pr(E)**

Similarly, the probability

$$Pr(\text{widowed}) = \frac{13362}{191596} = 0.0697$$

is an unconditional probability because it involves the original outcome set of the entire adult U.S. population, whereas

$$Pr(\text{widowed} \mid \text{female}) = \frac{11080}{99589} = 0.1113$$

is a conditional probability because it is a probability for the restricted sample set of adult females. The events "widowed" and "female" are dependent because the conditional and unconditional probabilities are not the same.

Note that

$$\begin{aligned}
Pr(\text{widowed and female}) &= \frac{11080}{191596} \\
&= \frac{99589}{191596} \cdot \frac{11080}{99589} \\
&= Pr(\text{female}) \cdot Pr(\text{widowed} \mid \text{female})
\end{aligned}$$

Basic Multiplication Rule

> **For any two events E and F, we have**
> $$\mathbf{Pr(E \text{ and } F) = Pr(E) \cdot Pr(F \mid E)}$$

Solving for $Pr(F \mid E)$ we obtain the following:

Formula for conditional probabilities

> $$\mathbf{Pr(F \mid E) = \frac{Pr(E \text{ and } F)}{Pr(E)}} \quad \textit{provided} \ \ \mathbf{Pr(E) \neq 0}$$

Applying the basic multiplication rule to independent events we obtain the following:

Multiplication Rule

> **If events E and F are independent, then**
> $$\mathbf{Pr(E \text{ and } F) = Pr(E) \cdot Pr(F)}$$

We can say that event E and F are independent if the probability of E is the same regardless of whether F did or did not occur. Formally, this rule takes the following form:

> **Events E and F are independent if**
> $$\mathbf{Pr(E \mid F) = Pr(E \mid \text{not} F)}$$

Example 3.29 **Draw two cards.** Consider drawing two cards from a standard deck of 52 cards. If we draw with replacement, we obtain the following probabilities:

$$Pr(\text{``first card is an ace''}) = 4/52$$

$$Pr(\text{``second card is either a queen or a king''} \mid \text{``first card is an ace''}) = 8/52$$

$$Pr(\text{``first card is an ace and second card is either a queen or a king''}) = 4/52 \cdot 8/52$$

If we draw without replacement, then the probabilities are

$$Pr(\text{``first card is an ace''}) = 4/52$$

$$Pr(\text{``second card is either a queen or a king''} \mid \text{``first card is an ace''}) = 8/51$$

$$Pr(\text{``first card is an ace and second card is either a queen or a king''}) = 4/52 \cdot 8/51$$

Now consider the *unconditional* event "second card is a queen or a king." For draws with replacement, we clearly have

$$Pr(\text{``second card is a queen or a king''}) = 8/52$$

What about the probability when draws are made without replacement? Surprisingly, that probability is the same as for draws with replacement:

$$Pr(\text{``second card is a queen or a king''}) = 8/52$$

To see this, imagine that the two cards are drawn face down without looking at them. The chance that the first card is an ace is 4/52. Because we have seen neither card, the chance that the second is an ace is also 4/52. In fact, if all of the cards were to be laid out face down, and you pointed to *any* one card, the chance that it's an ace is also 4/52. Similarly, the chance that the first card is either a queen or a king is 8/52. The chance that the second card is either a queen or a king is also 8/52.

Suppose two cards are drawn from a population box $\{\mathbf{0}, \mathbf{0}, \mathbf{1}\}$. Let E be the event that the first card is a zero, and let F be the event that the second card is a zero. Both unconditional probabilities $Pr(E) = 1/3$ and $Pr(F) = 1/3$ are equal. If the first card is replaced before the second is drawn, then the events are independent because $Pr(F \mid E) = 1/3$. But if the first card is not replaced, then we have $Pr(F \mid E) = 1/2$.

> - *For population box models, draws made with replacement are independent.*
>
> - *Draws from a finite box made without replacement are dependent.*
>
> - *If we consider boxes with an infinite number of cards, draws either with or without replacement are independent.*

One of the reasons for modeling with boxes is that it then becomes intuitively obvious whether or not events are independent. For example, people sometimes study recent pattern in the state lottery winnings, such as for the Illinois Pick Four, and play combinations of digits that are "overdue." A box model of the game eliminates extraneous information and makes it clear that on each draw, the chance of a particular digit is the same (theoretically 1/10), regardless of how long it has failed to come up.

Warning

> *The concepts of disjointness and independence are entirely different.*

To say that two events E and F are independent means that knowledge of the occurrence of E gives no information of the occurrence (hence, does not change the probability) of F. If they are disjoint, then the occurrence of E will tell you that the occurrence of F is impossible (hence, of zero probability). Suppose that $Pr(E)$ and $Pr(F)$ are positive. Then

> *Events* **E** *and* **F** *independent means:* $\mathbf{Pr(F \mid E) = Pr(F)}$
> *Events* **E** *and* **F** *disjoint means:* $\mathbf{Pr(F \mid E) = 0}$

Similarly,

> *Events* **E** *and* **F** *independent means:* $\mathbf{Pr(F \cap E) = Pr(F) \cdot Pr(E)}$
> *Events* **E** *and* **F** *disjoint means:* $\mathbf{Pr(F \cap E) = 0}$

If we consider three events E_1, E_2, and E_3, then we can extend the multiplication rule as follows:

$$Pr(E_1 \text{ and } E_2 \text{ and } E_3) = Pr(E_1) \cdot Pr(E_2 \mid E_1) \cdot Pr(E_3 \mid E_1 \text{ and } E_2)$$

Similarly, it can be extended to any number of events.

Events $E_1, E_2, E_3, \ldots$ are said to be **independent** if for each $i = 1, 2, \ldots$, the unconditional probability of the event E_i is the same as the conditional probability of E_i, given that E_1 and E_2 and E_3 and $\cdots$ and E_{i-1} have occurred. Thus, $Pr(E_2) = Pr(E_2 \mid E_1)$, $Pr(E_3) = Pr(E_3 \mid E_1 \text{ and } E_2)$, $Pr(E_4) = Pr(E_4 \mid E_1 \text{ and } E_2 \text{ and } E_3)$, and so on. These multiplication rule formulas hold for every ordering of the events. For independent events, conditional and unconditional probabilities are the same. This gives the multiplication rule an important simplicity.

Multiplication Rule

> If $\mathbf{E_1, E_2, E_3, \ldots}$ *are independent, then*
>
> $\mathbf{Pr(E_1} \text{ and } \mathbf{E_2} \text{ and } \mathbf{E_3} \text{ and } \cdots) = \mathbf{Pr(E_1) \cdot Pr(E_2) \cdot Pr(E_3)} \cdots$

Example 3.30   *Tosses of a fair coin.* Toss a fair coin six times. The tosses are independent. If we let E_i be the event of getting a head on the ith toss in the formula above, then $Pr(\text{"all heads"}) = (1/2)^6$. The opposite event has probability $Pr(\text{"at least one tail"}) = 1 - (1/2)^6$.

Remember that the addition rule has three key words:

disjoint, or, add

Also, the multiplication rule has three key words:

independent, and, multiply

The three key words for each rule:

disjoint	*independent*
or	*and*
add	*multiply*

All three key words have to match. If not, the rule will give you the wrong answer. However, the following De Morgan's Laws of logic can be used, along with the rule of opposites, to change events containing *or* to those containing *and*, and vice versa.

De Morgan's Laws

$$\text{not}(E \text{ or } F) = (\text{not } E) \text{ and } (\text{not } F)$$
$$\text{not}(E \text{ and } F) = (\text{not } E) \text{ or } (\text{not } F)$$

Example 3.31   *Risk of being shot down.*   Tom Clancy, in his novel *Red Storm Rising*, writes that "A pilot may think a 1 percent chance of being shot down in a given mission acceptable, then realizes that fifty such missions make it a 40 percent chance." The addition rule is not appropriate because we do not have disjointness; that is, we cannot add up 1% 50 times. It seems that Clancy assumed independence (was it reasonable?) to obtain

$$Pr(\text{"surviving 50 missions"}) = (0.99)^{50} = 0.6065$$

and then he used the rule of opposites.

EXERCISES FOR SECTION 3.4

- **3.3** If a fair coin is tossed four times, find the probabilities of the following events:

  a. Exactly three of the four tosses come up heads.
  b. The last two tosses come up heads.
  c. The third toss comes up heads.
  d. All tosses come up the same.

3.4 If a fair die is rolled three times, find the probabilities of the following events:

   a. All of the rolls show an even number of dots.
   b. The last two rolls show an even number of dots.
   c. The third roll shows an even number of dots.
   d. Every roll show a single dot.
   e. Every roll show the same number of dots.

3.5 Sketch a Venn diagram and label the probabilities of the regions $E \cap F$, $E \cap F'$, $F \cap E'$, and $(E \cup F)'$, if $Pr(E) = 0.35, Pr(F) = 0.55$, and $Pr(E \cup F) = 0.75$.

3.6 Sketch a Venn diagram and label the probabilities of the regions $E \cup F$, $E \cap F'$, $F \cap E'$, and $(E \cup F)'$, if $Pr(E) = 0.40, Pr(F) = 0.55$, and $Pr(E \cap F) = 0.15$.

3.7 Use the census tables in Section 3.4.2 on page 97 to find the probabilities of the unconditional events "female" and "married with spouse absent." Find the probabilities of the events "female | married with spouse absent" and "female and married with spouse absent" and explain why they are different.

3.8 Suppose $Pr(E) = 0.55, Pr(F) = 0.40$, and $Pr(F \mid E) = 0.20$. Find
   a. $Pr(E \cap F)$ b. $Pr(E' \cup F')$ c. $Pr(E' \cap F')$ d. $Pr(E \mid F)$

3.9 Suppose $Pr(E) = 0.50, Pr(F) = 0.20$ and $Pr(F \mid E) = 0.30$. Find
   a. $Pr(E \cap F)$ b. $Pr(E \cup F)$ c. $Pr(E' \cap F')$ d. $Pr(E \mid F)$

3.10 Suppose that a survey of married couples in a certain city shows that 20% of the husbands watched the 2003 Superbowl football game and 8% of the wives. Also, if the husband watched, then the probability that the wife watched increased to 25%. Find the probabilities of the following events: (Hint: See Exercise 3.9.)

   a. The couple both watched
   b. At least one watched
   c. Neither watched
   d. The husband watched given that the wife watched

3.11 Roll a fair die four times. What is the probability of at least one ace? Why is the answer not $\frac{1}{6} + \frac{1}{6} + \frac{1}{6} + \frac{1}{6} = \frac{4}{6} = \frac{2}{3}$?

3.12 A company makes optical lenses under contract to the U.S. military. The lenses are ground to precise specifications and are shipped in lots of 100. Military inspectors check 2 different lenses out of each lot of 100. Let E_1 be the event that the first lens inspected fails inspection and E_2 be the event that the second fails. If either one of the two fails, then the entire shipment of 100 lenses is returned. Suppose that 3 lenses in the shipment are bad. Find the probability that the shipment is rejected.

- **3.13** The event "snake eyes" is a pair of aces in the roll of two dice. In 24 rolls of a pair of fair dice, what is the chance of getting the event snake eyes at least once?

3.14 **Problem of de Méré.**[8] Which is more likely: at least one ace in four rolls of a fair die or at least one pair of aces in 24 rolls of a pair of fair dice? Be sure to find both probabilities.

- **3.15** (Proof) Suppose that $Pr(E)$ and $Pr(F)$ are positive. Recall that event E is **independent** of F, if $Pr(E \mid F) = Pr(E)$. Show that if E is independent of F, then F is independent of E, that is, $Pr(F \mid E) = Pr(F)$.

3.16 (Proof) Show that if events E and F are independent, then the events E and *not* F are also independent.

3.5 TREE DIAGRAMS

Whenever you deals with experiments that involve several stages, difficult problems can generally be represented by tree diagrams.

Example 3.32 *Draw two cards.* If you draw a card from a standard 52-card deck, the chance that it is a heart is $13/52 = 1/4$. Suppose a second card is drawn without replacing the first. What is the chance that the second card is a heart? If you know whether or not the first card was a heart, then the chance that the second one is a heart has the conditional probability: $12/51$ provided the first card was a heart, and $13/51$ provided the first card was not a heart. On the other hand, what if you do not see the first card and do not know whether it is or is not a heart? If you look at the problem in the correct way, it becomes clear that the chance that the second card is a heart is $13/52 = 1/4$, the same as the chance that the first card is a heart. This is clear from the independence of the draws. The chance that the first card is a heart is the same as the chance that the second card is a heart, is the same as the chance that the third card is a heart, and so on. The chance is $13/52$ for all cards. Sometimes one cannot "see it the right way" but a tree will reveal the truth. The tree is shown in Figure 3.4.

---

[8] In the seventeenth century, because of a misuse of the Addition Rule, it was commonly thought that the two events described here were equally likely. The Chevalier de Méré, an experienced gambler, observed that the two events were not equally likely and claimed a fallacy in the theory of numbers. He mentioned this to Blaise Pascal, who wrote a solution to this paradox in a letter to Pierre de Fermat in 1654.

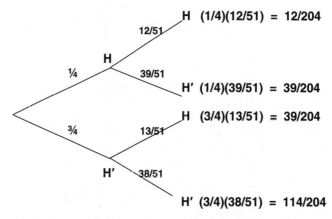

Figure 3.4 Tree diagram for hearts in a hand of two cards

Here are some rules for tree diagrams. The branches at each stage represent the possible outcomes of that stage. The probability written on each branch is the conditional probability of that outcome at that stage. The conditional probabilities at each branching (fork) must add up to 1. The terminal probabilities consist of the products to the conditional probabilities leading to that outcome. The sum of all of the terminal probabilities will always be equal to 1.

Notice that the chance that the second card is a heart is the sum of the terminal probabilities ending in heart, which is

$$(1/4) \cdot (12/51) + (3/4) \cdot (13/51) = 1/4\{12/51 + 39/51\} = 1/4$$

Example 3.33 *Birthday paradox.* Suppose a dozen people are attending a meeting. What is the probability that at least two of the people share the same birthday? Here we assume that people's birth date events are independent and that each of the 365 days is equally likely. Imagine the people are lined up in a row. The first person is asked for his or her birthday. Given that first birthday, the probability that the second person does not share the same birthday is 364/365. Given the first two birthdays, the probability the third does not share a birthday with the first two is 363/365. Given the first b birthdays, the probability the next person does not share a birthday with any that went before is $(365 − b)/365$. So the probability that none share a birthday is the following:[9]

$$\prod_{b=0}^{11} \frac{365-b}{365} = \left(\frac{364}{365}\right)\left(\frac{363}{365}\right)\cdots\left(\frac{354}{365}\right) = \frac{364!}{353! \cdot 365^{11}} = 0.83298$$

So the chance that at least two share a birthday is $1 - 0.83298 = 0.16702$.

EXERCISES FOR SECTION 3.5

- **3.17** Consider a two-stage experiment, where the event E is in the first stage and F is in the second stage, and the following probabilities:

  $Pr(E) = 0.30 \quad Pr(F \mid E) = 0.80 \quad Pr(F \mid E') = 0.60$

  Make a tree diagram including all branch probabilities branches and all terminal probabilities.

- **3.18** Consider a two-stage experiment, where the event E is in the first stage and F is in the second stage, and the following probabilities.

  $Pr(E') = 1/3 \quad Pr(F \mid E') = 1/4 \quad Pr(E \cap F) = 1/12$

  Make a tree diagram including all branch probabilities branches and all terminal probabilities:

- **3.19** Suppose that 10 people attend a meeting. What is the probability that at least two people share a common birthday?

- **3.20** Suppose that 16 people attend a meeting. What is the probability that at least two people share a common birthday?

- **3.21** How many people would have to attend a meeting so that there is at least a 50% chance that two people share a birthday?

3.6 BAYES' METHOD

Bayes' method is a way of updating the probability of an event when you are given additional information.[10] In a two-stage experiment, you can sometimes

---

[9] Here the symbol $\prod_{b=0}^{11}$ is used to denote the product of factors, analogous to how the symbol $\sum_{b=0}^{11}$ is used to denotes the sum of terms.

[10] Thomas Bayes (1702–1761) was a British clergyman and amateur mathematician. One of his papers, published in 1763 after his death, gave rise to *Bayesian inference*, whereby prior estimates of probabilities (often based on subjective probabilities) are updated using empirical data. In this section, we consider a special case of his approach. His biography can be found at the website www-gap.dcs.st-and.ac.uk/~history/Mathematicians/Bayes.html.

108   CHAPTER 3.   PROBABILITY

observe the outcome of the second stage without knowing the outcome of the first. Under such circumstances you can use Bayes' method to compute the chance some event occurred in the first stage after the second stage has been observed. We illustrate this with medical screening tests.

First some definitions. The **prevalence** of a disease is the probability of having the disease; it is found by dividing the number of people with the disease by the number of people in the population under study.

Screening tests for diseases are never completely accurate. Such tests generally produce some false negatives and some false positives. The **sensitivity** of a screening test refers to the probability that the test is positive given that the person has the disease. The **specificity** of a screening test is the probability that the test is negative when the person does not have the disease. See Figure 3.5.

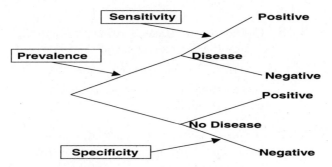

Figure 3.5 Tree diagram for medical testing

If you know nothing about a subject, then the probability that subject has the disease is just the prevalence of the disease. Now suppose this subject tests positive. This does not necessarily mean that the person has the disease because the test may be a true positive or a false positive. How does the information about the positive test change the probability that the subject actually has the disease? The **predictive value positive** is the probability that the person has the disease given that the test is positive:

$$PV^+ = Pr\,(\text{disease} \mid \text{positive}) = \frac{Pr\,(\text{disease and positive})}{Pr\,(\text{positive})}$$

The numerator is a terminal probability of the tree. Applying to the numerator, the multiplication rule suggested by the tree, we have the following:

$$Pr\,(\text{disease and positive}) = Pr\,(\text{disease}) \cdot Pr\,(\text{positive} \mid \text{disease})$$

Similarly, the denominator $Pr(\text{positive})$ is the sum of two terminal probabilities of the tree. If we partition the denominator into these mutually exclusive parts and apply a multiplication rule to each part, we obtain

Bayes' formula for screening tests

$$PV^+ = \frac{Pr(\text{disease}) \cdot Pr(\text{positive} \mid \text{disease})}{Pr(\text{disease}) \cdot Pr(\text{positive} \mid \text{disease}) + (1 - Pr(\text{disease})) \cdot (1 - Pr(\text{negative} \mid \text{not disease}))}$$

Similarly, the **predictive value negative** is the probability that the person does not have the disease given that the test is negative:

$$PV^- = Pr(\text{no disease} \mid \text{negative}) = \frac{Pr(\text{no disease and negative})}{Pr(\text{negative})}$$

Example 3.34 *Mammography.* Consider mammography as a screening test for breast cancer. The prevalence of breast cancer in women between 20 and 30 years of age is about 1 in 2500 or 0.04%. The sensitivity of mammography is 80% and the specificity is 90%. Find the predictive value positive: that is, if a woman in this population has a positive mammogram, find the chance that she has breast cancer.

Solution See Figure 3.6.

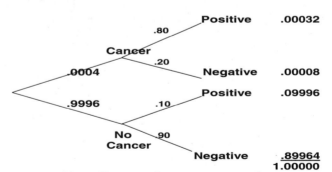

Figure 3.6 Tree diagram for mammography

$$PV^+ = Pr(\text{breast cancer} \mid \text{positive}) = \frac{0.00032}{0.00032 + 0.09996} = \frac{32}{10028} = 0.003191$$

Notice that a woman in her twenties who receives a positive mammogram has less than 1/3 of 1% chance of actually having breast cancer. How is this possible for a test with a sensitivity of 80% and a specificity of 90%? It is due to the small risk of breast cancer among young women, which has the consequence that most of the errors of the test are false positives. The fact that almost 99.7% of the positive mammograms turn out to be false alarms is the reason that most experts do not advise mammography before the age of 40.

However, notice that the risk of breast cancer was $0.0004 = 4/10000$ before a mammogram and rose almost eight-fold to $0.003191 = 32/10028$ after a positive mammogram. This is an example of how Bayes' method can be used to update the risk as a result of a test.

Similarly,

$$PV^- = Pr(\text{no breast cancer} \mid \text{negative}) = \frac{0.89964}{0.00008 + 0.89964} = \frac{89964}{89972} = 0.99991$$

Notice that a negative mammogram is very predictive of no breast cancer.

So far we have used Bayes' method only in the case of medical testing, using a binary tree. However, it applies to any two-stage experiment. All of these problems can be solved without a formal formula by using a tree diagram, sometimes with multiple branchings at each stage. However, we give a formal statement of Bayes' formula below.

Definition

> *A partition of an outcome set is a collection of events $\{\mathbf{E}_1, \mathbf{E}_2, \ldots\}$ that are pairwise disjoint and whose union is the entire outcome set.*

THEOREM 3.1 (**Partition Theorem**) *If $\{E_1, E_2, \ldots\}$ is a partition of the outcome set and E is any event, then*

$$Pr(E) = Pr(E_1)Pr(E \mid E_1) + Pr(E_2)Pr(E \mid E_2) + \cdots$$

The proof is left as Exercise 3.30 on page 113. It uses the basic multiplication rule on page 99 and the addition rule on page 96.

3.6 BAYES' METHOD

> **Bayes' Formula:** Suppose $\{E_1, E_2, \ldots\}$ partitions the outcome set and E is any event with nonzero probability. Then for any E_i we have
>
> $$\Pr(E_i \mid E) = \frac{\Pr(E_i \cap E)}{\Pr(E)} = \frac{\Pr(E_i)\Pr(E \mid E_i)}{\Pr(E_1)\Pr(E \mid E_1) + \Pr(E_2)\Pr(E \mid E_2) + \cdots}$$

Example 3.35 *Estimating the prevalence of HIV.*[11] In 1986, two tests were used for HIV, an enzyme-linked immunosorbent assay (ELISA) and an immunoblot assay called western blot (WB).

The Red Cross found that approximately 1% of donated blood tested ELISA positive. Because positive test results include both true positives and false negatives, one cannot simply conclude that the prevalence of HIV in donated blood was 1% in 1986.

In order to estimate the prevalence, the blood that initially tested ELISA positive was given two more ELISA tests. If either of these were positive, the blood is considered *repeat reactive*; and it turned out that 30% to 35% of the initial reactives were repeaters. Repeat reactive blood was then given the WB test. Roughly 8% of the repeat reactives tested WB positive. This resulted in about 0.025% of Red Cross donors in the United States being notified by the Red Cross as testing positive for HIV.

Max Essex, an AIDS researcher at Harvard University, said, "Ninety to 95 percent of the people who test positive do not have the virus." In other words, the *predictive value positive* is between 5% and 10%. He also stated that some percentage of the people infected with the HIV virus do not have detectable antibodies, "the best figure used is 5 percent"; that is, the *sensitivity* was approximately 95%.

The Centers for Disease Control and Prevention estimated that the number of healthy people in the United States who were then antibody positive for HIV was between 1 and 1.5 million. If we assume that there were roughly 240 million HIV-free people in the United States in 1986, then the *specificity* of these tests was between $1 - 1.5/240 = 99.375\%$ and $1 - 1/240 = 99.583\%$.

Using a predictive value positive of 10%, a specificity of $1 - 1/240$, substituting these values into Bayes' Formula, and then solving for $x =$ prevalence,

$$0.10 = \frac{0.95x}{0.95x + \frac{1}{240}(1-x)}$$

we get $x = 0.00049$. This results in the prevalence of HIV of approximately 0.049%.

---

[11] Based on a report by Joanne Silberner, *AIDS Blood Screens: Chapters 2 and 3*, Science News **130** (July 26, 1986), pp. 56–57.

EXERCISES FOR SECTION 3.6

3.22 The American Cancer Society as well as the medical profession recommend that people have themselves checked annually for any cancerous growths. If a person has cancer, then the probability is 0.99 that it will be detected by a test. Furthermore, the probability that the test results will be positive when no cancer actually exists is 0.10. Government records indicate that 8% of the population in the vicinity of a paint manufacturing plant has some form of cancer. Find PV^+ and PV^-.

- **3.23** The prevalence of breast cancer in the population of women between 40 and 50 years of age is 1/63. Assume the same sensitivity and specificity of mammography as in Example 3.34 above.

    a. Find the predictive value positive.

    b. Compare the risk of breast cancer before the test with the risk after a positive mammogram.

    c. Compare the risk of breast cancer before the test with the risk after a negative mammogram.

3.24 You go to a beach party. Two of you are bringing coolers with sandwiches. Your cooler contains 10 ham sandwiches and 5 cheese sandwiches. Your friend's cooler has 6 ham and 9 cheese sandwiches. At the beach, someone takes a sandwich at random from one of the coolers. It turns out to be a ham sandwich. What is the probability that the sandwich came from your cooler? First, sketch a tree diagram. Then use Bayes' method to compute the probability.

- **3.25** In a factory that manufactures silicon wafers, machines 1 and 2 produce, respectively, 40% and 60% of the output. Suppose 2 out of every 100 wafers produced by machine 1 are defective. Machine 2 is not only faster, but only 3 out of every 200 wafers it produces are defective. Make a tree of this two-stage experiment, the first stage of which is binary with possibilities machine 1 and machine 2. One wafer is found to be defective.

    a. What is the probability it was produced by machine 1? **b.** Machine 2?

3.26 A certain financial market experiences growth (G) 20% of the time, stagnation (S) 70% of the time, and recession (R) 10% of the time. A consulting firm is hired to forecast this market. In order to assess the reliability of this firm's forecasts, we investigate past performance of this consulting firm. Records show that whenever there was growth, the firm had correctly forecasted it 80% of the time but had erroneously forecasted stagnation 10% of the time, and recession 10% of the time. The firm's accuracy was the same for periods of stagnation and recession; that is, it forecasted correctly 80% of the time and made each of the other two erroneous forecasts 10% of the time. The tree for this two-stage experiment has three branchings at each stage. Suppose that now the firm forecasts growth. Given its forecast and past performance, find the probability of

a. growth b. stagnation c. recession

- **3.27** Messages arrive at a computer from three different sources A, B, and C, with probabilities 0.2, 0.3, and 0.5, respectively. The probability that messages from sources A, B, or C exceed 500 bytes in length are 0.8, 0.6, and 0.4, respectively.

    a. What is the probability that the next message will exceed 500 bytes?

    Suppose that a message exceeds 500 bytes.

    b. What is the probability that it was sent from source A?

    c. Source B? d. Source C?

3.28 A cab driver was involved in a deadly hit-and-run accident at night.[12] Two cab companies, the Green and the Blue, operate in the city; 85% of the cabs are Green and 15% are Blue. A witness identifies the cab as Blue. The court tests the reliability of the witness under the same circumstances that existed on the night of the accident and concludes that the witness can correctly identify the color of the cab 80% of the time. Use Bayes' methods to find the probability that the cab involved in the accident was actually Blue.

* * * * * * * *

- **3.29** (Proof) When studying screening tests, a partially completed 2-by-2 table relating the various events and probabilities would look something like the following. Here x is the prevalence of the disease in the population.

Events	Test positive	Test negative	Row sum
Infected	$x \cdot$ sensitivity		$x = Pr(\text{infected})$
Not infected		$(1-x) \cdot$ specificity	$1 - x = Pr(\text{not infected})$
Column sum	$Pr(\text{test positive})$	$Pr(\text{test negative})$	1

a. Fill in the blanks in the 2-by-2 table.
b. Use this notation to find a formula for $Pr(\text{test positive})$.
c. Use this notation to find a formula for the predictive value positive.
d. Let the specificity and sensitivity be fixed constants (begin with 0.90 and 0.95 and experiment with different values) and graph the predictive value positive as a function of x. Do you notice any patterns? Does the predictive value positive always exceed the prevalence of the disease?

3.30 (Proof) Prove the Partition Theorem 3.1 on page 110.

---

[12] From Massimo Piattelli-Palmarini, *Probability Blindness, Neither Rational nor Capricious*, Bostonia (March/April 1991), pp. 28–35

REVIEW EXERCISES FOR CHAPTER 3

3.31 **a.** What is an *outcome set*?   **b.** What is an *event*?
  c. What is the complement A' of an event A?
  d. Let A and B denote events. Show $Pr(A \cap B') = Pr(A) - Pr(A \cap B)$.

3.32 State which of the following equalities are always true and which are not. Here A, B, C are events and A', B', C' are their complements.

  a. $A' \cap B' = (A \cap B)'$
  b. $A \cup B) \cap C = A \cup (B \cap C)$
  c. $A \cup B = A \cup (B \cap A')$

• **3.33** Which of the following statements are always correct and which are not?

  a. $Pr(A) + Pr(A') = 1$
  b. $Pr(A) + Pr(B) = Pr(A \cup B) + Pr(A \cap B)$
  c. $Pr(A \cap B') = Pr(A) - Pr(B)$

3.34 Suppose E and F are disjoint events with $Pr(E) = 0.20$ and $Pr(F) = 0.50$.

  a. Find the probability that both E and F occur.
  b. Find the probability that either E or F occur.
  c. Find the probability that F occurs and E does not occur.
  d. Find the probability that neither occurs.

• **3.35** Suppose E and F are independent events with $Pr(E) = 0.20$ and $Pr(F) = 0.50$.

  a. Find the probability that both E and F occur.
  b. Find the probability that either E or F occur.
  c. Find the probability that F occurs and E does not occur.
  d. Find the probability that neither occurs.

3.36 If a pair of fair dice are rolled, find the probability of getting the following events:

  a. Both faces 4.   **b.** At least one face 4.
  c. No faces 4.   **d.** A total of 4 dots.

• **3.37** A fair coin is tossed four times. Find the probability of the following events:

  a. The sequence H-T-H-H.   **b.** No heads.
  c. At least one head.   **d.** Four heads.

REVIEW EXERCISES FOR CHAPTER 3

3.38 A fair coin is tossed six times. Find the probability of the following events:

  a. No heads. **b.** At least one head.

  c. Exactly one head. **d.** Six heads.

• **3.39** A box contains 2 red and 4 white balls. Two balls are chosen randomly with replacement. Find probabilities of the following events:

  a. Both balls are red. **b.** Both are white.
  c. Both are the same color. **d.** The balls are different colors.
  e. Answer the same questions above if both balls are drawn without replacement.

3.40 A box contains 2 red and 3 white balls and a second box contains 2 red and 4 white balls. A ball is chosen randomly from each box. Find probabilities of the following events.

  a. That both balls are red. **b.** That both are white.

  c. That both are the same color. **d.** Different color.

• **3.41** Two letters are selected at random without replacement from the word *statistical*.

  a. What is the probability that both letters are *s*?
  b. What is the probability that both letters are the same?
  c. If you know both letters are the same, what is the probability both letters are *s*?

3.42 Five cards are dealt at random to each of two players. If one player has no aces, what is the probability that the other player has no aces?

• **3.43** Three fair dice are rolled. Find the probability that the top faces show three different numbers.

3.44 A survey of students at a certain college showed that 60% of them read a daily newspaper and 40% read a weekly news magazine. Also, if a student reads a daily newspaper, then the chance of reading a weekly news magazine rose to 50%.

  a. What percentage of students read both?
  b. What percentage of students read at least one?
  c. What percentage of students read neither?
  d. What percentage of the readers of a news magazine also read a newspaper?

- **3.45** Fill in the blanks to make two true sentences.

    a. If two events are _____ and you want to find the probability that _____ will happen, you can _____ the probabilities.

    b. If two events are _____ and you want to find the probability that _____ will happen, you can _____ the probabilities.

3.46 Suppose that the birth of boys and girls is equally likely and independent. In a family of six children, find the probabilities of the following events:

a. There will be three boys and three girls.

b. There will be a 4:2 split (four boys and two girls, or vice versa).

- **3.47** Suppose that 20% of all personal computers of a certain brand break down in the first year of operation. In an office with 10 such computers, find the probability that at least one breaks down.

3.48 In a box of five widgets, one of them is defective. An inspector checks three at random (without replacement).

a. What is the probability that the first widget checked is defective?

b. What is the probability that the second widget checked is defective?

c. What is the probability that first widget checked is okay and the second one is defective?

d. What is the probability that the inspector finds no defective widgets among the three checked?

e. What is the probability that the inspector finds the defective widget among the three checked?

- **3.49** Ten people are in a room wearing badges marked 1 through 10. Three persons are selected at random and their badge numbers are recorded. What is the probability that the smallest of these badge numbers is 6?

3.50 An experiment has six possible outcomes. An outcome set for this experiment is $S = \{o_1, o_2, o_3, o_4, o_5, o_6\}$. Suppose the probabilities assigned to the single possible outcome are $Pr(\{o_1\}) = 0.04$, $Pr(\{o_2\}) = 0.20$, $Pr(\{o_3\}) = 0.10$, $Pr(\{o_4\}) = 0.40$, $Pr(\{o_5\}) = 0.20$, and $Pr(\{o_6\}) = 0.06$. Let $A = \{o_1, o_4, o_6\}$, $B = \{o_3, o_4, o_5\}$, and $C = \{o_2, o_5\}$ denote three events.

a. Which of these events are mutually exclusive?

b. Which are independent?

c. Compute $Pr(B \cup C)$.

- **3.51** The Cubs and White Sox are playing in the World Series. The series ends when one of these teams wins four games. Assume the teams are evenly matched and the outcome of any game is independent of the results of the previous games.

a. How many different ways can the Cubs win in a 6-game series? (For example, CSCCSC is a six-game series with the Cubs winning the first, third, fourth, and sixth games.)

b. How many different ways can the Cubs win in a 5-game series?

c. What is the probability the Sox win the first game?

d. What is the probability the Sox win the series?

e. What is the conditional probability that the Cubs win the series if the Sox win the first game?

3.52 A box contains 3 white and 2 blue balls. A second box contains 1 white and 4 blue balls. A box is chosen at random and a ball is selected at random from it.

a. Sketch the tree diagram for this experiment.

b. Find the probability that the ball is blue.

c. If the ball came from the first box, what is the probability it is blue?

d. If the ball is blue, what is the probability it came from the first box?

3.53 A box contains four fair coins and one two-headed coin.

• a. A coin is chosen at random from this box and tossed three times. It comes up heads each time. What is the probability that it is a fair coin?

b. Now more tosses. A coin is chosen at random from the same box and tossed six times. It comes up heads each time. What is the probability that it is a fair coin?

3.54 To reduce theft among employees, a company subjects all employees to lie-detector tests and then fires all employees who fail the test. In the past, the test has been proven to correctly identify guilty employees 90% of the time; however, 4% of the innocent employees also fail the test. Suppose that 5% of the employees are actually guilty.

a. What percentage of the employees fail the test?

b. What percentage of those fired were innocent?

• **3.55** The board of education of a certain city decides to give an aptitude test to all elementary school teachers. It is noted that of the 60 percent who pass the test, 70 percent of them are rated as good teachers by their principals, whereas of the 40 percent who failed the test only 50 percent are rated as good.

a. What percentage of the teachers are rated as good?

b. What is the probability that a teacher rated good fails the test?

3.56 Consider mammography as a test for breast cancer. The prevalence of breast cancer in women between 60 and 70 years of age is about 1/28. The sensitivity of mammography is 80% and the specificity is 90%. Find the predictive value positive.

- **3.57** Three manufacturing plants, C_1, C_2, and C_3 produce, respectively 10%, 50% and 40% of a company's output. Although plant C_1 is a small plant, its manager believes in high quality and only 1% of its products are defective. The other two, C_2, and C_3, are worse and produce items that are 3% and 4% defective, respectively. All products are sent to a central warehouse. One item is selected at random and observed to be defective.

    a. Find the probability that the item came from plant C_1.

    b. Find the probability that the item came from plant C_2.

    c. Find the probability that the item came from plant C_3.

3.58 A certain car model comes in the L model and the LX model. For the past few years, 70% of all the cars sold were L models. Of those buying the L model, 30% also purchased the extended warranty, whereas of those buying the LX model, 50% did so. Of those covered by an extended warranty, what percentage have the L model?

- **3.59** A blood test to screen for a certain disease is not completely reliable, and medical officials are unsure as to whether the test should be routinely given. Suppose that 99.5% of those with the disease will show positive on the test, but that 0.2% of those who are free of the disease also show positive on the test. If 0.1% of the population actually has the disease, find PV^+ and PV^-.

3.60 Repeat Exercise 3.59 but for a population with prevalence of 1/25.

- **3.61** In a population of 10,000 males and 10,000 females, 1060 of the males and 780 of the females are left-handed. A left-handed person is selected at random from this population. What is the probability this person is male?

3.62 Suppose that in a population with an equal number of males and females, 5% of the males and 0.25% of the females are color-blind. A randomly chosen person is found to be color-blind. What is the probability that the person is female?

- **3.63** A box contains the keys to a safe as well as six other keys that do not fit the safe. You remove one key at a time, without replacement, until you find the correct key. What is the probability that the correct key is selected on the second draw?

3.64 A multiple-choice quiz consists of five questions, each with four possible answers, only one of which is correct. You randomly guess each answer. Find probabilities of the following events.

    a. Getting them all wrong.   **b.** All right.

    c. At least one right.   **d.** At least one wrong.

- **3.65** The Senate of the 107th Congress consists of 50 Democrats, 49 Republicans, and 1 Independent. What is the probability that a randomly selected committee of five members consists of all Democrats?

3.66 What is the probability that a family of four children has at least one girl? Assume births of boys and girls are independent and equally likely.

- **3.67** A mechanical device has five components, all of which must function correctly for the device to work. If each component has a 99% chance of functioning correctly, what is the probability that the device fails to work.

3.68 A string of 20 Christmas lights is wired in series (that is, if one bulb fails, then the entire string fails to light). If the probability of any particular bulb failing sometime during the holiday season is 0.005, and if the failures are independent events, what is the probability that the lights will fail sometime during the holiday season?

- **3.69** Tomatoes grown in two valleys in California are sometimes infested with hornworms. Valley A is infested 20% of the time, valley B is infested 50% of the time, and at least one is infested 60% of the time. If state inspectors find hornworms on a shipment from valley B, what is the probability that farmers in valley A are experiencing the same problem?

3.70 During a power blackout, 100 persons were arrested on suspicion of looting. Each is given a polygraph test. It is known that the polygraph is 90% reliable when administered to a guilty suspect and 96% reliable when given to someone who is innocent. Suppose that of the 100 persons taken into custody only 12 were actually involved in any wrongdoing. What is the probability that a given suspect is innocent given that the polygraph says he is guilty?

- **3.71** A computer network consists of 10 computers, and the system can remain up so long as at least 4 computers are functioning. What is the number of computer combinations in which the system will be up?

3.72 A study was conducted to assess the accuracy of a new Department of Defense Test for Espionage and Sabotage (TES) with an older Counterintelligence Scope Polygraph (CSP).[13] In the study, subjects were either *programmed innocent* (PI) or *programmed guilty* (PG). The PG subjects were involved in a mock espionage scenario. Eighteen certified government examiners conducted blind examinations. In the CSP test, of the 131 PI subjects, 120 were correctly identified, 6 were rated as "guilty," and 5 were rated as inconclusive; and of the 60 PG subjects, 32 were correctly identified, 24 were found "not guilty," and 4 were inconclusive.

---

[13] Department of Defense Polygraph Institute report DODP194-R-008, *A comparison of Psychophysiological Detection of Deception Accuracy Rates Obtained Using the Counterintelligence Score Polygraph and the Test for Espionage and Sabotage Question Formats*, Polygraph **26** (1997), pp. 79–106.

a. For the CSP test, what percentage of the subjects were rated "guilty" by the examiners?

b. Of those who were rated "guilty," what percentage were PG subjects?

- **3.73** This is a continuation of Exercise 3.72. In the TES study, there were 56 PI subjects and 25 PG subjects. Of the 56 PI subjects, 48 were correctly identified, 6 were rated as "guilty," and 2 were rated as inconclusive. Of the 25 PG subjects, 20 were correctly identified, 5 were rated as "not guilty," and none were rated inconclusive.

  a. For the TES test, what percentage of the subjects were rated "guilty" by the examiners?

  b. Of those who were rated "guilty," what percentage were PG subjects?

3.74 In addition to the blood types O, A, B, and AB, as listed in Exercise 2.1 on page 25, there are two Rh factors. In the United States 85% of the population is Rh-positive and 15% is Rh-negative. Blood type and Rh factor are independent.

Blood type	O	A	B	AB
Percentage	45.4	39.5	11.1	4.0

Use this distribution of blood types in the the United States to make a 2 by 2 table giving the percentages of blood types and Rh factors for the U.S. population, rounded to one decimal place. Do your percentages add up to 100.0%?

- **3.75** An American man's risk of developing prostate cancer at some time in his life if 30%.[14] High levels of prostate specific antigen (PSA) in blood is used as a test for prostate cancer. Approximately 20% of men with prostate cancer have normal PSA levels and two out of three men with high PSA levels do not have cancer. For the PSA blood test find the following:

  a. Prevalence

  b. Sensitivity

  c. Specificity

  d. Predictive value positive

  e. Predictive value negative

3.76 Explain in your own words the difference between *disjointness* and *independence*. Give good examples of each.

---

[14]Men's Health: Controversies and Clear Thinking, *Newsweek*, June 16, 2003, pp. 64–65. (www.msnbc.com/news/923671.asp)

Chapter 4

DISCRETE RANDOM VARIABLES

4.1 INTRODUCTION

4.2 BASIC PROPERTIES OF DISCRETE RANDOM VARIABLES

4.3 PROBABILITY HISTOGRAMS

4.4 EXPECTED VALUE OR MEAN OF DISCRETE RANDOM VARIABLES

4.5 FUNCTIONS OF RANDOM VARIABLES

4.6 VARIANCE AND STANDARD DEVIATION OF DISCRETE RANDOM VARIABLES

4.7 DISCRETE UNIFORM RANDOM VARIABLES

4.8 LAW OF LARGE NUMBERS (optional)

4.1 INTRODUCTION

This chapter is a continuation of the topic of probability. We have considered general events in random experiments. In order to take full advantage of the power of mathematical analysis, it is most useful to consider numerical outcomes (measurements). To this end, we now introduce the concept of random variables. A **random variable** is a function that assigns a real number to each outcome of an experiment. For example, in a hand of five cards from a standard deck, we might be interested in the number of spades.

Or in the roll of a pair of dice we might be interested in the sum of the dots on the two top faces—or their product or their difference. For a given random experiment, a random variable is a number of interest that depends on the outcome of an experiment. By convention, random variables are denoted by capital letters such as T, U, V, W, X, Y, and Z.

Example 4.1 **Some random variables.** Consider a population of college athletes and choose one at random. This is a random experiment. Let V be the height of the athlete, W the weight, X the number of college credits earned by the athlete, and Y the grade point average. If we code the gender of each athlete 0 for female and 1 for male, this assignment is also a random variable—say, U.

In Chapter 10, we will consider relationships between random variables. For example, there is a relationship between an athlete's height V and gender U. Similarly, an athlete's height V and weight W are also related. A measure of the strength of the relationship of two variables is called *correlation*. The subject of how to predict the value of one variable, such as weight, if you know the value of another variable, such as height, is called *regression theory*, which is considered in Chapter 14. For now, we deal with one random variable at a time.

A random variable is said to be **discrete** if it assumes a finite or an infinite sequence of values. A random variable that can take any value in an interval of the real numbers is said to be **continuous**. In Example 4.1 above, the random variable X is discrete. The random variable V is continuous because height, as a concept, can be any positive real number. A note of caution: The recorded *measurements* of height are discrete if, for example, the measurements are rounded to the nearest 1/4 inch. Yet the *concept* of height is continuous. There is sometimes an ambiguous distinction between a concept and its measurement. A concept X may or may not be discrete, yet the measurement of that concept is generally discrete because of either rounding or limitations in precision of measurement. Rounding is a form of grouping data.

Example 4.2 **Cap size.** Consider the cap size of a randomly selected member of the graduating class of a large university. Cap size, which is the head circumfer-

EXERCISES FOR SECTION 4.1

- **4.1** In Example 4.1 above, which of U, V, W, X, and Y are discrete?

4.2 Which of the following random variables are discrete and which are continuous? A person's age, IQ score, weight, average number of cigarettes smoked per day, annual income. What about a person's blood type?

4.2 BASIC PROPERTIES OF DISCRETE RANDOM VARIABLES

Example 4.3   *Family planning.* Consider families of three children, observing whether each child is a boy or a girl. There are eight possibilities for each family: GGG, GGB, GBG, BGG, GBB, BGB, BBG, BBB, where each outcome is listed in order of birth. We can define the random variable X to be the function that assigns to each outcome the number of boys; and we can define Y to be the number of girls. If we are interested in the number of times there is a switch from girl to boy or from boy to girl, we can define the random variable Z to be the number of changes in sequence. In functional notation, we have the following:

$$\begin{array}{lll}
X(\text{GGG}) = 0 & Y(\text{GGG}) = 3 & Z(\text{GGG}) = 0 \\
X(\text{GGB}) = 1 & Y(\text{GGB}) = 2 & Z(\text{GGB}) = 1 \\
X(\text{GBG}) = 1 & Y(\text{GBG}) = 2 & Z(\text{GBG}) = 2 \\
X(\text{BGG}) = 1 & Y(\text{BGG}) = 2 & Z(\text{BGG}) = 1 \\
X(\text{GBB}) = 2 & Y(\text{GBB}) = 1 & Z(\text{GBB}) = 1 \\
X(\text{BGB}) = 2 & Y(\text{BGB}) = 1 & Z(\text{BGB}) = 2 \\
X(\text{BBG}) = 2 & Y(\text{BBG}) = 1 & Z(\text{BBG}) = 1 \\
X(\text{BBB}) = 3 & Y(\text{BBB}) = 0 & Z(\text{BBB}) = 0
\end{array}$$

Notice that $X + Y$ is the constant random variable 3.

For each variable, we can consider the events that range over the values of the random variable. For example:

"$X = 0$" = \{GGG\}

"$X = 1$" = \{BGG, GBG, GGB\}

"$X = 2$" = \{GBB, BGB, BBG\}

"$X = 3$" = \{BBB\}

For a discrete variable X, the function f_X that assigns to the real number x the probability of the event "$X = x$" is called the **probability mass (density) function** or just **probability function** or **density function**. In other words,

Definition

> **Discrete Probability Function:**
>
> $$\mathbf{f_X(x) = Pr(X = x)}$$

In a discussion for which the random variable is implicitly understood, it may be omitted from the notation; we then write simply $f(x)$ instead of $f_X(x)$. The notation $p(x) = Pr(X = x)$ is common, but we will use f instead because the symbol p is overused in probability theory.

The points x where f_X is nonzero are called the **mass points**; here we have four mass points: $x = 0, 1, 2, 3$. Note that X denotes a random variable, and x denotes a real variable. Consider the model where the sex of each child is independent of the sex of the previous children, where the chance of a boy birth is 0.52, and the chance of a girl is 0.48. Then for a family of three children, we have the values

$$\begin{aligned}
f_X(0) &= Pr(X = 0) = Pr(\text{GGG}) = (.48)^3 \\
f_X(1) &= Pr(X = 1) = Pr(\text{BGG or GBG or GGB}) \\
&= Pr(\text{BGG}) + Pr(\text{GBG}) + Pr(\text{GGB}) \\
&= (.52)(.48)(.48) + (.48)(.52)(.48) + (.48)(.48)(.52) = 3(.52)(.48)^2 \\
f_X(2) &= Pr(X = 2) = Pr(\text{GBB}) + Pr(\text{BGB}) + Pr(\text{BBG}) = 3(.52)^2(.48) \\
f_X(3) &= Pr(X = 3) = Pr(\text{BBB}) = (.52)^3
\end{aligned}$$

4.2 BASIC PROPERTIES OF DISCRETE VARIABLES

A table of mass points of a random variable X and their probabilities, such as the one below, is called the **probability table** of X. See Figure 4.1 for the graph of f_X.

x	$f_X(x)$		
0	$(0.48)^3$	=	0.1106
1	$3(0.52)^1(0.48)^2$	=	0.3594
2	$3(0.52)^2(0.48)^1$	=	0.3894
3	$(0.52)^3$	=	0.1406
	$\sum f_X(x)$	=	1

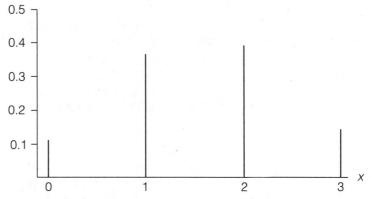

Figure 4.1 Probability mass function for the number of boys in a 3-child family

The **cumulative distribution function** F_X is defined by the following formula.

Definition

Cumulative Distribution Function:

$$\mathbf{F_X(x) = Pr(X \leq x)}$$

A table of values of the cumulative distribution function is given below, and Figure 4.2 shows the graph of F_X.

x	$F_X(x)$	=	$Pr(X \leq x)$
$-\infty < x < 0$			0
$0 \leq x < 1$	$f(0)$	=	0.1106
$1 \leq x < 2$	$f(0) + f(1)$	=	0.4700
$2 \leq x < 3$	$f(0) + f(1) + f(2)$	=	0.8594
$3 \leq x < \infty$			1

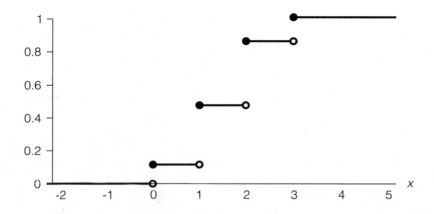

Figure 4.2 Cumulative distribution function for the number of boys in a 3-child family

Notice that f_X is zero except at the mass points, whereas F_X is a nondecreasing function with steps at the mass points.

The term **distribution** has been traditionally used to denote the cumulative distribution function F. Recently, some authors have used the term *distribution* informally to denote the probability mass function f or even the random variable X. We will emphasize the distinction by using the terms *mass function* or *density function* to denote f and the term *cumulative distribution* to denote F.

Here is an example of a discrete random variable that takes on countably infinite values.

Example 4.4 *Waiting for heads.* Toss a fair coin repeatedly until it comes up heads. The outcome set is $S = \{H, TH, TTH, TTTH, TTTTH, \ldots\}$. Let X be the number of tails before the first head appears:

x	Event	$f_X(x)$
0	H	1/2
1	TH	1/4
2	TTH	1/8
3	TTTH	1/16
4	TTTTH	1/32
$\ldots$	$\ldots$	$\ldots$
k		$(1/2)^k$
$\ldots$	$\ldots$	$\ldots$

4.2 BASIC PROPERTIES OF DISCRETE VARIABLES

See Figure 4.3 for the graph of f, and Figure 4.4 for the graph of F.

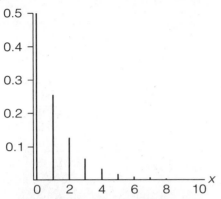

Figure 4.3 Probability mass function for the number of tosses before the first head

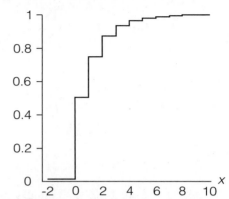

Figure 4.4 Cumulative distribution function for the number of tosses before the first head

Basic properties of f and F for discrete random variables

(a) $F(x) = Pr(X \leq x)$, for $-\infty < x < +\infty$

(b) $0 \leq F(x) \leq 1$

(c) F is nondecreasing

(d) F is continuous from the right; that is, $F(x) = \lim_{t \to x^+} F(t)$

(e) $\lim_{x \to -\infty} F(x) = 0$

(f) $\lim_{x \to \infty} F(x) = \sum_{\text{all } t} f(t) = 1$

(g) $f(x) = Pr(X = x)$

(h) $0 \leq f(x) \leq 1$

(i) $\sum_{\text{all } t} f(t) = 1$

(j) $f(x) = F(x) - \lim_{t \to x^-} F(t)$

(k) $F(x) = \sum_{t \leq x} f(t)$

(l) for any two numbers $a < b$, $Pr(a < X \leq b) = F(b) - F(a) = \sum_{a < t \leq b} f(t)$

EXERCISES FOR SECTION 4.2

- **4.3** Find the cumulative distribution function for the random variable X given by the following table:

x	$f_X(x)$		
0	$(0.75)^3$	=	27/64
1	$3(0.25)^1(0.75)^2$	=	27/64
2	$3(0.25)^2(0.75)^1$	=	9/64
3	$(0.25)^3$	=	1/64
	$\sum f_X(x)$	=	1

4.4 Find the cumulative distribution function for the random variable X given by the following table:

x	$f_X(x)$
0	1/27
1	6/27
2	12/27
3	8/27
	$\sum f_X(x) = 1$

- **4.5** Find the probability mass function for the following cumulative distribution function:

$$F(x) = \begin{cases} 0 & : \text{ for } x < 0 \\ 0.75 & : \text{ for } 0 \leq x < 1 \\ 1 & : \text{ for } x \geq 1 \end{cases}$$

4.6 Find the probability mass function for the following cumulative distribution function:

$$F(x) = \begin{cases} 0 & : \text{ for } x < 1 \\ 0.25 & : \text{ for } 1 \leq x < 3 \\ 0.75 & : \text{ for } 3 \leq x < 5 \\ 1 & : \text{ for } x \geq 5 \end{cases}$$

4.3 PROBABILITY HISTOGRAMS

Recall that histograms in Chapter 2 were used to graph data. The areas of the rectangles represented frequencies. Here we consider a different kind of histogram called a **probability histogram** or **theoretical histogram**.

They are similar except that areas of rectangles represent *probabilities* instead of *frequencies*.

For example, let X be the number of boys in a family for three children in Example 4.3. The probability histogram is constructed from the probability table on page 125, just as if it were a table of relative frequencies (see Figure 4.5).

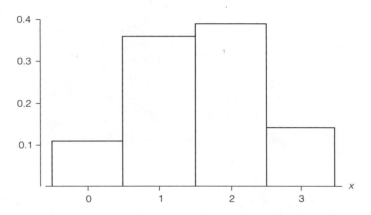

Figure 4.5 Histogram for the number boys in a family of three children

Because $\sum f_X(x) = 1$, the total area under a probability histogram must be 1.

To emphasize the distinction between the two types of histograms, histograms of data will be called **empirical histograms**.

Example 4.5 **John Kerrich.** John Kerrich was a South African mathematician who became a prisoner of war of the Germans in an internment camp in Denmark during the Second World War. While a prisoner, he tossed a coin 10,000 times and recorded the results.[1] The table below gives a partial list of these results.

---

[1]This is from his book *An Experimental Introduction to the Theory of Probability*, University of Witwatersrand Press, 1964.

John Kerrich's coin-tossing experiment

Number of tosses	Number of heads
10	4
50	25
100	44
500	255
1000	502
2000	1013
3000	1510
4000	2029
5000	2533
6000	3009
7000	3516
8000	4034
9000	4538
10,000	5067

Figure 4.6 shows the empirical histograms for the first n tosses for various values of n.

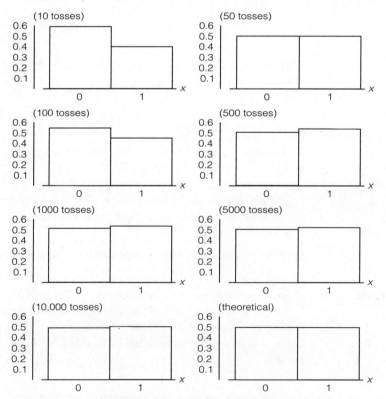

Figure 4.6 Empirical histograms of Kerrich's coin tosses

Notice that as n gets large, the empirical histograms look more and more like the theoretical histogram at the bottom right. The reason that the probability histogram is called the *theoretical histogram* is that this is what the empirical histogram would tend to if trials continued to infinity. This is the essence of the **law of large numbers**. We first state it in informal terms.

Law of large numbers

> *Suppose that an experiment is repeated* n *times. As the number of trials* n *tends to infinity, the empirical histogram of the data tends to look like the probability histogram.*

Notice that the law uses the "fuzzy" phrases "tends to" and "looks like." The phrase "tends to" is a term of calculus, and here it means "$\lim_{n \to \infty}$." Although it is possible to understand elementary calculus with only an intuitive understanding of limit, the concept of limit has a rigorous definition generally given with the use of notation of ϵ, δ, and N. The other "fuzzy" phrase is "looks like." It is a term of probability theory. A rigorous discussion of the law of large numbers can be found in Section 4.8 at the end of this chapter.

Percentiles

Informally, the rth percentile is the value x for which $F(x) = \frac{r}{100}$. However, sometimes there are many choices for x that satisfy this equation, and it could be that no value of x satisfies this equation. For this reason, the definition is usually expanded to: the rth percentile is the *smallest* value of x for which $F(x) \geq \frac{r}{100}$. The **median** is the 50th percentile.

Consider Example 4.4, where X denotes the number of tails of a fair coin until the first head appears. Because $F(x) = 0$ for $-\infty < x < 0$ and $F(x) = \frac{1}{2}$ for $0 \leq x < 1$, the median of the random variable X is $x = 0$. And the 80th percentile is $x = 2$ because $F(x) = \frac{3}{4}$ for $1 \leq x < 2$ and $F(x) = \frac{7}{8}$ for $2 \leq x < 3$.

Mode

The **mode** of a random variable X is defined to be the value of x where $f(x)$ attains its maximum value. For the number of boys in a 3-child family, the mode is $x = 2$. For the number of tosses of a fair coin until the first head, the mode is $x = 1$.

EXERCISES FOR SECTION 4.3

4.7 Sketch the probability histogram of the random variable of Exercise 4.3 on page 128:

x	$f_X(x)$		
0	$(0.75)^3$	=	27/64
1	$3(0.25)^1(0.75)^2$	=	27/64
2	$3(0.25)^2(0.75)^1$	=	9/64
3	$(0.25)^3$	=	1/64
	$\sum f_X(x)$	=	1

4.8 Sketch the probability histogram of the random variable of Exercise 4.4 on page 128:

x	$f_X(x)$
0	1/27
1	6/27
2	12/27
3	8/27
	$\sum f_X(x) = 1$

4.4 EXPECTED VALUE OR MEAN OF DISCRETE RANDOM VARIABLES

For grouped *data* with classes $1, 2, \ldots, k$, class marks $x_1, x_2, \ldots, x_k$, and class *relative* frequencies $f_1, f_2, \ldots, f_k$, the formula for the (empirical) mean $\overline{x}$ is

$$\overline{x} = \sum_{j=0}^{k} x_j f_j$$

For a discrete random variable X with mass points $x_1, x_2, \ldots$, the **(theoretical) mean** μ_X or **expected value** $E(X)$ is defined similarly, except that probabilities $f(x_j) = Pr(X = x_j)$ are used instead of frequencies f_j:

$$\mu_X = E(X) = \sum_j x_j f(x_j)$$

The mean is the weighted mean of all the values of X weighted by the probabilities. Values x of X that have highest probabilities are weighted the heaviest. In case of an infinite sum, the series *must be absolutely convergent*; that is, we must have $\sum_{j=1}^{\infty} |x_j| f(x_j) < \infty$. Otherwise, we say that $E(X)$ is *not defined*.

If the random variable is implicitly understood, the subscript X may be omitted: $\mu_X = \mu$. Because $f(x) = 0$ except at the mass points $x = x_j$, we can suppress the subscripts in the notation to write

$$\mu = E(X) = \sum_{\text{all } x} x f(x)$$

A consequence of the law of large numbers is that, if μ is defined, as the sample size n tends to ∞, the sample mean $\bar{x}$ tends toward the theoretical mean μ. However, without absolute convergence of the sum in the definition of $E(X)$, the law of large numbers does not apply. We will illustrate this difficulty when we deal with continuous random variables in Chapter 7.

Caution

> **Do not forget the factor x in the definition of μ:**
>
> $$\mathbf{Pr}(-\infty < \mathbf{X} < \infty) = \sum \mathbf{f(x)} = 1 \quad \text{but} \quad \mu = \sum \mathbf{x f(x)}$$

Example 4.6 **Dots on a fair die.** Roll a fair die and let X be the number of dots on the top face. Operationally, for a discrete random variable, in order to find the mean from the probability table, multiply the x column by the $f(x)$ column, and then find the sum:

x	$f(x)$	$xf(x)$
1	1/6	$1 \cdot 1/6 = 1/6$
2	1/6	$2 \cdot 1/6 = 2/6$
3	1/6	$3 \cdot 1/6 = 3/6$
4	1/6	$4 \cdot 1/6 = 4/6$
5	1/6	$5 \cdot 1/6 = 5/6$
6	1/6	$6 \cdot 1/6 = 6/6$
	$\sum = 1$	$\mu = 21/6 = 3.5$

Note that the expected number of dots $E(X) = 3.5$ cannot be the outcome of any trial of this experiment. The expected value is a theoretical average number of dots one would get if the experiment were to be repeated indefinitely.

4.5 FUNCTIONS OF RANDOM VARIABLES

Consider a change of variable as in the following example.

Example 4.7 **Dots on a fair die.** Roll a fair die, where X is the number of dots on the top face, as in Example 4.6. Consider the game of chance where you win W dollars, where $W = (X-3)^2$. The random variable W is a function of X—say, $W = g(X) = (X-3)^2$. One way of finding $E(W)$ is by making a table for W as we did for the variable X in Example 4.6:

w	$f_W(w)$			$wf_W(w)$
0	$Pr(X=3)$	=	1/6	$0 \cdot 1/6 = 0/6$
1	$Pr(X=2 \text{ or } X=4)$	=	2/6	$1 \cdot 2/6 = 2/6$
4	$Pr(X=1 \text{ or } X=5)$	=	2/6	$4 \cdot 2/6 = 8/6$
9	$Pr(X=6)$	=	1/6	$9 \cdot 1/6 = 9/6$
	$\sum f_W(w)$	=	1	$\mu_W = 19/6 = 3.1667$

It is operationally easier to expand this table in the following way:

x	$w = g(x)$	$f_X(x)$	$g(x)f_X(x)$
1	4	1/6	$4 \cdot 1/6 = 4/6$
2	1	1/6	$1 \cdot 1/6 = 1/6$
3	0	1/6	$0 \cdot 1/6 = 0/6$
4	1	1/6	$1 \cdot 1/6 = 1/6$
5	4	1/6	$4 \cdot 1/6 = 4/6$
6	9	1/6	$9 \cdot 1/6 = 9/6$
		$\sum = 1$	$\mu_W = 19/6 = 3.1667$

We write
$$E((X-3)^2) = \sum (x-3)^2 f_X(x)$$

In general, for a discrete random variable X and a change of variable $W = g(X)$, we define

$$E(g(X)) = \sum g(x) f_X(x)$$

Note that for $W = X$, this formula reduces to the definition of

$$E(X) = \sum x f_X(x)$$

For linear functions, we have the following theorem. The proof is left as an exercise.

THEOREM 4.1 *Suppose that X is a random variable with expected value $E(X)$. Then, for any real constants a and b, we have*

$$\boxed{E(aX + b) = aE(X) + b}$$

Warning In general, for any random variable X, we have $E(X^2) > E^2(X)$, except for trivial discrete cases where X has only one mass point of probability 1.

4.6 VARIANCE AND STANDARD DEVIATION OF DISCRETE RANDOM VARIABLES

The theoretical variance is defined for random variables analogous to the way the population variance is defined for data sets.

Definition

Variance and Standard Deviation

$$\sigma_X^2 = \text{Var}(X) = E\{(X-\mu)^2\} \text{ and}$$

$$\sigma_X = \text{SD}(X) = \sqrt{\text{Var}(X)} = \sqrt{E(X-\mu)^2}$$

Just as the standard deviation of a data set is a measure of the spread of the values in the data set, the standard deviation of a random variable is a measure of the spread of the values of the random variable.

THEOREM 4.2 *For any random variable X with expected value $E(X)$ and variance $\text{Var}(X)$, we have*

$$\text{Var}(X) = E(X^2) - E^2(X) = E(X^2) - \mu^2$$

Proof By definition,

$$\text{Var}(X) = E(X-\mu)^2 = E(X^2 - 2\mu X + \mu^2)$$

This equals $E(X^2) - 2\mu E(X) + \mu^2$ by Theorem 4.1. Because $E(X) = \mu$, we have

$$\text{Var}(X) = E(X^2) - 2\mu \cdot \mu + \mu^2 = E(X^2) - \mu^2$$

Hence, we have a useful formula for computing the standard deviation of a random variable:

$$\sigma = \sqrt{E(X^2) - \mu^2}$$

Compare the formula of Theorem 4.2 with the cautionary remark after Theorem 4.1 concerning the inequality $E(X^2) \geq E^2(X)$. Equality is possible only if $\text{Var}(X) = 0$ which, for a discrete random variable, means that X has only one mass point of probability 1.

Operationally, to find the variance of a discrete random variable, we subtract the square of the mean from the sum of the $x^2 f(x)$ column, as shown by the following example.

Example 4.8 *Dots on a fair die.* Roll a fair die and observe the number of dots on the top face:

x	$f(x)$	$xf(x)$	$x^2 f(x)$
1	1/6	$1 \cdot 1/6 = 1/6$	$1^2 \cdot 1/6 = 1/6$
2	1/6	$2 \cdot 1/6 = 2/6$	$2^2 \cdot 1/6 = 4/6$
3	1/6	$3 \cdot 1/6 = 3/6$	$3^2 \cdot 1/6 = 9/6$
4	1/6	$4 \cdot 1/6 = 4/6$	$4^2 \cdot 1/6 = 16/6$
5	1/6	$5 \cdot 1/6 = 5/6$	$5^2 \cdot 1/6 = 25/6$
6	1/6	$6 \cdot 1/6 = 6/6$	$6^2 \cdot 1/6 = 36/6$
	$\sum = 1$	$\sum = 21/6$	$\sum = 91/6$

$$\sigma^2 = \text{Var}(X) = E(X^2) - E(X)^2 = \left(\frac{91}{6}\right) - \left(\frac{21}{6}\right)^2 = \frac{182}{12} - \frac{147}{12} = \frac{35}{12}$$

and

$$\sigma = \sqrt{\frac{35}{12}} = 1.70763$$

A translation of a random variable from X to $X + b$ has no effect on the variance and standard deviation: $\text{Var}(X + b) = \text{Var}(X)$. Also, multiplication by -1 has no effect: $\text{Var}(-X) = \text{Var}(X)$ and $\text{SD}(-X) = \text{SD}(X)$. More generally, we have the following. The proof is left as an exercise.

THEOREM 4.3 *For any random variable X, with variance $\text{Var}(X)$, and any real constants a, b, we have*

$$\text{Var}(aX + b) = a^2 \text{Var}(X)$$

Corollary *Under the conditions of Theorem 4.3,*

$$\text{SD}(aX + b) = |a| \text{SD}(X)$$

4.6 VARIANCE AND STANDARD DEVIATION

EXERCISES FOR SECTION 4.6

- **4.9** Compute the mean μ and standard deviation σ of the random variable in Exercise 4.3 on page 128:

x	$f_X(x)$		
0	$(0.75)^3$	=	27/64
1	$3(0.25)^1(0.75)^2$	=	27/64
2	$3(0.25)^2(0.75)^1$	=	9/64
3	$(0.25)^3$	=	1/64
	$\sum f_X(x)$	=	1

4.10 Compute the mean μ and standard deviation σ of the random variable in Exercise 4.4 on page 128:

x	$f_X(x)$
0	1/27
1	6/27
2	12/27
3	8/27
	$\sum f_X(x) = 1$

- **4.11** Compute the mean μ and standard deviation σ of the random variables X, Y, and $X+Y$ of the family planning Example 4.3 on page 123. Assume a simple model whereby boys and girls are equally likely.

4.12 Repeat the above exercise except for a family of four children, where X is the number of boys and Y is the number of girls. Assume boys and girls are equally likely.

- **4.13** Compute the mean μ and standard deviation σ of the random variables X, Y, and $X+Y$ of the family planning Example 4.3 on pages 123–125. Assume the probability of a boy is .52 and the probability of a girl is .48.

4.14 Repeat the above exercise except for a family of four children, where X is the number of boys and Y is the number of girls. Assume the probability of a boy is .52 and the probability of a girl is .48.

- **4.15** Compute the mean μ and standard deviation σ of the random variable given by the following table:

x	$f_X(x)$
0	1/8
1	3/8
2	3/8
3	1/8
	$\sum f_X(x) = 1$

4.7 DISCRETE UNIFORM RANDOM VARIABLES

If the possible values of a discrete random variable are equally likely, then the random variable is said to be **uniform**. A simple example is the number of dots U in the roll of a single fair (six-sided) die. The probability mass function is $f(u) = 1/6$ for $u = 1, 2, 3, 4, 5, 6$. Here the mean is

$$E(U) = \frac{1}{6} + \frac{2}{6} + \frac{3}{6} + \frac{4}{6} + \frac{5}{6} + \frac{6}{6} = 3.5$$

and the variance is

$$\text{Var}(U) = E(U^2) - E^2(U) = \sum u^2/6 - \left(\sum u/6\right)^2 = 35/12 = 2.9167$$

So the standard deviation is $\text{SD}(U) = \sqrt{2.9167} = 1.7078$.

See Figure 4.7 for the probability histogram.

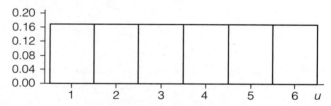

Figure 4.7 Probability histogram for discrete uniform random variable ($n = 6$)

In general, if a uniform random variable U has n equally likely values $u_1, u_2, u_3, \ldots, u_n$, then $f(u) = 1/n$ for $u = u_1, u_2, u_3, \ldots, u_n$.

Standard box model   This experiment is like the draw of one card from a box filled with one of each of the numbers $u_1, u_2, u_3, \ldots, u_n$.

Mean   In the particular case where $u_k = k$ for $k = 1, 2, \ldots, n$, as in the example of dots on a fair die, the expected value is

$$\begin{aligned}\mu = E(U) &= \frac{1 + 2 + 3 + \cdots + n}{n} \\ &= \frac{n(n+1)}{2n} \\ &= \frac{n+1}{2}\end{aligned}$$

Variance   Using Theorem 4.2, the variance is

$$\begin{aligned}\sigma^2 = \mathrm{Var}(U) &= E(U^2) - \mu^2 \\ &= \frac{1^2 + 2^2 + 3^2 + \cdots + n^2}{n} - \frac{(n+1)^2}{4} \\ &= \frac{n(n+1)(2n+1)}{6n} - \frac{(n+1)^2}{4} \\ &= \frac{(n+1)(2n+1)}{6} - \frac{(n+1)^2}{4} \\ &= \frac{n^2 - 1}{12}\end{aligned}$$

Standard deviation   Hence, $\sigma = \sqrt{\frac{n^2-1}{12}}$.

EXERCISES FOR SECTION 4.7

4.16 Subjects of a clinical trial of an experimental drug treatment are randomly and uniformly assigned to seven dosage levels U labeled 1 through 7.

   a. Find the probability function and sketch the probability histogram.

   b. Find μ and σ.

   c. Find $Pr(\mu - \sigma \leq U \leq \mu + \sigma)$.

• **4.17** Consider a discrete random variable U with probability function

$$f(u) = \tfrac{1}{10} \text{ for } u = 1, 2, 3, 4, 5, 6, 7, 8, 9, 10 \quad \text{and} \quad f(u) = 0 \text{ otherwise}$$

   a. Find the expected value μ of U.

   b. Find the standard deviation σ of U.

4.18 Consider the number of dots U appearing on the top face of a fair five-sided die. Find the mean μ and standard deviation σ of the random variable U.

• **4.19** Let U be the number of heads in a single toss of a fair coin. Find the mean μ and standard deviation σ of the random variable U.

4.8 LAW OF LARGE NUMBERS (optional)

Recall that the law of large numbers states that, for experiments that are repeated n times, the empirical histogram of the data tends to look like the probability histogram, as n tends to infinity. In this section, we consider a rigorous statement, which we include for sake of completeness.

THEOREM 4.4   **Rigorous statement of a law of large numbers for histograms.**
Suppose an experiment is conducted and a random variable X observed. Let Y_j indicate whether, on the jth replication of the experiment, X is in the interval $(a, b]$. This means $Y_j = 1$ if, on the jth independent replication of the experiment, $a < X \leq b$; otherwise, $Y_j = 0$. Then

$$Pr\left(\lim_{n \to \infty} \frac{1}{n} \sum_{j=1}^{n} Y_j = F_X(b) - F_X(a)\right) = 1$$

There are analogous results for cumulative distribution functions. The empirical cumulative distribution function was defined in Section 2.12 on page 52. Here we restate the definition when the data are generated by independent replications of a random experiment. The jth replication results is a random variable X_j, each with cumulative distribution function $F_X(x)$. A random experiment replicated n times generates a sequence of (independent) random variables $X_1, X_2, X_3, \ldots, X_n$. Define step functions $S_j(x) = 1$ if $X_j \leq x$ and $S_j(x) = 0$ if $X_j > x$. The empirical cumulative distribution function is

$$F_n(x) = \frac{1}{n} \sum_{j=1}^{n} S_j(x)$$

As the number of trials tends to infinity, the empirical distribution functions of the data tend to look like the cumulative distribution function; that is, for each fixed x,

$$Pr\left(\lim_{n \to \infty} F_n(x) = F_X(x)\right) = 1$$

To put it more strongly, imagine the number of independent trials tending to infinity. It is almost certain (the probability is 1) that the limit of the sequence of empirical cumulative distribution functions is the cumulative distribution function, uniformly in x:

$$Pr\left(\lim_{n \to \infty} \sup_{-\infty < x < \infty} |F_n(x) - F_X(x)| = 0\right) = 1$$

REVIEW EXERCISES FOR CHAPTER 4

4.20 Consider a discrete random variable X with probability function
$$f(x) = \frac{x+1}{10} \text{ for } x = 0, 1, 2, 3 \text{ and } f(x) = 0 \text{ otherwise}$$
    a. Find the expected value of X.   **b.** Find the standard deviation of X.

• **4.21** A pair of fair dice are rolled. Let S be the total number of dots shown. It may be helpful to imagine one of the dice red and the other green, and then to make a 6×6 matrix of the 36 possible outcomes of the experiment.

    a. Find the probability function or table.
    b. Sketch the probability histogram.
    c. Find μ_S.   **d.** Find σ_S.

4.22 A fair die is rolled three times. Let X be the number of different faces showing: 1, 2, or 3. For example, if the top faces of the three dice have 4, 6, and 4 dots, respectively, then 2 different faces are showing. Hint: First compute $f(3)$ and $f(1)$, and then subtract these from 1 to obtain $f(2)$.

    a. Sketch the mass density function.
    b. Sketch the cumulative distribution function.
    c. Find μ_X.   **d.** Find σ_X.

• **4.23** A lab technician knows that two of the eight pints of blood in the blood bank contain Type A Rh-positive blood. The technician selects two of the pints at random (without replacement). Let K be the number of pints of Type A Rh-positive blood obtained.

    a. Find the probability function or table of K.
    b. Graph the cumulative distribution function of K.
    c. Find the expected value μ_K.   **d.** Find the standard deviation σ_K.

4.24 Repeat Exercise 4.23, but this time suppose three pints of blood are selected.

• **4.25** Find the probability mass function for the discrete random variable X with the following cumulative distribution function:
$$F(x) = \begin{cases} 0 & : \text{ for } x < 0 \\ \frac{1}{8} & : \text{ for } 0 \leq x < 1 \\ \frac{4}{8} & : \text{ for } 1 \leq x < 2 \\ \frac{7}{8} & : \text{ for } 2 \leq x < 3 \\ 1 & : \text{ for } x \geq 3 \end{cases}$$

4.26 Consider families of four children. Assume that boys and girls are equally likely and independent. Let X be the absolute value of the difference between the number of boys and girls.

a. Find the probability table of X.
b. Find the cumulative distribution function of X and sketch the graph.
c. Find the mean.   d. Find the standard deviation.

- **4.27** A sample of size 4 is taken from a process that produces 5% defective items.

    a. Find the probability function or table for the number of defective items in the sample.
    b. Find the mean.   c. Find the standard deviation.

4.28 Let U be the number of dots on the roll of a fair die. Find the following:

   a. $E(U)$ b. $E(2U-7)$ c. $E(U^2+4)$ d. $\text{Var}(U)$

- **4.29** Consider blood types of the 40 children mentioned in Exercise 2.1 on page 25, where five of them had Type AB blood. In contrast, for the entire U.S. population, 4% of the children have Type AB blood.

    a. For a sample of size 2 chosen with replacement from the 40, find the probability function or table for the number of children selected with Type AB blood.
    b. For a sample of size 2 chosen with replacement from the U.S. population, find the probability function or table for the number of children selected with Type AB blood.

4.30 A pair of fair dice is rolled. Let X the smaller of the two number rolled.

   a. Find the probability table.
   b. Find the mean of X.   c. Find the standard deviation of X.

   Hint: Suppose one die is red and the other green. It may be helpful to make a 6×6 matrix of all 36 possible outcomes, such as the one on page 85.

* * * * * * * * *

4.31 Repeat Exercise 4.30, but this time for a pair of n-sided dice instead of the usual 6-sided dice.

4.32 Consider Example 4.4 on page 126 where a fair coin is tossed until it comes up heads. Here X be the number of tails before it comes up heads. Show:

   a. $Pr(X \text{ is even}) = Pr(X = 0, 2, 4, \ldots) = 2/3$
   b. $Pr(X \text{ is odd}) = Pr(X = 1, 3, 5, \ldots) = 1/3$

4.33 (Proof) Prove Theorem 4.1 on page 134.

4.34 (Proof) Prove Theorem 4.3 on page 136 and the Corollary.

4.35 Let $f(x) = \frac{1}{2^x}$ for $x = 1, 2, 3, \ldots$. Show that $\sum_{x=1}^{\infty} f(x) = 1$ and $\mu_X = 2$.

Chapter 5

RANDOM VARIABLES FOR SUCCESS/FAILURE EXPERIMENTS

5.1 BERNOULLI RANDOM VARIABLES

5.2 BINOMIAL RANDOM VARIABLES

5.3 HYPERGEOMETRIC RANDOM VARIABLES

In this chapter, we develop standard probability models for three families of discrete random variables. Our approach is to translate a real-world statistics problem into a standard model of drawing from a box of numbered cards. Then, from an analysis of the model we arrive at a model solution and then translate back from the model solution to a solution of the original problem. We call such models **population box models**. Classically they are known as **urn models**. The three standard models that we consider in this chapter all stem from experiments of drawing cards from a box of 0's and 1's. Drawing a card numbered 0 represents failure and drawing a card numbered 1 represents success.

5.1 BERNOULLI RANDOM VARIABLES

If a random experiment has only two possible outcomes, it is called a **dichotomous experiment** or **binary experiment**. One of the outcomes we call **success** and the other **failure**. If we code success with the number 1 and failure with the number 0, the resulting random variable is called a **Bernoulli random variable**.[1]   Here are some examples:

- Toss a coin: heads or tails
- Birth of a child: boy or girl.
- Is this traffic light green? yes or no.
- Are you employed now? yes or no
- Has your mother ever smoked? yes or no
- Are you at least 18 years of age? yes or no
- Did you pass the test? yes or no
- Will this person vote? yes or no

A Bernoulli random variable is an indicator of whether or not an event has occurred. It is always possible to design such an experiment by the use of a question whose answer is either yes (success) or no (failure). The two possible outcomes are not necessarily equally likely.

This is perhaps the simplest type of random variable. There are only two values: value 1 (success) with probability π and value 0 (failure) with probability $1 - \pi$. The probability table looks like this:

x	$f(x)$	$xf(x)$	$x^2 f(x)$
0	$1-\pi$	0	0
1	π	π	π
	$\Sigma = 1$	$\Sigma = \pi$	$\Sigma = \pi$

The mean is the sum of the $xf(x)$ column

$$\mu = \sum xf(x) = \pi$$

and the variance is

$$\text{Var}(X) = E(X^2) - \mu^2 = \sum x^2 f(x) - \pi^2 = \pi - \pi^2 = \pi(1-\pi)$$

so the standard deviation is

$$\sigma = \sqrt{\pi(1-\pi)}$$

---

[1]Named after Jakob Bernoulli (1654–1705), one of a family of famous Swiss mathematicians and author of the first treatise on probability, *Ars conjectandi*, written in 1713.

5.1 BERNOULLI RANDOM VARIABLES

Standard population box model

A single card is drawn from a box consisting of cards marked with 0's and 1's. The proportion of 1's in the box corresponds to the probability of success π. The proportion of 0's in the box corresponds to the probability of failure $1 - \pi$.

Example 5.1

Box model for $\pi = 0.80$. As a box model for the Bernoulli random variable with $\pi = 0.80$, we could have four cards numbered 1 and one card numbered 0 ($\pi = 4/5 = 0.80$); or we could have 40 cards with 1's and 10 cards with 0's ($\pi = 40/50 = 0.80$); or 80 cards with 1's and 20 cards with 0's ($\pi = 80/100 = 0.80$), and so on. **The number of cards in the box is not important, but the proportion of 1's in the box must be 0.80. Success corresponds to getting a card with the number 1; failure is getting a card with the number 0.** A box model consisting of 0's and 1's is called a **Bernoulli box**. See Figure 5.1 for a probability histogram of the contents of this box.

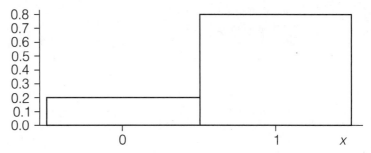

Figure 5.1 Probability histogram for Bernoulli random variable with $\pi = 0.80$

Example 5.2

Assembly line. Consider a refrigerator coming off the assembly line of a manufacturing plant. A quality-control inspector examines it by submitting the refrigerator to various tests. Success means that the refrigerator passed all tests. Failure means the refrigerator is sent back for repairs or adjustments. The Bernoulli random variable corresponds to the response to the question, Does the refrigerator pass inspection? The parameter π is the probability of success.

Example 5.3

Draw a card. Draw a card from a standard deck; suppose success is getting an ace. A box model for this experiment may contain 52 cards, four cards with the number 1, and 48 cards with the number 0; or the box model may contain 13 cards, one card with the number 1, and twelve cards with the number 0.

Example 5.4 **Genetics.** The work of Gregor Mendel (1822–1884) is the basis of the modern theory of genetics. The theory states that hereditary traits, such as flower color, are carried by genes, which appear in pairs. Each plant inherits one copy of the gene from each parent. The table below shows some of the data obtained by Mendel from his experiments with garden pea plants. He studied one single hereditary trait at a time. For each trait, he asked the question, Does the second generation of garden pea plant have the same trait as the first generation?

Parental trait	First generation	Second generation
yellow seeds × green seeds	all yellow	6022 yellow and 2001 green
round seeds × wrinkled seeds	all round	5474 round and 1850 wrinkled
green pods × yellow pods	all green	428 green and 152 yellow
long stems × short stems	all long	787 long and 277 short
axial flowers × terminal flowers	all axial	651 axial and 207 terminal
inflated pods × constricted pods	all inflated	882 inflated and 299 constricted
red flowers × white flowers	all red	705 red and 224 white

We can view the second-generation data as the summarization of a large number of experiments and describe these results in terms of many Bernoulli random variables. For each second-generation plant, assign a random variable X that assumes the value 1 if the plant has the same hereditary trait as its first-generation parents, and the value 0 otherwise. For example, the experiments involving crossing red flower peas with white flower peas, the second generation can be viewed as the realization of 929 Bernoulli random variables of which 705 were 1 and 224 were 0.

EXERCISES FOR SECTION 5.1

- **5.1** Find μ and σ for Example 5.3 on page 145.

  5.2 Consider Mendel's experiment as described in Example 5.4. Encode a red-flowered plant with 1 and yellow with 0. Suppose that when a second-generation plant is selected from the crossing of a pea plant with red flowers and one with white flowers, the probability of obtaining a pea plant with red flowers is 3/4. Find μ and σ.

- **5.3** Consider Mendel's experiment as described in Example 5.4.
  Suppose one of the 8023 second-generation plants is chosen from Mendel's yellow seeds × green seeds trials. Encode the trait of yellow seeds with 1 and green seeds with 0. Find μ and σ.

- **5.4** Let X be a Bernoulli random variable with parameter π.
  Find $E[(X-\mu)^i]$ for $i = 1, 2, 3$.

- **5.5** The **coefficient of skewness** of a random variable X is defined to be:
  $$\frac{E[(X-\mu)^3]}{\sigma^3}$$
  Show that the coefficient of skewness of a Bernoulli random variable is
  $$\frac{1-2\pi}{\sqrt{\pi(1-\pi)}}$$

5.2 BINOMIAL RANDOM VARIABLES

A binomial random variable is obtained by counting successes when a Bernoulli experiment is repeated. Each repetition is called a **trial**. The number of successes we obtain is called the **success count** or simply **count**, which we denote by the symbol K.

Rules for a binomial random variable K

Rule 1. The number of Bernoulli trials n is predetermined.

Rule 2. The random variable K is the number of successes in the n trials.

Rule 3. The trials are independent.

Rule 4. The probability of success π is the same for every trial.

Example 5.5 *Roll a fair die.* Roll a fair die four times and count the number of times a top face of the die shows a single dot (ace). Let K be the number of aces you get. The random variable K is binomial. Here $n = 4$, the rolls are independent, and $\pi = 1/6$.

Standard box model

We use the same 0–1 population box as for a Bernoulli random variable. Instead of drawing one card, we draw a random sample of n cards *with replacement* and count the number of 1's (successes) in the sample.

One often understands a concept better by considering counterexamples:

Counterexample 1. A box contains five cards; three of them numbered 1 and two numbered 0. Draw cards with replacement until you obtain three 0's; then count the number of 1's. This violates Rule 1 above. The number of draws depends on the outcome of the experiment. Rules 2, 3, and 4 clearly hold.

Counterexample 2. A box contains six cards numbered 1 through 6. Draw three cards with replacement and add the numbers drawn. This experiment violates Rule 2 because we are not counting how many times some event occurs.

Counterexample 3. Use the same box as in counterexample 1. Draw three cards *without replacement* and count the number of 1's. This violates Rule 3 because the draws are dependent. Rules 1 and 2 clearly hold. Also, note that Rule 4 holds because the (unconditional) probability of success on each draw is the same, namely, $\pi = 3/5$.

Counterexample 4. Draw one card from each of two boxes and count the number of 1's. The first box is the same as in Counterexample 1, the second contains two cards numbered 0 and two cards numbered 1. This violates Rule 4. For the first draw, the probability of success is $3/5 = 0.60$, and for the second it's $2/4 = 0.50$. However, Rules 1, 2, and 3 hold.

Example 5.6 *Roll a fair die.* Roll a fair die four times. On each roll, let "A" denote the event of getting an ace (an ace occurs when the top face has only one dot) and let "N" denote the opposite event of not getting an ace. If success is getting an ace, the standard box model consists of four draws with replacement from a box with one card labeled 1 and five cards labeled 0. In four trials, there are as many outcomes as there are four-letter words using the letters A and N: $2^4 = 16$. With the right emphasis, you can pronounce all 16 possible words in the table below.

If we count the number of aces in four rolls of a die we obtain the following table.

k	Event "$K = k$"	$f_K(k)$
0	{NNNN}	$1 \cdot (1/6)^0 \cdot (5/6)^4$
1	{ANNN, NANN, NNAN, NNNA}	$4 \cdot (1/6)^1 \cdot (5/6)^3$
2	{AANN, ANAN, ANNA, NAAN, NANA, NNAA}	$6 \cdot (1/6)^2 \cdot (5/6)^2$
3	{AAAN, AANA, ANAA, NAAA}	$4 \cdot (1/6)^3 \cdot (5/6)^1$
4	{AAAA}	$1 \cdot (1/6)^4 \cdot (5/6)^0$

The coefficients 1, 4, 6, 4, 1 appearing in the column for $f_K(k)$ are the number of words you can make with k A's and $(4-k)$ N's, for $k = 0, 1, 2, 3, 4$, respectively.

In general, the number of n-letter words with k A's and $(n-k)$ N's is called a **binomial coefficient**, denoted $\binom{n}{k}$, and read "n choose k" or "n take k." Thus, the binomial probabilities $f_K(k)$ take the form

$$f_K(k) = \begin{cases} \binom{n}{k}\pi^k(1-\pi)^{n-k} & : \quad \text{for } k = 0, 1, 2, \ldots, n \\ 0 & : \quad \text{elsewhere} \end{cases} \quad (5.1)$$

where

π is the probability of success (in the above example, $\pi = \frac{1}{6}$)

$1 - \pi$ is the probability of failure (in the above example, $1 - \pi = \frac{5}{6}$)

k is the number of successes

$n - k$ is the number of failures

Figure 5.2 shows histograms of various binomial random variable.

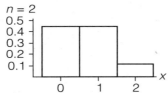
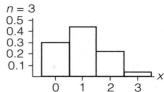
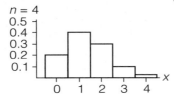

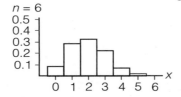
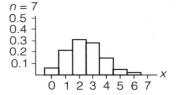

Figure 5.2 Probability histograms for binomial random variables with $\pi = \frac{1}{3}$ and $n = 2, 3, 4, 5, 6,$ and 7

We can use the following formula for the binomial coefficients[2]

$$\binom{n}{k} = \frac{n!}{k!(n-k)!} \qquad (5.2)$$

or we can use **Pascal's Triangle**[3]:

```
                    1
                 1     1
              1     2     1
           1     3     3     1
        1     4     6     4     1
     1     5    10    10     5     1
  1     6    15    20    15     6     1
```

Each entry is the sum of the two entries diagonally above it. In each row, the first entry is 1, and the second entry is the number of trials n. Example 5.6 uses the row

$$1 \quad 4 \quad 6 \quad 4 \quad 1$$

whose entries are the numbers $\binom{4}{k}$ for $k = 0, 1, 2, 3, 4$. If one needs all of the binomial coefficients for a particular small value of n, then it's often easier to use Pascal's triangle than Formula 5.2.

A binomial random variable is a sum of n independent and identical Bernoulli random variables. Let's say that the outcome of the first Bernoulli trial is denoted X_1, the outcome of the second is X_2, and so on. Then

$$K = X_1 + X_2 + \cdots + X_n$$

For each $k = 1, 2, \ldots, n$, we have

$$E(X_k) = \pi \quad \text{and} \quad \text{SD}(X_k) = \sqrt{\pi(1-\pi)}$$

---

[2]Using elementary counting techniques, one can see that there are $n!$ words with n letters, provided all letters are different. So the word SAT can be rearranged $3! = 6$ ways. If, however, a letter is repeated k times, one must divide by $k!$. So STAT can be rearranged $\frac{4!}{2!}$ ways, and STATS can be rearranged $\frac{5!}{2!2!}$ ways. In general, an n-letter word with k A's and $(n-k)$ N's can be rearranged $\frac{n!}{k!(n-k)!}$ ways.

[3]Named after Blaise Pascal (1623–1662), a French mathematician and philosopher.

In the language of error analysis (Section 1.3 on page 5), we say that, for each of the k Bernoulli trials, we expect $E(X_k) = \pi$ successes, give or take $SD(X_k) = \sqrt{\pi(1-\pi)}$ successes or so. Or, using the notation of Section 1.3, we have for each k,

$$X_k = \pi \pm SE_{X_k} = \pi \pm \sqrt{\pi(1-\pi)}$$

After n repetitions, we expect

$$E(K) = E(X_1) + E(X_2) + \cdots + E(X_n) = n\pi \tag{5.3}$$

successes, give or take

$$SD(K) = \sqrt{SD(X_1)^2 + SD(X_2)^2 + \cdots + SD(X_n)^2} = \sqrt{n\pi(1-\pi)} \tag{5.4}$$

successes or so. Thus,

$$K = n\pi \pm \sqrt{n\pi(1-\pi)}$$

Although Equation 5.3 is intuitive, we sketch a direct proof in the Exercise 5.34 on page 162. However, Theorem 10.1 on page 280 will show that, in general, when one adds random variables, the means also add. Thus,

$$E(K) = \mu_K = n\pi \tag{5.5}$$

Also, in Theorem 10.8 on page 286 we will show that, when one adds *independent* random variables, then the variances add. Thus, for a binomial random variable, which is a sum of n Bernoulli trials, each with variance $\pi(1-\pi)$, we have

$$\text{Var}(K) = \sigma_K^2 = n\pi(1-\pi) \tag{5.6}$$

or

$$SD(K) = \sigma_K = \sqrt{n\pi(1-\pi)}$$

A direct proof of 5.6 is also given in Exercise 5.35 on page 162.

Note that the variances are proportional to the number of draws n. Standard deviations are *not* proportional to the number of draws; rather, they are proportional to the *square root* of the number of draws.

Example 5.7 ***Roll a fair die.*** In one roll of a fair die, we expect $\pi = 1/6$ aces, give or take $\sqrt{\pi(1-\pi)} = 0.373$ aces or so. In $n = 4$ rolls of the die, we expect $n\pi = 4/6$ aces, give or take $\sqrt{n\pi(1-\pi)} = \sqrt{4} \cdot 0.373 = 0.745$ aces or so. In $n = 180$ rolls, we expect 30 aces, give or take $\sqrt{180} \cdot \sqrt{\pi(1-\pi)} = 5$ aces or so. Notice that the standard deviation is proportional to the square root of the number of rolls.

Example 5.8 ***Genetics.*** In the simplest form of Mendel's genetic theory, each gene of a pair can assume two forms called **alleles**. For example, in garden peas the allele R may carry the trait for red flowers and the allele w the trait for white flowers. In this case, one of the alleles R is **dominant** (denoted by capital letters) and the other w is **recessive** (denoted by lowercase letters). This means that the plants that have one or two dominant R alleles will have red flowers, and that only the plants with two w alleles will have white flowers. Each parent contributes one gene, which is randomly selected from that parent's two copies. Mendel, in one of his experiments, started with pure type red-flower parents with the pair of genes RR and pure type white-flower parents with the pair ww. Because each parent gives one copy of the gene to its offspring, all the first-generation plants had the gene pattern Rw and had red flowers. But in the second generation, each plant inherits one gene from each of its (first-generation) Rw parents. The possible gene pairs are RR, Rw, and ww (the genes are not ordered, so Rw is the same as wR). If we assume each parent is equally likely to contribute a copy of either allele to its offspring, then the number of R alleles that a pea plant will inherit from its parents is a binomial random variable with $n = 2$ and $\pi = 1/2$. According to this theory, the probability that a second-generation plant will have red flowers is

$$f_K(0) + f_K(1) = \frac{1}{4} + \frac{2}{4} = \frac{3}{4}$$

(We assume the trait has no effect on survival to maturity.) If each second-generation plants is viewed as an independent trial, then the number of red-flowered plants grown from a random sample of N second-generation plants will be a binomial random variable with $\pi = 3/4$. In this case, if Mendel's cross-fertilization experiment were repeated and 929 of the second-generation plants were randomly selected, then the number of red-flower plants would be a binomial random variable with mean

$$\mu_K = n\pi = 929 \cdot \frac{3}{4} = 696.75$$

and standard deviation

$$\sigma_K = \sqrt{n\pi(1-\pi)} = \sqrt{929 \cdot \frac{3}{4} \cdot \frac{1}{4}} = 13.20$$

EXERCISES FOR SECTION 5.2

5.6 Use the ideas of footnote 2 on page 150 to count the number of rearrangements of the word

MISSISSIPPI

5.7 Sketch a tree diagram for the number of aces in four rolls of a fair die as described in Example 5.6 on page 148. This is a four-stage experiment. Label the 2^4 terminal outcomes and their probabilities.

5.8 In a binomial experiment, find μ_K and σ_K for

    a. $n = 100$ and $\pi = 0.40$ **b.** $n = 400$ and $\pi = 0.40$

    c. $n = 100$ and $\pi = 0.60$ **d.** $n = 400$ and $\pi = 0.60$

5.9 In a binomial experiment, find μ_K and σ_K for

    a. $n = 100$ and $\pi = 1/3$ **b.** $n = 900$ and $\pi = 1/3$

    c. $n = 100$ and $\pi = 2/3$ **d.** $n = 900$ and $\pi = 2/3$

5.10 In a test for extrasensory perception (ESP), the subject has to guess which one of three cards lying face down on a table is the ace of spades. Suppose that the subject has no ESP and is just guessing. Find the probability that in 12 guesses, the number of cards guessed correctly is

    a. 4   **b.** 5   **c.** 6   **d.** less than 6   **e.** at least 6   **f.** more than 6

5.11 Only about 10% of all people survive an infection of the Ebola virus. In 10 independent cases of Ebola, what is the probability that the number of survivors is

    a. 0,   **b.** 1   **c.** 2   **d.** less than 2   **e.** at least 2   **f.** more than 2

5.12 In one toss of a fair coin, we expect _____ heads, give or take _____ heads or so.
In $n = 100$ tosses, we expect _____ heads, give or take _____ heads or so.
In $n = 400$ tosses, we expect _____ heads, give or take _____ heads or so.
Note that the standard deviation is proportional to the square root of the number of tosses.

5.13 A machine makes faulty widgets 1% of the time. The expected count of faulty widgets in a box of 100 is $E(K) = n\pi = 100(0.01) = 1$. What is the chance that a box of 100 contains the following?

    a. Exactly one faulty widget.

    b. No faulty widgets.

    c. At least one faulty widget.

    d. In a crate of 16 boxes, how many faulty widgets do you expect? What is the chance of getting exactly what you expect?

5.14 A multiple-choice quiz consists of five questions, each with four possible answers, only one of which is correct. You randomly guess each answer. What is the probability of getting the following?

a. Exactly one right. **b.** Exactly three right.
c. More right than wrong. **d.** At least one wrong.

- **5.15** Assuming boys and girls are equally likely, among families with four children, what percentage have more girls than boys?

* * * * * * * * *

⋆ **5.16** Refer to Mendel's garden pea experiment of Example 5.8, where he mated pure type red-flower RR plants with white-flower ww plants. If only second-generation plants with red flowers are parents for a third-generation experiment, what is the probability that a third-generation plant will have red flowers? Begin by computing the probability of genotype ww in two steps. Under the assumption of random mating, the number of parents of a third-generation plant that has genotype Rw is a binomial random variable with $n = 2$ and $\pi = 2/3$, so the probability the parents are both Rw is 4/9. Use a tree diagram to show that the probability they are both Rw is $\frac{4}{9}$ and the conditional probability of ww is $\frac{1}{4}$. So the probability of ww is $\frac{4}{9} \cdot \frac{1}{4} = \frac{1}{9}$. Now use the rule of opposites to compute the probability that a third-generation plant will have red flowers.

NOTES ON TECHNOLOGY

Binomial Probabilities

This is how to use technology to find $f_K(k) = \binom{20}{k}(0.42)^k(0.58)^{20-k}$ (and the cumulative distribution $F_K(k)$) for $n = 20$, $\pi = 0.42$, and a list of k values.

Minitab: Enter the list of values k in column C1. Select **Calc→Probability Distributions→Binomial**. In the display boxes, enter 20 for the number of trial, .42 for the probability of success, and C1 for the input column. Click **OK**. For the cumulative distribution, do the same except click the cumulative probability button in the Binomial Distribution display.

Excel: Enter the list of values k in column A. Click on cell B1 and type **=Binomdist(A1,20,.42,0)**. Copy cell B1 to the remainder of the desired list in column B. For the cumulative distribution $F_K(k)$, type **=Binomdist(A1,20,.42,1)**. If you select the function through the icon **f$_\mathbf{x}$**, you will be prompted for the arguments of the functions.

TI-83 Plus: To find the probability for a single value of k, such as $k = 4$, use the **DISTR** function, which is can be reached at **2nd Vars**. Select

binomialpdf. Then complete the entry to show *binomialpdf*(20, .42, 4) and hit **ENTER**. The display should show .0247251446. To store the probabilities for all values of k in list L1, you can enter **binompdf(20,.42)→STO→L1**. The value of the cumulative distribution $F_K(4)$ can be found by using *binomialcdf*(20, .42, 4) to display the value .0348955147.

5.3 HYPERGEOMETRIC RANDOM VARIABLES

A hypergeometric random variable also counts successes in repeated Bernoulli trials, except that sampling is done *without replacement*.

Standard box model

We use the same Bernoulli box of 0's and 1's as for the binomial random variables. The hypergeometric random variable K is the number of 1's (successes) we obtain in drawing a random sample of n cards *without replacement*.

Population size is N, and the number of 1's in the population is $N_1 = \pi N$, where π is the (unconditional) probability of drawing a 1. The number of 0's is thus $N_0 = N - N_1 = N - \pi N$. The sample size (number of draws) is n.

Example 5.9

A hand of cards. We draw a random hand of five cards from a standard deck of cards. In each draw, let K be the number of aces drawn. Let "A" denote the event of getting an ace (success) and "M" the opposite event; that is, the event of not getting an ace (failure). A box model might have $N = 52$ cards of which $N_1 = 4$ have the letter A and $N - N_1 = 48$ have the letter M. In a sample of five, there are as many outcomes as there are five-letter words using the letters A and M with no more than four A's. The table below lists them:

k	Event "$K = k$"
0	{MMMMM}
1	{AMMMM,MAMMM,MMAMM,MMMAM,MMMMA}
2	{AAMMM,AMAMM,AMMAM,AMMMA,MAAMM, MAMAM,MAMMA,MMAAM,MMAMA,MMMAA}
3	{AAAMM,AAMAM,AAMMA,AMAAM,AMAMA, AMMAA,MAAAM,MAAMA,MAMAA,MMAAA}
4	{AAAAM,AAAMA,AAMAA,AMAAA,MAAAA}
5	{ }

In this case, the probability mass function takes the form:

k	$f_K(k)$
0	$(48 \cdot 47 \cdot 46 \cdot 45 \cdot 44)/(52 \cdot 51 \cdot 50 \cdot 49 \cdot 48)$
1	$5(4 \cdot 48 \cdot 47 \cdot 46 \cdot 45)/(52 \cdot 51 \cdot 50 \cdot 49 \cdot 48)$
2	$10(4 \cdot 3 \cdot 48 \cdot 47 \cdot 46)/(52 \cdot 51 \cdot 50 \cdot 49 \cdot 48)$
3	$10(4 \cdot 3 \cdot 2 \cdot 48 \cdot 47)/(52 \cdot 51 \cdot 50 \cdot 49 \cdot 48)$
4	$5(4 \cdot 3 \cdot 2 \cdot 1 \cdot 48)/(52 \cdot 51 \cdot 50 \cdot 49 \cdot 48)$
5	0

In general, the probability mass function] $f_K(k)$ takes the form

$$f_K(k) = \begin{cases} \frac{\binom{N_1}{k}\binom{N-N_1}{n-k}}{\binom{N}{n}} = \frac{\binom{\pi N}{k}\binom{N-\pi N}{n-k}}{\binom{N}{n}} & : \text{ for } k = 0, 1, 2, \ldots, n \\ 0 & : \text{ elsewhere} \end{cases}$$

This formula follows from the following principle given in Chapter 3: When all outcomes are equally likely, we have

$$Pr(\text{Event}) = \frac{\text{count of favorable outcomes}}{\text{count of possible outcomes}}$$

For a box model of N_1 1's and $N - N_1$ 0's, with draws made without replacement, this is equal to

$$\frac{(\text{number of subsets of size } k \text{ from the 1's })(\text{number of subsets of size } n-k \text{ from the 0's })}{(\text{number of possible subsets of size } n \text{ from the population of size } N)}$$

$$= \frac{\binom{N_1}{k}\binom{N-N_1}{n-k}}{\binom{N}{n}}$$

Just as for a binomial random variable, a hypergeometric random variable is a sum of Bernoulli random variables. Because for each draw the expected value of the Bernoulli trial is π, it turns out that for a sum of random variables, the expected values add. Thus, for $n \leq N$,

$$E(K) = \mu_K = n\pi \qquad (5.7)$$

Note that the expected value of K is the same regardless of whether draws are made with or without replacement. The variance of K is smaller when draws are made without replacement because the variability decreases as cards are removed from the box. The variance is reduced by a factor $\frac{N-n}{N-1}$:

$$\text{Var}(K) = \sigma_K^2 = \frac{N-n}{N-1} \, n\pi(1-\pi) \tag{5.8}$$

The derivation of this formula is left as Exercise 10.16. Taking the square root we obtain the equation

$$\sigma_K = \sqrt{\frac{N-n}{N-1}} \, \sqrt{n\pi(1-\pi)} \tag{5.9}$$

Example 5.10 **A hand of cards.** In a hand of five cards from a standard deck, the count of aces is expected to be

$$\mu_K = n\pi = \frac{5}{13} \quad \text{give or take}$$

$$\sigma_K = \sqrt{\frac{N-n}{N-1}} \sqrt{n\pi(1-\pi)} = \sqrt{\frac{52-5}{52-1}} \sqrt{\frac{5}{13}\frac{12}{13}} = 0.5720 \quad \text{or so}$$

Reduction factor

> The same **Reduction Factor** $\sqrt{\frac{N-n}{N-1}}$ *will appear whenever we compare the standard deviation for draws with replacement to the standard deviation for draws without replacement.*

Observe that, if we consider the extreme case of taking all cards from the box, the sample consists of the entire population, or $n = N$. Because we have emptied the box and there were N_1 1's in the box, N_1 is the only possible value of K. Thus, $E(K) = N_1 = \pi N$ and $\sigma_K = 0$. This is consistent with Equation 5.9 because the reduction factor is $\sqrt{\frac{N-n}{N-1}} = \sqrt{\frac{N-N}{N-1}} = 0$. At the other extreme, if we take just one card from the box, the reduction factor is 1, which is not surprising because, for $n = 1$, there is no distinction between sampling with or without replacement.

If a population size N is large in comparison to the sample size n, then the reduction factor is close to 1. In those situations, the reduction factor can generally be ignored.

General rule Commonly, a population is considered large if $N \geq 20n$; that is, if the sample consists of no more than 5% of the population.

When a population is large, binomial probabilities are good approximations of hypergeometric probabilities.

In the extreme case when the Bernoulli population box is infinite, the hypergeometric random variable is equal to the binomial. This is the situation when one is modeling a production line, a human population that includes future generations, and so on.

Small Population

> When sampling more than 5% of a population, we say that we are sampling from a **small population**. *The reduction factor is also know as the* **Small Population Reduction Factor** $\sqrt{\frac{N-n}{N-1}}$ *because it can generally be ignored unless* $N < 20n$.

For example, in the Harris survey, a sample of 1256 was drawn. Although the sample was drawn without replacement, the population of all adult Americans is considered large in comparison to the sample. We can thus assume the simpler formulas of the binomial random variable.

EXERCISES FOR SECTION 5.3

- **5.17** Find μ_K and σ_K, when $N = 500$, $\pi = 0.50$, and 5% of the population is sampled, for the following cases:

    a. draws with replacement **b.** draws without replacement

5.18 In your pocket, you have eight red marbles and three green ones. You reach in and take a handful of six marbles. Find the probability that you get the following:

   a. no green marbles **b.** one green marble
   c. all three green marbles
   d. at least one green marble
   e. Find the expected number of green marbles.
   f. Find the standard deviation.

- **5.19** Consider a (bridge) hand of 13 cards drawn from a well-shuffled deck of 52 cards.

    a. Find the probability of getting 7 spades.

    b. Find the mean and standard deviation of the spade count.

    c. Find the probability of getting 7 cards of one suit.

5.20 In a hand of 13 cards drawn from a well-shuffled deck of 52 cards, what is the probability that there are:

a. 5 spades? **b.** no spades?
c. less than 5 9 spades? **d.** no more than 2 spades?

NOTES ON TECHNOLOGY

Hypergeometric Probabilities

This is how to use technology to find $f_K(k) = \frac{\binom{14}{k}\binom{10}{9-k}}{\binom{24}{9}}$ and the cumulative distribution $F_K(k)$ for $N = 24, N_1 = 14, n = 9$, and $k = 0, 1, \ldots, 9$.

Minitab: Enter the list of 10 values of k in column C1. Select
Calc→Probability Distributions→Hypergoemetric
In the display boxes, enter 20 for **Population size** (N), 14 for **Successes in population** (X), 9 for **Sample size** (n), C1 for **Input column**, and C2 for **Optional storage**. For the probability, click the **Probability** button, and for cumulative distribution, click the **Cumulative probability** button. Then click **OK**.

Excel: Enter the list of 10 values of k in column A. Go to cell B1. Select icon **f$_x$** and choose the **Statistical** category and then the function **HYPERGEOMDIST**. Enter A1 for the number of successes in the sample, 9 for the size of the sample, 14 as the number of successes in the population, and 24 for the population size. Hit **OK**. Copy cell B1 to the remainder of the list in column B. For the cumulative distribution, use the properties of Excel to make a running sum of values in column B and place them in column C.

TI-83 Plus: This is somewhat involved but can be done.[4] Enter
seq$(14 \, nCr \, X \, * \, 10 \, nCr \, (9 - X) \, \div \, 24 \, nCr \, 9, X, 0, 9)$ and press **STO→** L_1. The function **seq** can be found under **LIST → OPS** and nCr can be found under **MATH → PRB**. The values will be entered in list L1. Note that $f_K(k) = L_1(k+1)$ for $k = 0, 1, \ldots, 9$.

To obtain the cumulative distribution in list L2, enter **cumSum**(L_1) and press **STO→** L_2. The function **cumSum** can be found under **LIST → OPS**. Note that $F_K(k) = L_2(k+1)$ for $k = 0, 1, \ldots, 9$.

---

[4] As an alternative, programs can easily be written for the TI-83 that will find the hypergeometric probabilities and the cumulative distribution. The website http://www.stat.luc.edu has such a program.

REVIEW EXERCISES FOR CHAPTER 5

- **5.21** The West Nile virus is spread to humans by mosquitoes. The probability that an infected person gets sick is less than 1%. But of the 877 people who were sickened by West Nile in Illinois in the year 2002, 63 died. Using 63/877 as the probability of a sick person dying from the disease, find the probability that, in a sample of 10 independent cases,

    a. all will recover
    b. exactly 8 persons will recover
    c. at least 8 persons recover

5.22 The St. Louis encephalitis virus is also spread by mosquitoes. The probability that a person infected with the virus gets encephalitis is 1 in 150. But of the 70 who developed St. Louis encephalitis in and around Monroe, La., in 2001, three died. Using 3/70 as the probability of a sick person dying from the disease, find the probability that in a sample of 10 independent cases:

    a. all will recover
    b. exactly 9 persons will recover
    c. at least 9 persons recover

- **5.23** Suppose that 40% of all students at a large university live in dormitories. Four students are chosen at random. Of those selected, let K be the number who live in dormitories.

    a. Find the probability function of K.
    b. Sketch a histogram of K.

5.24 Suppose that 20% of all personal computers of a certain brand break down in the first year of operation. In an office with 10 such computers, find the probability that:

    a. none break down
    b. exactly two break down
    c. exactly five break down
    d. all break down
    e. at least one breaks down

- **5.25** Suppose that the birth of boys and girls is equally likely. In a family of five children, what is the chance that there will be a 3–2 split (three boys and two girls, or vice versa)?

5.26 Consider drawing a random sample of size 4 from a box of 100 cards; 50 cards marked 1 and 50 cards marked 0.

    a. Compute $f_K(2) = Pr(K = 2)$ for draws with replacement.
    b. Compare with $f_K(2) = Pr(K = 2)$ for draws without replacement.

REVIEW EXERCISES FOR CHAPTER 5

- **5.27** A department store has to decide whether to accept a shipment of 20 kitchen cabinets. Five are randomly inspected for damage by the store. If 4 of the 20 are damaged, find the probability that the number of damaged cabinets in the random sample is

    a. 0   **b.** 4   **c.** 1   **d.** at least 1

    The expected number of damaged cabinets found in the inspection is

    e. _____, give or take   **f.** _____ or so.

- **5.28** A lot of 12 television sets includes three that are defective. Two of the sets are shipped to a hotel.

    a. Find the probability function for the number of defective TV's shipped to the hotel.
    b. Find the cumulative distribution function.
    c. How many defective sets can the hotel expect?
    d. What is the variance?

- **5.29** Among a store's 16 delivery trucks, 5 have worn brakes. If 3 of the trucks are randomly selected for a regular maintenance, what is the probability that at least one of the selected trucks will have worn brakes?

- **5.30** I suspect that a coin is not fair and comes up heads too many times. I decide to test the fairness by tossing it 20 times. For a fair coin, the expected number of head is **a.**_____ and the standard deviation is **b.** _____. Suppose I am prepared to decide that the coin is not fair if it comes up heads 16 or more times. For a fair coin, the probability of getting 16 or more heads is **c.** _____.

- **5.31** Suppose that one out of 48 chicken eggs have fragile shells that are likely to be damaged in shipment. Also suppose that eggs are randomly packaged into cartons, each carton containing a dozen eggs. Find the probability that a carton containing 12 eggs has at least one such fragile egg.

- **5.32** Among 50 math majors at a university, 30 are women. If five of these majors are selected at random, what is the probability that exactly two of the five are women?

    a. Use a hypergeometric probability.
    b. Use a binomial probability as an approximation of the hypergeometric probability.
    c. Is this considered to be sampling from a small population?

- **5.33** Among 120 statistics majors at a university, 80 have computers with Internet access at home. If five of these majors are selected at random, what is the probability that exactly two of the five have Internet access at home?

a. Use a hypergeometric probability.
b. Use a binomial probability as an approximation of the hypergeometric probability.
c. Is this considered to be sampling from a small population?

5.34 (Proof) Prove Equation 5.5 on page 151. Use the fact that

$$E(K) = \sum_{k=0}^{n} k f_K(k) = \sum_{k=0}^{n} k \binom{n}{k} \pi^k (1-\pi)^{n-k}$$
$$= n\pi \sum_{k=1}^{n} \frac{(n-1)!}{(k-1)!(n-k)!} \pi^{k-1}(1-\pi)^{n-k}$$

recognizing that the last sum is the sum of the probabilities of a binomial random variable for $(n-1)$ trials.

5.35 (Proof) Prove Equation 5.6 on page 151. First use the method of the previous exercise to show that $E(K(K-1)) = n(n-1)\pi^2$. Then show that

$$\begin{aligned} \mathrm{Var}(K) &= E(K^2) - E^2(K) \\ &= E(K(K-1)) + E(K) - E^2(K) \\ &= n(n-1)\pi^2 + n\pi - n^2\pi^2 \end{aligned}$$

5.36 The coefficient of skewness of a random variable X is defined to be:

$$\frac{E[(X-\mu)^3]}{\sigma^3}$$

Use the method of Exercise 5.34 to show that, for a binomial success count K of n trials, $E(K(K-1)(K-2)) = n(n-1)(n-2)\pi^3$. Then show that the coefficient of skewness is

$$\frac{1-2\pi}{\sqrt{n\pi(1-\pi)}}$$

Chapter 6

INTRODUCTION TO HYPOTHESIS TESTING

6.1 OVERVIEW

6.2 TWO TYPES OF ERROR

6.3 THE SIGN TEST

6.4 BINOMIAL EXACT TEST

6.5 TRIVIAL EFFECT CAN BE SIGNIFICANT

6.6 FISHER'S EXACT TEST (optional)

Now we are ready for some statistical inference. We start with situations in which we use probability theory to analyze a random sample from a population to either confirm or deny a hypothesis about the population.

In this chapter, we will consider three tests based on the binomial and hypergeometric random variables that we studied in Chapter 5. Later in Chapter 12 we will return to more tests of hypotheses after we examine the normal random variable and the Central Limit Theorem in Chapter 8 and the law of averages and sampling theory in Chapter 11.[1]

---

[1] Instructors preferring a more traditional ordering of topics may choose to cover the present chapter after Chapter 11.

6.1 OVERVIEW

In American criminal justice, a judge or jury must decide on the guilt or innocence of a defendant based on evidence presented by the prosecution. The procedural rules of evidence say that the defendant must be assumed innocent until proven guilty beyond a reasonable doubt. The judge or jury must decide on the level of evidence needed to find a person guilty beyond a reasonable doubt.

This is also the model for statistical hypothesis testing. With every hypothesis that says that something is true, there is the opposite hypothesis that says it is false. But the hypotheses are not symmetric. One of them, called the **null hypothesis**, and denoted H_0, is the default hypothesis, which corresponds to innocence in criminal law. The other is called the **alternative hypothesis** and is denoted H_a. According to statistical rules of evidence, we must assume that the null hypothesis is true until proven false beyond a reasonable doubt.

The burden of proof lies in proving the alternative hypothesis beyond a reasonable doubt. If we succeed, then we **reject the null hypothesis** or **accept the alternative hypothesis**. If we fail, then we **retain the null hypothesis**; this does not mean that the null hypothesis has been proven, only that there is insufficient evidence to disprove it. One should *not* say *accept the null hypothesis* because this implies that evidence has been provided in its favor.

6.2 TWO TYPES OF ERROR

A hypothesis can be either true or false. An error occurs if a true null hypothesis is rejected. This is called a **Type I Error**. On the other hand, an error also occurs if a false null hypothesis is retained. This is called a **Type II Error**.

Example 6.1   *Two-headed coin.*   Suppose that a friend has a coin that you suspect may have been altered to have heads on both sides. To test whether it is an altered coin, we ask her to toss it $n = 10$ times. Then we make a decision based on the results. If the coin comes up tails at least once, then it's clearly not two-headed. On the other hand, if the coin comes up heads *every* time, then there is evidence that the coin has been altered (because it is unusual for a normal coin to come up heads 10 times in a row), although it is still possible that the coin is normal. This leads to two hypotheses:

H_0 : It's a normal coin. H_a : It's a two-headed coin.

Let the parameter π be the probability of heads for the coin in question. In terms of π, the hypotheses are

$$H_0 : \pi = 0.50 \qquad H_a : \pi = 1.00$$

Suppose that we use the following **decision rule:**

Reject H_0 if the coin comes up heads every time.

Retain H_0 otherwise.

There are four possibilities, of which two lead to errors as shown by Figure 6.1.

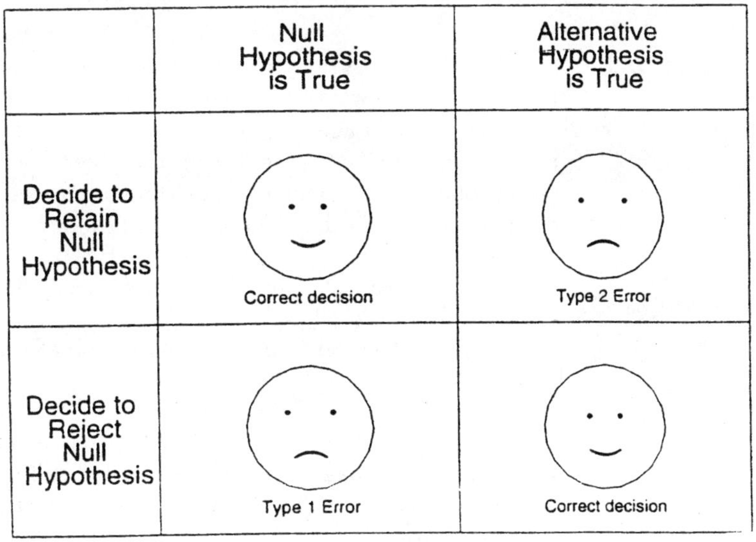

	Null Hypothesis is True	Alternative Hypothesis is True
Decide to Retain Null Hypothesis	Correct decision	Type 2 Error
Decide to Reject Null Hypothesis	Type 1 Error	Correct decision

Figure 6.1 Four ways an experiment may come out, including two types of error

Type I Error: This occurs if the null hypothesis is true but we reject it.

Type II Error: This occurs if the null hypothesis is false but we retain it.

If H_0 is true, then the probability of making the Type I error is denoted by α. In this example,

$$\alpha = Pr(H_0 \text{ is rejected} \mid H_0 \text{ is true}) = Pr(10 \text{ heads} \mid \text{coin is normal})$$
$$= Pr(10 \text{ heads} \mid \pi = 0.50) = \frac{1}{2^{10}} = \frac{1}{1024} \approx 0.001$$

On the other hand, if H_0 is false, then the probability of making the Type II error is denoted by β. In this example,

$$\beta = Pr(H_0 \text{ is retained} \mid H_0 \text{ is false}) = Pr(\text{at least one tail} \mid \text{coin is two-headed})$$
$$= Pr(\text{at least one tail} \mid \pi = 1.00) = 0$$

There is a $0.001 = 0.1\%$ risk of error if H_0 is true, and there is a 0% risk of error if H_0 is false.

Example 6.2 **Medical treatments.** Suppose that a new medical treatment is compared to a placebo. Twenty-four subjects are put into $n = 12$ pairs chosen for their similarity in age, sex, and severity of illness. In each pair, one is chosen randomly to receive the treatment, the other a placebo. The test is conducted double blind. After 6 months, doctors examine the patients and, for each pair, decide who is healthier. The hypotheses follow.

H_0 : The treatment makes no difference in the health of the subjects.

H_a : The treatment improves the health of the subjects.

Let the parameter π be the *population* proportion of pairs in which the treated subjects does better than the placebo subject, and let the count K be the number of pairs in which the treated subject is declared healthier than the placebo subject. The hypotheses can be restated as follows:

$H_0 : \pi = 0.50$

$H_a : \pi > 0.50$

Note that the hypotheses are statements about the population, not statements about the sample of 12 pairs of subjects. The decision rule deals with the sample of 12 pairs of subjects. Suppose we make the decision rule as follows:

Retain H_0 if $K \leq 9$ and reject H_0 if $K \geq 10$

For this decision rule, let's find α.

Solution   This is the risk of an error in the case when H_0 is true.

$$\begin{aligned}\alpha &= Pr(H_0 \text{ is rejected} \mid H_0 \text{ is true}) = Pr(K \geq 10 \mid \pi = 0.5) \\ &= \binom{12}{10}(.5)^{10}(1-.5)^2 + \binom{12}{11}(.5)^{11}(1-.5)^1 + \binom{12}{12}(.5)^{12}(1-.5)^0 \\ &= 0.0193\end{aligned}$$

Note that the null hypothesis is stated as an *equality* $\pi = 0.50$; this makes it a **simple hypothesis**. As we calculated above, for a given decision rule, α is a constant. Unlike the preceding Example 6.1, the alternative hypothesis here is an *inequality* $\pi > 0.50$, which makes it a **compound hypothesis**. Thus,

$$\beta = Pr(\text{retaining } H_0 \mid H_a \text{ is true})$$

depends on the true value of the parameter π in H_a. This makes β a function of the value of the true value of π.

As the value of π increases, the value of β, the risk of retaining H_0, decreases.

> *The power of a test is $1 - \beta$. The power is the probability that the test (correctly) rejects the null hypothesis when the null hypothesis is false.*

See Figure 6.2 for a graph of β and Figure 6.3 for a graph of the power curve $1-\beta$, both for the preceding Example 6.2 on testing medical treatments.

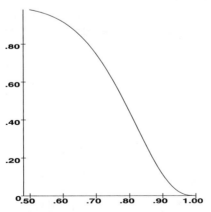

Figure 6.2   Graph of β

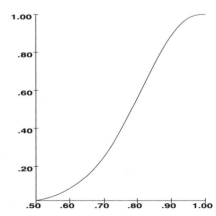

Figure 6.3   Graph of power curve $1 - \beta$

168   CHAPTER 6. INTRODUCTION TO HYPOTHESIS TESTING

Example 6.3   *Medical treatments.* Let's now compute β for Example 6.2 on testing medical treatments in the case when the treatment is so effective that the true value of π is

   a. $\pi = 0.80$

and

   b. $\pi = 0.95$

Solution   a. Recall that the decision rule says that the null hypothesis should be retained when $K \leq 9$.

$$\begin{aligned}
\beta &= \beta(0.80) = Pr(H_0 \text{ is retained} \mid \pi = 0.80) \\
&= Pr(K \leq 9 \mid \pi = 0.80) = 1 - Pr(K \geq 10 \mid \pi = 0.80) \\
&= 1 - \left\{ \binom{12}{10}(.8)^{10}(1-.8)^2 + \binom{12}{11}(.8)^{11}(1-.8)^1 + \binom{12}{12}(.8)^{12}(1-.8)^0 \right\} \\
&= 1 - 0.5583 = 0.4417
\end{aligned}$$

Instead of computing all of these terms, one can find the value in Table 2 on page 505 by looking up $\pi = 0.80, n = 12, k = 9$.

   b. Now suppose that the treatment is so effective that $\pi = 0.95$. Using Table 2, we have

$$\begin{aligned}
\beta &= \beta(0.95) = Pr(H_0 \text{ is retained} \mid H_0 \text{ is false}) \\
&= Pr(K \leq 9 \mid \pi = 0.95) = 0.0196
\end{aligned}$$

The power of a test can generally be increased by increasing α, the risk of a Type I error. For example, in Example 6.2 on testing medical treatments, we could decrease the risk of a Type II error β, and hence increased the power $1 - \beta$, by lowering the critical rejection value from 10 to 9:

Retain H_0 if $K \leq 8$ and   reject H_0 if $K \geq 9$

But then α, the risk of a Type I error, will increase to

$$\alpha = Pr(K \geq 9 \mid \pi = 0.5) = 0.0730$$

Also, for a fixed α, a test can be made more powerful by increasing the sample size n.

EXERCISES FOR SECTION 6.2

- **6.1** In the test for a two-headed coin in Example 6.1 on page 164, use the decision rule

    Reject H_0 if the coin comes up heads every time.   Retain H_0 otherwise.

    Find α and β for the following number of tosses:

    a. $n = 1$ **b.** $n = 4$ **c.** $n = 25$ **d.** $n = 100$

6.2 A box of dice contains one that has been altered by changing the side with one dot (ace) to five dots. Thus, the altered die has no aces but two sides with five dots. A die is chosen at random and rolled 24 times. A fair die is expected to come up ace 4 times in 24 rolls. Consider the hypotheses

H_0: It's a good die. H_a: It's the altered die.

Use the following decision rule: Reject H_0 if an ace never comes up in the 24 rolls, and retain H_0 otherwise.

a. Find α, the probability of a Type I error.

b. Find β, the probability of a Type II error.

- **6.3** A coin has been bent in such a way that it comes up heads 75% of the time. This bent coin is placed in a box with 9 fair ones. A friend selects one coin randomly and tosses it 12 times. Consider the hypotheses:

    H_0: It's a fair coin. H_a: It's the altered coin.

    Use the following decision rule: Reject H_0 if the coin comes up heads 12 times in the 12 tosses, and retain H_0 otherwise.

    a. Find α, the probability of a Type I error.

    b. Find β, the probability of a Type II error.

6.4 Repeat Exercise 6.3 for the decision rule: Reject H_0 if the coin comes up heads 11 or 12 times in the 12 tosses, and retain H_0 otherwise.

- **6.5** In Example 6.2 on page 166, let's make it harder to prove the treatment effective. For the same $n = 12$ pairs of patients, use the decision rule:

    Reject H_0 if $K \geq 11$.   Retain H_0 otherwise.

    a. Find α as in Example 6.2.

    b. Find β for $\pi = 0.80$ as in Example 6.3 part a.

    c. Find β for $\pi = 0.95$ as in Example 6.3 part b.

6.6 Repeat Exercise 6.5 with $n = 20$ pairs of patients and the decision rule:

Reject H_0 if the $K \geq 16$ Retain H_0 otherwise.

- **6.7** A Pepsi soft drink distributor is planning to test the consumer preference of the soft drinks Coke and Pepsi. Ten subjects are asked to taste each of the two soft drinks in unlabeled cups and then say which they prefer. The distributor is prepared to declare Pepsi the winner if at least 6 out of the 10 prefer Pepsi. Let π be the population proportion of subjects who prefer Pepsi over Coke. This test is similar to Example 6.2.

    a. State the null and alternative hypotheses, the decision rule, and find α.

    b. Find β under the condition that $\pi = 0.60$. Table 2 on page 504 in the back of the book may be helpful.

    c. Find β under the condition that $\pi = 0.70$.

6.3 THE SIGN TEST

This is a test about the population median. In a population box of numbered cards, there are as many values above the median as there are below the median. To test the hypothesis that a specific value m is the median of the box, we can draw a random sample from a population box and assign the plus sign $(+)$ for every value that the number drawn is larger than m, and assign the minus sign $(-)$ for every one that is below m. If the null hypothesis that m is the median is true, then there should be about the same number of $(+)$ signs as $(-)$ signs. If not, this is evidence against the null hypothesis.

The results look like the tosses of a coin—say, $(+)$ for heads and $(-)$ for tails. Although the sign test does not have much power in comparison with some other tests we will learn (that is, one needs rather strong evidence against a null hypothesis to reject it) it is simple to use and easy to understand. If the population is symmetric—that is, if the median and mean are the same—then it can also be used as a test of the mean of the population.

Example 6.4 *Heights of men.* It is hypothesized that the median height of men in the United States has increased over the last few decades, perhaps as a result of better nutrition and health care. In 1970, the median height among men in the age group from 18 to 24 was 69.5 inches. To test this hypothesis, the heights of a random sample of 20 men in this age group were taken last year, with the results given below:

69.5	72.4	74.0	74.5	72.0	68.0	68.8	70.0	69.5	70.5
74.0	72.0	71.0	71.0	72.0	69.0	69.8	69.0	71.8	73.0

We use the sign test on the hypotheses

H_0: The median height has not changed ($Med = 69.5$)

H_a: The median height has increased ($Med > 69.5$)

Assign plus (+) if the height is above 69.5 inches, minus (−) if below, and zero (0) if tied:

(0)	(+)	(+)	(+)	(+)	(−)	(−)	(+)	(0)	(+)
69.5	72.4	74.0	74.5	72.0	68.0	68.8	70.0	69.5	70.5
(+)	(+)	(+)	(+)	(+)	(−)	(+)	(−)	(+)	(+)
74.0	72.0	71.0	71.0	72.0	69.0	69.8	69.0	71.8	73.0

Because we are only interested in the values either above or below the median, we discard the $k_{(0)} = 2$ ties. This leaves us with $n = 18$ signs: $k_{(+)} = 14$ plus signs and $k_{(-)} = 4$ minus signs.

Presuming the null hypothesis, the proportion of men in the population who are above 69.5 inches is $\pi = 0.50$. We can restate the hypotheses in terms of the proportion π of men who are above 69.5 inches tall:

$H_0: \quad \pi = 0.50$

$H_a: \quad \pi > 0.50$

In particular, presuming the null hypothesis, we expect only $n\pi = (18)(0.50) = 9$ plus signs instead of the observed 14. **This means that the data set is evidence for the alternative hypothesis.**

> *It is always a good idea to check whether the data set is evidence for the alternative hypothesis. If it were not the case (for example, if we had observed nine or fewer (+) signs), then we would simply retain the null hypothesis without further calculations.*

For the sign test, if the evidence is in favor of the alternative hypothesis, we denote $k_{(+)} = 14$ as the **test statistic**:

Test statistic $k_s = 14$.

P-value
> *A measure of the strength of the evidence against the null hypothesis is the **P-value**, which is the probability of obtaining data as extreme as or more extreme than was observed (of course, under the presumption that the null hypothesis is true).*[2]

Because the alternative hypothesis is $\pi > 0.50$, the P-value is the probability of obtaining as many as (**as extreme as**) or more than (**more extreme than**) k_s (+) signs in n trials (under the assumption of the null hypothesis $\pi = 0.50$).[3] Here the P-value computes as follows:

$$P\text{-value} = Pr(K \geq k_s) = Pr(K \geq 14) = 1 - Pr(K \leq 13) = 1 - F_K(13) = 0.0154$$

---

[2] The notation P-value is not related to the symbol p used for the sample percentage in Chapter 1.

[3] If the alternative hypothesis had been $\pi < 0.50$, then the P-value would have been the probability of obtaining k_s or fewer (+) signs.

The term $F_K(13)$ can be obtained using technology (as shown on page 154) or from the Cumulative Binomial Distribution Table 2 on page 506 in the back of this book.

> *The smaller the **P**-value, the greater the evidence against the null hypothesis. How small should the **P**-value be in order to reject the null hypothesis? If you set α, the risk of a Type I error, to be a specific value, such as $\alpha = 0.02$, then we reject the null hypothesis when **P**-value $< \alpha$ and retain $\mathbf{H_0}$ otherwise.*

For a given test of a hypothesis, if the **P-value < 0.05**, it is commonly said that **the evidence is significant**. And if the **P-value < 0.01**, it is commonly said that **the evidence is highly significant**. As it turns out, the evidence that the median height of men has increased, as presented in Example 6.4, is significant, but not highly significant.

These criteria for rejection are just rough guidelines. In a specific test, one must carefully weigh the consequences of false rejection or false retention of a null hypothesis. Clearly, the burden of proof needed to convict someone of murder must be much higher than for a parking violation.

Suppose we decide to set $\alpha = 0.02$. We say that we are testing at the $\alpha = 0.02$ **level of significance**, or that the **critical P-value** is 0.02. We then report that a null hypothesis was either rejected or retained at the $\alpha = 0.02$ level of significance, depending of whether the P-value is less than, or greater than, 0.02. If the null hypothesis is rejected, we also say that **the evidence is significant at the $\alpha = 0.02$ level of significance**.

We should decide on the level of significance α needed to reject the null hypothesis before actually computing the P-value. Otherwise, seeing the evidence before deciding on the level (called **data snooping**) may bias our decision.

> *Reporting the **P**-value, with or without a decision to reject or retain, permits readers to choose their own level of significance.*

NOTES ON TECHNOLOGY

Sign Test

Although there are software and calculator procedures for the sign test, it is generally simpler to use the procedures for computing the binomial cumulative distribution as described in the Notes on Technology on page 154. If the evidence does not support the alternative hypothesis, then there is nothing to test (we simply retain the null hypothesis and state that P-value ≥ 0.50). If the evidence supports the alternative hypothesis, then use the procedure described on page 154 with $\pi = 0.50$, n being the number of $(+)$ and $(-)$ signs, and k_s, the **test statistic**, being the number of $(+)$ signs. If the alternative hypothesis is $\pi < 0$, then P-value $= F_K(k_s)$. For example, for the TI-83 Plus, this would be the function P-value $= binomcdf(n, .50, k_s)$ when specific values of n and k_s are entered. But, if the alternative hypothesis is $\pi > 0$, then P-value $= Pr(K \geq k_s) = 1 - F_K(k_s - 1) = 1 - binomcdf(n, .50, k_s - 1)$.

In scientific work, the results of tests are generally reported by stating the following six items:

1. *The null hypothesis.*
2. *The alternative hypothesis.*
3. *The level of significance α.*
4. *The results, including the sample size and test statistic.*
5. *The P-value.*
6. *The conclusion in plain English in context of the problem.*

Example 6.5 *More male births.* In 1710, the Scottish scientist John Arbuthnot suspected that there are more male births than female births. He tested this hypothesis as follows. He examined 82 years of London birth records. For each year, he compared the number of male births to the number of female births. He assigned the plus sign $(+)$ if there were more male than female births, and the minus sign $(-)$ if there were more female births. If the chance for a male birth were the same as for a female birth, you would expect about the same number of pluses and minuses. Well, he observed no ties, no minus signs, and 82 plus signs. This is like tossing a coin 82 times and coming up with no tails and 82 heads (probability P-value $= (0.5)^{82} = 2.07 \times 10^{-25}$). It is easy to see the convincing evidence against the null hypothesis of equal chance.

Example 6.6   ***Weight gains.*** Do women recover their weight after childbirth? Below are the weights in pounds of a random sample of 20 women measured during an office visit when a pregnancy was first diagnosed. Ten years later, the same women were weighed again. This sample was taken from a population of 1296 women who were part of the Kaiser Foundation Health Plan during the period 1961–1972.[4]

Subject:	1	2	3	4	5	6	7	8	9	10
Before:	112	119	130	115	151	137	162	116	130	121
After:	125	130	139	132	155	146	167	130	153	118

Subject:	11	12	13	14	15	16	17	18	19	20
Before:	112	140	136	129	115	102	138	133	118	185
After:	112	157	155	136	136	111	175	172	121	163

Test whether women gain weight in the 10 years after a pregnancy at the $\alpha = 5\%$ level of significance.

Solution   Here we use the sign test, listing the six steps mentioned in the box above. Let π be the proportion of women in the population of $N = 1296$ who increase in weight over the 10-year period.

Step 1. H_0 : The median weight has remained the same ($\pi = 0.5$).

Step 2. H_a : The median weight has increased ($\pi > 0.5$).

Step 3. $\alpha = 0.05$

Step 4. Because there is one tie, which is discarded, it leaves us with a sample of $n = 19$. There are 17 increases in weight, so 17 signs (+) and 2 signs (−). The test statistic is $k_s = 17$.

Step 5. The *P*-value is calculated using the $\pi = 0.5$ column of Table 2 for the group $n = 19$ on page 507. *P*-value $= Pr(K \geq k_s) = 1 - 0.9996 = 0.0004$. Because *P*-value $< \alpha$, the null hypothesis is rejected.

Step 6. The conclusion in English: There is strong evidence (*P*-value $= 0.0004$) to conclude that the median weight of women increased 10 years after pregnancy.

Warning   Because we are not comparing these women with a group of women who have not been pregnant, we cannot conclude that a pregnancy **causes** an increase in weight over a 10-year interval.

---

[4]Source: J. L. Hodges Jr., David Krech & Richard S. Crutchfield, *StatLab*, McGraw-Hill, 1975.

EXERCISES FOR SECTION 6.3

6.8 Use the sign test at the $\alpha = 0.05$ level of significance on the hypotheses

$$H_0 : Med = 25 \quad H_a : Med > 25$$

for a population from which the following sample came:

26.8   29.2   27.6   27.4   20.9   30.1   23.7   28.3   25.0   31.8

6.9 A three-month program of physical fitness is conducted on five subjects. The at-rest pulse is measured before and after the program with the following results:

Subject	1	2	3	4	5
Before	73	77	68	62	72
After	68	72	64	62	71

Test whether the median at-rest pulse has gone down. Use the sign test at the $\alpha = 0.10$ level of significance.

6.10 Below are the cash receipts in dollars for ice cream vendors at two locations at a city beach:

Day	1	2	3	4	5	6	7	8	9
North	290.15	183.75	205.35	193.86	256.89	201.15	207.99	299.23	235.18
South	188.86	202.45	215.35	180.77	200.45	205.65	157.84	295.46	201.35

Assuming the differences of daily receipts are independent, test the hypothesis that cash receipts for the north location are usually greater than for the south location against the null hypothesis that the median difference is 0. Use the sign test at the $\alpha = 0.10$ level of significance. Follow Example 6.6.

6.11 Below are the reported and measured heights in inches of 12 men. Test at the $\alpha = 10\%$ level of significance whether men tend to exaggerate their height.

Subject	1	2	3	4	5	6	7	8	9	10	11	12
Reported	67.0	68.0	70.0	68.0	73.0	73.0	71.0	70.0	76.0	69.0	70.0	67.0
Measured	66.3	67.5	69.3	67.5	73.3	73.0	70.8	69.5	73.3	68.8	69.0	66.0

6.12 A brochure claims that the average daily high temperature in Paris, France, is 18° Celsius in May. Suppose that last May the daily high temperature exceeded 18°C on 8 of the 31 days of the month, equaled 18°C on 2 days, and fell short of 18°C on 21 days. Test whether the temperature was significantly below the claimed average last May. Use $\alpha = 0.05$. Is it appropriate to use a sign test here? Explain.

6.13 A double-blind study was performed on consumers comparing a new formulation of a diet soft drink with the original formulation. Of 100 subjects,

57 preferred the new formulation, 38 preferred the original formulation, and 5 were undecided or could not tell the difference. Test whether consumers prefer the new formulation at the 5% level.

6.4 BINOMIAL EXACT TEST

We start this section with a test of the effectiveness of the Salk polio vaccine in preventing *death* from polio in Example 6.7 below. The null hypothesis is that the vaccine is ineffective in preventing death. Considering the data derived from the 1954 Salk vaccine trial, it will turn out that the evidence against the null hypothesis is too weak to reject it.

Then in Example 6.8 we will look at data of the same trial to judge the effectiveness of the vaccine in preventing *paralytic* polio. We will see that, in this case, the evidence against the null hypothesis is overwhelming; so the null hypothesis is rejected.

Example 6.7   ***Does the Salk vaccine prevent deaths from polio?*** In the randomized control double-blind study conducted by the U.S. Public Health Service in 1954 as summarized in Table 1.1 on page 12, there were 200,745 children who received the vaccine and 201,229 who received the placebo, randomly assigned. It turned out that 4 children in the placebo group died of polio and none of the children in the vaccine group died of polio.

	Fatal polio	Other	Row total
Vaccine	0	200,745	200,745
Placebo	4	201,225	201,229
Column total	4	401,970	401,974

Because we are trying to prove the effectiveness of the Salk vaccine in preventing death, the opposite of this hypothesis is the null hypothesis, which states that the vaccine is not effective.

H_0: the vaccine is not effective against fatal polio.

H_a: the vaccine is effective against fatal polio.

The federal Food and Drug Administration (FDA) does not take these decisions lightly. The FDA does not declare effectiveness of drugs based solely on a belief that they might work. Effectiveness has to be proven beyond a reasonable doubt. In testing the effectiveness of drugs, we must work under the assumption that the drug is not effective until the null hypothesis is discredited.

Let's consider only the children who die of polio; that is, consider only the fatal polio column of the table. Let π be the probability that a fatal polio case turns up in the vaccine group. *Under the assumption of the null hypothesis* that the treatment makes no difference, we have $\pi = 0.50$ because the outcome is not affected by the vaccine. Under the assumption of the alternative hypothesis that the vaccine is effective, we have $\pi < 0.50$ because the vaccine reduces the risk of fatal polio. The hypotheses can be restated in terms of π:

Step 1. H_0: the vaccine is not effective against fatal polio ($\pi = 0.50$).

Step 2. H_a: the vaccine is effective against fatal polio ($\pi < 0.50$).

Step 3. And let's choose $\alpha = 0.05$.

After stating H_0, H_a and α, we are ready to consider the evidence. Looking at the fatal polio column, we find that in the vaccine group, 0 of the polio cases were fatal, and in the placebo group, 4 of the polio cases were fatal. It's completely one-sided. There is evidence that the vaccine works. Because π refers to the probability of a fatal polio case falling in the vaccine group, the test statistic is the number of fatal polio cases in the vaccine group:

Step 4. Test statistic: $k_s = 0$

How strong is this evidence? Expressing the null hypothesis in terms of a box model, the experiment is like drawing four cards with replacement from a Bernoulli box with $\pi = 0.50$ and obtaining a sample count of $k_s = 0$. Although this is evidence as extreme as possible in favor of H_a, the P-value is not less than α:

Step 5. P-value $= F_K(k_s) = Pr(K \leq 0) = 0.0625$

Can the *observed* difference be explained by sample variability? The evidence that the Salk vaccine prevents polio deaths measured by the P-value $= 0.0625$. The P-value is greater than our chosen $\alpha = 0.05$; therefore, the evidence is not significant at this level of the test. This means that the null hypothesis H_0 is retained. After all, tossing a coin four times and getting no heads is not convincing evidence that the coin is in some way unbalanced.

Step 6. Conclusion in English: The effectiveness of the Salk vaccine in preventing polio deaths is *not* proven beyond a reasonable doubt.

However, recall what we said in Chapter 1. This does not show that the vaccine is ineffective in preventing death from polio. It just shows that there is insufficient evidence to conclude that the vaccine prevents death from polio.

Absence of evidence is not evidence of absence.[5]

178 CHAPTER 6. INTRODUCTION TO HYPOTHESIS TESTING

Example 6.8 **Does the Salk polio vaccine prevent paralytic polio?** Again using data from the same Salk vaccine trials, as summarized in Table 1.1 on page 12, it turned out that 115 children in the placebo group and 33 children in the vaccine group came down with paralytic polio.

	Paralytic polio	Other	Row total
Vaccine	33	200,712	200,745
Placebo	115	201,114	201,229
Column total	148	401,826	401,974

Because the study was designed to prove beyond a reasonable doubt that the Salk vaccine is effective against polio, this is the alternative hypothesis. The null hypothesis it that the vaccine is not effective. Considering only the paralytic polio cases, let π be the probability that a paralytic polio case turns up in the vaccine group:

Step 1. H_0: the vaccine is not effective against paralytic polio ($\pi = 0.50$).

Step 2. Ha: the vaccine is effective against paralytic polio ($\pi < 0.50$).

Step 3. This time, let's choose $\alpha = 0.01$.

Even though there were more than 400,000 children involved in the study, we can never be completely certain of our decision. Yet if we wait for complete certainty, we will wait forever. Based on the evidence, we must decide either that the effectiveness of the vaccine is proven beyond a reasonable doubt (hence, reject H_0) or that the effectiveness of the vaccine is not proven (hence retain the default H_0). There are risks involved in erring either way. If we decide that the drug is effective, when it really is not, we risk the expense and trouble of administering an ineffective drug as well as the risk of possible side effects of the vaccine (which might even be deadly). If, on the other hand, we decide that the drug is ineffective, when it really does work, then we risk needless illness, including the paralysis and death of many children.

The evidence By the rules of hypothesis testing, we have to assume the vaccine is ineffective until disproved beyond a reasonable doubt. Working under the assumption of the null hypothesis that the vaccine is worthless means that all of the $115 + 33 = 148$ children would have come down with paralytic polio regardless or whether they were assigned to the placebo group or vaccine group. The decision of group placement was made at random, as by the toss of a fair coin. Of the 148 children who came down with paralytic polio, we would expect that the count K of children placed into the vaccine group to be approximately $E(K) = n\pi = 74$. But if the vaccine is effective against paralytic polio,

---

[5]Douglas G. Altman, J. Martin Bland, *Statistical Notes: Absence of evidence is not evidence of absence*, British Journal of Medicine **311** (1995), p. 485. http://bmj.com/cgi/content/full/311/7003/485.

6.4 BINOMIAL EXACT TEST

then we would anticipate fewer cases in the vaccine group. The fact that we observe only $k_s = 33$ of these children in the vaccine group is evidence for the alternative hypothesis.

Step 4. Test statistic: $k_s = 33$

How strong is this evidence?

The *P*-value is $F_K(k_s) = Pr(K \leq 33 \mid \pi = 0.50)$, which is the probability of obtaining data at least as extreme as those observed. Equivalently, what is the chance that in $n = 148$ tosses of a coin, we get $k_s = 33$ or fewer heads? The calculation of this probability needs the use of technology as described on page 154 (our Table 2 at the back of the book does not go beyond $n = 20$). For example, the function *binomcdf* on the *Texas Instruments* TI-83 graphing calculator gives us the value

Step 5. $P\text{-value} = Pr(K \leq 33) = F_K(33) = \textit{binomcdf}\,(148, .50, 33)$
$= 3.91930486 \times 10^{-12}$

This probability is less then 1 in 250 billion. The smaller the *P*-value, the greater the evidence against the null hypothesis. Because the evidence against the null hypothesis is extremely strong, retaining it would be outrageous. We reject H_0 and accept the effectiveness of the vaccine H_a in preventing paralytic polio.

Step 6. Conclusion in English: There is overwhelming evidence (P-value $< 1/250,000,000,000$) to conclude that Salk vaccine is effective in preventing paralytic polio.

Remark

Because of the large number of cases of paralytic polio, the *P*-value is difficult to compute without the use of technology. In Chapter 8, we will show how to use the normal density curve to find good approximations in cases where n is large. Then in Chapter 12 we will revisit hypothesis testing making use of this approximation.

Example 6.9

Drug testing of DynaCirc versus placebo. Just as in the Salk vaccine trials, this was a large **randomized controlled experiment**; that is, subjects were randomly assigned either to a treatment group or a control group, as if by the toss of a fair coin. The treatment was supposed to reduce the risk of suffering a stroke using a new drug *DynaCirc* in addition to the standard treatment. The control group received the same standard treatment but received a placebo instead of DynaCirc. In the course of the experiment, nine patients suffered strokes; of these nine, three were in the treatment group.

Is this evidence that DynaCirc is effective medication in reducing the risk of stroke? There are two possibilities. One is the null hypothesis H_0 that DynaCirc is not effective in reducing the risk of stroke.

> **The null hypothesis is retained if the observed difference between the two groups can reasonably be explained by sample variability.**

If not, then the null hypothesis is rejected in favor of the alternative hypothesis H_a that DynaCirc is effective in preventing stroke.

Solution Let's look at the nine subjects in the study who suffered stroke. If the null hypothesis were true, then the chance that any stroke victim came from the placebo or the treatment group is the same, namely, $\pi = 0.50$. On the other hand, if the alternative hypothesis were true that the DynaCirc treatment had some positive effect on the risk of stroke, then the chance π of any one of the nine stroke victims being in the treatment group would be less than 0.50. We can summarize the two hypotheses as follows:

Step 1. $H_0: \quad \pi = 0.50$

Step 2. $H_a: \quad \pi < 0.50$

Step 3. $\alpha = 0.05$

The fact that only three of the nine stroke victims were in the treatment group is evidence for the alternative hypothesis.

Step 4. Test statistic: $k_s = 3$

According to the rules of evidence we have to assume H_0 to compute the P-value. Under the null hypothesis, we expect $\mu = n\pi = 4.5$ of the nine stroke victims to come from the treatment group. How likely then is it that $k_s = 3$ or fewer came from the treatment group? It is equivalent to the chance of tossing a fair coin $n = 9$ times and obtaining $K \leq 3$ heads. The probability is the P-value:

Step 5. P-value $= Pr(K \leq k_s) = F_K(3) = Pr(K \leq 3)$

$$= \sum_{k=0}^{3} f_K(k) = \sum_{k=0}^{3} \binom{9}{k}(0.50)^k(1-0.50)^{9-k} = 0.2539$$

This sum can also be obtained from Table 2 on page 504 for $\pi = 0.50, n = 9$, and $k = 3$. If DynaCirc is not effective against stroke, the results of getting three or fewer stroke victims in the treatment group is not at all surprising; it has probability 0.2539. The results of the experiment are consistent with the null hypothesis. The alternative hypothesis is not proved beyond a reasonable doubt. Our decision must be to retain H_0.

Step 6. In plain English: There is insufficient evidence (P-value = 0.2539) to conclude that DynaCirc reduces the risk of stroke.

6.4 BINOMIAL EXACT TEST

Example 6.10 **Do women who smoke cigarettes during pregnancy have more low-birthweight babies?** In a study done in 1971 of 5466 pregnancies[6] among women who gave birth at age 25 or less, 2076 of the pregnant women smoked during their pregnancies, whereas 3390 did not. Of the 2076 smokers, 185 had low-birthweight babies (defined as weighing less than 2.5 kg), and of the 3390 nonsmokers, 193 had low-birthweight babies.

	Low birthweight	Normal	Row total
Smokers	185	1891	2076
Nonsmokers	193	3197	3390
Column total	378	5088	5466

Does this show that smoking causes low birthweight? This example needs careful consideration. This was *not* a randomized controlled experiment, which means that the two groups, smokers and nonsmokers, were not chosen by a random process. The women themselves decided whether or not to smoke. If the subjects entered the study already members of the treatment groups, it's called an **observational study** instead of a controlled experiment.

One problem, in contrast to the previous examples, is that there is no probability model that uses a probability π to assign subjects to the groups. This is not serious; it can be fixed by using the proportion of women found in the study who smoked, 2076/5466.

> We can model the experiment as drawing 378 cards from a population box consisting of 5466 cards; 2076 of them are marked 1 and 3390 are marked 0.

The more serious matter is that membership in the two groups may be influenced by some (perhaps hidden) factors other than smoking during pregnancy. For example, there may be personality differences that influence whether people smoke or not. The two groups may differ in educational level, income, lifestyle, alcohol consumption, and so on. All of these factors may influence birthweight. This makes it impossible to prove that smoking causes low birthweight. A statistical test can assess the **association** between smoking and birthweight, not whether smoking **causes** low birthweight. Indeed, there is a surprising follow-up that we will discuss in Example 6.11.

Solution Let's now perform the binomial exact test at the $\alpha = 1\%$ level of significance. Among low-birthweight babies, let π be the proportion of those whose mothers

---

[6] J. Yerushalmy, *The Relationship of Parents' Cigarette Smoking to Outcome of Pregnancy: Implications as to the Problem of Inferring Causation from Observed Associations*, American Journal of Epidemiology **93** (1971), pp. 443–456, as quoted in J. L. Hodges Jr., David Krech & Richard S. Crutchfield, *StatLab*, McGraw-Hill, 1975.

smoked. If there is no association between low birthweight and smoking, this proportion should be the same as the proportion of *all* women in the study who smoked, namely, 2076/5466 = 0.3798.

Step 1. $H_0: \pi = 2076/5466 = 0.3798$

Step 2. $H_a: \pi > 0.3798$

Step 3. $\alpha = 0.01$

Step 4. The number of low-birthweight babies is $n = 185 + 193 = 378$. Because π is the population proportion of mothers who smoked, the test statistic here must be the number of mothers in the sample who smoked, namely, $k_s = 185$.

> *The result of the experiment can be modeled as drawing 378 cards from the population box and getting 185 cards marked 1.*

Step 5. Because the alternative hypothesis is $\pi > 0.3798$, evidence against the null hypothesis lies with large values of k. The P-value is the probability of obtaining data as extreme or more extreme than was observed. The "as extreme as was observed" part results in $K = k_s$ and the "more extreme than was observed" results in $K > k_s$. Combining $K = k_s$ and $K > k_s$ results in P-value $= Pr(K \geq k_s)$. Here $n = 378, k_s = 185$ and $\pi = 0.3798$. Thus, P-value $= Pr(K \geq 185) = 1 - Pr(K \leq 184) = 1 - F_K(184) = 0.00000929$. The null hypothesis is rejected.

Step 6. The conclusion in English: There is strong evidence (P-value $= 9.29 \times 10^{-6}$) to decide on an association between smoking during pregnancy and low birthweight of babies.

Outline of the binomial exact test

Here is an outline for the binomial exact test.
We start with a treatment-and-response table.

	Success	Failure	Row total
Treatment	k	$N_1 - k$	N_1
Control	$n - k$	$(N - N_1) - (n - k)$	$N - N_1$
Column total	n	$N - n$	N

If you find that $N - n$ is smaller than n, you may want to reverse the role of "success" and "failure" in order to work with smaller numbers.

In order to obtain the full power of the binomial exact test, n (or $N - n$, if the role of "success" and "failure" are reversed) should be no more than

5% of N. Otherwise, we are sampling from a small population as described on page 157, which means that we should be using hypergeometric probabilities instead of binomial probabilities. In these cases, Fisher's exact test as described in Section 6.6 has more power because it uses hypergeometric probabilities.

Care must be taken to write the treatments in rows and the outcomes in columns. The condition that n should be no more than 5% of N usually prevents us from interchanging the roles of rows and columns.

Step 1. If we know that the experiment used a probability, such as $\pi = 0.50$, in allocating subjects to the treatment and control, use that probability in the null hypothesis. Otherwise, use $H_0: \pi = N_1/N$.

Step 2. If we are trying to prove that the treatment increases "successes," then the alternative hypothesis should be of the form $H_a: \pi > \cdots$. Otherwise, if we are trying to prove that the treatment decreases "successes," then reverse the inequality $H_a: \pi < \cdots$.

Step 3. Choose a value of α. If a lot of evidence is needed to convince us of the effectiveness of the treatment, choose a small value of α.

Step 4. Consider your sample to be the n subjects in the success column and let $k_s = k$ be the number found in the treatment group.

> *The experiment can be modeled as drawing* **n** *cards from the population box with probability of getting a card marked 1 given by* π *specified by* $\mathbf{H_0}$. *The results are modeled as getting* $\mathbf{k_s}$ *cards marked 1.*

Step 5. If the alternative hypothesis is of the form $\pi > \cdots$, then P-value = $Pr(K \geq k_s)$. If the alternative hypothesis is of the form $\pi < \cdots$, then P-value = $Pr(K \leq k_s)$. Reject the null hypothesis if P-value $< \alpha$, and retain the null otherwise.

Step 6. Write the conclusion in context of the problem in common English.

Example 6.11   *Follow-up of birthweight and smoking.*   A follow-up study was done that looked at 210 of the women who did not smoke during their pregnancies but began smoking after their babies were born.[7] It turns out that 20 of their babies had low birthweight. Test whether women who did not smoke during their pregnancies but started smoking after their babies were born

---

[7] J. Yerushalmy, *Infants with Low Birth Weight Born before Their Mothers Started to Smoke Cigarettes*, American Journal of Obstetrics and Gynecology **112** (1972), pp. 227–284.

have a lower risk of low birthweight than women who smoked during their pregnancies.

	Low birthweight	Normal	Row total
Future smokers	20	190	210
Smokers	185	1891	2076
Column total	205	1081	2286

Solution The null hypothesis says that the probability of low birthweight for "future" smokers is $\pi = 210/2286$. The alternative says that the probability is smaller. Here $n = 205$ and $k_s = 20$. Evidence for the alternative lies with small values of k. Thus, P-value $= Pr(K \leq 20) = 0.667$. Comparing the "future" smokers to the smokers shows no significant difference in the risk of low-birthweight babies. Actually, the percentage of low-birthweight babies among the "future" smokers is slightly higher than that among mothers who smoked during their pregnancies (for this reason the P-value is greater than 0.50). Clearly, low birthweight is greatly influenced by some factor in the group of smokers other than smoking itself!

Yet the binomial exact test comparing the "future" smokers to nonsmokers shows a significant higher risk of low-birthweight babies among the "future" smokers. This is Exercise 6.18.

EXERCISES FOR SECTION 6.4

6.14 Are men more likely than women to be ambidextrous? In a study of 1000 men and 1000 women, 19 men and 7 women were classified as ambidextrous. Use the binomial exact test at the 1% level of significance.

• **6.15** Are women less likely than men to be color-blind? In a study of 1000 women and 1000 men, 4 of the women and 50 of the men were color-blind. Use the binomial exact test at the 1% level of significance.

6.16 In a *Newsweek* poll, 250 adults under the age of 30 and 500 over the age of 30 were asked "Do you believe Americans today are as willing to work hard at their jobs to get ahead as they were in the past?" Conduct a binomial exact test at the $\alpha = 5\%$ level to test whether fewer adults under 30 feel that Americans are as willing to work hard to get ahead than they were in the past. Use $\pi = 1/3$.

	Yes	No	Row total
Under 30	58	192	250
Over 30	145	355	500
Column total	203	547	750

- **6.17** Are people who skip breakfast more likely to suffer late morning fatigue? The table below gives the results of a random sample 100 adults who skipped breakfast and 200 who did not. Conduct a binomial exact test at the $\alpha = 5\%$ level.

	Fatigue	No fatigue	Row total
Skipped breakfast	38	62	100
Ate breakfast	45	155	200
Column total	83	217	300

6.18 This is a continuation of Examples 6.10 and 6.11. Perform a binomial exact test at the $\alpha = 5\%$ level of significance to show that "future" smokers have a higher risk of low-birthweight babies than nonsmokers. Use $\pi = 210/3390$.

	Low birthweight	Normal	Row total
Future smokers	20	190	210
Nonsmokers	173	3007	3180
Column total	193	3197	3390

- **6.19** In a double-blind test, 120 subjects were given a drug to reduce stomach acidity and 120 were given a placebo. In assessing adverse reactions, it was noted that 9 of the subjects in the drug group and 3 in the placebo group complained of headaches. Use the binomial exact test at the 5% level to test whether headaches are more frequent among subjects treated with the drug.

6.20 In a larger study of the same drug as in Exercise 6.19, 900 subjects were treated with the drug and compared with 100 in a placebo group. This time, 60 of the subjects in the drug group and 4 in the placebo group complained of headaches. Use the binomial exact test at the 5% level to test whether headaches are more frequent among subjects treated with the drug. (Do not use $\pi = 0.50$ here.)

6.5 TRIVIAL EFFECT CAN BE SIGNIFICANT

Alzheimer's disease is a slow and progressive degeneration of mental ability, starting with orientation and memory. It affects mostly the elderly, but some victims are as young as 40. The drug tacrine is thought to increase the concentration of a key neurotransmitter in the brain. Because Alzheimer's patients have bad days followed by good days, the placebo effect of any treatment may be quite strong. Anecdotal evidence would be of little value. A double-blind study of the drug in 1992 showed that significantly more treated patients improved compared to those on placebo. Improvement was measured by a test of cognitive performance. This was an important result because, up to that time, no drug treatment had been found that had any effect on the progression of the disease.

Significance should be understood in the sense of a statistical test of hypothesis. The mental improvement was detectable by tests and could not be explained by sample variability. It proved for the first time that the progression of the disease could be altered with drug treatment. This drug opened the door to further scientific investigations.

As a part of the trials, family members and physicians were asked to evaluate the subjects of the experiment. Those evaluations showed no difference between the treated and control patients.

Although the treatment was found *significant* as a treatment for patients, it was not *sufficiently* effective to be of clinical value. The effect was so small that it was not noticeable by family members or physicians, and there were many side effects. In summary, this particular treatment was trivial but statistically significant.

The opposite can happen. There are medical procedures that have been shown to have no significant value, but are performed nevertheless because the physician wants to do everything possible for the patient. Clinical studies show that in certain slow growing forms of prostate cancer, surgery and other aggressive treatments do not lead to improvement in either the quality or length of life in elderly patients when compared to less aggressive hormone treatments, which merely impede the growth of the cancer. Playing the role of scientist, the physician may realize that surgery in such cases will not likely lead to any change in the prognosis of the patient. Yet playing the role of care giver, the physician may recommend it, especially when the patient wants to "get all of the cancer out."

6.6 FISHER'S EXACT TEST (optional)

Fisher's exact test of hypotheses has more power than the binomial exact test in small studies. Say N patients are randomly assigned to treatment and control groups. Given that n of these N patients turn out to be "successes," we test the hypothesis that the treatment works (against the null hypothesis that the treatment had no effect) by considering the number of these "successes" that came from the treatment group. This is also what the binomial exact test does; however, if N is small [general rule: n and $N - n$ are both larger than $N/20$], then we obtain a more powerful test by using hypergeometric probabilities instead of binomial probabilities.

	Success	Failure	Row total
Treatment	k	$N_1 - k$	N_1
Control	$n - k$	$(N - N_1) - (n - k)$	$N - N_1$
Column total	n	$N - n$	N

We can model the population by a box consisting of N cards, N_1 of them marked 1 and $N - N_1$ marked 0. The probability of drawing a card marked

1 is $\pi = N_1/N$. Given that n of the subjects were "successes," let K be the random variable that counts the number of these that came from the treatment group. It turns out that the result of the experiment is like drawing n cards from the box, without replacement, and getting k cards marked 1.

THEOREM 6.1 $Pr(K = k) = \dfrac{\binom{N_1}{k}\binom{N-N_1}{n-k}}{\binom{N}{n}}$

Proof Consider, as in the table above, N repeated independent Bernoulli trials, with N_1 patients assigned to the treatment group, and the rest to the control group. Assuming the null hypothesis that the treatment makes no difference, the probability of success is the same—say, π—for all patients regardless of whether they are in the treatment or control group. Suppose that for n of the N patients the outcome was success.

Looking only at the N_1 patients in the treatment group, the number of successes is k. This is a draw of k successes out of N_1. Similarly, for the control group, we have $n - k$ successes out of $N - N_1$. Define the events

E : "k successes among the N_1 patients in the treatment group"

F : "$n - k$ successes among the $N - N_1$ patients in the control group"

G : "n successes among the N patients in the study"

with unconditional probabilities

$$Pr(E) = \binom{N_1}{k} \pi^k (1-\pi)^{N_1-k} \tag{6.1}$$

$$Pr(F) = \binom{N-N_1}{n-k} \pi^{n-k} (1-\pi)^{(N-N_1)-(n-k)} \tag{6.2}$$

$$Pr(G) = \binom{N}{n} \pi^n (1-\pi)^{N-n} \tag{6.3}$$

Given that for n of the N patients the outcome was success, the probability that k of them came from the treatment group is the condition probability

$$Pr(K = k) = Pr(E \mid G) = \frac{Pr(E \text{ and } G)}{Pr(G)}$$

Notice that the event "E and G" is the same as "E and F." The events E and F are independent because they refer to different groups of patients. Using the multiplication rule for independent events, we have $Pr(E \text{ and } F) = Pr(E) \cdot Pr(F)$. Now substituting equations 6.1, 6.2, and 6.3 into

$$Pr(E \mid G) = \frac{Pr(E) \cdot Pr(F)}{Pr(G)}$$

we obtain the hypergeometric probability

$$f_K(k) = \frac{\binom{N_1}{k}\binom{N-N_1}{n-k}}{\binom{N}{n}}$$

Fisher's exact test can be tedious to compute without technology such as described in the **Notes on Technology: Hypergeometric Probabilities** on page 159. According to the general rule given near the end of Chapter 5 (on page 157), a hypergeometric sample count K is close to binomial if $20n < N$. If this is the case, then we can use the simpler binomial exact test.[8] The same rule would apply if we reversed the role of the n "successes" and $N - n$ "failures" and we had $20(N - n) < N$.

Example 6.12 **New cancer treatment.** A new drug for the treatment of cancer is being tested. Fifteen patients are given the new drug, and they are compared to 10 who are given the old drug. The 15 patients in the new drug group are selected at random from among the 25 total number of patients in the study. All patients are followed for one year to see if the cancer stays in remission for that long. The results are summarized in the table below.

	Remission	Cancer returns	Row total
New drug	7	8	15
Old drug	2	8	10
Column total	9	16	25

Solution We enumerate the six steps to test the hypothesis that the new drug works better than the old drug at the $\alpha = 0.05$ level of significance.

Step 1. H_0: New drug is not better than the old one.

Step 2. H_a: New drug is better than the old one.

Step 3. $\alpha = 0.05$

Note that 47% of the patients in the new drug group stayed in remission for at least one year, whereas only 20% of the patients in the old drug group did so. This is evidence in favor of the alternative hypothesis.

---

[8]By this criterion, the study of Example 6.10 on page 181 has a small N, so the binomial exact test is not as strong as Fisher's exact test. However, the binomial exact test was strong enough to reject the null hypothesis; so, Fisher's exact test is not needed.

6.6 FISHER'S EXACT TEST (optional)

> *Always check to see that the evidence supports the alternative hypothesis. If the evidence went the other way, there would be nothing to test; we would have **P**-value > 0.50. Blindly testing without first checking whether there is evidence to support H_a may result in mistakenly reversing the procedure and computing $1 - (\mathbf{P}\text{-value})$ instead of **P**-value.*

Because $20n = 20(9) > 25 = N$, this case can be considered a small-population study. Indeed, the binomial exact test results in P-value = 0.2317. In hopes of obtaining a sufficiently small P-value to reject the null hypothesis, we use the more powerful Fisher's exact test. Given that 9 patients went into remission for at least a year, let K be the number of those found in the new drug group.

Step 4. The test statistic is $k_s = 7$. Also $N = 25, N_1 = 15$, and $n = 9$.

Step 5. P-value $= Pr(K \geq k_s) = Pr(K \geq 7)$. This is the probability of obtaining data at least as extreme as those observed. We have

$$\begin{aligned} Pr(K \geq 7) &= Pr(K = 7) + Pr(K = 8) + Pr(K = 9) \\ &= \frac{\binom{15}{7}\binom{10}{2}}{\binom{25}{9}} + \frac{\binom{15}{8}\binom{10}{1}}{\binom{25}{9}} + \frac{\binom{15}{9}\binom{10}{0}}{\binom{25}{9}} \\ &= 0.1417 + 0.0315 + 0.0025 = 0.1757 \end{aligned}$$

Because the P-value ≥ 0.05 the null hypothesis is retained.

Step 6. Even with Fisher's exact test, the evidence is insufficient to conclude that the new drug is better than the old one.

EXERCISES FOR SECTION 6.6

- **6.21** Does the Salk vaccine prevent death from polio? Use the data in Example 6.7, but use Fisher's exact test to compute the P-value. Compare this P-value with that of the binomial exact test as found in Example 6.7.

- **6.22** A lawsuit was brought against Transco Services of Milwaukee charging it with age discrimination in the termination of 11 employees.[9] Use the data below

---

[9] This example is from *StatXact-3 for Windows User Manual*, Cytel Software Corp., Cambridge, MA, 1995, p. 434. It is also cited in A. Tamhane and D. Dunlop, *Statistics and Data Analysis from Elementary to Intermediate*, Prentice Hall, 2000.

to determine whether a significantly larger percentage of old employees were terminated in comparison to younger employees. Use Fisher's exact test to obtain P-value $= 0.0037$. Compare with results of the binomial exact test.

	Terminated	Retained	Row total
Old employees	10	17	27
Young employees	1	24	25
Column total	11	41	52

- **6.23** A large bank hires 14 equally qualified employees. Ten of them are assigned jobs as tellers and four as loan officers. It is suspected that women are less likely than men to receive the more desirable loan officer jobs. Use the data given by the table below to test this hypothesis at the $\alpha = 0.05$ level of significance.

    a. Use the binomial exact test.
    b. Use Fisher's exact test.

	Loan officer	Teller	Row total
Women	1	9	10
Men	3	1	4
Column total	4	10	14

6.24 It has been observed that low-calorie diets in mice are associated with fewer cases of cancer. In an attempt to test this in primates, researchers divided 120 rhesus monkeys into two equal groups; over the span of more than 10 years, one group ate without limit and the other groups consumed only 70% as many calories as the first group.[10] In the low-calorie group 2 monkeys developed cancer, and in the well-fed group, 7 monkeys had cancer. Perform Fisher's exact test at the $\alpha = 0.05$ level to decide whether a low-calorie diet is associated with reduced risk of cancer.

	Cancer	No cancer	Total
Low-cal	2	58	60
Well-fed	7	53	60
Total	9	111	120

- **6.25** Continuing with the study of Exercise 6.24, researchers found that only 1 monkey in the low-calorie group developed endometriosis compared with 6 in the well-fed group. Perform Fisher's exact test at the $\alpha = 0.05$ level to decide whether a low-calorie diet is associated with a reduced risk of endometriosis.

	Endometriosis	No endometriosis	Total
Low-cal	1	59	60
Well-fed	6	54	60
Total	7	113	120

---

[10] J. Travis, *Low-cal diet may reduce cancer in monkeys*, Science News **158** (2000), p. 341.

REVIEW EXERCISES FOR CHAPTER 6

6.26 To test a new drug, a group of 10 matched pairs of volunteers is used. In each pair, one volunteer is treated with the drug while the other is treated with a placebo as a control. At the end of the double-blind experiment, a physician examines each pair and declares which of the two is healthier. Let K be the number of pairs in which the treated patient was declared healthier than the control. The null hypothesis H_0 is that the drug is absolutely ineffective (neither good nor bad). It has been decided to reject H_0 whenever $K > 8$.

    a. Find α, the probability of falsely rejecting H_0.

    b. Suppose that the drug is so effective that for each matched pair, the probability is 90% that the treated patient will be declared healthier. Find β, the probability of falsely retaining H_0.

6.27 It is hypothesized that in 1989 the median family income in Wilmette, Illinois, was above $50,000. A random sample of 20 families was taken from the U.S. census data (see Example 2.12). The results are given below. The units are in $1000s. Use the sign test at the $\alpha = 0.05$ level of significance to test the null hypothesis H_0: $Med = \$50,000$ against the alternative hypothesis H_a: $Med > \$50,000$.

Family incomes in 1989 of 20 families in zip code 60091

25	70	28	93	47	83	72	41	130	248
43	36	76	35	42	79	161	79	282	78

6.28 Body mass index (BMI) is computed as weight in kilograms divided by the square of the height in meters. The following data are the percentile rank BMI for 59 Hispanic boys from the fall of 1998 kindergarten class. Percentiles were computed using the Centers for Disease Control (CDC) Growth Charts. These percentiles are age- and gender-specific. These boys are a random sample from the 1233 Hispanic boys in the Early Childhood Longitudinal Study-K (ECLS-K) who had their height, weight, and age measured in the fall of 1998.[11]

77.73	0.98	77.87	71.08	62.79	99.85	94.85	19.74	67.38
66.03	97.17	99.09	29.85	79.35	85.37	67.37	74.15	55.12
1.78	99.79	95.35	85.59	99.54	88.60	12.56	46.22	46.80
79.81	39.61	40.16	78.21	29.38	91.83	99.89	98.01	60.25
23.98	89.83	70.77	75.99	82.95	62.50	85.39	3.62	35.76
84.48	33.84	99.96	91.75	81.14	48.95	11.25	86.60	97.53
85.47	98.17	1.80	18.90	94.90				

---

[11] For more information about the CDC Growth Charts, visit the website www.cdc.gov/growthcharts. For more information about ECLS-K visit the National Center for Education Statistics at http://nces.ed.gov.

The sample mean is 67.37, and the sample standard deviation is 29.91. Epidemiologists are concerned that there is a trend in the last decade toward children in America being increasingly overweight. Is the median percentile rank of the Hispanic boys in the ECLS-K study above the 50th percentile of the CDC Growth Charts? Use a sign test at the 1% significance level. State the null and alternative hypotheses. Compute a P-value and state your conclusion in common English.

- **6.29** Repeat the sign test at the 1% significance level as in the above exercise, but this time for white, non-Hispanic girls. The following data are the percentile rank BMI for 34 white, non-Hispanic girls from the fall of 1998 kindergarten class. These girls are a random sample from the 5084 white, non-Hispanic girls in the Early Childhood Longitudinal Study-K (ECLS-K) who had their height, weight, and age measured in the fall of 1998.

8.27	81.41	15.40	40.84	86.70	28.45	51.85	82.36	21.38
85.70	62.50	92.74	85.32	91.49	48.79	11.18	36.05	87.92
25.68	73.40	34.92	17.65	0.16	92.70	57.11	46.72	66.86
70.91	98.20	64.79	65.81	67.14	62.34	8.06		

6.30 Continue with Exercise 6.28. A BMI percentile rank of 95% or more on the CDC growth charts is an indication of obesity. Do more than 5% of the Hispanic boys in the ECLS-K study rank at or above the 95th percentile on the CDC growth charts? Use the 1% level of significance. State the null and alternative hypotheses and compute a P-value and state your conclusion in common English.

- **6.31** A certain kind of computer chip has a failure rate of 8%. During quality-control inspection of a batch of 100, 15% are found to be defective. Is this within historical limits or is the failure rate out of control? Use the binomial exact test at the 0.05 level of significance.

6.32 An osteoporosis drug, raloxifen, was tested against a placebo in a large double-blind experiment to see whether it could prevent breast cancer in women as well as help osteoporosis. The average age of the women was 66, and the number of cases of breast cancer after four years of treatment is given in the table below. Use the binomial exact test (with technology) at the $\alpha = 0.01$ level of significance. Assume the random selection for treatment was done with $\pi = 2/3$.

	Breast cancer	No breast cancer	Row total
Raloxifen	22	5107	5129
Placebo	39	2537	2576
Column total	61	7644	7705

- **6.33** In the 1969 trial of Dr. Benjamin Spock for conspiracy to violate the Military Service Act,[12] a jury was drawn from panel of 350 persons, consisting of 102 women and 248 men. The judge selected 100 potential jurors out of the panel of 350 persons. The potential jurors included only 9 women. If 100 potential jurors were randomly chosen from a panel of 102 women and 248 men, what is the chance of getting 9 or fewer women? Was the judge biased against women in choosing the panel? Set up and conduct a formal test of hypothesis using the binomial exact test, with $\alpha = 0.01$ and $\pi = 102/350$.

	Chosen	Not chosen	Row total
Women	9	93	102
Men	91	157	248
Column total	100	250	350

6.34 In a controlled experiment of 63 leukemia patients, a new drug, VCR, to be used in combination with the standard drug Prednisone, was tested against the Prednisone alone.[13] Use the data given in the table below to test whether the new treatment is better than the standard treatment. Use the binomial exact test at the $\alpha = 0.01$ level of significance. Use $\pi = 2/3$. Use the failure column to work with a smaller value $n = 11$.

	Success	Failure	Row total
New drug	38	4	42
Standard treament	14	7	21
Column total	52	11	63

- **6.35** Repeat Exercise 6.34, except this time use Fisher's exact test at $\alpha = 0.01$.

6.36 To test vitamin C as a treatment for the prevention of colds, the Nobel Prize–winning chemist Linus Pauling conducted a double-blind randomized experiment on 279 French skiers. It turned out that 17 of the 140 skiers in the vitamin C group, and 31 of the 139 skiers in the placebo group, came down with colds. Use Fisher's exact test at the $\alpha = 0.05$ level of significance to test for the effectiveness of vitamin C in preventing colds.

	Cold	No cold	Row total
Vitamin C	17	123	140
Placebo	31	108	139
Column total	48	231	279

- **6.37** Is the risk of getting the flu lower among students who get a flu shot? Consider the following table for 48 college students. Use the binomial exact test at the

---

[12]Hans Zeisel, *Dr. Spock and the case of the vanishing women jurors*, University of Chicago Law Review **37** (1969), pp. 1–18.

[13]P. C. O'Brien and T. R. Fleming, *A multiple testing procedure for clinical trials*, Biometrics **35** (1979), pp. 549–56.

$\alpha = 0.05$ level of significance.

	Flu	No flu	Row total
Shot	3	12	15
No shot	18	15	33
Column total	21	27	48

6.38 Repeat the flu shot Exercise 6.37, except this time use Fisher's exact test at $\alpha = 0.05$.

• **6.39** Acupuncture was tried as a therapy for cocaine addiction. In a double-blind randomized control experiment, 28 addicted subjects were treated with an eight-week course of acupuncture consisting of needles inserted into four specific points in the outer ear five times per week. In the placebo group, 27 addicted subjects were treated with a course of a sham treatment consisting of needles inserted into four nonacupuncture ear points. A third group of 27 was treated with relaxation therapy. Of those who completed the eight-week course, 7 of the acupuncture subjects, 4 of the placebo controls, and 2 of the relaxation controls tested free of cocaine. Use the binomial exact test to decide whether the acupuncture treatment is better than the placebo. Use $\alpha = 0.05$, and $\pi = 0.50$.

	Success	Failure	Row total
Acupuncture	7	21	28
Placebo	4	23	27
Column total	11	44	55

6.40 In the acupuncture experiment of Exercise 6.39, only 13 of the 28 in the acupuncture group, 17 of the 27 in the sham treatment group, and 22 of the 27 in the relaxation group completed the entire eight-week course of treatments. Repeat the binomial exact test comparing acupuncture with placebo at $\alpha = 0.05$, but this time consider only the subjects who completed the eight-week treatments. In this case, do not use $\pi = 0.50$.

	Success	Failure	Row total
Acupuncture	7	6	13
Placebo	4	13	17
Column total	11	19	30

• **6.41** Repeat Exercise 6.40 on acupuncture, except this time use Fisher's exact test at $\alpha = 0.05$.

6.42 Repeat Exercise 6.39 on acupuncture, except this time use Fisher's exact test at $\alpha = 0.05$.

Chapter 7

CONTINUOUS RANDOM VARIABLES

7.1 BASIC PROPERTIES

7.2 PERCENTILES AND MODES

7.3 EXPECTED VALUE OR MEAN

7.4 FUNCTIONS OF RANDOM VARIABLES

7.5 VARIANCE AND STANDARD DEVIATION

7.6 CHEBYSHEV'S INEQUALITY

Probability is a **measure** of an event, in the same sense that area is measure of a two dimensional region. Because of this abstract connection, integration theory is important in the understanding of probability theory. As a result of the Fundamental Theory of Calculus, which relates the derivative and the integral, we will see in this chapter that both differential calculus and integral calculus are natural tools in probability theory.

7.1 BASIC PROPERTIES

Let's return to Example 3.10 on page 81: highway accidents that occur on Interstate 94 somewhere between mile marker 3 (New Buffalo exit) and mile marker 6 (Union Pier exit). The random variable X is the distance between the place of the accident and the Michigan–Indiana border measured in miles along the highway. Note that X is not a discrete random variable. In particular, $Pr(X = x) = 0$ for *all* values of x, because the chance of an accident occurring at any *one* of a continuum of points on the highway is zero.

> *This phenomenon that the accident must have occurred at some point, yet* $\Pr(X = x) = 0$ *for all values of* x, *is called the* **Continuum Paradox**.

This means that defining, for a continuous random variable, the probability function f_X to be $Pr(X = x)$, as we did for discrete random variables, would not be useful. However, the cumulative distribution function of a continuous random variable can be defined the same as before:

$$\mathbf{F_X(x) = Pr(X \leq x)} \quad \textit{for} \quad -\infty < \mathbf{x} < +\infty$$

Notice that the only *possible* values of X are between the 3 and 6; that is,

$$F_X(x) = Pr(X \leq x) = 0 \text{ whenever } -\infty < x < 3$$

$$F_X(x) = Pr(X \leq x) = 1 \text{ whenever } 6 \leq x < +\infty$$

Because we assume that no part of the three-mile stretch of highway $3 \leq x \leq 6$ is more prone to accidents than any other, the probability $Pr(X \leq x)$ there is proportional to the distance traveled between New Buffalo and Union Pier:

$$F(x) = Pr(X \leq x) = \frac{x-3}{6-3} \quad \text{for } 3 \leq x \leq 6$$

Thus, the graph of F is piecewise linear, starting at 0 for $-\infty < x \leq 3$, increasing to 1 for $3 \leq x \leq 6$, and remaining at 1 for $6 \leq x < \infty$ (see Figure 7.1).

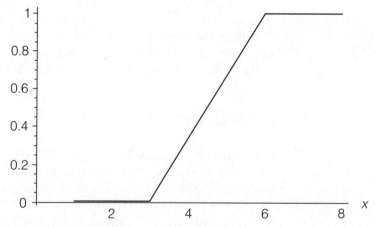

Figure 7.1 Cumulative distribution function $F(x)$ for distance from Michigan–Indiana border

Notice that the cumulative distribution function F of a continuous random variable is not a step function as it is in the discrete case. Here F is a continuous nondecreasing function. We want to define the **probability density function** f to have the following property.

> **For any two numbers $a < b$,**
> $$\mathbf{Pr(a < X \leq b) = F(b) - F(a) = \int_a^b f(t)\,dt.}$$

This is analogous to property **(1)** for discrete f and F on page 127, replacing summation with integration. The importance of the above formula is that, once you have the probability density function f, probabilities can be computed without the use of F. By the fundamental theorem of calculus, we have
$$f(x) = F'(x)$$
for all points where F' exists.

Note that, for the accident on the road example, f is undefined at the points $x = 3$ and $x = 6$ because F is not differentiable there. For x in the interval $(3, 6)$, the probability density function f is constant with value $f(x) = 1/3$ and $f(x) = 0$ elsewhere. See Figure 7.2.

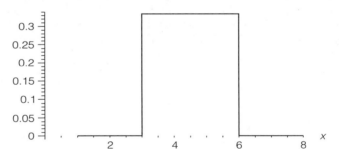

Figure 7.2 Probability density function $f(x)$ for distance from Michigan–Indiana border

Analogous to property **(k)** for discrete f and F in Chapter 4 on page 127, we have
$$F(t) = \int_{-\infty}^{t} f(x)\,dx$$

Note. There are random variables that are combinations of discrete and continuous random variables. For example, let Y be the number of passengers in the car involved in the accident on Interstate 94. Clearly, Y is discrete. However, the random variable Z defined by $Z = X \cdot Y$ (which is the number of passenger-miles traveled between the Michigan–Indiana border and Union Pier before the accident occurred) is a combination of a discrete and a continuous random variable. Another example is the random variable $X + Y$, which does not have a simple natural interpretation. The theoretical tool of the *Riemann-Stieltjes Integral* can deal with both discrete and continuous random variables and combinations of them. The Riemann-Stieltjes Integral (and the more general Lebesgue-Stieltjes Integral) are used in rigorous studies of random variables, as taught in advanced courses of mathematical probability and statistics. In this book, we consider only purely discrete and purely continuous random variables and treat them separately. In the discrete case, when F is a step function, we use summation, and, when F is continuous, we use Riemann integration.

Example 7.1 *Failure of air conditioning.* Consider the failure of air conditioning on commercial aircraft. Let T be the accumulated flight time after each major overhaul until the air conditioning again needs repair. The random variable T measures the waiting time for failure of air conditioning. Certainly, $Pr(T < 0) = 0$, so $f_T(t) = 0$ whenever $t < 0$. In Chapter 9, we will show that random variables for certain types of waiting time have density functions of the following exponential form:

$$f_T(t) = \frac{1}{\theta} e^{-\frac{t}{\theta}}, \text{ whenever } t \geq 0, \text{ and } f_T(t) = 0, \text{ whenever } t < 0 \qquad (7.1)$$

for some real number θ. The parameter θ turns out to be the average waiting time. The frequency polygon in Figure 7.3 follows such a curve.[1]

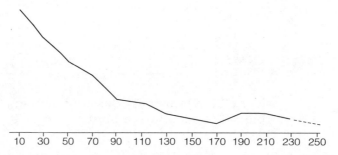

Figure 7.3 Hours between 213 failures of air conditioning on a fleet of airplanes

Let's say that the average waiting time for failure is $\theta = 80$. What percentage of the aircraft need air conditioning repairs within the first 40 hours?

---

[1] From the article Frank Proschan, *Theoretical explanation of observed decrease failure rate*, Technometrics **5** (1973), pp. 375–383. Graph adapted from J. L. Hodges Jr., David Krech & Richard S. Crutchfield, *StatLab*, McGraw-Hill, 1975.

7.1 BASIC PROPERTIES

Solution   Because

$$F(t) = \int_{-\infty}^{t} f(x)\,dx = \int_{-\infty}^{0} 0\,dx + \int_{0}^{t} \frac{1}{80} e^{-\frac{x}{80}}\,dx = -e^{-\frac{x}{80}}\Big|_{0}^{t} = 1 - e^{-\frac{t}{80}}$$

we have $Pr(T \le 40) = F(40) = 1 - e^{-\frac{40}{80}} = 1 - e^{-\frac{1}{2}} = 0.3935$.

Example 7.2   ***Two accidents on a road.***   Suppose that two accidents occur independently on Interstate 94 between New Buffalo and Union Pier. When there are two accidents, the emergency plan calls for the New Buffalo paramedics to handle the one closer to New Buffalo with the other one handled by Union Pier paramedics. Let Y mark the spot where the accident closer to New Buffalo occurred. Let's find the functions $F(y)$ and $f(y)$.

Solution   Let X_1 and X_2 mark the spots where the two accidents occur. The random variables X_1 and X_2 are different, but they have the same cumulative distribution function $F(x) = 0$ for $x < 3$, $F(x) = \frac{1}{3}x - 1$ for $3 \le x < 6$, and $F(x) = 1$ for $x \ge 6$. The random variable Y, being the spot where the accident closer to New Buffalo occurred, is the smaller of X_1 and X_2.

So, we have

$$\begin{aligned} Pr(Y > y) &= Pr(X_1 > y \text{ and } X_2 > y) \\ &= Pr(X_1 > y) \cdot Pr(X_2 > y) \\ &= (1 - F(y)) \cdot (1 - F(y)) = (2 - \frac{1}{3}y)^2 \text{ for } 3 \le y < 6 \end{aligned}$$

Thus, the cumulative distribution function can be computed as follows:

$$\begin{aligned} F_Y(y) &= 0 \text{ for } y < 3 \\ F_Y(y) &= 1 \text{ for } y > 6 \\ F_Y(y) &= Pr(Y \le y) \\ &= 1 - Pr(Y > y) \\ &= 1 - (2 - \frac{1}{3}y)^2 \qquad \text{for } 3 \le y < 6 \\ &= -3 + \frac{4}{3}y - \frac{1}{9}y^2 \end{aligned}$$

We then can find the density function f:

$$\begin{aligned} f_Y(y) &= F_Y'(y) = \frac{4}{3} - \frac{2}{9}y \quad \text{for } 3 \le y < 6 \\ f_Y(y) &= F_Y'(y) = 0 \qquad\qquad \text{elsewhere} \end{aligned}$$

Basic properties of f and F for continuous random variables

(a) $F(x) = Pr(X \leq x)$, for $-\infty < x < +\infty$

(b) $0 \leq F(x) \leq 1$

(c) F is nondecreasing

(d) F is continuous

(e) For all x, $Pr(X = x) = 0$ and $Pr(X \neq x) = 1$

(f) $\lim_{x \to -\infty} F(x) = 0$

(g) $\lim_{x \to \infty} F(x) = 1$

(h) $0 \leq f(x)$

(i) $\int_{-\infty}^{\infty} f(t)\, dt = 1$

(j) $f(x) = F'(x)$

(k) $F(x) = \int_{-\infty}^{x} f(t)\, dt$

(l) for any two numbers $a < b$, $Pr(a < X \leq b) = F(b) - F(a) = \int_{a}^{b} f(t)\, dt$. Furthermore, the inequalities $<$ and $\leq$ may be interchanged with any combination of $<$ and $\leq$.

EXERCISES FOR SECTION 7.1

- **7.1** Find the probability density function f for the continuous random variable X with cumulative distribution function

$$F(x) = \begin{cases} 0 & : \text{ for } x < 0 \\ \frac{x^2}{4} & : \text{ for } 0 \leq x < 2 \\ 1 & : \text{ for } x \geq 2 \end{cases}$$

7.2 Find the probability density function f for the continuous random variable X with cumulative distribution function

$$F(x) = \begin{cases} 0 & : \text{ for } x < 5 \\ \frac{x}{5} - 1 & : \text{ for } 5 \leq x < 10 \\ 1 & : \text{ for } x \geq 10 \end{cases}$$

7.3 Find each of the following for the density function

$$f(x) = \begin{cases} (3/2)(2x - x^2) & : \text{ for } 0 < x < 1 \\ 0 & : \text{ elsewhere} \end{cases}$$

a. The cumulative distribution function F
b. $F(0.25)$ **c.** $Pr(X \geq 0.25)$
d. $F(0.75)$ **e.** $Pr(0.25 < X \leq 0.75)$

7.4 Find each of the following for the density function

$$f(x) = \begin{cases} |x| & : \text{ for } -1 \leq x \leq 1 \\ 0 & : \text{ elsewhere} \end{cases}$$

a. The cumulative distribution function F
b. $F(-0.50)$ **c.** $Pr(X \geq 0)$
d. $F(0.75)$ **e.** $Pr(-0.50 < X \leq 0.75)$

7.5 Find each of the following for the density function

$$f(x) = \begin{cases} x + 1 & : \text{ for } -1 < x < 0 \\ -x + 1 & : \text{ for } 0 < x < 1 \end{cases}$$

a. The cumulative distribution function F
b. $F(-0.50)$ **c.** $Pr(X \geq 0)$
d. $F(0.50)$ **e.** $Pr(-0.50 < X \leq 0.50)$

7.6 Find each of the following for the density function

$$f(x) = \begin{cases} e^{-x} & : \text{ for } 0 \leq x < \infty \\ 0 & : \text{ for } x < 0 \end{cases}$$

a. The cumulative distribution function F
b. $F(0)$ **c.** $F(1)$
d. $Pr(X \geq 1)$ **e.** $Pr(1 < X \leq 2)$

7.7 A book is on one-hour reserve at a college library. Let X be the time it is checked out (in hours) by a random student. Suppose

$$f(x) = \begin{cases} 2x & : \text{ for } 0 < x < 1 \\ 0 & : \text{ elsewhere} \end{cases}$$

a. Sketch f and F (note that $F(x) = 1$ for $x \geq 1$) and find
b. $F(0.25)$ **c.** $Pr(X \geq 0.25)$
d. $F(0.75)$ **e.** $Pr(0.25 < X \leq 0.75)$

7.8 While driving a country road, the time T between oncoming cars is a continuous random variable that follows Equation 7.1 on page 198. Let's say that the average time between oncoming cars is $\theta = 2$ minutes. Find the following:

a. $F(2)$ **b.** $Pr(T > 2)$ **c.** $Pr(2 < T \leq 6)$

7.2 PERCENTILES AND MODES

For a random variable X, the rth percentile, which we denote $x_{\frac{r}{100}}$, is the value x for which $F(x) = \frac{r}{100}$. In particular, the **median** is the 50th percentile $x_{.50}$.

Example 7.3 *Failure of air conditioning.* Continuing with Example 7.1 for air conditioning failure on aircraft, the median is the solution Med, of the following equation:

$$0.50 = \int_{-\infty}^{Med} f(t)\,dt = \frac{1}{80}\int_{0}^{Med} e^{-\frac{t}{80}}\,dt = 1 - e^{-\frac{Med}{80}}$$

or

$$Med = 80\ln 2 = 55.45. \quad \text{See Figure 7.4.}$$

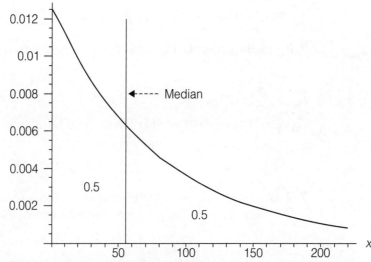

Figure 7.4 Exponential probability density function with $Med = 80\ln 2$ shown

7.2 PERCENTILES AND MODES

Example 7.4 *Normal curves.* The **normal density functions** are given by the formula

$$f(x) = \frac{1}{\sigma\sqrt{2\pi}} e^{-\frac{1}{2}\left(\frac{x-\mu}{\sigma}\right)^2}, \text{ for } -\infty < x < \infty$$

where μ is any real number and σ is any positive real number as graphed in Figure 7.5.

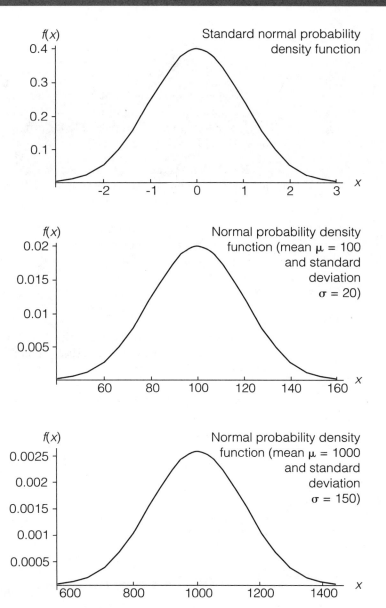

Figure 7.5 Normal density functions for various values of μ and σ

For any of these curves, we have

$$\Pr(a < X < b) = \frac{1}{\sigma\sqrt{2\pi}} \int_a^b e^{-\frac{1}{2}(\frac{x-\mu}{\sigma})^2} \text{ for any values a and b.}$$

Regardless of the choice of parameters μ and $\sigma > 0$, it turns out that

$$\int_{-\infty}^{\infty} f(x)\,dx = \frac{1}{\sigma\sqrt{2\pi}} \int_{-\infty}^{\infty} e^{-\frac{1}{2}(\frac{x-\mu}{\sigma})^2}\,dx = 1$$

IQ scores are discrete, but a convenient model is a continuous one given by the normal density curve with $\mu = 100$ and $\sigma = 15$, which we will denote by f_{IQ}. See Figure 8.9 on page 235 for a frequency polygon of the IQ scores of 2904 children.

Recall that the **mode** of a random variable X is the value of x where $f(x)$ attains its maximum. To find the mode of a continuous random variable, we find the maximum of the function f.

Although it is evident from the graph that the mode of a normal random variable occurs at the point $x = \mu$, we can use the techniques of calculus to find the maximum of a function. The only critical points are the solutions of the equation

$$f'(x) = \frac{1}{\sigma\sqrt{2\pi}} e^{-\frac{1}{2}(\frac{x-\mu}{\sigma})^2} \left(\frac{\mu - x}{\sigma}\right) = 0$$

and the only solution of the equation $f'(x) = 0$ is $x = \mu$. The maximum of f is attained at this unique critical point. For IQ scores, we thus have $Mode_{IQ} = 100$.

EXERCISES FOR SECTION 7.2

- **7.9** Find the 25th percentile and median for the random variable given in Exercise 7.1 on page 200:

$$F(x) = \begin{cases} 0 & : \text{ for } x < 0 \\ \frac{x^2}{4} & : \text{ for } 0 \leq x < 2 \\ 1 & : \text{ for } x \geq 2 \end{cases}$$

7.10 Find the 80th percentile and median for the random variable given in Exercise 7.2 on page 200:

$$F(x) = \begin{cases} 0 & : \text{for } x < 5 \\ \frac{x}{5} - 1 & : \text{for } 5 \leq x < 10 \\ 1 & : \text{for } x \geq 10 \end{cases}$$

- **7.11** Find the interquartile range (IQR = 75th percentile $- 25^{th}$ percentile) for air-conditioning failure in Example 7.1 on page 198.

  7.12 Find the interquartile range (IQR = 75th percentile $- 25^{th}$ percentile) of the random variable Y given in Example 7.2 (on page 199) of two accidents on a road.

- **7.13** Find the median of the random variable Y given in Example 7.2 (on page 199) of two accidents on a road.

  7.14 Find the IQR, the median, and the two modes of the random variable given in Exercise 7.4 on page 201:

  $$f(x) = \begin{cases} |x| & : \text{for } -1 \leq x \leq 1 \\ 0 & : \text{elsewhere} \end{cases}$$

- **7.15** Find the mode, median, and IQR of the random variable given in Exercise 7.5 on page 201:

  $$f(x) = \begin{cases} x + 1 & : \text{for } -1 < x < 0 \\ -x + 1 & : \text{for } 0 < x < 1 \end{cases}$$

  7.16 Find the median of the random variable given in Exercise 7.6 on page 201:

  $$f(x) = \begin{cases} e^{-x} & : \text{for } 0 \leq x < \infty \\ 0 & : \text{for } x < 0 \end{cases}$$

7.3 EXPECTED VALUE OR MEAN

Recall that for a *discrete* random variable X, the mean is defined as

$$\mu = E(X) = \sum_{\text{all } x} x f(x)$$

For a *continuous* random variable X, the mean is defined analogously, replacing the sum by an integral:

$$\mu = E(X) = \int_{-\infty}^{\infty} x f(x) \, dx$$

206   CHAPTER 7.   CONTINUOUS RANDOM VARIABLES

This formula compares with the integral Formula (2.4) on page 65 for computing the mean of grouped data. Here, as in the discrete case, we need absolute convergence $\int_{-\infty}^{\infty} |x| f(x)\, dx < \infty$. Otherwise, we say that the expected value is undefined.

Warning

> As with discrete random variables, do not forget the factor **x** in the definition of μ. We have
>
> $$\mathbf{Pr}(-\infty < \mathbf{X} < \infty) = \int_{-\infty}^{\infty} \mathbf{f(x)\, dx} = 1, \quad \text{but} \quad \mu = \int_{-\infty}^{\infty} \mathbf{x f(x)\, dx}$$

Example 7.5   *Normal curves.* Let's return to the normal density curves of Example 7.4:

$$f(x) = \frac{1}{\sigma\sqrt{2\pi}} e^{-\frac{1}{2}\left(\frac{x-\mu}{\sigma}\right)^2}$$

The fact that one of the parameters is denoted by the symbol μ suggests that it is the mean. Let's actually show that.

Solution   Using the definition of mean, we have the following:

$$\begin{aligned}
E(X) &= \int_{-\infty}^{\infty} x f(x)\, dx \\
&= \frac{1}{\sqrt{2\pi}} \int_{-\infty}^{\infty} \frac{x}{\sigma} e^{-\frac{1}{2}\left(\frac{x-\mu}{\sigma}\right)^2} dx \\
&= \frac{1}{\sqrt{2\pi}} \int_{-\infty}^{\infty} \frac{x-\mu}{\sigma} e^{-\frac{1}{2}\left(\frac{x-\mu}{\sigma}\right)^2} dx + \frac{1}{\sqrt{2\pi}} \int_{-\infty}^{\infty} \frac{\mu}{\sigma} e^{-\frac{1}{2}\left(\frac{x-\mu}{\sigma}\right)^2} dx \\
&= \frac{1}{\sqrt{2\pi}} \int_{-\infty}^{\infty} \frac{x-\mu}{\sigma} e^{-\frac{1}{2}\left(\frac{x-\mu}{\sigma}\right)^2} dx + \mu \frac{1}{\sigma\sqrt{2\pi}} \int_{-\infty}^{\infty} e^{-\frac{1}{2}\left(\frac{x-\mu}{\sigma}\right)^2} dx \\
&= \frac{1}{\sqrt{2\pi}} \left\{ \int_{-\infty}^{0} \frac{x-\mu}{\sigma} e^{-\frac{1}{2}\left(\frac{x-\mu}{\sigma}\right)^2} dx + \int_{0}^{\infty} \frac{x-\mu}{\sigma} e^{-\frac{1}{2}\left(\frac{x-\mu}{\sigma}\right)^2} dx \right\} + \mu \int_{-\infty}^{\infty} f(x)\, dx
\end{aligned}$$

With a change of variable $z = \frac{x-\mu}{\sigma}$, $dz = \frac{dx}{\sigma}$, we obtain

$$E(X) = \frac{1}{\sqrt{2\pi}} \left\{ \int_{-\infty}^{0} z e^{-\frac{1}{2}z^2} \sigma\, dz + \int_{0}^{\infty} z e^{-\frac{1}{2}z^2} \sigma\, dz \right\} + \mu \cdot 1$$

With another change of variable $u = -\frac{1}{2}z^2$, $du = -z\, dz$ we obtain

$$E(X) = \frac{\sigma}{\sqrt{2\pi}} \left\{ \int_{-\infty}^{0} -e^u\, du + \frac{1}{\sqrt{2\pi}} \int_{0}^{-\infty} -e^u\, du \right\} + \mu$$

$$= \frac{\sigma}{\sqrt{2\pi}}\{-1+1\} + \mu = \mu$$

In particular, for IQ scores we have $E(IQ) = \mu_{IQ} = 100$.

7.4 FUNCTIONS OF RANDOM VARIABLES

Consider a change of variable $W = g(X)$ for a continuous random variable X. We have

$$E(g(X)) = \int_{-\infty}^{\infty} g(x) f_X(x) \, dx$$

Note that for $W = X$, this formula reduces to the definition of the mean

$$E(X) = \int_{-\infty}^{\infty} x f_X(x) \, dx$$

Example 7.6   *Standard normal curve.*   For the *standard* normal random variable Z with parameters $\mu = 0$ and $\sigma = 1$, we show that $E(Z^2) = 1$.

Solution   The random variable Z has the probability density function

$$f_Z(z) = \frac{1}{\sqrt{2\pi}} e^{-\frac{1}{2}z^2}$$

Then

$$\begin{aligned} E(Z^2) &= \int_{-\infty}^{\infty} z^2 f_Z(z) \, dz \\ &= \frac{1}{\sqrt{2\pi}} \int_{-\infty}^{\infty} z^2 e^{-\frac{1}{2}z^2} \, dz \end{aligned}$$

Using the integration-by-parts formula

$$\int u \, dv = uv - \int v \, du$$

with $u = z$ and $dv = ze^{-\frac{1}{2}z^2}\,dz$, we obtain the following:

$$\begin{aligned}
E(Z^2) &= \frac{1}{\sqrt{2\pi}}\left\{-ze^{-\frac{1}{2}z^2}\Big|_{-\infty}^{\infty} + \int_{-\infty}^{\infty} e^{-\frac{1}{2}z^2}\,dz\right\} \\
&= \frac{1}{\sqrt{2\pi}}\left\{0 + \int_{-\infty}^{\infty} e^{-\frac{1}{2}z^2}\,dz\right\} \\
&= \frac{1}{\sqrt{2\pi}} \int_{-\infty}^{\infty} e^{-\frac{1}{2}z^2}\,dz \\
&= \int_{-\infty}^{\infty} f_Z(z)\,dz = 1
\end{aligned}$$

Remark Note that $E(Z) = 0$ but $E(Z^2) = 1$. In general, for any continuous random variable X, we have $E(X^2) \neq E^2(X)$.

However, just as with discrete random variables, we do have linear formulas given by the following theorem. The proof is left as an exercise.

THEOREM 7.1 *Suppose that X is a random variable with expected value $E(X)$. Then, for any real constants a and b, we have*

$$E(aX + b) = aE(X) + b$$

7.5 VARIANCE AND STANDARD DEVIATION

In the continuous case, the variance and standard deviation are defined the same as in the discrete case.

$$\sigma_X^2 = \text{Var}(X) = E\{(X - \mu)^2\}$$

$$\sigma_X = \text{SD}(X) = \sqrt{\text{Var}(X)} = \sqrt{E(X - \mu)^2}$$

7.5 VARIANCE AND STANDARD DEVIATION

Example 7.7 *Normal curves.* In Example 7.5, we showed that for each normal density function

$$f(x) = \frac{1}{\sigma\sqrt{2\pi}} e^{-\frac{1}{2}\left(\frac{x-\mu}{\sigma}\right)^2}$$

the parameter μ in the formula is the mean. Now let's show that the parameter σ in the formula turns out to be the standard deviation of the random variable.

Solution Using the definition of variance, we have

$$\begin{aligned}\mathrm{Var}(X) &= \int_{-\infty}^{\infty} (x-\mu)^2 f(x)\, dx \\ &= \frac{1}{\sigma\sqrt{2\pi}} \int_{-\infty}^{\infty} (x-\mu)^2 e^{-\frac{1}{2}\left(\frac{x-\mu}{\sigma}\right)^2} dx\end{aligned}$$

Then, using a change of variable $z = \frac{x-\mu}{\sigma}$, $dz = \frac{dx}{\sigma}$, we obtain

$$\begin{aligned}\mathrm{Var}(X) &= \frac{1}{\sigma\sqrt{2\pi}} \int_{-\infty}^{\infty} (x-\mu)^2 e^{-\frac{1}{2}\left(\frac{x-\mu}{\sigma}\right)^2} dx \\ &= \sigma^2 \frac{1}{\sqrt{2\pi}} \int_{-\infty}^{\infty} z^2 e^{-\frac{1}{2}z^2} dz = \sigma^2 E(Z^2)\end{aligned}$$

It was shown in Example 7.6 that $E(Z^2) = 1$. Thus, $\mathrm{Var}(X) = \sigma^2$.

In particular, for the continuous model of IQ, we have $\mathrm{SD}(IQ) = \sigma_{IQ} = 15$.

THEOREM 7.2 *For any random variable X with expected value $E(X)$ and variance $\mathrm{Var}(X)$, we have*

$$\boxed{\mathrm{Var}(X) = E(X^2) - E^2(X) = E(X^2) - \mu^2}$$

The proof is the same as for the discrete case Theorem 4.2 and is omitted.

As with the discrete case, this leads to a useful formula for computing the standard deviation of a random variable:

$$\boxed{\sigma = \sqrt{E(X^2) - \mu^2}}$$

210   CHAPTER 7.   CONTINUOUS RANDOM VARIABLES

As in the discrete case, a translation of a random variable from X to $X + b$ has no effect on the variance and standard deviation: $\text{Var}(X + b) = \text{Var}(X)$. Also, multiplication by -1 has no effect: $\text{Var}(-X) = \text{Var}(X)$ and $\text{SD}(-X) = \text{SD}(X)$. More generally, we have the following. The proof is left as an exercise.

THEOREM 7.3 *For any random variable X, with variance $\text{Var}(X)$ and real constants a, b, we have*

$$\text{Var}(aX + b) = a^2 \text{Var}(X)$$

Corollary *Under the conditions of Theorem 7.3:*

$$\text{SD}(aX + b) = |a|\text{SD}(X)$$

The "standard" center of a random variable is the mean μ. And the "standard" unit of measure is the standard deviation σ. This results in the **z-scale** or **standardized scale** for statistics, which is a measure of the number of standard deviations above the mean. The z-scale is illustrated in Figure 7.6.

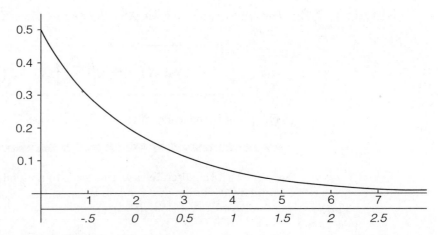

Figure 7.6 Exponential density curve (with $\mu = 2, \sigma = 2$) with underneath z-scale

7.5 VARIANCE AND STANDARD DEVIATION

A recoding to standard units of a random variable X results in the **standardized random variable**

$$Z = \frac{X - \mu_X}{\sigma_X}$$

We have, for all standardized random variables, $E(Z) = \mu_Z = 0$ and $SD(Z) = \sigma_Z = 1$.

z-score For a particular value x, the standardized value

$$z = \frac{x - \mu}{\sigma}$$

is called the **z-score** of x. For example, because $\mu_{IQ} = 100$ and $\sigma_{IQ} = 15$, we say that, for $IQ = 127$, the z-score is $z = 1.8$ because an IQ of 127 is 1.8 SD's above average.

EXERCISES FOR SECTION 7.5

- **7.17** Compute μ, $E(X^2)$, and σ for the random variable given in Exercise 7.3 on page 201:

$$f(x) = \begin{cases} (3/2)(2x - x^2) & : \text{ for } 0 < x < 1 \\ 0 & : \text{ elsewhere} \end{cases}$$

7.18 Compute μ, $E(X^2)$, and σ for the random variable given in Exercise 7.4 on page 201:

$$f(x) = \begin{cases} |x| & : \text{ for } -1 \leq x \leq 1 \\ 0 & : \text{ elsewhere} \end{cases}$$

- **7.19** Compute μ and σ for the random variable given in Exercise 7.1 on page 200:

$$F(x) = \begin{cases} 0 & : \text{ for } x < 0 \\ \frac{x^2}{4} & : \text{ for } 0 \leq x < 2 \\ 1 & : \text{ for } x \geq 2 \end{cases}$$

7.20 Compute μ and σ for the random variable given in Exercise 7.2 on page 200.

$$F(x) = \begin{cases} 0 & : \text{ for } x < 5 \\ \frac{x}{5} - 1 & : \text{ for } 5 \leq x < 10 \\ 1 & : \text{ for } x \geq 10 \end{cases}$$

- **7.21** Compute μ and σ for the accident-on-the-road example shown in Figure 7.2 on page 197.

  7.22 Compute the mean and standard deviation of the two-accident random variable Y of Example 7.2 on page 199.

- **7.23** Compute μ and σ for the book-on-reserve Exercise 7.7 on page 201:

$$f(x) = \begin{cases} 2x & : \quad \text{for} \quad 0 < x < 1 \\ 0 & : \quad \text{elsewhere} \end{cases}$$

  7.24 The coefficient of skewness of a random variable X is defined to be:

$$\frac{E[(X-\mu)^3]}{\sigma^3}$$

  Find the the coefficient of skewness of the two-accident random variable Y of Exercise 7.22.

- **7.25** As in the Exercise 7.24 above, compute the coefficient of skewness, except for the book-on-reserve Exercise 7.23.

7.6 CHEBYSHEV'S INEQUALITY

In Chapter 2, we considered three **common rules** that we used to understand standard deviations of data. The 68% rule, 95% rule, and 99.7% rules are based on the normal curve. If we apply them to areas under general probability histograms and density curves, we obtain approximations that are reasonable for many random variables.

Recall that, for any random variable X, the standardized random variable Z is defined as follows:

$$\boxed{Z = \frac{X-\mu}{\sigma}}$$

68% rule | *Approximately 68% of the area under a probability histogram falls between $Z = -1$ and $Z = +1$.*

95% rule | *Approximately 95% of the area under a probability histogram falls between $Z = -2$ and $Z = +2$.*

99.7% rule | *Approximately 99.7% of the area under a probability histogram falls between $Z = -3$ and $Z = +3$.*

Because these rules are approximations, there are exceptions. Now we state **Chebyshev's Inequality**[2], which has no exceptions. This necessarily gives results that are weaker than the common rules.

THEOREM 7.4   **Chebyshev's Inequality.** *Let X be any random variable and let Z be the standardized random variable of X. Then, for all positive values of r, we have*

$$Pr(-r < Z < +r) > 1 - \frac{1}{r^2}$$

Comparing Chebyshev's Inequality with the common rules, we have the following:

r	Chebyshev's Inequality	Common Rules
1	$Pr(-1 < Z < +1) > 0$	$Pr(-1 < Z < +1) \approx 0.68$
2	$Pr(-2 < Z < +2) > \frac{3}{4} = 0.75$	$Pr(-2 < Z < +2) \approx 0.95$
3	$Pr(-3 < Z < +3) > \frac{8}{9} = 0.889$	$Pr(-3 < Z < +3) \approx 0.997$

Other forms of Chebyshev's Inequality are:

$$\begin{array}{rcl} Pr(|X - \mu| < r\sigma) & > & 1 - \frac{1}{r^2} \\ Pr(-r\sigma < X - \mu < r\sigma) & > & 1 - \frac{1}{r^2} \\ Pr(|Z| \geq r) & \leq & \frac{1}{r^2} \\ Pr(|X - \mu| \geq r\sigma) & \leq & \frac{1}{r^2} \end{array}$$

In order to prove Chebyshev's Inequality, we first prove **Markov's Lemma**.[3]

Lemma   *(Markov's Lemma) If Y is a random variable which takes only positive values, then for any positive number a, we have*

$$Pr(Y \geq a) \leq \frac{E(Y)}{a}$$

---

[2] Named after P. L. Chebyshev (1821–1894), one of the founders of the Russian school of probability.
[3] A. A. Markov (1856–1922), another member of the Russian school of probability.

Proof If Y is a continuous random variable that takes only positive values, then we have $F_Y(y) = Pr(Y \leq y) = 0$ for $y < 0$. Hence, $f_Y(y) = F'_Y(y) = 0$ for $y < 0$. By definition of expected value,

$$\begin{aligned} E(Y) &= \int_{-\infty}^{\infty} yf(y)\,dy = \int_{-\infty}^{0} yf(y)\,dy + \int_{0}^{\infty} yf(y)\,dy \\ &= \int_{0}^{\infty} yf(y)\,dy \geq \int_{a}^{\infty} yf(y)\,dy \\ &\geq a \int_{a}^{\infty} f(y)\,dy = aPr(Y \geq a) \end{aligned}$$

Dividing by the positive number a results in the lemma. A proof for the discrete case is similar.

Proof of Chebyshev's Inequality Let Z be the standardized random variable of X. Then $E(Z^2) = \text{Var}(Z) = 1$. By Markov's Lemma, we have $Pr(Z^2 \geq a) \leq \frac{1}{a}$. Letting $r^2 = a$, we have

$$Pr(Z^2 \geq r^2) \leq \frac{1}{r^2}$$

or

$$Pr(Z^2 < r^2) > 1 - \frac{1}{r^2}$$

But $Z^2 < r^2$ is equivalent to $|Z| < r$, or to $-r < Z < r$. This results in

$$Pr(-r < Z < r) > 1 - \frac{1}{r^2}$$

which is Chebyshev's Inequality.

EXERCISES FOR SECTION 7.6

7.26 American women's weights have a mean of 145 lb and standard deviation of 30 lb.

   a. What do the common rules say about women's weights between 115 and 175 lb? What does Chebyshev's Inequality show about the same interval of weights?

   b. For women's weights between 85 and 205 lb?

   c. For women's weights between 55 and 235 lb?

- **7.27** Suppose IQ scores have a mean $\mu = 100$ and standard deviation $\sigma = 15$.

  a. What do the common rules say about IQ scores between 85 and 115? What does Chebyshev's Inequality say about the same interval of IQ scores?
  b. For IQ scores between 70 and 130?
  c. For IQ scores between 55 and 145?

REVIEW EXERCISES FOR CHAPTER 7

7.28 Given the probability density function

$$f(x) = x/8 \text{ for } 0 < x < 4 \text{ and } f(x) = 0 \text{ elsewhere}$$

  a. Find the cumulative distribution function $F(x)$.
  b. Compute $Pr(1 < X \leq 6)$.
  c. Find $E(X)$.   d. Find Var(X).

- **7.29** Given the probability density function

  $$f(x) = x/24 \text{ for } 1 \leq x \leq 7 \text{ and } 0 \text{ elsewhere}$$

  a. Sketch the cumulative distribution function $F(x)$.
  b. Compute $Pr(1 < X \leq 5)$
  c. Find $E(X)$.   d. Find Var(X).
  e. Find the median.   f. Find the mode.

7.30 Given the probability density function

$$f(x) = 20x^3(1-x) \text{ for } 0 < x < 1 \text{ and } 0 \text{ elsewhere}$$

find the following:

  a. The cumulative distribution function $F(x)$
  b. $E(X)$ c. $E(X^2)$ d. SD(X)
  e. Find $Pr(0.5 < X < 2)$.

- **7.31** Given the probability density function

  $$f(x) = 3x^2/8 \text{ for } 0 < x < 2 \text{ and } 0 \text{ elsewhere}$$

  find the following:

  a. The cumulative distribution function $F(x)$
  b. $Pr(X > 1)$ c. $E(X)$ d. SD(X)
  e. Compute $Pr(\mu - 2\sigma < X < \mu + 2\sigma)$.

7.32 A **continuous uniform random variable** U is one for which $f(u)$ is a constant c over an interval $[a, b]$, such as in the accident on the highway example. Because the total area under the probability density curve must be 1, it follows that $c = 1/(b-a)$; that is,

$$f(u) = \begin{cases} \frac{1}{b-a} & : \text{ for } a < u < b \\ 0 & : \text{ for } u < a \text{ and } u < b \end{cases}$$

Show that
$$F(u) = \begin{cases} 0 & : \text{ for } u \leq a \\ \frac{u-a}{b-a} & : \text{ for } a \leq u \leq b \\ 1 & : \text{ for } b \leq u \end{cases}$$

7.33 Continuing with Exercise 7.32, find a formula for the qth percentile $u_{\frac{q}{100}}$.

7.34 Continuing with Exercise 7.32, show that

a. $E(U) = \dfrac{a+b}{2}$ **b.** $\text{Var}(U) = \dfrac{(b-a)^2}{12}$

- **7.35** Consider the random variable of Exercise 7.5 on page 201.

$$f(x) = \begin{cases} x+1 & : \text{ for } -1 < x < 0 \\ -x+1 & : \text{ for } 0 < x < 1 \end{cases}$$

a. Find μ. **b.** Find $E(X^2)$. **c.** Find σ.

7.36 Consider the random variable of Exercise 7.6 on page 201.

$$f(x) = \begin{cases} e^{-x} & : \text{ for } 0 \leq x < \infty \\ 0 & : \text{ for } x < 0 \end{cases}$$

a. Find μ. **b.** Find $E(X^2)$. **c.** Find σ.

- **7.37** Two numbers are randomly chosen between 0 and 1. Let Y be the larger of the two.

  a. Follow Example 7.2 on page 199 to show that $F(y) = y^2$ for $0 < y < 1$.
  b. Find the density function.
  c. Find $Pr(0.25 < Y < 0.75)$.
  d. Compute the expected value of Y.
  e. Compute $\text{Var}(Y)$.

7.38 Let X denote a random variable with cumulative distribution function

$$F(x) = \begin{cases} 0 & : \text{ for } x \leq -1 \\ \frac{x^3}{2} + \frac{1}{2} & : \text{ for } -1 < x \leq 1 \\ 1 & : \text{ for } 1 < x \end{cases}$$

a. Sketch f and F.
b. Find $Pr(0.5 < X < 1.5)$.
c. Find $E(X)$.
d. Find $E(X^2)$.
e. Find Var(X).
f. Find SD(X).

• **7.39** Suppose that the duration of a phone call in minutes to the reservation department of Midwest Express Airlines is a random variable T with probability density function
$$f(t) = .2e^{-0.2t} \text{ for } t \geq 0 \text{ and } f(t) = 0 \text{ for } t < 0$$

a. What is the expected duration of a call?
b. What is the standard deviation?
c. What is the mode?
d. What is the median?

7.40 For a 2-year-old laptop computer, suppose that the total number of years it will be in service is a random variable X with cumulative distribution function
$$F(x) = 1 - \frac{4}{x^2} \quad \text{for } x > 2 \quad \text{and} \quad F(x) = 0 \quad \text{elsewhere}$$

a. Find an equation of the probability density function.
b. Find the probability that a 2-year-old laptop computer will be in service less than a total of 5 years.
c. Find the expected value μ.

• **7.41** The lifetime of a light bulb, measured in hours, may be represented by a continuous random variable T with cumulative distribution function
$$F(t) = \begin{cases} 0 & \text{for } t \leq 0 \\ 1 - e^{-\frac{t}{1000}} & \text{for } t > 0 \end{cases}$$

a. What is the probability of the event "$T > 1000$"?
b. Find the density function of T.
c. Show that the expected value of T is $\mu = 1000$. Hint: Use integration by parts.

7.42 Sixteen-ounce boxes of shredded wheat cereal packed automatically by machine are sometimes overweight and sometimes underweight. The actual weight in ounces over or under 16 is a random variable X whose probability density function is
$$f(x) = \tfrac{3}{32}(4 - x^2) \text{ for } -2 < x < 2 \quad \text{and} \quad 0 \text{ elsewhere}$$

Negative values pertain to ounces under 16. Find the probability that a box of cereal will be:

a. more than one ounce underweight.
b. neither underweight nor more than 1 ounce overweight.
c. exactly 16 ounces.

- **7.43** The number of minutes that a flight from Chicago to Springfield is early or late is a random variable whose probability density function is given by
  $$f(x) = \frac{36-x^2}{288} \text{ for } -6 \leq x \leq 6, \text{ and } f(x) = 0 \text{ elsewhere}$$
  Negative values are indicative of the flight being early, positive values are indicative of being late. Find the probability that one of these flights is:

  a. at least one minute late.
  b. anywhere from 1 to 3 minutes early.
  c. exactly on time.
  d. Find the mean and standard deviation.

7.44 The length of time necessary to perform a specified operation in a manufacturing plant is a random variable T. The mean and standard deviation of T have been computed to be 12 and 2 minutes, respectively. Use Chebyshev's Inequality to find a lower bound for $Pr(6 < T < 18)$.

- **7.45** The mean June daytime temperature in Stockholm is 14 degrees Celsius (C) and the variance is 50 degrees Celsius. Find the mean and variance in degrees Fahrenheit (F) if $F = (9/5)C + 32$.

* * * * * * * * *

7.46 For people with a certain deadly disease, let T be the number of years between the onset of the disease and death. The density function of such a random variable is
$$f(t) = kte^{-ct}$$
for $t \geq 0$, where c and k are positive real numbers.

a. Find the cumulative distribution function F in terms of c and k.
b. Show that $k = c^2$.
c. If 50% of the people with this disease die in 4 years, find c and k.

7.47 (Proof) Prove Theorem 7.1 on page 208.

7.48 (Proof) Prove Theorem 7.2 on page 209.

7.49 (Proof) Prove Theorem 7.3 on page 210 and the Corollary.

7.50 The coefficient of skewness of a random variable X is a measure of the lack of symmetry. The coefficient is defined to be:
$$\frac{E[(X-\mu)^3]}{\sigma^3}$$

Show that a random variable that is symmetric about μ has skewness equal to zero.

Chapter 8

NORMAL RANDOM VARIABLES

8.1 INTRODUCTION

8.2 NORMAL APPROXIMATION OF BINOMIAL

8.3 CONTINUITY CORRECTION

8.4 CENTRAL LIMIT THEOREM

8.5 PROCESSES THAT FOLLOW THE NORMAL CURVE

8.1 INTRODUCTION

Normal random variables, as introduced in Examples 7.4 on page 203 through 7.7 on page 209, have the famous bell-shaped density curves. For example, heights of adult American men follow a normal curve with mean of 175 cm and standard deviation of 6 cm. The first graph in Figure 8.1 shows this density function. Similarly, heights of adult American women follow the normal curve with mean of 162 cm and standard deviation of 5.5 cm. This density function is the second graph of Figure 8.1. Recall that, for normal curves, the mean, mode, and median coincide. Also, the inflection points occur one standard deviation on either side of the mean.

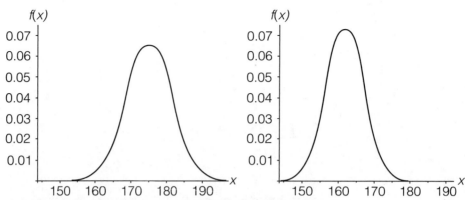

Figure 8.1 Normal density functions; for $\mu = 175$ and $\sigma = 6$ on the left, and for $\mu = 162$ and $\sigma = 5.5$ on the right

The ideal normal curves, discovered by Gauss and Newton, have the density functions given by the equation

$$f(x) = \frac{1}{\sigma\sqrt{2\pi}} e^{-\frac{1}{2}\left(\frac{x-\mu}{\sigma}\right)^2}$$

for various values of μ and σ.

Note that the normal curve is symmetric about the mode, which occurs at the mean $x = \mu$. The curve increases from $-\infty$ up to the mean $x = \mu$, and decreases from there to $+\infty$. The second derivative shows that the curve is concave up from $-\infty$ to an inflection point at $x = \mu - \sigma$. Then it turns concave down to the other inflection point at $x = \mu + \sigma$, on the other side of the maximum at $x = \mu$. After that point, it becomes concave up again to ∞. This describes the characteristic bell shape.

As shown in Examples 7.5 and 7.7, the parameters μ and σ are actually the mean and standard deviation, respectively, of the normal random variable.

All normal density curves have the same shape. Although there are many normal curves, one for each pair of values μ and σ, once you change to standard units $z = \frac{x-\mu}{\sigma}$, all become the standard normal curve

$$f_Z(z) = \frac{1}{\sqrt{2\pi}} e^{-\frac{1}{2}z^2}$$

This is accomplished by a change in the horizontal scale so that the mean occurs at $z = 0$; the inflection points occur one standard deviation away, at $z = -1$ and $z = +1$, as shown in Figure 8.2.

8.1 INTRODUCTION   221

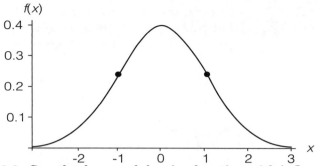

Figure 8.2 Standard normal density function with inflection points shown

We will use the symbol $\Phi(z) = F_Z(z)$ to denote the cumulative distribution function of the standard normal random variable. Unfortunately, Φ cannot be expressed as an explicit formula because it is not possible to find an antiderivative of the normal density function in terms of elementary functions. To compute the cumulative distribution Φ, we must resort to calculators, computer programs, or tables. A table of the values of Φ is given in Table 3 on pages 508–509.

The common 68%, 95%, and 99.7% rules come from the normal curve. For example, we have $\Phi(-1) = 0.1587$. This means that 15.87% of the normal curve falls to the left of $z = -1$; that is, approximately $100\% - 2(15.87\%) = 68.26\% \approx 68\%$ of the curve falls within one standard deviation of the mean. Figure 8.3 illustrates this in terms of the cumulative distribution function and the density function.

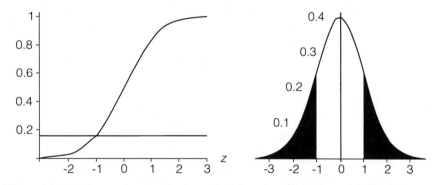

Figure 8.3 Normal cumulative distribution function $\Phi(z)$ with cut at $z = -1$ and normal density curve with middle 68% unshaded

Similarly, we have $\Phi(-2) = 0.0228$, which means that 2.28% of the normal curve falls to the left of $z = -2$; that is, approximately $100\% - 2(2.28\%) = 95.44\% \approx 95\%$ of the curve is within two standard deviations of the mean.

> *A more accurate value for the 95%-rule is $z = 1.95996 \approx 1.96$; that is, approximately 95% of the normal curve falls within 1.96 standard deviations of the mean.*

We use the notation z_q for the qth quantile (or $100q^{th}$ percentile) of the z curve, such as $z_{.025} = -1.96$ or $z_{.975} = 1.96$. For the 99.7% rule, we have $\Phi(-3) = 0.0013$, or $z_{.0013} = -3$, which means that 0.13% of the normal curve falls to the left of $z = -3$; or approximately $100\% - 2(0.13\%) = 99.74\% \approx 99.7\%$ of the curve falls within three standard deviations of the mean. See Figure 8.4.

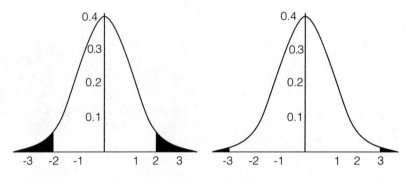

Figure 8.4 The 95% rule and the 99.7% rule

Among the first scientists to use the normal curve to describe the variability of behavioral and physiological data were L. A. J. Quetelet (1796–1884) and Francis Galton (1822–1911). Their work was followed by many other studies in the early 1900s. Pearson and Lee (1903) examined family data and found that physical measurements such as height, arm span, and length of forearm follow the normal curve.

Although the ideal normal density curve goes from $-\infty$ to $+\infty$, curves that follow the normal curve in nature have fatter and shorter tails. This makes the normal curve unreliable as a model beyond 3 σ's or so from μ.

Psychometric tests of general and special mental abilities often yield distributions that closely conform to the normal distribution. The IQ scores from the Stanford-Binet Intelligence Test follow the normal curve with a mean of 100 and a standard deviation of 15 for people in general. Data from the 1919 edition of L. M. Terman's book consist of IQ scores based on the Stanford revision of the Binet-Simon Intelligence Scale for 65 boys and 47 girls in five kindergarten classes in San Jose and San Mateo, California. They fit a normal curve with mean 104.5 and standard deviation 16.24 to Terman's data. They also found that other observed frequencies differ only slightly from the theoretical frequencies.

8.1 INTRODUCTION

Example 8.1   **Brain weight.** In 1905, Pearl[1] studied variations in brain weights of 416 Swedish men between the ages of 20 and 80. He obtained a sample mean of $\bar{x} = 1400.48$ and a sample standard deviation of $s = 106.328$. Comparing his empirical frequency table with a theoretical one derived from the normal curve with mean $\mu = 1400.48$ and standard deviation $\sigma = 106.328$, we find that the normal curve is a useful summary of these data.

Brain weight (grams)	Frequency	Theoretical
$-\infty$–1100	0	0.980456
1100–1150	1	2.86471
1150–1200	10	8.50259
1200–1250	21	20.3077
1250–1300	44	39.0336
1300–1350	53	60.3827
1350–1400	86	75.1791
1400–1450	72	75.3360
1450–1500	60	60.7614
1500–1550	28	39.4425
1550–1600	25	20.6061
1600–1650	12	8.66361
1650–1700	3	2.93117
1700–1750	1	0.797964
1750–∞	0	0.210483
Totals	416	416.0001

What proportion of brains weigh between 1300 and 1500 grams?

Let X denote a brain weight from this population. If we assume the distribution of brain weights (in grams) is normal with mean 1400.48 and standard deviation 106.328, then 1300 grams is $z = \frac{1300-1400.48}{106.328} = -0.945$ in standard units, and 1500 grams is $z = \frac{1500-1400.48}{106.328} = 0.936$ in standard units. So the probability that X is between 1300 and 1500 grams is $\Phi(0.936) - \Phi(-0.945) = 0.8254 - 0.1723 = 0.6531$. In a random sample of 416 subjects from this population, we expect $416 \times 0.6531 = 271.7$ cases to be in this interval. Pearl's data actually included 271 brain weights in this interval! This procedure was actually used to create the Theoretical column of the above table.

What is the 31st percentile of brain weights?

The 31st percentile of a standard normal variable Z can be found by using Table 3 on page 508 *backward*. The 31st percentile occurs about halfway between between $\Phi(-0.50) = 0.30854$ and $\Phi(-0.49) = 0.31207$. So the 31st percentile is $z_{.31} \approx -0.495$. Or we can use a programmable calculator such as the TI-83 where we get $invNorm(.31) = -0.4959$. We write $z_{.31} = -0.04959$.

The 31st percentile of a normal curve with mean 1400.48 and standard deviation 106.328 is $z = 0.4959$ standard deviations below the mean, which

---
[1] Described in Ingram Olkin, Leon J. Gleser, and Cyrus Derman, *Probability Models and Applications*, Macmillan Publishing Co. Inc., 1980.

What proportion of brains in the above study exceed 1600 grams?

is $1400.83 - 0.4959 \times 106.328 = 1348.1$ grams. For Pearl's data, the 31st percentile is 1350 grams.

The value $x = 1600$ grams is $z = \frac{1600 - 1400.48}{106.328} = 1.876$ in standard units. So the probability that X exceeds 1600 grams is $1 - \Phi(1.876) = 0.0303$. This means that, in a random sample of 416 from this population, we expect $416 \times 0.0303 = 12.6$ cases to exceed 1600 grams. Pearl's data included 16 brain weights in excess of 1600 grams.

> *It is not unusual for the normal curve to be a better approximation for intervals near the mean than for intervals far from the mean.*

NOTES ON TECHNOLOGY

Normal distribution

We find the area under a normal curve with $\mu = 1400.48, \sigma = 106.328$ for various intervals, as used for Example 8.1, and we find the 31st percentile.

Minitab: Minitab gives values of the cumulative distribution just as in Table 3 in the back of the book. Enter the values 1300 and 1500 in column C1. Select **Calc→Probability Distributions→Normal**. In the display boxes, enter 1400.48 for the mean, 106.318 for the standard deviation, and C1 for the input column. Be sure that Cumulative probability has been selected and then click **OK**. The two values can be subtracted to obtain the probability $0.8256 - 0.1726 = 0.6530$. To find the 31st percentile, enter .31 in column C1. Repeat **Calc→Probability Distributions→Normal** as above, except select the Inverse Cumulative Probability button. The result should show $x = 1.35\text{E}+03$, which means $x = 1350$.

Excel: Excel gives values of the cumulative distribution just as Table 3 in the back of the book. Type **=Normaldist(1500,1400.48,106.328,1)** for the value of the cumulative distribution at $x = 1500$. Repeat this for $x = 1300$, and subtract the results. To find the 31st percentile, use the function **=NormInv(.31,1400.48,106.328)**. The result should be 1347.757294.

TI-83 Plus: Use **DISTR** and select **normalcdf**. Complete the entry to show *normalcdf*$(1300, 1500, 1400.48, 106.328)$. The result should be .6530268174. This differs slightly from the value obtained using Table 3 in the back of the book for Example 8.1. If z scores are used, then the last two arguments can be omitted. Here the z-score for 1300 is -0.94500, and for 1500 it's 0.93597. So *normalcdf*$(-.94500, .93597)$

results in the same probability. To find the 31st percentile, enter the argument .31 in the **invNorm** option of **DISTR**. This results in a z-score of $invNorm(.31) = -.49580307$. The 31st percentile is $x = \mu - 0.49580307\sigma = 1400.48 - 0.49580307(106.328) = 1347.757224$. Alternatively, you can use the function $invNorm(.31, 1400.48, 106, .328) = 1347.757224$.

EXERCISES FOR SECTION 8.1

- **8.1** Consider a normal population with $\mu = 135$ and $\sigma = 5$. Convert the following values to z-scores:

    a. 142 b. 135 c. 102 d. 140 e. 130

- **8.2** Consider a normal population with $\mu = 1.6$ and $\sigma = 2.1$. Convert the following values to z-scores:

    a. 3.7 b. 1.6 c. -1.3 d. 1.8 e. 0

- **8.3** Consider a normal population with $\mu = -50$ and $\sigma = 6$. Convert the following z-scores to original units:

    a. -1 b. 0 c. 1.2 d. -2.5 e. 4.5

- **8.4** Consider a normal population with $\mu = 5.6$ and $\sigma = 1.2$. Convert the following z-scores to original units:

    a. -1.5 b. -2.1 c. 0 d. 1.2 e. 3.0

- **8.5** Calculate the area under the standard normal curve between

    a. $z = 0.00$ and $z = 1.25$ b. $z = 0.25$ and $z = 1.25$
    c. $z = -1.25$ and $z = -0.25$ d. $z = -1.25$ and $z = 0$

- **8.6** Consider a normal variable X with $\mu = 35$ and $\sigma = 20$.

    a. Find $Pr(X < 40)$ b. Find $Pr(X > 45)$
    c. Find $Pr(0 < X < 35)$ d. Find $Pr(35 < X < 45)$
    e. Find $Pr(-5 < X < 45)$ f. Find $Pr(40 < X < 45)$

- **8.7** Consider a normal variable X with $\mu = 52$ and $\sigma = 8$.

    a. Find $Pr(X < 60)$ b. Find $Pr(52 < X < 60)$
    c. Find $Pr(44 < X < 52)$ d. Find $Pr(46 < X < 60)$
    e. Find $Pr(60 < X < 62)$ f. Find $Pr(X > 62)$.

8.8 Consider a normal variable X with $\mu = 35$ and $\sigma = 20$. Find the following percentiles:

    a. 50th **b.** 90th **c.** 10th **d.** 80th **e.** 20th

- **8.9** Suppose Z is a standard normal random variable with probability density function $f(z) = \frac{1}{\sqrt{2\pi}} e^{-\frac{z^2}{2}}$, $-\infty < z < \infty$. Compute $E(Z)$, $E(Z^2)$, and $E(Z^3)$. (These are called the *first three moments* of Z.)

8.10 Verify the Theoretical value 75.3360 for the row $1400 - 1450$ of the table in Example 8.1 on page 223.

- **8.11** American women's weights follow the normal curve with mean 145 lb and standard deviation of 30 lb. Find the quartiles Q_1, $med = Q_2$, Q_3, and the interquartile range, IQR.

8.12 American women's height follow the normal curve with mean 63.5 inches and standard deviation of 2.5 inches. If the police force sets standard so the shortest 1% are rejected, what is the minimum height requirement?

- **8.13** Serum cholesterol levels in 17-year-olds follow the normal curve with mean 176 mg/dl and standard deviation of 30 mg/dl. What percentage of 17-year-olds have the following cholesterol levels?

    a. below 166   **b.** above 166   **c.** below 186

    d. above 220   **e.** above 260

8.14 Durations of pregnancies (gestation periods) in humans vary according to the normal curve with a mean of $\mu = 266$ days and a standard deviation of $\sigma = 16$ days. What percentage of pregnancies have the following durations?

    a. less than 258 days   **b.** more than 274 days   **c.** less than 255 days

    d. more than 286 days   **e.** more than 310 days

- **8.15** Continuing with Exercise 8.13, find the 25th, 90th, and 95th percentile serum cholesterol level for 17-year-olds.

8.16 Continuing with Exercise 8.14, find the 30th, 70th, and 99th percentile gestation periods for humans.

- **8.17** American men's heights follow the normal curve with mean 69 and standard deviation of 2.5 inches. If the Marine Corps accepts only men between 64 inches and 78 inches tall, what percentage are not acceptable because of height?

8.18 For a normal curve, the 90th percentile is how many standard deviations above the mean?

- **8.19** The ACT and SAT tests are used to predict success in college. Scores for college-bound high school students on the SAT follow the normal curve with mean of 500 and standard deviation of 100. Scores on the ACT follow the normal curve with mean 18 and standard deviation 6. Although the means and standard deviations of the two tests are based on different populations of college-bound students, let's assume that both tests measure the same kind of ability. Suppose that Aras scores 700 on the mathematics part of the SAT test and Rimas scores 24 on the ACT test of mathematical ability. Who did better?

8.20 Suppose that the average high temperature in July in a certain city is a normal random variable with parameters $\mu = 25$ (degrees Celsius) and $\sigma = 2$. Find the probability that on a randomly chosen day in July, the high temperature will be:

   a. above 28   **b.** below 27   **c.** between 24 and 26

8.2 NORMAL APPROXIMATION OF BINOMIAL

For fixed π, as n gets larger, binomial random variables tend to look like normal random variables, as seen by the probability histograms in Figure 8.5.

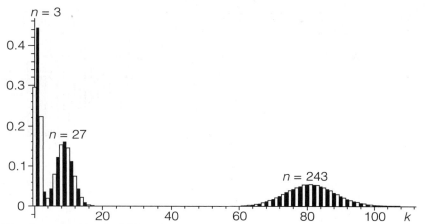

Figure 8.5 Probability histograms for binomial random variables with $\pi = \frac{1}{3}$ and n increasing from 3 to 27 to 243

The tendency is masked by the fact that, as n increases, the means $n\pi$ and standard deviations $\sqrt{n\pi(1-\pi)}$ increase as well; that is, as n tends to ∞, the probability histograms not only move to the right but also flatten out. If we change scale so as to keep the mean and standard deviation fixed—that is, if we standardize the binomial random variables—then the trend toward the normal curve is apparent. The binomial variable K for n trials will standardize to Z_n as follows:

$$Z_n = \frac{K - \mu}{\sigma} = \frac{K - n\pi}{\sqrt{n\pi(1-\pi)}}$$

Using the equations $E(aX + b) = aE(X) + b$ and $\text{SD}(aX + b) = |a|\text{SD}(X)$, we have

$$\mu_{Z_n} = E(\frac{K-\mu}{\sigma}) = \frac{1}{\sigma}E(K - \mu) = 0$$

and

$$\sigma_{Z_n} = |\frac{1}{\sigma}|\text{SD}(K - \mu) = \frac{1}{\sigma}\text{SD}(K) = 1$$

For each n, the standardized binomial variable Z_n is not itself binomial, but it has the same shape as can be seen in Figure 8.6.

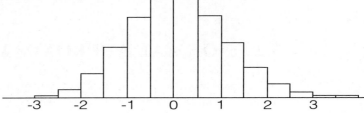

Standardized binomial random variable histogram, $n = 20$, $\pi = \frac{1}{3}$

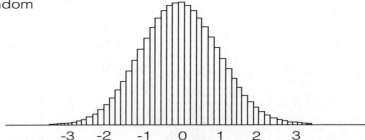

Standardized binomial random variable histogram, $n = 200$, $\pi = \frac{1}{3}$

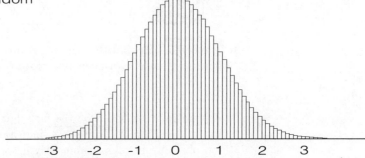

Standardized binomial random variable histogram, $n = 500$, $\pi = \frac{1}{3}$

Figure 8.6 Standardized binomial random variables with $\pi = \frac{1}{3}$ and n increasing from 20 to 200 to 500

8.2 NORMAL APPROXIMATION OF BINOMIAL

The result that the binomial random variable tends to the normal random variable is attributed to De Moivre and Laplace. Their observation can be stated as follows:

Theorem 8.1 (Central Limit Theorem for the Binomial Variable[2])
Let Z_n be the standardized binomial random variable for n trials, where π is fixed:

$$Z_n = \frac{K - \mu}{\sigma} = \frac{K - n\pi}{\sqrt{n\pi(1-\pi)}}$$

For all real numbers z, we have

$$\lim_{n \to \infty} F_{Z_n}(z) = \Phi(z)$$

where Φ is the cumulative distribution function of the standard normal random variable. In particular, for real numbers $a < b$, we have

$$\lim_{n \to \infty} Pr(a < Z_n \leq b) = \Phi(b) - \Phi(a)$$

Example 8.2 *Code errors.* Radio signals transmitted to earth from deep space probes are very weak against a noisy background. Suppose there is a 3% chance a transmitted bit will be misread. If 1000 bits are transmitted, what is the chance that no more than 35 of the bits were misread?

Solution Let K be the number of bits that are misread (here K is binomial with $n = 1000$ and $\pi = 0.03$). The exact probability of no more than 35 errors is

$$Pr(K \leq 35) = F_K(35) = \sum_{k=0}^{35} \binom{1000}{k}(0.03)^k(1 - 0.03)^{1000-k} = 0.84608$$

The value $k = 35$ has a z-score:

$$z = \frac{k - \mu}{\sigma} = \frac{k - n\pi}{\sqrt{n\pi(1-\pi)}} = \frac{35 - 30}{\sqrt{1000(0.03)(1 - 0.03)}} = 0.92688$$

We thus obtain the normal approximation

$$Pr(K \leq 35) = F_K(35) = F_{Z_n}(0.92688) \approx \Phi(0.92688) = 0.82301$$

This normal approximation is fairly close to the exact probability 0.84608.

---

[2]Historically, this is the first central limit theorem in probability. A version of this theorem appeared in Abraham De Moivre's (1667–1754) book *Doctrine of Chance*, published in 1733. The distribution function version of this result is due to Pierre S. Laplace (1749–1827) in his book *Théorie Analytique des Probabilités*, which appeared in 1812.

8.3 CONTINUITY CORRECTION

Let us consider 25 tosses of a fair coin. The binomial variable K counting the number of heads is discrete. But the Central Limit Theorem uses a continuous normal random variable to approximate it. If we want to use the normal approximation to estimate the probability K takes on a specific value—say, $K = 13$—the normal approximation breaks down. That is due to the fact that for any *continuous* random variable X, the probability it takes on a single value $Pr(X = x)$ is always zero. Yet we know that the chance of obtaining 13 heads in 25 tosses of the coin is not zero:

$$Pr(K = 13) = f_K(13) = \binom{25}{13}(0.5)^{13}(1 - 0.5)^{12} = 0.15498$$

In Figure 8.7 we have the probability histogram of K, on which is superimposed the density curve f_X of the normal random variable X, with $\mu_X = n\pi = 12.5$ and $\sigma_X = \sqrt{n\pi(1 - \pi)} = 2.5$.

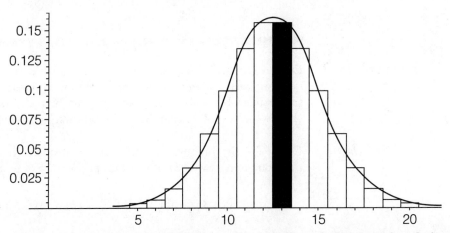

Figure 8.7 Continuity correction: Probability of $K = 13$ heads in $n = 25$ tosses of a fair coin, shaded; and a normal curve with $\mu = 12.5$ and $\sigma = 2.5$ superimposed

The exact probability of getting 13 heads is the area of the shaded rectangle. The rectangle, centered at 13, goes from 12.5 to 13.5 and has height $f_K(13) = \binom{25}{13}(0.5)^{13}(0.5)^{12} = 0.1550$.

Notice that the area of the shaded rectangle is approximately the area under the normal curve from $x = 12.5$ ($z = \frac{12.5 - 12.5}{2.5} = 0$) to $x = 13.5$ ($z = \frac{13.5 - 12.5}{2.5} = 0.40$), which $\Phi(0.40) - \Phi(0) = 0.1554$. Comparing this approximation with the true value we see that this is within 0.0004 of the exact value.

8.3 CONTINUITY CORRECTION

Going halfway between possible values of discrete random variables better approximates the area under the probability histograms. This is called the **continuity correction**.

> *General Rule.* *If we have a binomial random variable* **K** *with parameters* **n** *and* π, *and if the sample size* **n** *is so large that the expected number of successes* $\mathbf{n}\pi$ *is at least 5, and the expected number of failures* $\mathbf{n}(1-\pi)$ *is at least 5, then the normal random variable* **X** *with mean* $\mu_\mathbf{X} = \mathbf{n}\pi$ *and* $\sigma_\mathbf{X} = \sqrt{\mathbf{n}\pi(1-\pi)}$ *is considered a good approximation of the binomial random variable, provided that the continuity correction is used.*

Example 8.3 **Code errors.** Let's repeat Example 8.2 on page 229 but this time with the continuity correction. We are looking for $Pr(K \leq 35)$ in the case where $n = 1000$ and $\pi = 0.03$.

Solution The event "$K \leq 35$" is the set $\{0, 1, \ldots, 35\}$ and the complement is the set $\{36, 37, \ldots, 1000\}$. Halfway between the two sets is the number 35.5. The continuity correction is the step

$$Pr(K \leq 35) = Pr(K < 35.5)$$

> *The continuity correction always goes halfway between the included set and the excluded set. So the continuity correction for the event "***K < 35***" would be* **Pr(K < 35) = Pr(K < 34.5)**

$$\begin{aligned}
Pr(K \leq 35) &= Pr(K < 35.5) = F_K(35.5) \\
&= F_{Z_n}\left(\frac{35.5 - 30}{\sqrt{1000(0.03)(0.97)}}\right) \\
&= F_{Z_n}(0.83419) \\
&\approx \Phi(0.83419) = 0.84603
\end{aligned}$$

This normal approximation with the continuity correction is accurate to within four decimal places of the exact value 0.84608. Recall that the normal approximation without the continuity correction was 0.82301 in Example 8.2.

Example 8.4 *IQ scores.* It was noted in the last chapter that IQ scores, which are discrete, follow the normal curve with mean 100 and standard deviation 15. Using the continuity correction, we can make the following approximations:

$$Pr(IQ = 100) = Pr(99.5 < IQ < 100.5) \approx \Phi(\frac{100.5 - 100}{15}) - \Phi(\frac{99.5 - 100}{15}) = 0.0266$$

$$Pr(IQ > 105) = Pr(IQ > 105.5) \approx 1 - \Phi(\frac{105.5 - 100}{15}) = 0.3570$$

$$Pr(IQ \geq 105) = Pr(IQ > 104.5) \approx 1 - \Phi(\frac{104.5 - 100}{15}) = 0.3821$$

$$Pr(100 \leq IQ < 105) = Pr(99.5 < IQ < 104.5) \approx \Phi(\frac{104.5 - 100}{15}) - \Phi(\frac{99.5 - 100}{15}) = 0.1312$$

> *The continuity correction can be used whenever a discrete random variable is being approximated by a continuous one.*

EXERCISES FOR SECTION 8.3

- **8.21** Suppose that IQ scores follow the normal curve with a mean of 100 and standard deviation of 15. Use the continuity correction to estimate the percentage of people with IQ scores of:

    a. exactly 101   **b.** over 130

    c. at least 130   **d.** between 98 and 102, inclusively

  8.22 Use the normal approximation with continuity correction to find the probability of getting fewer than 40 heads in 100 tosses of a fair coin.

- **8.23** Use the normal approximation to find the probability of getting exactly 50 heads in 100 tosses of a fair coin.

  8.24 Use the normal approximation to find the probability of getting exactly 20 aces in 120 rolls of fair die.

- **8.25** Use the normal approximation to find the probability of getting between 8 and 10 heads, inclusive, in 15 tosses of a fair coin. Then find the exact value using binomial probabilities.

8.26 Use the normal approximation to find the probability of getting between 4 and 10 aces, inclusive, in 36 rolls of a fair die. Then find the exact value using binomial probabilities.

• **8.27** Assume that the births of boys and girls are equally likely. Use the normal approximation to approximate the probability, in 100 births, of getting

    a. at least 54 boys   **b.** more than 54 boys

8.28 Consider the blood types in the sample of Exercise 2.1. Given that 4% of all Americans have Type AB blood, find the probability of getting at least 5 people with Type AB blood in a sample of 40:

    a. using the normal approximation   **b.** using the binomial probabilities

    c. Explain why the normal approximation is not recommended in this case.

• **8.29** Consider the blood types in the sample of Exercise 2.1. Given that 39.5% of all Americans have Type A blood, find the probability of getting 11 or fewer people with Type A blood in a sample of 40:

    a. using the normal approximation   **b.** using the binomial probabilities

    c. Explain why the normal approximation is recommended in this case.

8.30 A college can admit up to 480 freshmen. The college accepts 650 students because it is known from past experience that approximately 70% of the accepted students actually enroll. What is the probability that the college ends up with more students than it can accommodate?

8.4 CENTRAL LIMIT THEOREM

The Central Limit Theorem for Binomial Variables says that sample counts K from a Bernoulli population box (that is, a box with cards of 0's and 1's) tend to follow a normal curve as the sample size n gets large. It turns out that the Central Limit Theorem can be extended to population box models that are not Bernoulli boxes. All that is required is that the numbers in the box have a finite mean μ and standard deviation σ, and that we draw a random sample of n value $X_1, X_2, \cdots X_n$ with replacement (independent draws).

Theorem 8.2 (Central Limit Theorem) *Given n independent random draws $X_1, X_2, \cdots X_n$ from a box with mean μ and nonzero standard deviation σ, let the sum of the draws be given by the random variable*

$$\text{SUM} = X_1 + X_2 + \cdots + X_n$$

Then

$$\mu_{\text{SUM}} = n\mu \quad \text{and} \quad \sigma_{\text{SUM}} = \sqrt{n}\sigma$$

Furthermore, let

$$Z_n = \frac{\text{SUM} - n\mu}{\sqrt{n}\sigma}$$

be the standardized random variable SUM. *Then, for all real numbers z, we have*[3]

$$\lim_{n \to \infty} F_{Z_n}(z) = \Phi(z)$$

8.5 PROCESSES THAT FOLLOW THE NORMAL CURVE

The Central Limit Theorem is to statistics what the Fundamental Theorem of Calculus is to calculus. Whenever a random variable is the result of the summing or the averaging of many different factors, with no dominating factor, then the probabilities tend to follow the normal curve. Because many things in nature are the result of such averaging, the normal distribution is one of the most important distributions. For example, let's consider people's heights. Most factors that determine a person's height have about the same importance. However, there are some genetic ones that predominate. One of them is the sex of the person. For *each* sex, the heights follow a normal curve as shown in the two curves of Figure 8.1. The distribution of adult heights is a mixture of these two normal distribution. The result is shown in Figure 8.8, which does not follow the normal curve.

Example 8.5   *Intelligence quotient.*   Many factors, such as general health, blood circulation, and genetic factors determine intelligence. The IQ test is designed to measure intelligence. If intelligence is the result of an additive effect of many factors, with no dominating factor, then the Central Limit Theorem argues for IQ's following a normal curve. The actual frequency polygon (Figure 8.9)[4] is slightly skewed to left because of physical conditions that result in brain damage (e.g., auto accidents, birth defects, lead poisoning). In their

---

[3] In fact, for all $-\infty < t < \infty$ and all n, we have $|F_{Z_n}(t) - F_Z(t)| \leq \frac{c}{\sqrt{n}}$, where $c = 3E(|\frac{X-\mu}{\sigma}|^3)$. This version is a Berry-Esséen Theorem as stated in William Feller, *An Introduction to Probability Theory and its Applications*, volume II, John Wiley & Sons, 1971.

book, *The Bell Curve Intelligence and Class Structure in American Life*, Richard J. Herrnstein and Charles Murray discuss the g factor, which they define as "a general capacity for inferring and applying relationships drawn from experience." If the g factor were dichotomous, and we divided the general population into two large groups based on the g factor, then IQ scores would tend to be bimodal as was the distribution of adult heights in Figure 8.8. The normal shape of the IQ distribution is evidence against the hypothesis that there are two kinds of people: those with the g factor and those without.

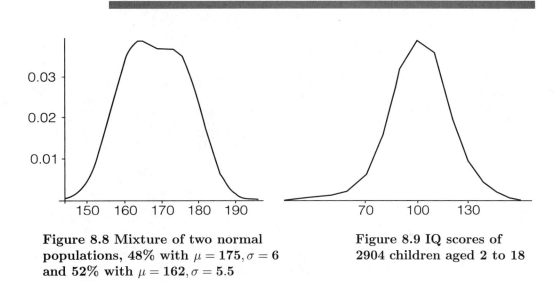

Figure 8.8 Mixture of two normal populations, 48% with $\mu = 175, \sigma = 6$ and 52% with $\mu = 162, \sigma = 5.5$

Figure 8.9 IQ scores of 2904 children aged 2 to 18

EXERCISES FOR SECTION 8.5

- **8.31** Herrnstein and Murray in the above-mentioned book describe the ongoing National Longitudinal Survey of Labor Market Experience of Youth that began in 1979. As part of the survey, the youths took the Armed Forces Qualification Test. They converted the scores on the test to the IQ metric with mean 100 and standard deviation 15 and refer to these as "IQ scores." They report that 12,686 young men and women were included in the survey, and that these "IQ scores" followed the normal curve. Find the 5th, 25th, 75th and 95th percentiles of these "IQ scores." The authors use these cut points to break the population into 5 classes:

    Very Dull   Dull   Normal   Bright   Very Bright

---

[4] Graph from J. L. Hodges Jr., David Krech & Richard S. Crutchfield, *StatLab*, McGraw-Hill, 1975.

REVIEW EXERCISES FOR CHAPTER 8

8.32 The following data are provided by the College Board APT summary report, 1979–80.[5]

SAT math scores for 1979–80 college-bound male high school students

Score	Frequency
200–249	3,423
250–299	18,434
300–349	39,913
350–399	51,603
400–449	61,691
450–499	72,186
500–549	72,804
550–599	58,304
600–649	46,910
650–699	30,265
700–749	16,246
750–799	6,414
Total	478,193

The reported mean score was 491 with standard deviation 120. Let X denote a test score from this population. Assuming the scores follow the normal with mean 491 and standard deviation 120, find the following probabilities:

   a. $Pr(400 \leq X < 600)$ b. $Pr(X \geq 700)$ c. $Pr(X = 491)$

   d. What is the 75th percentile?

   e. Follow Example 8.1 on page 223 to compute the theoretical expected frequencies based on the assumption that X is normal with mean 491 and standard deviation 120. How well do these compare with the actual frequencies of the 478,193 scores above?

• **8.33** A "three-minute" egg timer measures times that turn out to be normally distributed with $\mu = 3$ minutes and $\sigma = 0.2$.

   a. What is the probability that such an egg timer will measure a time in excess of 3.3 minutes?

   b. What is the probability that such an egg timer will measure a time accurate to within 0.1 minutes of 3 minutes?

   c. What is the first percentile time of such an egg timer?

---

[5] George C. Canavos, *Applied Probability and Statistical Methods*, Little, Brown and Company, 1984.

REVIEW EXERCISES FOR CHAPTER 8

8.34 IQ scores are normally distributed with $\mu = 100$ and a $\sigma = 15$.

    a. Find the 95th percentile IQ score.

If an individual is randomly selected, find:

    b. the probability that he or she will have an IQ score above 130.

If two individuals are chosen at random, find:

    c. the chance that both will have a score above 130.

    d. the chance that at least one will have a score above 130.

• **8.35** IQ scores are normally distributed with $\mu = 100$ and $\sigma = 15$.

    a. Find the 99th percentile IQ score.

If an individual is randomly selected, find:

    b. the probability that she will have an IQ score above 120.

If four individuals are chosen at random, find:

    c. the probability that all will have a score above 120.

    d. the probability that at least one will have a score above 120.

    e. the probability that exactly three of the four will have scores above 120.

8.36 Breakfast food packages are listed as containing 12 ounces of cereal. The filling machine is subject to errors that follow a normal curve with a standard deviation of 0.1 ounce.

    a. If the machine is set to fill a package with 12.1 ounces of cereal, what percentage of packages will be at least 0.1 ounce short of the listed 12-ounce weight?

    b. If the producer wishes to have at most 2% of the packages to have a shortage of 0.1 ounce or more (that is, weight under 11.9 ounces), what mean filling weight should be used?

• **8.37** The amount of cola put into "16-ounce" bottles by an automatic filling machine is normally distributed with a mean μ and standard deviation $\sigma = 0.05$ ounces. The value of μ can be adjusted, but σ is the same regardless of the adjustment.

    a. If the machine is adjusted so that $\mu = 16.15$, what is the probability that a randomly selected bottle has more than 16.2 ounces of soda?

    b. If the machine is adjusted so that $\mu = 16.05$, what is the probability that a randomly selected bottle is underfilled (that is, has less than 16 ounces of soda)?

    c. The machine operator wants to adjust the machine so that exactly 5% of the bottles will be underfilled. The operator should adjust μ equal to what value?

8.38 A coffee vending machine is set to fill a cup with an average of 6.4 ounces of soft drink. The amount of fill varies. Sometimes the machine overfills the cup until it overflows and sometimes it fills it under the legal minimum. Suppose the amount of fill follows the normal curve with an average of 6.4 ounces and a standard deviation of 0.3 ounces.

  a. What percentage of the times will a 7-ounce cup overflow?
  b. What percent of the times will it fill less than the legal minimum of 6 ounces?

• **8.39** A civil service exam has been designed so that the time needed to finish it will average 45 minutes. Examinations of this type in the past produced a standard deviation of 10 minutes. Assuming that the time to complete the test follows the normal curve, what percentage of the applicants taking the exam will not have finished after one hour?

8.40 Suppose that the sizes (diameters) of men's heads follow a normal curve with a mean of 7 inches and a standard deviation of 1 inch. A clothing store wants to stock hats in proportion to the sizes of customers' heads.

  a. What percentage of the customers will have head sizes between 6.5 and 7.5 inches?
  b. Ninety-nine percent of the customers will have head sizes between _____ and _____ inches.
  c. Two percent of the customers will have head sizes greater than _____ inches.

• **8.41** An insurance agent has 1000 policy holders of car insurance. Suppose that the probability of an insurance claim is 0.15 on each policy each year.

  a. Find the probability that there will be fewer than 120 claims in a given year.
  b. Find the 90th percentile for the number of claims.

8.42 An airline flies airplanes that hold 100 passengers. Typically, some 10% of the passengers with reservations do not show up for the flight. The airline generally overbooks flights in an attempt to fill them.

  a. Find the probability that a flight booked for 100 passengers will fly full.
  b. Find the probability that for a flight with 105 reservations, everyone will get a seat?
  c. If the flight is booked for 106 passengers, can the airline be at least 95% sure that everyone will get a seat?

REVIEW EXERCISES FOR CHAPTER 8

- **8.43** In a particular gubernatorial election, the incumbent won with 56% of the votes as opposed to 44% for her challenger. An exit poll is taken of a random sample of $n = 120$ voters as they leave the polling places. Assuming the voters told the truth, use the normal approximation to calculate the probability that less than a half in the sample voted for the incumbent (this means that the exit poll incorrectly predicts the winner).

- **8.44** In an election for mayor of a large city, 60% voted Democratic and 40% voted Republican. An exit poll is taken of a random sample of $n = 100$ voters as they leave the polling places. Assuming the voters told the truth, use the normal approximation to calculate the probability that the poll correctly predicts the election winner.

- **8.45** Use the normal approximation to find the probability of getting exactly 24 aces in 120 rolls of a fair die.

- **8.46** Use the normal approximation to find the probability of getting more than 15 aces in 60 rolls of a fair die.

- **8.47** Bricks are used to build a wall. The distance between the bricks is approximately 12 inches. This includes the length of the brick plus the mortar between the bricks. However, the length of bricks and the thickness of mortar vary. Thus, the distances between bricks vary with a standard deviation of 0.25 inches. Say that a section of the wall is 100 bricks long. Use the general Central Limit Theorem to find the following:

    a. the expected length of this section of the wall
    b. the standard deviation of the length of this section of the wall
    c. the probability that the wall is at least 5 inches longer than the expected length
    d. Is it necessary to assume that the distances between bricks follow the normal curve?

- **8.48** In some states, an automobile driver is legally drunk if the driver's blood alcohol concentration is at least 0.10%. A convenient test for alcohol is a breathalyzer test where the suspected driver blows into a balloon. The test is subject to error of approximately 0.004% or so. A comparison of breathalyzer results with blood results have shown that, for a given driver, breathalyzer readings X follow a normal curve with mean μ being the driver's true blood alcohol level as measured by a blood test, and $\sigma = 0.004\%$.

    a. Find the probability that a driver is declared legally drunk by a breathalyzer test when the true blood alcohol level is only 0.09%.
    b. If a breathalyzer reading shows 0.105%, how certain are you that the driver is really legally drunk?
    c. Find the breathalyzer reading needed to be 99% certain that a driver is legally drunk.

- **8.49** In the 1972 presidential election, 56% of all registered voters actually voted. A random sample of 900 registered voters were surveyed after the election. One of the questions was whether they had voted in the presidential election. It turned out that 573 of the 900 people said Yes. Use the normal approximation to estimate the probability that, in a sample of 900 registered voters, 573 or more of them had actually voted. What can explain such a low probability?

8.50 In the 2000 presidential election, 47.89% of the votes went for George W. Bush. In October 2001, in a random sample of 650 voters, 368 claimed that they had voted for Bush. What is the probability that in a sample of 650 voters, there would be 368 or more voters for Bush? What can explain such a low probability?

Chapter 9

WAITING TIME RANDOM VARIABLES

9.1 GEOMETRIC RANDOM VARIABLES

9.2 EXPONENTIAL RANDOM VARIABLES

9.3 POISSON RANDOM VARIABLES

9.4 POISSON APPROXIMATION OF BINOMIAL FOR RARE EVENTS

9.5 POISSON RANDOM VARIABLE AS INVERSE EXPONENTIAL

In this chapter, we study three random variables, all based on the concept of waiting for some event to occur.

9.1 GEOMETRIC RANDOM VARIABLES

Let us repeat independent Bernoulli trials until the first success occurs. The number of trials T, sometimes called the waiting time for a success to occur, is called a **geometric random variable**. For example, the number T of rolls of a die until the next ace appears is a geometric random variable. This is a discrete random variable with infinite values $k = 1, 2, 3, \ldots$.

Standard box model   We have a Bernoulli box of 0's and 1's. To model a geometric random variable T, draw cards with replacement until you obtain the first success. Let T be the number of draws.

Memoryless process A waiting time for repeated Bernoulli trials is a **memoryless process** in the sense that a long string of failures does not make the next ace come sooner. In symbols, for positive whole numbers s and t,

$$Pr(T > s + t \mid T > s) = Pr(T > t)$$

or

$$Pr(T > t \mid T > s) = Pr(T > t - s)$$

Every *discrete* memoryless process results is a geometric random variable.

We will now derive the probability mass function $f_T(k)$ for T, the number of rolls of a fair die until an ace comes up. You may want to draw a tree diagram to follow this example.

The box model consists of one card numbered 1 (success) and five cards numbered 0 (failure). The event "$T = 1$" means that we observe a success on the first draw. There is a 1/6 chance of this happening; thus

$$Pr(T = 1) = 1/6$$

The event "$T = 2$" means that we fail on the first draw and then succeed on the second draw:

$$Pr(T = 2) = Pr(\text{one failure followed by a success}) = (5/6)(1/6)$$

Now suppose that the first two draws fail but the third draw succeeds. Then "$T = 3$," and because the draws are independent, we have by the multiplication rule

$$Pr(T = 3) = Pr(\text{two failures followed by success}) = (5/6)^2(1/6)$$

and so on:

t	Event	$f_T(t) = Pr(T = t)$
1	$\{A\}$	$1/6$
2	$\{NA\}$	$(5/6)^1(1/6)$
3	$\{NNA\}$	$(5/6)^2(1/6)$
4	$\{NNNA\}$	$(5/6)^3(1/6)$
5	$\{NNNNA\}$	$(5/6)^4(1/6)$

$\cdot$
$\cdot$
$\cdot$

In the table above, "A" denotes getting "ace" and "N" denotes "no ace." Clearly, the probability mass function is

$$f_T(t) = (1/6)(5/6)^{t-1} \quad \text{for } t = 1, 2, 3, \ldots$$

Intuitively, in repeated draws with replacement, one gets a success about every 6th draw. This means that $E(T) = 6$.

Probability mass function for geometric random variable

In general, if π is the probability of success of the repeated Bernoulli experiment, then

$$f_T(t) = \begin{cases} (1-\pi)^{t-1}\pi & : \quad \text{for } t = 1, 2, 3, \ldots \\ 0 & : \quad \text{elsewhere} \end{cases}$$

Figures 4.3 and 4.4 on page 127 show the probability mass function and cumulative distribution function of X, the number of tails before the first head appears in repeated tosses of a fair coin. The random variable $T = X + 1$, which counts the number of tosses to obtain the first head, is a geometric random variable with $\pi = 1/2$. Figure 9.1 shows the probability histogram for the geometric random variable with $\pi = \frac{1}{6}$.

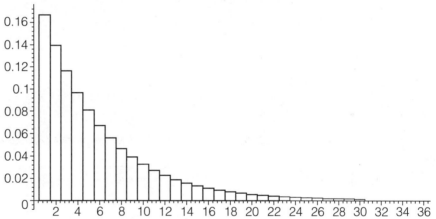

Figure 9.1 Probability histogram of geometric random variable T with $\pi = \frac{1}{6}$

Power series

In order to calculate formulas for $E(T)$ and $Var(T)$, we need some results about power series.

Let

$$s_n(x) = \sum_{k=0}^{n} x^k = 1 + x + x^2 + x^3 + \cdots + x^n$$

Then

$$x s_n(x) = x \sum_{k=0}^{n} x^k = x + x^2 + x^3 + x^4 + \cdots + x^{n+1}$$

When we subtract, the terms telescope, and we obtain

$$s_n(x) - xs_n(x) = (1 + x + x^2 + \cdots + x^n) - (x + x^2 + x^3 + \cdots + x^{n+1})$$
$$= 1 - x^{n+1}$$

or

$$s_n(x) = \frac{1 - x^{n+1}}{1 - x}$$

Letting n tend to ∞, for $|x| < 1$, we have

$$\sum_{k=0}^{\infty} x^k = \lim_{n \to \infty} s_n(x) = \lim_{n \to \infty} \frac{1 - x^{n+1}}{1 - x} = \frac{1}{1 - x} \quad \text{for } |x| < 1 \qquad (9.1)$$

If we define

$$g(x) = \sum_{k=0}^{\infty} x^k, \qquad \text{we have} \qquad g(x) = \frac{1}{1 - x}, \qquad \text{for } |x| < 1$$

Differentiating we obtain

$$g'(x) = \sum_{k=1}^{\infty} kx^{k-1} = \frac{1}{(1 - x)^2}$$

and

$$g''(x) = \sum_{k=2}^{\infty} k(k - 1)x^{k-2} = \frac{2}{(1 - x)^3}$$

THEOREM 9.1 *If T is a geometric random variable with parameter π, then*

$$\boxed{E(T) = \frac{1}{\pi}}$$

Proof Let the function g be defined as above. Then we have

$$E(T) = \sum_{t=1}^{\infty} t f_T(t) = \sum t(1 - \pi)^{t-1}\pi$$
$$= \pi \sum t(1 - \pi)^{t-1} = \pi g'(1 - \pi)$$
$$= \pi \frac{1}{\pi^2} = \frac{1}{\pi}$$

9.1 GEOMETRIC RANDOM VARIABLES

THEOREM 9.2 *If T is a geometric random variable with parameter π, then*

$$\text{Var}(T) = \frac{1-\pi}{\pi^2}$$

Proof Consider the random variable $T(T+1)$. Then

$$\begin{aligned}
E(T(T+1)) &= \sum_{t=1}^{\infty} t(t+1) f_T(t) = \sum_{t=1}^{\infty} t(t+1)(1-\pi)^{t-1}\pi \\
&= \pi \sum_{t=1}^{\infty} t(t+1)(1-\pi)^{t-1} = \pi \sum_{t=2}^{\infty} t(t-1)(1-\pi)^{t-2} \\
&= \pi g''(1-\pi) = \pi \frac{2}{\pi^3} = \frac{2}{\pi^2}
\end{aligned}$$

Or

$$E(T(T+1)) = E(T^2) + E(T) = \frac{2}{\pi^2}$$

Transposing and using Theorem 9.1, we have

$$E(T^2) = \frac{2}{\pi^2} - E(T) = \frac{2}{(\pi)^2} - \frac{1}{(\pi)} = \frac{2-\pi}{\pi^2}$$

Finally,

$$\text{Var}(T) = E(T^2) - E(T)^2 = \frac{2-\pi}{\pi^2} - \left(\frac{1}{\pi}\right)^2 = \frac{1-\pi}{\pi^2}$$

EXERCISES FOR SECTION 9.1

9.1 Let W be the number of failures until the first success; that is, $W = T - 1$. Show that W has mean $\frac{1-\pi}{\pi}$ and variance $\frac{1-\pi}{\pi^2}$.

9.2 Let W be the number of failures until the first success, that is, $W = T - 1$. Find the cumulative distribution function of W.

9.2 EXPONENTIAL RANDOM VARIABLES

In the previous section, we considered geometric random variables, which are discrete waiting times for repeated independent Bernoulli experiments such as counting the rolls of a die until an ace appears. Here we consider events that occur over continuous "time" in such a way that the waiting times between consecutive occurrences are independent; that is, knowledge of a certain waiting time between two events has no effect on the waiting time of the next event. In addition to independence, we assume the waiting times between any two events are identical random variables. Finally, we assume that the waiting time is memoryless.

Example 9.1 **Waiting for a phone call.** Consider the time between incoming phone calls midday to a receptionist in a busy office. For the sake of the example, assume that most incoming phone calls are the results of independent decisions made by many people. Let the waiting time T be the time from the end of one conversation to the ring of the next incoming phone call.

Clearly, this is a continuous random variable with possible values on the half line $t \geq 0$. Suppose that the phone has not rung for, say, five minutes. Because the calls are made independently, callers are not aware that the receptionist has been waiting for five minutes. This means that a call is no more likely to come within the next minute than during any other minute.

Such a random variable T is a **memoryless process** because it satisfies the property

$$Pr\,(T > s + t \mid T > s) = Pr\,(T > t), \text{ for all } s > 0,\, t > 0$$

as defined in the previous section, except that here the t and s need not be whole numbers.

A memoryless process of a continuous variable T is called a **Poisson process**.[1] Here are some examples.

Examples of Poisson processes

- A Geiger counter clicks every time an alpha particle is detected from an atomic disintegration in some radioactive material, say, radioactive iodine. The clicks are the result of independent atomic disintegrations (assuming no chain reaction) among a large number of atoms. Here T is the time between clicks.

- Time between oncoming cars while driving along a country road.

- Time between deaths of members of the U.S. Congress.

---
[1] Simeon Denis Poisson (1781–1840) was a French mathematician and physicist.

- Time between catastrophic air crashes.

- Space between typographical errors in the *Chicago Tribune*. Although space here is measured in number of typographical characters, which is discrete, a continuous model works well.

- Age of French children at first vaccination.[2]

Example 9.2 ***Not all waiting times are memoryless.*** The waiting time for a car to wear out is not memoryless. When looking to buy a used car, we know that a car with 70,000 miles on its odometer will generally wear out before one with 20,000 miles. In contrast, the waiting time for a click of a Geiger counter is memoryless. Whether you have waited 7 seconds or 2 seconds, will have no effect on the time of the next click.

The random variable of waiting times of a Poisson process is called an **exponential random variable**. The memoryless property

$$Pr(T > s+t \mid T > s) = Pr(T > t), \text{ for all } s > 0, t > 0$$

can be written in the equivalent form using the formula for conditional probability

$$\frac{Pr(T > s+t \text{ and } T > s)}{Pr(T > s)} = Pr(T > t), \text{ for all } s > 0, t > 0$$

Because the event "$T > s+t$ and $T > s$" is the same as "$T > s+t$," this reduces to the form

$$Pr(T > s+t) = Pr(T > s)Pr(T > t), \text{ for all } s > 0, t > 0$$

Waiting function Denoting the *waiting function* $w(t) = Pr(T > t)$, we have

$$w(s+t) = w(s)w(t), \text{ for all } s > 0, t > 0 \qquad (9.2)$$

The property (9.2) that addition $s+t$ inside a function w results in multiplication $w(s) \cdot w(t)$ outside is a basic property of exponential functions. We will state it formally as a lemma.

---

[2] G. Martin-Boyer, et al., *Situation Vaccinale en France*, Bulletin de L'Institute National de la Santé et de la Recherche Médicale **26** (1971), p. 362.

Lemma If a continuous function w is defined on the interval $[0, \infty)$ and satisfies Equation 9.2, then for some parameter $\alpha > 0$, we have

$$w(t) = \alpha^t \qquad \text{for all } t > 0$$

Proof Let $\alpha = w(1)$. Then by Equation 9.2, we have $w(2) = w(1+1) = w(1) \cdot w(1) = \alpha^2$, $w(3) = w(1+1+1) = w(1) \cdot w(1) \cdot w(1) = \alpha^3$, and so on. In general, for any whole number n, we have by repeated uses of Equation 9.2,

$$w(n) = w(1 + 1 + \cdots + 1) = w(1) \cdot w(1) \cdot \ldots \cdot w(1) = \alpha^n$$

Let r be a positive rational number $r = \frac{n}{m}$. The number n can be expressed as $\frac{n}{m}$ added to itself m times. Again using Equation 9.2 repeated times, we have $w(n) = w(\frac{n}{m} + \frac{n}{m} + \ldots + \frac{n}{m}) = w(\frac{n}{m})^m = \alpha^n$. Thus,

$$w\left(\frac{n}{m}\right) = \alpha^{\frac{n}{m}} \qquad \text{or} \qquad w(r) = \alpha^r$$

By the continuity of the function w, we can extend the result from all positive rational numbers r to all positive real numbers t.

Notice that the waiting function $w(t) = Pr(T > t)$ also satisfies the conditions

$$w(t) \text{ decreases on } [0, \infty)$$
$$\lim_{t \to \infty} w(t) = 0 \quad \text{and}$$
$$w(t) = 1 \quad \text{for } t \leq 0$$

We can take the natural log to convert the statement of the lemma to the natural base e, thereby defining a parameter θ satisfying $\alpha = e^{-\frac{1}{\theta}}$. Then we have

$$w(t) = \begin{cases} e^{-\frac{t}{\theta}} & : \quad \text{for } t > 0 \\ 1 & : \quad \text{for } t < 0 \end{cases}$$

Because $w(t) = 1 - F(t)$, there exists a parameter θ such that the cumulative distribution function F satisfies the following.

Cumulative distribution function for exponential random variable

$$F(t) = \begin{cases} 1 - e^{-\frac{t}{\theta}} & : \quad \textbf{for } t > 0 \\ 0 & : \quad \textbf{for } t < 0 \end{cases}$$

9.2 EXPONENTIAL RANDOM VARIABLES

And the density function f satisfies the following.

Density function for exponential random variable

$$f(t) = F'(t) = \begin{cases} \frac{1}{\theta} e^{-\frac{t}{\theta}}, & : \text{ for } t > 0 \\ 0 & : \text{ for } t < 0 \end{cases}$$

Figure 9.2 is a graph of the density function for an exponential random variable. It models the waiting time between clicks of a Geiger counter of the famous Rutherford and Geiger experiment mentioned in Exercise 2.48 on page 70 and Exercise 9.43 on page 265. Their mean waiting time was 1.9393 seconds.

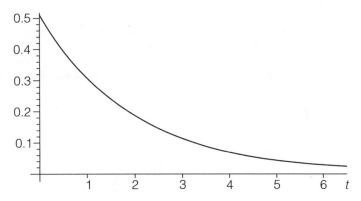

Figure 9.2 Density curve of exponential random variable with $\theta = 1.9393$

THEOREM 9.3 *If T is an exponential random variable with parameter θ, then*

$$E(T) = \theta$$

Proof We integrate by parts, using $u = t$ and $dv = \frac{1}{\theta} e^{-\frac{t}{\theta}}$.

$$\begin{aligned} E(T) &= \int_{-\infty}^{\infty} tf(t)\,dt = \frac{1}{\theta} \int_0^{\infty} t e^{-\frac{t}{\theta}}\,dt \\ &= -t e^{-\frac{t}{\theta}}\Big|_0^{\infty} + \int_0^{\infty} e^{-\frac{t}{\theta}}\,dt \\ &= -t e^{-\frac{t}{\theta}} - \theta e^{-\frac{t}{\theta}}\Big|_0^{\infty} = \theta \end{aligned}$$

We can find $E(T^2)$ by integrating by parts twice, which will lead us to the following theorem stating the variance. The proof is Exercise 9.5.

THEOREM 9.4 *If T is an exponential random variable with parameter θ, then*

$$\text{Var}(T) = \theta^2 \quad \text{and} \quad \sigma_T = \theta$$

Here are two cautionary examples of random variables of waiting times that are not memoryless.

Example 9.3 **Waiting for a train.** During rush hours, subway trains leave Washington Street station every 5 minutes. If you arrive at the station at a random time during rush hour, what is the probability that you will have to wait more than 2 minutes for a train?

Solution Although this is a waiting time, it is not an exponential random variable because it is not a memoryless process. We know that there is an upper bound of 5 minutes for the wait for the next train. Here the waiting time U is a continuous uniform random variable over the interval $[a,b] = [0,5]$ as described in Exercise 7.32.

$$F_U(u) = \begin{cases} 0 & : \quad \text{for } u < a \\ \frac{u-a}{b-a} & : \quad \text{for } a < u < b \\ 1 & : \quad \text{for } u > b \end{cases}$$

In particular, $Pr(U > 2) = 1 - F_U(2) = 1 - \frac{2-0}{5-0} = \frac{3}{5}$.

Example 9.4 **Waiting for a phone call.** Suppose the receptionist did not record the time of a call but did record the time of the one after that. What is the chance that this unrecorded call occurred in the first half of the wait until the next call? This is no longer an exponential waiting time because we know an upper bound for the wait for the first call. As in the previous example, it turns out that the conditional wait is a continuous uniform random variable on the interval between two calls, so the answer is $\frac{1}{2}$.

EXERCISES FOR SECTION 9.2

- **9.3** The time between arrivals of patients in an emergency ward follows an exponential curve with mean of 4.6 minutes. Find equations of the probability density function and the cumulative distribution function.

9.4 The distance a gas molecule travels before colliding with another molecule follows an exponential curve. For nitrogen at standard temperature and pressure, the mean distance traveled is $\mu = 0.006$ mm. Find equations of the probability density function and the cumulative distribution function.

9.5 (Proof) Prove Theorem 9.4.

9.6 (Proof) Let T be a memoryless continuous random variable and suppose
$$Pr(T > 1) = \alpha$$
Show that the mean and standard deviation of T are both equal to $-\dfrac{1}{\ln \alpha}$.

9.7 (Proof) Continuing with Exercise 9.6, show that the median is $\dfrac{\ln 0.5}{\ln \alpha}$.

9.3 POISSON RANDOM VARIABLES

A **Poisson random variable** counts the number of events that occur in a fixed unit of time of a Poisson process. Whereas an exponential random variable considers the time between events, a Poisson random variable considers the number of events occurring in a given time. A Poisson random variable is discrete, whereas an exponential random variable is continuous. Also, a Poisson random variable is infinite because there is no upper bound of the number of events in an interval.

For example, the number of Geiger counter clicks during a 15-second interval is a Poisson random variable. The time between clicks is an exponential random variable.

The Poisson random variables form an important family of random variables for another reason. They are good approximations of binomial random variables when sample sizes n are large (typically $n \geq 100$) and probabilities of success π are small (typically $n\pi < 10$). For example, they can be used to model the number of spectators who will need medical treatment at an Olympic sports event. The number of spectators n is large, the chance of any one of them needing medical treatment π is small. This approximation will be considered in Theorem 9.7, the Poisson Limit Theorem.

We will derive these interpretations after giving basic definitions and results.

Definition

> For each positive parameter λ, the Poisson random variable is a discrete random variable with probability mass functions
> $$f_Y(y) = \begin{cases} e^{-\lambda}\left(\dfrac{\lambda^y}{y!}\right) & : \text{ for } y = 0, 1, 2, \ldots \\ 0 & : \text{ elsewhere} \end{cases} \qquad (9.3)$$

Figure 9.4 shows the probability histograms for three Poisson random variables with $\lambda = 5$, 30, and 75.

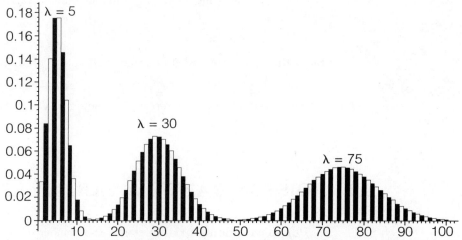

Figure 9.3 Probability histograms of three Poisson random variables with $\lambda = 5$, **30**, and **75**

Note that the histograms are skewed toward the right. As the the parameter λ gets larger, the skew tends to disappear and the histograms begin to more nearly follow a normal curve.

Recall the power series (Taylor series) expansion of the exponential function

$$e^x = 1 + x + \frac{x^2}{2!} + \cdots + \frac{x^k}{k!} + \cdots = \sum_{k=0}^{\infty} \frac{x^k}{k!} \qquad (9.4)$$

With this expansion for $x = \lambda$ and $k = y$, we can verify that

$$\sum_{y=0}^{\infty} f_Y(y) = e^{-\lambda} \sum_{y=0}^{\infty} \frac{\lambda^y}{y!} = e^{-\lambda} \cdot e^{\lambda} = 1$$

9.3 POISSON RANDOM VARIABLES

THEOREM 9.5 *If Y is a Poisson random variable with the parameter λ, then*

$$\mu = E(Y) = \lambda$$

Proof
$$\begin{aligned}
E(Y) &= \sum_{y=0}^{\infty} y f(y) = \sum_{y=0}^{\infty} y e^{-\lambda} \frac{\lambda^y}{y!} \\
&= e^{-\lambda} \sum_{y=1}^{\infty} \frac{\lambda^y}{(y-1)!} = \lambda e^{-\lambda} \sum_{y=1}^{\infty} \frac{\lambda^{y-1}}{(y-1)!} \\
&= \lambda e^{-\lambda} e^{\lambda} = \lambda
\end{aligned}$$

The last sum equals e^λ by Equation 9.4.

Lemma *If Y is a Poisson random variable with the parameter λ, then*

$$E(Y^2) = \lambda^2 + \lambda$$

Proof
$$\begin{aligned}
E(Y(Y-1)) &= \sum_{y=0}^{\infty} y(y-1) f(y) = \sum_{y=0}^{\infty} y(y-1) e^{-\lambda} \frac{\lambda^y}{y!} \\
&= e^{-\lambda} \sum_{y=2}^{\infty} \frac{\lambda^y}{(y-2)!} = \lambda^2 e^{-\lambda} \sum_{y=2}^{\infty} \frac{\lambda^{y-2}}{(y-1)!} \\
&= \lambda^2 e^{-\lambda} e^{\lambda} = \lambda^2
\end{aligned}$$

Thus, $E(Y^2) = E(Y(Y-1)) + E(Y) = \lambda^2 + \lambda$.

Combining Theorem 9.5 and its Lemma, we see that

$$\text{Var}(Y) = E(Y^2) - E^2(Y) = \{\lambda^2 + \lambda\} - \lambda^2 = \lambda$$

Hence, the following.

THEOREM 9.6 *If Y is a Poisson random variable with the parameter λ, then*

$$\text{Var}(Y) = \lambda \quad \text{and} \quad \sigma_Y = \sqrt{\lambda}$$

EXERCISES FOR SECTION 9.3

9.8 In 1996, there were 707 homicides in Los Angeles. Using the Poisson model with $\lambda = 707/365$, find the probability that on a randomly selected day in 1996 the number of homicides was: **a.** 0 **b.** 1 **c.** 2 **d.** more than 2.

• **9.9** In the past 100 years, there have been 31 earthquakes worldwide that are considered significant by the U.S. Geological Survey because of their social impact or geological framework. Using the Poisson model with $\lambda = 31/100$, find the probability that on a randomly selected year the number of such earthquakes is: **a.** 0 **b.** 1 **c.** 2 **d.** more than 2.

9.10 On weekend nights, a large urban hospital has an average of four emergency arrivals every hour. Using the Poisson model, find the probability that on a randomly selected hour, the number of arrivals is: **a.** 0 **b.** 1 **c.** 5 **d.** less than 2.

• **9.11** In 1979, the Eugene, Oregon, newspaper *Register-Guard* published 329 articles reporting homicides. Assuming a Poisson model, what is the probability that on a randomly selected day the number of articles reporting homicides was: **a.** 0 **b.** 1 **c.** at least 1.

NOTES ON TECHNOLOGY

Poisson Probabilities

This is how to use technology to find $f_Y(y)$ and $F_Y(y)$ for $\lambda = 6$ and a list of y values. This procedures are similar to finding the binomial probabilities discussed on page 154

Minitab: Enter the list of values y in column C1. Select **Calc→Probability Distributions→Poisson**. In the display boxes, enter 6 for the mean and C1 for the input column. Click **OK**. For the cumulative distribution, do the same except click the cumulative probability button in the Poisson Distribution display.

Excel: Enter the list of values y in column A. Click on cell B1 and type =**Poisson(A1,6,0)**. Copy cell B1 to the remainder of the desired list in column B. For the cumulative distribution $F_Y(y)$, enter 1 instead of 0 for the last argument of the Poisson function.

TI-83 Plus: To find the probability for a single value of y, such as $y = 10$, use the **DISTR** function. Select **poissonpdf**, complete the entry to show **poissonpdf(6,10)**, and hit **ENTER**. The value of the cumulative distribution $F_Y(10)$ can be found by using **poissoncdf(6,10)**.

9.4 POISSON APPROXIMATION OF BINOMIAL FOR RARE EVENTS

For large values of n, it is difficult to evaluate the binomial probabilities

$$f_K(k) = \binom{n}{k} \pi^k (1-\pi)^{n-k}$$

because the binomial coefficients $\binom{n}{k}$ involve multiplication and division by large numbers such as $n!$, $k!$, and $(n-k)!$.

Fortunately, in these cases we have *two* curves that can be used to approximate the binomial. In Section 8.2, we considered the normal approximation, which is useful for large $n\pi$ and $n(1-\pi)$. Here we consider the **Poisson approximation**, which is useful for small means $\mu = n\pi$.

Let K be a binomial random variable for n trials with probability of success π and expected value $E(K) = n\pi$. Let Y be the Poisson random variable with $E(Y) = \lambda = n\pi$:

$$f_Y(y) = \begin{cases} e^{-\lambda}\frac{\lambda^y}{y!} & : \quad \text{for } y = 0, 1, 2, \ldots \\ 0 & : \quad \text{elsewhere} \end{cases}$$

The Poisson Limit Theorem, which is Theorem 9.7 below, states that, for large n and small π, we have the approximation

$$f_K(k) \approx f_Y(k) \quad \text{for } \lambda = n\pi$$

General rule If $n \geq 100$ and $\lambda = n\pi < 10$, then the approximation

$$f_K(k) \approx f_Y(k)$$

is very good. And generally the approximation is excellent when $n \geq 100$ and $\lambda = n\pi \leq 5$.

Let's make a comparison.

Example 9.5 **First aid.** From previous experience of Olympic games, it is estimated that for any specific sports event one in a thousand spectators will require first aid treatment. Suppose that there are $n = 2000$ spectators for a particular event.

Let's find the probability that three spectators will require first aid treatment. Recall the probability mass function for the binomial random variable:

$$f_K(k) = \binom{n}{k} \pi^k (1-\pi)^{n-k}$$

Solution Here $n = 2000$, $\pi = 0.001$, and $k = 3$:

$$f_K(3) = \binom{2000}{3} (0.001)^3 (0.999)^{1997} = 0.18054$$

The probability mass function for the Poisson variable is

$$f_Y(k) = e^{-\lambda} \frac{\lambda^k}{k!}$$

Here $n = 2000$, $\lambda = n\pi = 2$, and $k = 3$, so the Poisson approximation is

$$f_Y(3) = e^{-2} \frac{2^3}{3!} = 0.18045$$

As can be seen from this result, the approximation is very close, and much easier to compute than the binomial.

Now let's find the probability that *at least* three spectators require first aid. The exact value of the probability is

$$\begin{aligned} Pr(K \geq 3) &= 1 - \{f_K(0) + f_K(1) + f_K(2)\} \\ &= 1 - \sum_{k=0}^{3} \binom{2000}{k} (0.001)^k (0.999)^{2000-k} = 0.32332356 \end{aligned}$$

Using the Poisson approximation, the value we obtain is

$$\begin{aligned} Pr(K \geq 3) &\approx 1 - e^{-2} \left\{ \frac{2^0}{0!} + \frac{2^1}{1!} + \frac{2^2}{2!} \right\} \\ &= 1 - e^{-2} \{1 + 2 + 2\} = 0.32332358 \end{aligned}$$

This is the same as the exact value to the first seven decimal places!

9.4 POISSON APPROXIMATION OF BINOMIAL

Recursive formula for Poisson

The Poisson probability mass functions can be computed recursively using the formulas

$$f_Y(0) = e^{-\lambda}$$
$$f_Y(k+1) = \frac{\lambda}{k+1} f_Y(k) \quad \text{for } k = 0, 1, 2, 3, \ldots \tag{9.5}$$

Recursive formula for binomial

We can also write a recursive formula for the binomial probability mass functions with $\pi = \frac{\lambda}{n}$ as follows:

$$f_K(0) = (1 - \frac{\lambda}{n})^n$$
$$f_K(k+1) = \frac{n-k-1}{n-\lambda} \frac{\lambda}{k+1} f_K(k) \quad for \; k = 0, 1, 2, 3, \ldots \tag{9.6}$$

THEOREM 9.7 **Poisson Limit Theorem.** *For a positive constant λ, let π and n vary so that $\lambda = n\pi$. Then, for $k = 0, 1, 2, \ldots$, we have*

$$\lim_{n \to \infty} f_K(k) = \lim_{n \to \infty} \binom{n}{k} \pi^k (1-\pi)^{n-k} = f_Y(k)$$

Proof First we consider the case $k = 0$. We see that

$$\lim_{n \to \infty} f_K(0) = \lim_{n \to \infty} \binom{n}{0} \pi^0 (1-\pi)^n = \lim_{n \to \infty} (1-\pi)^n$$
$$= \lim_{n \to \infty} (1 - \frac{\lambda}{n})^n = e^{-\lambda} = f_Y(0) \tag{9.7}$$

The last limit

$$\lim_{n \to \infty} (1 - \frac{\lambda}{n})^n = e^{-\lambda}$$

is one of the fundamental formulas for the exponential function.[3] We now use the recursive binomial Formula 9.6 for the case $k = 0$, followed by Equation

---

[3] One way of seeing it is to use L'Hospital's rule on $\ln(1-\frac{\lambda}{n})^n$. Another way is to take the natural logarithm and use the Taylor series expansion $\ln(1-t) = -\sum_{k=1}^{\infty} \frac{t^k}{k}$. Then $\ln(1-\frac{\lambda}{n})^n = n\ln(1-\frac{\lambda}{n}) = -n\sum_{k=1}^{\infty} \frac{\lambda^k}{kn^k} = -\lambda - \frac{\lambda^2}{2n} - \frac{\lambda^3}{3n^2} - \cdots = -\lambda - R$, where $0 \leq R = n\sum_{k=2}^{\infty} \frac{\lambda^k}{kn^k} \leq n\sum_{k=2}^{\infty} (\frac{\lambda}{n})^k = n(\frac{\lambda}{n})^2 \sum_{k=0}^{\infty} (\frac{\lambda}{n})^k = n(\frac{\lambda}{n})^2 \frac{1}{1-\frac{\lambda}{n}}$. This last equality comes from the power series formula (9.1). Then R is squeezed $0 \leq R \leq \frac{\lambda^2}{n-\lambda}$, and tends to 0 as $n \to \infty$. Thus, $\lim_{n \to \infty} \ln(1-\frac{\lambda}{n})^n = -\lambda$, or $\lim_{n \to \infty} (1-\frac{\lambda}{n})^n = e^{-\lambda}$.

9.7, and the recursive Poisson Formula 9.5 for the case $k = 1$.

$$\lim_{n \to \infty} f_K(1) = \lim_{n \to \infty} \frac{n-1}{n-\lambda} \frac{\lambda}{1} f_K(0) = \frac{\lambda}{1} f_Y(0) = f_Y(1)$$

Similarly, for $k = 1$, we have

$$\lim_{n \to \infty} f_K(2) = \lim_{n \to \infty} \frac{n-2}{n-\lambda} \frac{\lambda}{2} f_K(1) = \frac{\lambda}{1} f_Y(1) = f_Y(2)$$

and, in general,

$$\lim_{n \to \infty} f_K(k+1) = \lim_{n \to \infty} \frac{n-k-1}{n-\lambda} \frac{\lambda}{k+1} f_K(k) = \frac{\lambda}{k+1} f_Y(k) = f_Y(k+1)$$

EXERCISES FOR SECTION 9.4

9.12 Suppose that the probability that any one passenger on a subway train will be robbed is $\pi = 0.0001$. Use the Poisson approximation to find the probability that of the next 60,000 passengers:

a. three passengers will be robbed

b. that at least three will be robbed.

• **9.13** The probability is 0.003 that any one person attending a parade on a very hot day will suffer from heat exhaustion. What is the probability that 12 of the 4000 persons attending the parade will suffer from heat exhaustion? Use the Poisson approximation of the exact value.

9.14 Suppose that one prescription in every 1000 is filled in error. If a patient requires 10 prescriptions, find the probability that at least one is filled in error. Use the Poisson approximation of the exact value.

• **9.15** Consider the binomial count K with parameters $n = 10$ and $\pi = 0.5$.

a. Compute the exact value of $Pr(5 < K \leq 8)$.

b. Compute the Poisson approximation.

c. Compute the normal approximation with continuity correction.

9.16 Consider the binomial count K with parameters $n = 50$ and $\pi = 0.1$.

a. Compute the exact value of $Pr(3 < K \leq 6)$.

b. Compute the Poisson approximation.

c. Compute the normal approximation with continuity correction.

9.5 POISSON RANDOM VARIABLE AS INVERSE EXPONENTIAL

Consider the waiting times T between clicks of a Geiger counter, which follows an exponential curve with mean waiting time, say θ. The value of θ depends, of course, on the units of time that we use. For example, if we use seconds, then the value of θ will be sixty times larger than if we use minutes. Let's fix the unit of time, say 1 minute, and count the number of clicks Y that occur. We will show that Y will be a Poisson random variable with mean $\lambda = \frac{1}{\theta}$. We use the above Poisson Limit Theorem 9.7.

If we return to the first aid at an Olympic event Example 9.5 on page 256, it can be viewed in terms of waiting for an event. The event "needs first aid" occurs on average once every 1000 spectators. Picture all of the spectators arranged in a line. On the average, 1000 spectators pass between events. If we use stadiums as unit of measurement, the mean "waiting time" for an event is $\theta = \frac{1000}{2000} = \frac{1}{2}$ stadiums. After 2000 spectators (1 stadium) pass, we count the number of first aid events Y that occur. As we saw in the previous section using the Poisson Limit Theorem, Y is approximately a Poisson random variable. The mean count of events per stadium is $\lambda = \frac{1}{\theta} = 2$.

We can use the same idea for clicks of a Geiger counter except that we take the limit $n \to \infty$ to get precisely, not approximately, a Poisson random variable. For each whole number n, we break the unit interval of time $[0, 1]$ into subintervals:

$$[0, \frac{1}{n}], \ [\frac{1}{n}, \frac{2}{n}], \ [\frac{2}{n}, \frac{3}{n}], \ [\frac{3}{n}, \frac{4}{n}], \ \ldots, \ [\frac{n-1}{n}, 1]$$

Waiting one minute is the same as waiting for the n subintervals to pass. Under the assumption that the events occur independently, the number K of subintervals in which a click occurs is a binomial random variable. By doing this for increasing larger n, the Poisson Limit Theorem shows that the number of the intervals with clicks K approaches a Poisson random variable Y.

Because time is continuous, the probability of two clicks occurring at the same time is zero. So, as n increases, the subintervals become small, and the probability that any of them have more than one click approaches zero. Counting the subintervals with clicks K becomes the same as counting the number of clicks, which is the variable Y. Following this Geiger counter example, we have Theorem 9.8 below. We call the variable T a waiting time. However, a more formal statement of the theorem replaces the condition of "waiting time" by a continuity condition that, as n tends to infinity, the probability that two events occur in an interval of length $\frac{1}{n}$ tends to zero.

THEOREM 9.8 *Let T be an exponential random variable of waiting times for an event, with mean waiting time of θ units. Let Y the be count of events that occur in a unit of time $[0,1]$. Then*

$$Pr(Y = y) = f_Y(y) = e^{-\lambda}\frac{\lambda^y}{y!} \quad \text{for } y = 1, 2, 3, \ldots$$

Note that changing the units of time changes the mean waiting time and the mean for count of events, correspondingly. If the mean waiting time for a click of a Geiger counter is 1.5 seconds, then we expect 40 clicks per minute. But for a 15-second interval of time, we expect $\lambda = 10$ clicks. The probability that there are $y = 5$ clicks in a 15-second interval is thus

$$e^{-\lambda}\frac{\lambda^y}{y!} = e^{-10}\frac{10^5}{5!}$$

Similarly, the mean number of clicks per a 10-second unit of time is $\lambda = 10/1.5 = 20/3$. The probability of $y = 10$ clicks in 10 seconds is thus

$$e^{-\lambda}\frac{\lambda^y}{y!} = e^{-20/3}\frac{(20/3)^{10}}{10!}$$

Example 9.6 *Typographical errors.* The average number of errors per page in a manuscript of a certain book is thought to be two. Two proofreaders start Chapter 7 after lunch taking turns reading the book to each other. They find no errors for the first five pages. You suspect that the proofreaders are missing errors (perhaps they are drowsy after lunch). Assuming the Poisson process is an appropriate probability model, what is the chance of no errors between those pages?

Solution Letting Y be a Poisson random variable for five consecutive pages, we expect $\lambda = 10$ errors, but we observed 0 errors:

$$Pr(Y = 0) = e^{-10}\frac{10^0}{0!} = 0.0000453$$

In this case, we can also solve the problem using an exponential random variable T. The mean waiting time is $\theta = 1/\lambda = 1/2$, but we observe no errors for at least five pages. Then

$$Pr(T > 5) = e^{-\frac{5}{\theta}} = e^{-10} = 0.0000453$$

There is very strong evidence to conclude that either the proofreaders are missing errors or the assumption of a Poisson model with a mean of two errors per page is wrong.

EXERCISES FOR SECTION 9.5

- **9.17** In a statistical study of word usage, noncontextual words such as *the*, *and*, *of*, and so forth occur at different rates for different authors. A given author being investigated uses such words on the average 40% of the time. If a given passage of 50 words is examined, what is the probability of finding exactly 20 such noncontextual words?

9.18 For the Geiger counter example above, where the mean waiting time between clicks is 1.5 seconds, find the probability of:

   a. exactly 7 clicks in 30 seconds

   b. exactly 20 clicks in 30 seconds

   c. exactly 40 clicks in 60 seconds

REVIEW EXERCISES FOR CHAPTER 9

- **9.19** A fair fair coin is tossed five times. Two heads are observed. What is the conditional probability that the two heads appeared during the first three tosses of the coin?

9.20 Suppose a pair of fair dice are rolled several times. The rolls are independent, and the probability of getting a 7 on any roll is $\frac{1}{6}$. The probability histogram is shown in Figure 9.1 on page 243.

   a. What is the probability of getting three 7's in 10 rolls?

   b. What is the probability of getting the first 7 on the seventh roll?

   c. Given that there were exactly six 7's in 10 rolls, what is the conditional probability that exactly four 7's occurred during the first five rolls?

   d. If no 7's were observed during the first seven rolls, then what is the conditional probability of getting the first 7 on the fourteenth roll?

   e. Let X denote the number of 7's in 10 rolls. Find the mean of X.

   f. Let Y denote the number of rolls until the first 7 is observed. For example, $Y = 1$ if a 7 is observed on the first roll. Find the mean of Y.

9.21 Suppose the time between hits on a website is exponential with mean of 30 minutes. Find the probability that a hit is followed by another one in less than 20 minutes.

9.22 Let T have an exponential random variable with mean of 0.25.

   a. Sketch the graph of f_T.
   b. Sketch the graph of F_T.
   c. Compute $Pr(0.1 < T < 0.3)$.
   d. Compute $Pr(T > 0.5)$.
   e. Find the median.

• **9.23** Suppose T is an exponential random variable with mean 2 and variance 4.

   a. Compute the exact probability that T takes on a value more than three standard deviations above the mean.
   b. Use Chebyshev's Inequality to obtain an upper bound for this probability.

9.24 For certain heavy duty computer disk drives, the mean time to failure is a random variable whose probability function is given by the exponential random variable with a mean of 100,000 hours.

   a. Find the probability that a given disk drive will fail in the first 50,000 hours.
   b. Find the probability that of four such disk drives, at least one fails in the first 50,000 hours.

• **9.25** Suppose that the operating lifetime of a certain type of electronic device is an exponential random variable with mean of two years. Find the probability that:

   a. such a device will last over 4 years.
   b. such a device will last between 1 and 3 years.
   c. at least one out of 5 such devices will last over 4 years.

9.26 Use the Poisson approximation to estimate the probability of four or more leukemia cases in a population of size $n = 7076$ when the probability of getting leukemia is $\pi = 0.000248$.

• **9.27** An appliance repair shop receives an average of three calls per day for Maytag washing machine repairs. Assuming a Poisson process, find the probability of:

   a. no calls on a given day.
   b. three calls on a given day.
   c. waiting at least two days to receive a call.

REVIEW EXERCISES FOR CHAPTER 9

9.28 Suppose that the probability is 0.0001 that any one passenger on a subway train carries a gun. Let G be the probability that 10 of the next 50,000 passengers will have a gun.

    a. Find the binomial formula for G and evaluate it.

    b. Find the formula for the Poisson approximation of G and evaluate it.

• **9.29** In a large shipment of disposable cigarette lighters, 1% fail to meet federal safety standards. Use the Poisson approximation of the binomial to find the probability that among 200 lighters randomly selected from the shipment:

    a. exactly one will fail the federal standards.

    b. exactly two will fail the federal standards.

    c. more than one will fail the federal standards.

9.30 Suppose that the number of chocolate chips in a cookie follows the Poisson random variable with $\lambda = 4$.

    a. What is the expected number of chocolate chips per cookie?

    b. What is the standard deviation?

    c. What is the probability that a cookie will have less than three chocolate chips?

    d. What is the probability that a package of 9 cookies will have exactly 36 chips?

• **9.31** Suppose we roll a pair of fair dice 36 times counting the number of times we obtain a pair of aces (snake eyes).

    a. Use Poisson approximations to estimate the probabilities of getting zero, one, two, three, and four successes.

    b. Compare the Poisson approximations with the exact probabilities.

9.32 Suppose that on a particular stretch of a highway the time between automobile accidents is a Poisson process with an average of three accidents per week.

    a. Find the probability density function for the time T between automobile accidents.

    b. Find the probability mass function for the number of accidents Y per week.

    c. What is the probability that there is no accident for a period of two weeks?

    d. What is the probability that there is at least one accident this week?

• **9.33** Suppose the number of days to failure of a diesel locomotive has an exponential random variable with mean 43.3 days. Find the probability that there are exactly two failures in 30 days.

264   CHAPTER 9.   WAITING TIME RANDOM VARIABLES

9.34 During the morning rush, an average of 40 cars per hour go through an intersection located near an elementary school.

  a. Find the probability that during a four-minute period there will be no cars going through that intersection.
  b. Find the probability for one car.
  c. Find the probability for more than one car.

• **9.35** A restaurant finds that on weekends there are an average of five orders for lobster per day. If the restaurant has eight live lobsters on a particular Sunday, what is the probability that more than eight will be ordered by customers that day? Assume a Poisson model.

9.36 In the 1954 Salk vaccine trials, there were 33 cases of paralytic polio among 200,745 vaccinated children. What is the probability that among 10,000 vaccinated children, the number of cases of paralytic polio is:

  a. 0   b. 1   c. 2

• **9.37** An auto dealer plans to sell an average of 10 cars per week in May. Assuming the Poisson model, what is the probability that during a given week in May, there will be fewer than 5 cars sold?

9.38 In a study of automobile accidents on a heavily traveled bridge, it was found that there were an average of 5 accidents per week when the speed limit was set at 50 miles per hour. What is the probability of having fewer than 4 accidents per week for two consecutive weeks? Assume a Poisson model.

• **9.39** A hospital is expecting 500 births during the next year. Assuming the Poisson model, what is the probability that there will be more than 4 births on any given day?

9.40 Suppose that 1% of all items in a discount store are unmarked and require a price check. If there is a customer ahead of you in the express lane who has 10 items, what is the probability that you will be delayed because at least one of his items requires a price check? Use a Poisson model.

• **9.41** Ten percent of certain integrated circuits fail in the first year. Assume a Poisson process.

  a. Find the mean life of such an integrated circuit.
  b. What is the probability of such an integrated circuit failing in the second year?

9.42 Suppose that 60% of patients survive one year from the time they are first diagnosed with liver cancer. Assume a Poisson process.

  a. Find the mean survival time of such patients.
  b. What percentage of such patients die during their second year?

REVIEW EXERCISES FOR CHAPTER 9

- **9.43** The table below shows the result of the famous experiment of Rutherford and Geiger in which they observed the number Y of alpha particles emitted by a radioactive substance in $n = 2608$ periods of 7.5 seconds each:

y	f
0	57
1	203
2	383
3	525
4	532
5	408
6	273
7	139
8	45
9	27
10	16
$\sum = 2608$	

  a. Compute the mean number $\bar{y}$ of alpha particles per 7.5-second period.

  b. Use this value of $\bar{y}$ for the Poisson parameter λ to make a side-by-side table comparing the relative frequencies of the Rutherford–Geiger data, and the corresponding Poisson probabilities. Why do we expect these data to follow a Poisson random variable?

9.44 In 1898, L. Bortkiewicz studied data for the number of deaths caused by a kick of a horse from 10 Prussian cavalry corps for a period of 20 years as given in Exercise 2.46 on page 70.

Number of deaths y	Frequency f
0	109
1	65
2	22
3	3
4	1

For these data, $\bar{x} = 0.61$. If the number of deaths from the kick of a horse per cavalry corps per year followed the Poisson random variable with a mean of 0.61, find the following probabilities:

  a. no deaths from the kick by a horse during a one year period
  b. one death from the kick by a horse during a one year period
  c. more than one death from the kick by a horse during a one year period
  d. two deaths from the kick by a horse during a one year period
  e. three deaths from the kick by a horse during a one year period

f. Make a side-by-side table comparing the relative frequencies of Bortkiewicz's data as given in the table above with the probabilities found in this exercise.
   g. Why do we expect these data to follow a Poisson model?

   * * * * * * * * *

- **9.45** An airport security office spent three hours at Terminal A and five hours in Terminal B and confiscated six knives or other potential weapons. What is the conditional probability that she did not confiscate any of them in Terminal A? For this problem, suppose the number of items confiscated per hour is a Poisson process, and that there is no difference in rates between the two terminals; only the length of time affects the number of items confiscated.

  9.46 Suppose X is a continuous random variable with density function $f_X(x)$. Let $Q(x)$ be a monotone decreasing function with derivative $Q'(x) < 0$ on the interval (a, b). Suppose further that $Pr(a < X < b) = 1$. Define a new random variable $Q = Q(X)$ as the composite of Q and X. Show that the density function of Q is $f_Q(q) = -f_X(x_q)/Q'(x_q)$, where x_q is the unique solution to the equation $q = Q(x_q)$.

- **9.47** Refer to the previous Exercise 9.46. Let X be the random variable with density $f(x) = e^{-x}$ for $x \geq 0$ and $f(x) = 0$ for $x < 0$. Let the new random variable be $Q = e^{-X}$. Find the density function of Q.

Chapter 10

TWO OR MORE RANDOM VARIABLES

10.1 JOINT RANDOM VARIABLES AND DENSITIES

10.2 INDEPENDENCE OF RANDOM VARIABLES

10.3 EXPECTATION, COVARIANCE, AND CORRELATION

10.4 LINEAR COMBINATION OF RANDOM VARIABLES

10.1 JOINT RANDOM VARIABLES AND DENSITIES

So far we have considered univariate statistics; that is, we analyze single random variables (separately), such as the annual salary X of a person chosen at random from a certain community or the educational level Y of that person measured in years of school completed. However, when we want to consider how variables are related to each other—how the educational level of a person Y is related to the person's salary X—then we must deal with the two variables simultaneously as joint (or multivariate) variables (X, Y).

Some examples

- A region's rainfall X, with probability function f_X, and its wheat production Y, with probability function f_Y, are related and should be considered jointly.

- How is the high school GPA X of a student related to success in college Y?

- How is the world price of crude oil X related to inflation Y?

- In a study of photosynthesis, we may be interested in how the production of oxygen X in spinach chloroplasts is related to the concentration Y of a certain added chemical, the temperature Z, and the amount of light W.

As in the last case, we must often deal with more than two variables jointly. In this chapter, we will consider only two random variables X and Y jointly and their joint (bivariate) probability mass function $f_{XY}(x,y)$, which gives the likelihood of X and Y taking on values x and y, respectively. If the context is clear, we will use the simpler notation $f(x,y)$ instead of $f_{XY}(x,y)$.

We will consider the strength of relationships between variables. Two random variables that are not related are said to be independent. For example, suppose we roll two dice, a red one and a green one. The outcome on one die does not affect the outcome on the other. If X is the number of dots observed on the red die and Y the number on the green one, then it's clear that X and Y are independent. For variables that are related, we will define the correlation coefficient to measure the strength of the linear relationship between them. In this chapter, we will explore the mean and variance of linear combinations of random variables. In Chapter 15, we will consider statistical inference based on relationships between random variables to study, for example, how our knowledge of a girl's height at age 10 affects our estimate of her height at age 25.

In this chapter, our discussion will be restricted to discrete random variables. The study of joint discrete random variables is simpler than that of continuous ones. For two discrete variables X and Y, all joint information can be displayed in a table; whereas computations involving two or more continuous random variables require the use of partial differentiation and multiple integration. The good news is contained in the following box.

> *Even though our examples and proofs in this chapter are restricted to discrete cases, the main results of this chapter hold for both discrete and continuous random variables.*

We start with a simple discrete example.

Example 10.1 *Five tosses of a fair coin.* For five tosses of a fair coin, let X be the number of heads and let Y be the number of changes in sequence from H to T or from T to H. For example, the sequence HTTHH has $x = 3$ heads and

$y = 2$ changes in sequence. For the 32 possible outcomes of the experiment, we list the values of X and Y:

Outcome	Heads x	Changes y	Outcome	Heads x	Changes y
TTTTT	0	0	TTTTH	1	1
TTTHT	1	2	TTTHH	2	1
TTHTT	1	2	TTHTH	2	3
TTHHT	2	2	TTHHH	3	1
THTTT	1	2	THTTH	2	3
THTHT	2	4	THTHH	3	3
THHTT	2	2	THHTH	3	3
THHHT	3	2	THHHH	4	1
HTTTT	1	1	HTTTH	2	2
HTTHT	2	3	HTTHH	3	2
HTHTT	2	3	HTHTH	3	4
HTHHT	3	3	HTHHH	4	2
HHTTT	2	1	HHTTH	3	2
HHTHT	3	3	HHTHH	4	2
HHHTT	3	1	HHHTH	4	2
HHHHT	4	1	HHHHH	5	0

Considered separately, the variables X and Y have the following outcomes.

X: Number of heads

x	Event $\{X = x\}$
0	{TTTTT}
1	{TTTTH,TTTHT,TTHTT,THTTT,HTTTT}
2	{TTTHH,TTHTH,THTTH,HTTTH,TTHHT, THTHT, HTTHT,THHTT,HTHTT,HHTTT}
3	{HHHTT,HHTHT,HTHHT,THHHT,HHTTH, HTHTH, THHTH,HTTHH,THTHH,TTHHH}
4	{HHHHT,HHHTH,HHTHH,HTHHH,THHHH}
5	{HHHHH}

Y: Changes in sequence

y	Event $\{Y = y\}$
0	{TTTTT,HHHHH}
1	{TTTTH,TTTHH,TTHHH,THHHH, HTTTT,HHTTT,HHHTT,HHHHT}
2	{TTTHT,TTHTT,TTHHT,THTTT,THHTT,THHHT, HTTTH,HTTHH,HTHHH,HHTTH,HHTHH,HHHTH}
3	{TTHTH,THTTH,THTHH,THHTH, HTTHT,HTHTT,HTHHT,HHTHT}
4	{THTHT,HTHTH}

The probability mass functions are as follows:

x	$f_X(x) = Pr(X = x)$
0	1/32
1	5/32
2	10/32
3	10/32
4	5/32
5	1/32

y	$f_Y(y) = Pr(Y = y)$
0	2/32
1	8/32
2	12/32
3	8/32
4	2/32

In order to consider the discrete random variables X and Y *jointly*, the joint probability mass function must have two variables:

$$f_{XY}(x, y) = Pr(X = x \text{ and } Y = y)$$

We obtain the following values:

(x, y)	$\{X = x, Y = y\}$	$f_{XY}(x, y) = Pr(X = x \text{ and } Y = y)$
(0, 0)	{TTTTT}	1/32
(1, 0)	∅	0
(1, 1)	{TTTTH,HTTTT}	2/32
(1, 2)	{THTTT,TTHTT,TTTHT}	3/32
(2, 1)	{TTTHH, HHTTT}	2/32
.		
.		
.		

and so on.

10.1 JOINT VARIABLES AND DENSITIES

The joint (or bivariate) probability mass function can be summarized in a table as follows.

Table 10.1: Five tosses of a fair coin

	$f_{XY}(x,y)$	y=0	y=1	y=2	y=3	y=4
x	0	1/32	0	0	0	0
	1	0	2/32	3/32	0	0
	2	0	2/32	3/32	4/32	1/32
	3	0	2/32	3/32	4/32	1/32
	4	0	2/32	3/32	0	0
	5	1/32	0	0	0	0

Joint probability table

- The sum of all the entries in a joint probability table must be 1:

$$\sum_{\text{all } y} \sum_{\text{all } x} f_{XY}(x,y) = 1$$

- It is customary to show the row and column totals, which are called **marginal probabilities**. For each value of x, the row total is the value of the probability mass function $f_X(x)$

$$\sum_y f_{XY}(x,y) = f_X(x)$$

also called the **X marginal density function**.

- Similarly, for each value of y, the column total is the value of the probability mass function $f_Y(y)$

$$\sum_x f_{XY}(x,y) = f_Y(y)$$

also called the **Y marginal density function**.

So we have the following table with marginal probabilities, this time written in terms of percentages instead of fractions:

Table 10.2: Five tosses of a fair coin

	$f_{XY}(x,y)$	0	1	2	3	4	$f_X(x)$
	0	3.13	0	0	0	0	3.13
	1	0	6.25	9.38	0	0	15.63
	2	0	6.25	9.38	12.50	3.13	31.25
x	3	0	6.25	9.38	12.50	3.13	31.25
	4	0	6.25	9.38	0	0	15.63
	5	3.13	0	0	0	0	3.13
	$f_Y(y)$	6.25	25.00	37.50	25.00	6.25	100.00

(columns indexed by y)

Definition

> *As in the single variable case, we can define the cumulative distribution function as*
>
> $$\mathbf{F_{XY}(x,y) = \Pr(X \leq x \text{ and } Y \leq y)} \qquad (10.1)$$

For discrete random variables X and Y, we have[1]

$$F_{XY}(x,y) = \sum_{s \leq x} \sum_{t \leq y} f_{XY}(s,t)$$

Example 10.2   *Families of 5 children.* Now we will consider a similar discrete example, using empirical probabilities. The table below is obtained from 15,162 families of 5 (or more) children.[2] We consider the first 5 children only; their sexes are listed below. Here f denotes the frequency of each sequence, x denotes the number of girls, and y the number of changes in sequence.

---

[1] For *continuous* random variables X and Y, the cumulative distribution function is defined the same way as in the discrete case, but the density function f_{XY} is a double partial derivative

$$f_{XY}(x,y) = \frac{\partial^2}{\partial x \partial y} F_{XY}(x,y)$$

and we have

$$F_{XY}(x,y) = \int_{-\infty}^{y} \int_{-\infty}^{x} f_{XY}(s,t)\,ds\,dt$$

[2] K. O. Renkonen, O. Mäkelä, and R. Lehtovaara, *Factors affecting the human sex ratio*, Annales Medicinae Experimentalis et Biologiae Fenniae **39** (1961), pp. 173–84, and described by Bernard Rosner in his book *Fundamentals of Biostatistics*, fifth edition, PWS-Kent Publishing Company, 1999.

10.1 JOINT VARIABLES AND DENSITIES

Sequence	f	x	y	Sequence	f	x	y
MMMMM	549	0	0	MMMMF	516	1	1
MMMFM	478	1	2	MMMFF	495	2	1
MMFMM	524	1	2	MMFMF	469	2	3
MMFFM	476	2	2	MMFFF	509	3	1
MFMMM	472	1	2	MFMMF	441	2	3
MFMFM	456	2	4	MFMFF	453	3	3
MFFMM	496	2	2	MFFMF	473	3	3
MFFFM	495	3	2	MFFFF	440	4	1
FMMMM	486	1	1	FMMMF	447	2	2
FMMFM	479	2	3	FMMFF	452	3	2
FMFMM	492	2	3	FMFMF	430	3	4
FMFFM	469	3	3	FMFFF	429	4	2
FFMMM	502	2	1	FFMMF	440	3	2
FFMFM	445	3	3	FFMFF	427	4	2
FFFMM	455	3	1	FFFMF	461	4	2
FFFFM	488	4	1	FFFFF	518	5	0

This results in an empirical joint frequency table as follows:

Table 10.3: Counts of families of 5 children

	$f_{XY}(x,y)$	0	1	2	3	4	x totals
	0	549	0	0	0	0	549
	1	0	1002	1474	0	0	2476
	2	0	997	1419	1881	456	4753
x	3	0	964	1387	1840	430	4621
	4	0	928	1317	0	0	2245
	5	518	0	0	0	0	518
	y totals	1067	3891	5597	3721	886	15162

To obtain the empirical joint frequency table, we divide all entries by 15,162 to obtain the following table, written in percentages:

Table 10.4: Probabilities of families of 5 children

	$f_{XY}(x,y)$	0	1	2	3	4	$f_X(x)$
	0	3.62	0	0	0	0	3.62
	1	0	6.61	9.72	0	0	16.33
	2	0	6.58	9.36	12.41	3.01	31.35
x	3	0	6.36	9.15	12.14	2.84	30.48
	4	0	6.12	8.69	0	0	14.81
	5	3.42	0	0	0	0	3.42
	$f_Y(y)$	7.04	25.66	36.91	24.54	5.84	100.00

When comparing to Table 10.2 for 5 tosses of a fair coin shown below (copied from page 272), we notice that there are differences.

Table 10.2: Probabilities of 5 tosses of a fair coin

	$f_{XY}(x,y)$	0	1	2	3	4	$f_X(x)$
	0	3.13	0	0	0	0	3.13
	1	0	6.25	9.38	0	0	15.63
x	2	0	6.25	9.38	12.50	3.13	31.25
	3	0	6.25	9.38	12.50	3.13	31.25
	4	0	6.25	9.38	0	0	15.63
	5	3.13	0	0	0	0	3.13
	$f_Y(y)$	6.25	25.00	37.50	25.00	6.25	100.00

(with y as the column variable)

Note that values in the upper left of Table 10.4 are larger than corresponding entries of Table 10.2; and the ones in the lower right of Table 10.4 are smaller than those in Table 10.2.

There are other differences. Heads and tails are equally likely. Yet we see that 49.35% of the children are girls and 50.65% are boys. This is consistent with the findings of John Arbuthnot (as we saw in Example 6.5), who used the sign test to show that male births are more frequent that female births.

For the children, the chance of a change in sequence is 49.12%, whereas for each toss of a fair coin, the chance of a change in sequence is 50%. Indeed, for the four kinds of changes in sequence we have the following frequencies:

Table 10.5: Pairs in families of 5 children

Pair	f	$f\%$
F F	15,109	24.913%
F M	14,906	24.578%
M F	14,886	24.545%
M M	15,747	25.965%
	$\sum = 60,648$	$\sum = 100.001\%$

Is there enough evidence to show that the gender of the next child is not independent of the previous? We will return to this example in Section 16.3.

EXERCISES FOR SECTION 10.1

- **10.1** Suppose that X and Y are random variables with joint probability function given by the following table.

$f_{XY}(x,y)$	$y=2$	3	4	5
$x=0$	$\frac{1}{24}$	$\frac{3}{24}$	$\frac{1}{24}$	$\frac{1}{24}$
$x=1$	$\frac{2}{24}$	$\frac{2}{24}$	$\frac{6}{24}$	$\frac{2}{24}$
$x=2$	$\frac{2}{24}$	$\frac{1}{24}$	$\frac{2}{24}$	$\frac{1}{24}$

  a. Find $Pr(X = 1 \text{ and } Y = 3)$. b. Find $Pr(X + Y \leq 4)$.

10.2 Consider random variable X and Y whose joint probability function is

$$f_{XY}(x, y) = \frac{x + y}{30} \quad \text{for} \quad x = 0, 1, 2, 3 \text{ and } y = 0, 1, 2$$

  a. Find $Pr(X = 1 \text{ and } Y = 2)$.
  b. Find $Pr(X + Y < 2)$.
  c. Find $Pr(XY \leq 2)$.

10.2 INDEPENDENCE OF RANDOM VARIABLES

Recall that two *events* are independent if the knowledge of the occurrence of one does not change the probability of the other:

$$Pr(E \mid F) = Pr(E)$$

The formula for conditional probability, when $Pr(F) \neq 0$,

$$Pr(E \mid F) = \frac{Pr(E \text{ and } F)}{Pr(F)}$$

yields the multiplication rule:

> **Events E and F are independent if and only if**
> $$\mathbf{Pr(E \text{ and } F) = Pr(E)Pr(F).}$$

Applying this concept to discrete random variables X and Y, we say that random variables X and Y are **independent** if knowledge of the value of one random variable has no effect on the probability of the value of the other. Symbolically,
$$Pr(X = x \mid Y = y) = Pr(X = x)$$
Also, using the formula for conditional probability, when $Pr(Y = y) \neq 0$,
$$Pr(X = x \mid Y = y) = \frac{Pr(X = x \text{ and } Y = y)}{Pr(Y = y)}$$
results in the following characterization:

Discrete random variables **X** *and* **Y** *are independent if and only if, for all values* **x** *and* **y**, *we have*
$$\mathbf{Pr(X = x \text{ and } Y = y) = Pr(X = x) \cdot Pr(Y = y)}.$$

In terms of the probability functions, we have:

Random variables **X** *and* **Y** *are* **independent** *if and only if, for all values* **x** *and* **y**, *we have*
$$\mathbf{f_{XY}(x,y) = f_X(x) f_Y(y)}.$$

Recall that the probability functions $f_X(x)$ and $f_Y(y)$ are the marginal probabilities of a joint probability table. This results in the following rule:

Discrete random variables **X** *and* **Y** *are independent if and only if the joint probability table is a multiplication table of the marginal probabilities.*

In the fair coin toss Example 10.1 on page 268, we clearly do not have independence of the number of heads X and changes in sequence Y because, for example, $f_X(0) f_Y(1) = (\frac{1}{32})(\frac{2}{32}) = \frac{1}{512}$, which is *not* equal to $f_{XY}(0,1) = 0$. In games of chance, independence is often intuitively obvious.

10.2 INDEPENDENCE OF RANDOM VARIABLES

Example 10.3 *A pair of coins and a die.* We toss a pair of fair coins and roll a fair die. It is intuitively clear that the number of heads X and number of dots Y are independent. A look at the joint probability table, which is a multiplication table of the margins, confirms this.

	$f_{XY}(x,y)$	1	2	3	4	5	6	$f_X(x)$
	0	1/24	1/24	1/24	1/24	1/24	1/24	1/4
x	1	1/12	1/12	1/12	1/12	1/12	1/12	1/2
	2	1/24	1/24	1/24	1/24	1/24	1/24	1/4
	$f_Y(y)$	1/6	1/6	1/6	1/6	1/6	1/6	1

Example 10.4 *Sex and age distribution.* The U.S. Bureau of the Census figures for the resident population in Illinois as of July, 1995 (Series A projections), distributed by age and sex are summarized below (numbers in thousands):

Ages	0–4	5–17	18–24	25–64	65 and up	Totals
Females	450	1074	543	3108	892	6067
Males	471	1131	572	2997	592	5763
Totals	921	2205	1115	6105	1484	11830

Imagine selecting an Illinoisan at random. We create two random variables as follows. Let Y indicate the males; so $Y = 1$ if the person is male, and $Y = 0$ if the person is female. Let X be the lower bound of the age classifications; so if the person is 21 years of age, then $X = 18$. We can compute the joint probability table (and the marginal densities of X and Y) by dividing each entry in the above table by 11,830, the total population in thousands.

	$f_{XY}(x,y)$	0	5	18	25	65	f_X
x	0	0.03804	0.09079	0.04590	0.26272	0.07540	0.51285
	1	0.03981	0.09560	0.04835	0.25334	0.05004	0.48715
	f_Y	0.07785	0.18639	0.09425	0.51606	0.12544	1.00000

It is clear that the value of X and Y are not independent because the joint probability table is not a multiplication table of the margins. For example:

$$Pr\,(X = 65 \text{ and } Y = 0) = 0.07540$$
$$Pr\,(X = 65) \cdot Pr(Y = 0) = (0.12544) \cdot (0.51285) = 0.064332$$

Women live longer than men. Thus, the probability of getting a female when choosing from the "65 and up" age group is larger than the probability of getting a female when selecting from the general population because

$$Pr(Y = 0 \mid X = 65) = \frac{f_{X,Y}(65,0)}{f_X(65)} = \frac{0.07540}{0.12544} = 0.60108$$
$$Pr(Y = 0) = 0.51285$$

EXERCISES FOR SECTION 10.2

- **10.3** Continuing with Exercise 10.1, suppose that X and Y are random variables with joint probability function given by the following table:

$f_{XY}(x,y)$		y		
	2	3	4	5
x = 0	$\frac{1}{24}$	$\frac{3}{24}$	$\frac{1}{24}$	$\frac{1}{24}$
1	$\frac{2}{24}$	$\frac{2}{24}$	$\frac{6}{24}$	$\frac{2}{24}$
2	$\frac{2}{24}$	$\frac{1}{24}$	$\frac{2}{24}$	$\frac{1}{24}$

  a. Find $Pr(X = 1 \mid Y = 3)$.
  b. Are X and Y independent? Why?

10.4 Continuing with Exercise 10.2, consider random variable X and Y whose joint probability function is

$$f_{XY}(x,y) = \frac{x+y}{30} \quad \text{for} \quad x = 0, 1, 2, 3 \text{ and } y = 0, 1, 2$$

  a. Find $Pr(X = 2 \mid Y = 1)$.
  b. Are X and Y independent? Why?

10.3 EXPECTATION, COVARIANCE, AND CORRELATION

Consider random variables X and Y and joint probability function $f_{XY}(x,y)$. Let $g(x,y)$ be a function of two variables x and y. When we apply g to the random variables X and Y, the function g defines a random variable $W = g(X,Y)$.

10.3 EXPECTATION, COVARIANCE, & CORRELATION

Definition

> *The expected value or mean of a random variable* $W = g(X, Y)$ *is defined to be*
>
> $$\mu_W = E(W) = E(g(X, Y)) = \sum_{\text{all x}} \sum_{\text{all y}} g(x, y) f(x, y)$$

In other words, we can compute the expected value of the random variable $W = g(X, Y)$ by multiplying each entry $f(x, y)$ of the bivariate table by the number $g(x, y)$, and then adding these products.[3]

Example 10.5 **Gambling.** Suppose that there is a game in which one tosses a fair coin five times to win, in dollars, the number of heads minus the square of the number of changes in sequence. Using the random variables X and Y of the coin tossing Example 10.1, the function that computes the winnings $g(x, y) = x - y^2$ defines a random variable $W = g(X, Y) = X - Y^2$. Find the expected winnings.

Solution

$$\begin{aligned}
E(W) &= E(X - Y^2) = \sum_{\text{all } x} \left\{ \sum_{\text{all } y} (x - y^2) f(x, y) \right\} \\
&= \{(0 - 0^2)(1/32)\} + \{(1 - 1^2)(2/32) + (1 - 2^2)(3/32)\} \\
&\quad + \{(2 - 1^2)(2/32) + (2 - 2^2)(3/32) + (2 - 3^2)(4/32) + (2 - 4^2)(1/32)\} \\
&\quad + \{(3 - 1^2)(2/32) + (3 - 2^2)(3/32) + (3 - 3^2)(4/32) + (3 - 4^2)(1/32)\} \\
&\quad + \{(4 - 1^2)(2/32) + (4 - 2^2)(3/32)\} + \{(5 - 0^2)(1/32)\} \\
&= \{0/32\} + \{0/32 - 9/32\} + \{2/32 - 6/32 - 28/32 - 14/32\} \\
&\quad + \{4/32 - 3/32 - 24/32 - 13/32\} + \{6/32 + 0/32\} + \{5/32\} \\
&= \frac{0 - 9 - 46 - 36 + 6 + 5}{32} = \frac{-80}{32} = -2.50
\end{aligned}$$

We conclude that in this game you expect to lose $2.50 per play, on average.

As you see, this involves many computations. Fortunately, there are shortcuts in some cases. The following theorem shows that the expected value of

---

[3] In the continuous case, we replace the double sums by double integrals $\mu_W = E(W) = E(g(X, Y)) = \int_{-\infty}^{\infty} \int_{-\infty}^{\infty} g(x, y) f(x, y) \, dx dy$.

280   CHAPTER 10.   TWO OR MORE RANDOM VARIABLES

a sum of random variables is the sum of the expected values. For example, in five tosses of a fair coin we expect $E(X) = 2.5$ heads and $E(Y) = 2$ changes in sequence. Both can be computed from the probability functions of Example 10.1 as shown on page 270. Let $S = X + Y$ be the sum of the number of heads plus the number of changes in sequence. It is intuitive that

$$E(S) = E(X + Y) = E(X) + E(Y) = 2.5 + 2 = 4.5$$

THEOREM 10.1 Let X and Y be random variables with finite expectations. Then for any real constants a and b we have

$$E(aX + bY) = aE(X) + bE(Y)$$

Proof
$$\begin{aligned}
E(aX + bY) &= \sum_{\text{all } x}\sum_{\text{all } y}(ax + by)f_{XY}(x,y) \\
&= \sum_{\text{all } x}\left\{\sum_{\text{all } y}ax f_{XY}(x,y)\right\} + \sum_{\text{all } y}\left\{\sum_{\text{all } x}by f_{XY}(x,y)\right\} \\
&= \sum_{\text{all } x}ax\left\{\sum_{\text{all } y}f_{XY}(x,y)\right\} + \sum_{\text{all } y}by\left\{\sum_{\text{all } x}f_{XY}(x,y)\right\} \\
&= a\sum_{\text{all } x}x f_X(x) + b\sum_{\text{all } y}y f_Y(y) \\
&= aE(X) + bE(Y)
\end{aligned}$$

The result can easily be extended to a linear combination of n random variables. For a random sample of size n from a Bernoulli box, the sample count K is the sum of n Bernoulli random variables, each with expected value π. This is stated in the following corollary.

Corollary For the binomial and hypergeometric variable count K, we have $E(K) = n\mu$.

Unfortunately, Theorem 10.1 does not help in evaluating $E(Y^2)$ of Example 10.5. However, it can be evaluated separately using the probability function for Y on page 270 to obtain $E(Y^2) = 5$. This eliminates some of the computations used above to evaluate $E(W) = E(X - Y^2) = 2.5 - 5 = -2.5$.

Fortunately, for *independent* random variables there is a multiplication rule:

THEOREM 10.2   *If X and Y are* independent *random variables with defined expectations, then*

$$E(XY) = E(X)E(Y)$$

Proof   If X and Y are independent, then $f_{XY}(x,y) = f_X(x)f_Y(y)$. We have

$$\begin{aligned}
E(XY) &= \sum_{\text{all } x} \left\{ \sum_{\text{all } y} xy f_{XY}(x,y) \right\} \\
&= \sum_{\text{all } x} \left\{ \sum_{\text{all } y} xy f_X(x) f_Y(y) \right\} \\
&= \sum_{\text{all } x} x f_X(x) \left\{ \sum_{\text{all } y} y f_Y(y) \right\} \\
&= \sum_{\text{all } x} x f_X(x) E(Y) \\
&= \left\{ \sum_{\text{all } x} x f_X(x) \right\} E(Y) = E(X)E(Y)
\end{aligned}$$

This result leads to a measure of the strength of a linear relationship between X and Y. The **correlation** between the random variables X and Y is defined as follows:

Let

$$Z_X = \frac{X - \mu_X}{\sigma_X} \text{ and } Z_Y = \frac{Y - \mu_Y}{\sigma_Y}$$

be the standardized random variables of X and Y, respectively. Recall that the standardized random variables Z_X and Z_Y describe X and Y in terms of standard units; that is, for each variable, the change of scale moves the

origin to the mean, and a unit of measure becomes a standard deviation:
$E(Z_X) = 0$, $\mathrm{SD}(Z_X) = 1$, $E(Z_Y) = 0$, $\mathrm{SD}(Z_Y) = 1$.

Definition

> *The correlation coefficient of* **X** *and* **Y** *is defined to be*
>
> $$\rho_{\mathbf{XY}} = \mathrm{Corr}(\mathbf{X}, \mathbf{Y}) = E(\mathbf{Z_X} \cdot \mathbf{Z_Y})$$
>
> *provided the expected value exists.*

THEOREM 10.3 *If X and Y are independent random variables, then $\mathrm{Corr}(X,Y) = 0$, whenever it exists.*

Proof If X and Y are independent, then so are the standardized random variables Z_X and Z_Y. By Theorem 10.2, if X and Y are independent, then $E(Z_X Z_Y) = E(Z_X)E(Z_Y) = 0 \cdot 0 = 0$, and hence $\mathrm{Corr}(X,Y) = 0$.

At the opposite extreme from independence, X and Y may be **functionally** or **deterministically related**. For example, if, for some constants m and b, we have $Y = mX + b$, then X and Y are said to be **linearly related**. In this case, $Z_X = Z_Y$ if the slope m is positive; or $Z_X = -Z_Y$ if the slope m is negative. This results in $\mathrm{Corr}(X,Y) = +1$ or $\mathrm{Corr}(X,Y) = -1$, respectively. If X and Y are not deterministically related, we say they are **statistically** or **stochastically** related.

> *The correlation coefficient ρ measures how close* **X** *and* **Y** *are to being linearly related.*

- If $\mathrm{Corr}(X,Y) > 0$, we say that X and Y are **positively correlated**.

- If $\mathrm{Corr}(X,Y) < 0$, we say that X and Y are **negatively correlated**.

- If $\mathrm{Corr}(X,Y) = 0$, we say that X and Y are **uncorrelated**.

- If $\mathrm{Corr}(X,Y) = +1$ or -1, then X and Y are linearly related; that is, $Y = mX + b$ for some constants m and b.

10.3 EXPECTATION, COVARIANCE, & CORRELATION

Warning There are other functional relationships between X and Y that result in $\text{Corr}(X,Y) = 0$. So the converse of Theorem 10.3 is not true.

Even statistically related X and Y can be uncorrelated without being independent. Indeed, in Example 10.1, we have noted that the number of heads is not independent of the number of changes in sequence. Yet if we work out the formula, we see that $\text{Corr}(X,Y) = 0$.

Definition

> The covariance of the random variables **X** and **Y** is, defined as follows:
>
> $$\sigma_{\mathbf{XY}} = \text{Cov}(\mathbf{X},\mathbf{Y}) = \mathbf{E}\{(\mathbf{X} - \mu_{\mathbf{X}}) \cdot (\mathbf{Y} - \mu_{\mathbf{Y}})\}$$

Because $Z_X = \frac{X - \mu_X}{\sigma_X}$ and $Z_Y = \frac{Y - \mu_Y}{\sigma_Y}$, this leads immediately to the following theorem.

THEOREM 10.4 For random variables X and Y, we have

$$\text{Corr}(X,Y) = \frac{\text{Cov}(X,Y)}{\sigma_X \sigma_Y}$$

The proof of the following theorem is left as Exercise 10.7.

THEOREM 10.5 For random variables X and Y, we have

$$\text{Cov}(X,Y) = E(XY) - E(X)E(Y) = E(XY) - \mu_X \mu_Y$$

Example 10.6 **Two tosses of a fair coin.** Toss a fair coin twice. Let X be the number of heads on the first toss and Y be the number of heads on the second toss. The probability table is below.

	$f_{XY}(x,y)$	y = 0	y = 1	$f_X(x)$
x	0	0.25	0.25	0.50
	1	0.25	0.25	0.50
	$f_Y(y)$	0.50	0.50	1.00

We have $E(X + Y) = 0.5 + 0.5 = 1$.

Because X and Y are independent, we have $E(XY) = 0.5 \cdot 0.5 = 0.25$.

Also, $\text{Cov}(X, Y) = 0$.

EXERCISES FOR SECTION 10.3

- **10.5** Show that the random variables X and Y of Example 10.1 on page 268 are not independent. Show, however, that they are uncorrelated.

10.6 Let X be a Bernoulli random variable with probability of success π and let $Y = 1 - X$. Note that X is the function that indicates a success; Y indicates a failure. Make a joint probability table. Find the covariance and the correlation coefficient.

10.7 (Proof) Prove Theorem 10.5.

10.4 LINEAR COMBINATION OF RANDOM VARIABLES

As was noted in Theorem 10.1, we have the relationship

$$E(aX + bY) = aE(X) + bE(Y)$$

Also, because we know that for any constants c we have $E(X+c) = E(X)+c$, it follows that:

> **For any random variables X and Y and any constants a, b, c, we have $\mathbf{E(aX + bY + c) = aE(X) + bE(Y) + c}$.**

This may be extended to any number of random variables as follows.

THEOREM 10.6 If $X_1, X_2, \ldots X_n$ are random variables, and if $a_1, a_2, \ldots, a_n, c$ are real numbers, then

$$E(a_1 X_1 + a_2 X_2 + \cdots + a_n X_n + c) = a_1 E(X_1) + a_2 E(X_2) + \cdots + a_n E(X_n) + c$$

10.4 LINEAR COMBINATION OF VARIABLES

The variance of a sum of two random variables follows the formula given by the following theorem.

THEOREM 10.7 *For any random variables X and Y, we have*

$$\text{Var}(X+Y) = \text{Var}(X) + 2\text{Cov}(X,Y) + \text{Var}(Y)$$

Proof By the variance formula Theorem 4.2, we have

$$\text{Var}(X+Y) = E\left((X+Y)^2\right) - E^2(X+Y)$$

Expanding $(X+Y)^2 = X^2 + 2XY + Y^2$ results in

$$E(X+Y)^2 = E(X^2 + 2XY + Y^2) = E(X^2) + 2E(XY) + E(Y^2)$$

Similarly,

$$E^2(X+Y) = \{E(X) + E(Y)\}^2 = E^2(X) + 2E(X)E(Y) + E^2(Y)$$

Subtracting the two, we have

$$\begin{aligned}
\text{Var}(X+Y) &= \{E(X^2) + 2E(XY) + E(Y^2)\} - \{E^2(X) + 2E(X)E(Y) + E^2(Y)\} \\
&= \{E(X^2) - E^2(X)\} + 2\{E(XY) - E(X)E(Y)\} + \{E(Y^2) - E^2(Y)\} \\
&= \text{Var}(X) + 2\{E(XY) - E(X)E(Y)\} + \text{Var}(Y)
\end{aligned}$$

Finally, by Theorem 10.5, the middle term of this expression is $2\text{Cov}(X,Y)$, which gives us the statement of the theorem.

Using Theorems 4.3, this can easily be extended to

$$\text{Var}(aX + bY + c) = a^2\text{Var}(X) + 2ab\text{Cov}(X,Y) + b^2\text{Var}(Y)$$

Theorem 10.7 can also be extended to more than two random variables as follows.

> **If** $a_1, a_2, a_3, \ldots, a_n$ **are real numbers and** $X_1, X_2, X_3, \ldots, X_n$ **are random variables, then**
> $$\operatorname{Var}\left(\sum_{j=1}^n a_j X_j\right) = \sum_{j=1}^n a_j^2 \operatorname{Var}(X_j) + 2 \sum_{1 \leq i < j \leq n} a_i a_j \operatorname{Cov}(X_i, X_j)$$

For independent random variables, the covariances are all zero. This yields the following important result.

THEOREM 10.8 If $X_1, X_2, X_3, \ldots X_n$ are independent random variables, and $a_1, a_2, \ldots, a_n, c$ are real constants, then

$$\operatorname{Var}(a_1 X_1 + a_2 X_2 + \cdots + a_n X_n + c) = a_1^2 \operatorname{Var}(X_1) + a_2^2 \operatorname{Var}(X_2) + \cdots + a_n^2 \operatorname{Var}(X_n)$$

Recall that for a sample of size n from a Bernoulli box, the sample count K is the sum of n independent Bernoulli random variables, each with variance $\pi(1-\pi)$. The corollary follows.

Corollary For the binomial variable count K, we have $\operatorname{Var}(K) = n\pi(1-\pi)$.

Independence of random variables is analogous to orthogonality of lines in Euclidean geometry: If X and Y are independent random variables, then

$$\operatorname{SD}^2(X+Y) = \operatorname{SD}^2(X) + \operatorname{SD}^2(Y)$$

Following the same analogy, Theorem 10.7 is like the law of cosines.

For differences of random variables, we have the following results:

$$E(X - Y) = E(X) - E(Y)$$

And because $\operatorname{Var}(-Y) = (-1)^2 \operatorname{Var}(Y) = \operatorname{Var}(Y)$, we have

$$\operatorname{Var}(X - Y) = \operatorname{Var}(X) - 2\operatorname{Cov}(X, Y) + \operatorname{Var}(Y)$$

> **For any independent random variables X and Y, we have**
> $$\operatorname{Var}(X - Y) = \operatorname{Var}(X + Y) = \operatorname{Var}(X) + \operatorname{Var}(Y)$$

EXERCISES FOR SECTION 10.4

10.8 Compute $\text{Cov}(X, Y)$ of the random variables X and Y described in Exercise 10.2 on page 275:

$$f_{XY}(x, y) = \frac{x+y}{30} \quad \text{for} \quad x = 0, 1, 2, 3 \text{ and } y = 0, 1, 2$$

Then compute the mean and variance of $X - Y$.

- **10.9** Let X be the binomial count for n trials with probability of success π and let $Y = n - X$. Note that X counts the number of successes, and Y counts the number of failures. Use Theorem 10.7 to find the covariance and the correlation coefficient. This is an extension of Exercise 10.6.

10.10 In Example 10.3 on page 277, we tossed a pair of fair coins and rolled a fair die. Let $X + Y$ be the sum of the number of heads and the number of dots.

   a. Find the expected value and variance of $X + Y$.
   b. Is $\rho(X, Y) = 0$?

- **10.11** In Example 10.1 on page 268, we tossed a fair coin five times. Let $X + Y$ be the sum of the number of heads and the number of changes of sequence.

   a. Find the expected value and variance of this sum.
   b. Is $\rho(X, Y) = 0$?

10.12 Consider discrete random variables X and Y that take on values $-1, 0,$ and 1, and have the following joint probability function:

$f_{XY}(x,y)$	-1	0	1
-1	$\frac{1}{8}$	$\frac{1}{8}$	$\frac{1}{8}$
0	$\frac{1}{8}$	0	$\frac{1}{8}$
1	$\frac{1}{8}$	$\frac{1}{8}$	$\frac{1}{8}$

(with y across top, x down the side)

   a. Show that X and Y have identical marginal density functions.
   b. Show that X and Y are dependent random variables.
   c. Show that X and Y are uncorrelated.
   d. Show $\text{Var}(X + Y) = \text{Var}(X - Y)$

10.13 For the random variables X and Y described in Example 10.4 on page 277, show that the correlation coefficient is negative but close to zero.

10.14 A box contains 10 cards of which six are labeled 1 and four are labeled 0. A sample of 10 cards is drawn at random from the box *with replacement*. Let

$$X_j = \begin{cases} 1 \text{ if the } j^{th} \text{ card drawn is marked } 1 \\ 0 \text{ if the } j^{th} \text{ card drawn is marked } 0 \end{cases}, \quad \text{for } j = 1, \ldots 10$$

Let $K = X_1 + X_2 + \ldots + X_{10}$. The X's are Bernoulli random variables, and K is binomial. Use the formula in Theorem 10.1 on page 280 to compute the expected value of K. Use the formula in Theorem 10.8 on page 286 to compute the variance of K.

- **10.15** A box contains 10 cards of which six are labeled 1 and four are labeled 0. A sample of 5 cards is drawn at random from the box *without replacement*. Let

$$X_j = \begin{cases} 1 \text{ if the } j^{th} \text{ card drawn is marked } 1 \\ 0 \text{ if the } j^{th} \text{ card drawn is marked } 0 \end{cases}, \text{ for } j = 1, \ldots 5$$

Let $K = X_1 + X_2 + \ldots + X_5$. The X's are Bernoulli random variables and K is a hypergeometric random variable with $N = 10$, $N_1 = 6$, and $n = 5$. Use the formula in Theorem 10.1 on page 280 to compute the expected value of K. The joint probability function of X_1 and X_2 is

	$f_{X_1 X_2}(x_1, x_2)$	$x_2 = 0$	$x_2 = 1$	f_{X_1}
x_1	0	$\frac{2}{15}$	$\frac{4}{15}$	$\frac{4}{10}$
	1	$\frac{4}{15}$	$\frac{5}{15}$	$\frac{6}{10}$
	f_{X_2}	$\frac{4}{10}$	$\frac{6}{10}$	1

a. Compute the covariance of X_1 and X_2.
b. Use the formula in Theorem 10.7 to compute the variance of $X_1 + X_2$.
c. Noting that the covariance of X_i and X_j, for distinct i and j, is the same as the covariance of X_1 and X_2, compute the variance of K.

* * * * * * * * *

10.16 (Proof) Generalize Exercise 10.15 to find the variance of a hypergeometric random variable (Equation 5.8 on page 157) for any choice of N, N_1, and n.

REVIEW EXERCISES FOR CHAPTER 10

- **10.17** In a community of working couples, the husband's income X and the wife's income Y, in thousands of dollars, have the following joint probability function f defined by the values below. Construct a joint probability table with marginal probabilities.

$$f(10, 5) = 0.20 \quad f(10, 10) = 0.05 \quad f(10, 15) = 0.02$$
$$f(15, 5) = 0.10 \quad f(15, 10) = 0.20 \quad f(15, 15) = 0.04$$
$$f(20, 5) = 0.05 \quad f(20, 10) = 0.30 \quad f(20, 15) = 0.04$$

a. Find $E(X+Y)$.   b. Find $E(2X-3Y)$.   c. Find $Pr(18 < X+Y < 27)$.

REVIEW EXERCISES FOR CHAPTER 10

10.18 For the joint probability table of Exercise 10.1 on page 275.
  a. Find $E(X^2)$. b. Find $E(Y^2)$. c. Find $E(XY)$.
  d. Find Cov(X,Y). e. Find Corr(X,Y).
  f. Are X and Y independent? Why?

• **10.19** For the joint probability table of Exercise 10.17, find:
  a. Find $E(X^2)$. b. Find $E(Y^2)$. c. Find $E(XY)$.
  d. Find Cov(X,Y). e. Find Corr(X,Y).
  f. Are X and Y independent? Why?

10.20 Let X_1, X_2, and X_3 be independent random variables with means $-2, 0, 3$ and variances $4, 2, 1$, respectively. Let $S = X_1 + X_2 + X_3$ and $T = 2X_1 - 3X_2 - X_3 + 4$.

  a. Find $E(S)$. b. Find Var(S). c. Find $E(T)$. d. Find Var(T).

• **10.21** Suppose you have a cup of pills containing one aspirin, two sleeping tablets, and three laxatives. You choose three tablets randomly (without replacement). Let X be the number of laxatives and Z the number of sleeping tablets in the sample. Make a joint probability table with marginal probabilities. Compute the following:

  a. $E(X+Z)$ b. $E(X \cdot Z)$ c. Cov(X,Z) d. Var$(X+Z)$

10.22 A store has a regular checkout line and an express checkout line. Let X be the number of customers in the regular line and Y the number in the express line:

$f_{XY}(x,y)$	0	1	2	3
0	0.03	0.07	0.00	0.00
1	0.10	0.10	0.08	0.02
2	0.05	0.10	0.10	0.05
3	0.00	0.10	0.10	0.10

  a. Find $E(X+Y)$. b. Find Var$(X+Y)$. c. Find Cov(X,Y).
  d. Find $P(X>Y)$. e. Find $E(X-Y)$. f. Find Var$(X+Y)$.

• **10.23** The personnel department of a large corporation gives two aptitude tests to job applicants. One measures verbal ability; the other, quantitative ability. From many years' experience, the company found that the verbal scores tend to be normally distributed with a mean of 50 and standard deviation of 10. The quantitative scores tend to be normally distributed with a mean of 100 and standard deviation of 20. The covariance between the two scores is 75. A composite score C is assigned to each applicant, where

$$C = 3(\text{verbal score}) + 2(\text{quantitative score})$$

a. Find the mean and standard deviation of the composite score.
b. If the composite score is also normal, what percentage of the applicants get a composite score of at least 375?

10.24 (Proof) Suppose Y_1, Y_2 is a simple random sample of two elements from a population with probability function $f(y)$. What is the joint probability function of Y_1 and Y_2?

• **10.25** (Proof) Suppose Y_1, Y_2, Y_3 is a simple random sample of three elements from a population with probability function $f(y)$. What is the joint probability function of Y_1, Y_2, Y_3?

10.26 Suppose Z_1 and Z_2 are independent standardized random variables (so each has a mean of 0 and standard deviation of 1). Let $X = Z_1 + Z_2$ and $Y = Z_1 - Z_2$. Compute these correlations:

a. $\text{Corr}(Z_1, Z_2)$ b. $\text{Corr}(Z_1, Y)$ c. $\text{Corr}(X, Y)$

• **10.27** Suppose Z_1 and Z_2 are independent standardized random variables (so each has a mean of 0 and standard deviation of 1). Let $X = Z_1$ and $Y = aZ_1 + bZ_2$. Compute these correlations: a. $E(Y)$ b. $\text{Var}(Y)$ c. $\text{Corr}(X, Y)$

$$* \ * \ * \ * \ * \ * \ * \ * \ *$$

10.28 (Proof) Use multiple integration to prove Theorem 10.1 for the case when the random variables X and Y are continuous.

• **10.29** Let X and Y be *continuous* random variables with joint cumulative distribution function

$$F(x,y) = \begin{cases} xy & \text{if } 0 \leq x \leq 1 \text{ and } 0 \leq y \leq 1 \\ x & \text{if } 0 \leq x \leq 1 \text{ and } y > 1 \\ y & \text{if } x > 1 \text{ and } 0 \leq y \leq 1 \\ 1 & \text{if } x > 1 \text{ and } y > 1 \\ 0 & \text{otherwise} \end{cases}$$

a. Use partial differentiation to show that the joint density function of X and Y is
$$f(x,y) = \begin{cases} 1 & \text{if } 0 \leq x \leq 1 \text{ and } 0 \leq y \leq 1 \\ 0 & \text{otherwise} \end{cases}$$

b. Find the marginal density function of X.
c. Are X and Y independent random variables? Explain.
d. Compute the mean and standard deviation of X.
e. Compute $Pr(0 < X \leq \frac{1}{2} \text{ and } \frac{1}{2} < Y \leq 1)$.

10.30 (Proof) Use multiple integration to prove Theorem 10.2 for the case when the random variables X and Y are continuous.

Chapter 11

SAMPLING EXPERIMENTS AND THE LAW OF AVERAGES

11.1 POPULATIONS AND PARAMETERS

11.2 SAMPLES AND STATISTICS

11.3 LAW OF AVERAGES FOR THE SAMPLE COUNT

11.4 LAW OF AVERAGES FOR THE SAMPLE SUM

11.5 LAW OF AVERAGES FOR THE SAMPLE PROPORTION

11.6 LAW OF AVERAGES FOR THE SAMPLE MEAN

11.7 THE Z STATISTIC

11.8 THE T STATISTIC

11.9 ESTIMATORS OF PARAMETERS; ACCURACY AND PRECISION

In this chapter, we consider the behavior of samples from a known population. A **sampling experiment** is the draw of a sample from a population, the outcome set being the collection of all possible samples. In this context, a statistic (such as the sample mean, sample percentage, sample sum, or sample standard deviation) is a random variable for the sampling experiment.

In a **box model**, the population is represented by the contents of the box, a sampling experiment is the draw of cards from the box, and a sample is the cards drawn in the sampling experiment.

Because we are considering a known population, this analysis is still in the area of deductive reasoning. However, sampling theory will prepare us for the goal of inference theory, where we reverse the analysis; that is, we use data obtained from a random sample to make judgments and estimates about the unknown population from which it came.

11.1 POPULATIONS AND PARAMETERS

First, we consider randomly drawing a *single* element of a population, which is a sample of size $n = 1$. This process results in some observation. We restrict our attention to numerical observations (measurements) X. It permits us to model our population by a box of numbered cards.

Taking a *single* element at random from the population box and observing the number generates a random variable X. The expected value $E(X) = \mu_X$ of the random variable X is the mean μ of the population, and the standard deviation σ_X of the random variable X is the standard deviation σ of the population. The population and the random variable X have the same mean, standard deviation, and so on, so generally there is no confusion in identifying the population itself with the random variable X of drawing a single element of the population.

We can summarize the results of a single draw as follows: Consider a population box with mean μ and standard deviation σ. If we draw a single card from the box, we expect the number on the card to be approximately μ, give or take σ or so.

Example 11.1 **Height of an athlete.** Suppose that the mean height of all male athletes at a large university is $\mu = 185$ cm and the standard deviation is $\sigma = 6$ cm. One of these athletes (a random sample of size 1) is waiting in another room for a regular physical examination. If you had to guess the athlete's height X, it is reasonable to use the mean height of all athletes, namely, $\mu = 185$ cm.

It is unlikely that your guess or estimate of 185 cm would be the athlete's exact height. Actually, because height is continuous, the probability is zero that that any athlete's height is exactly 185 cm. The difference between μ and the athlete's actual height is $X - \mu$ and is called the **random sampling error** (or the **random error** of the experiment). The random sampling error will be different for different athletes.

11.1 POPULATIONS AND PARAMETERS

The standard deviation of all possible random sampling errors is called the **standard error**. In this case ($n = 1$), the standard error is the standard deviation $\sigma = 6$ cm. It is reasonable to use it as an estimate of the size of the random sampling error. We can summarize this estimate as follows:

The athlete's height is estimated to be 185 cm, give or take 6 cm or so.

Next we consider the special case of a Bernoulli box consisting of cards numbered 0 and 1. The mean of the box is the proportion of 1's. So, if the proportion of 1's in the box is π, we have

$$\mu = \pi$$

Also,

$$\sigma = \sqrt{\pi(1-\pi)}$$

Example 11.2 *Getting ace on a fair die.* We model the roll of a fair die to count the number of aces. An ace is the outcome with a single dot on the top face. There are only two ways the experiment can come out; either we succeed in getting an ace or we fail. This is a Bernoulli experiment and thus the box will contain 0's and 1's. Because the probability of a success is 1/6, the simplest model is the one consisting of one card numbered 1 and five cards numbered 0. Here $\mu = \pi = \frac{1}{6}$ and $\sigma = \sqrt{\pi(1-\pi)} = 0.3727$.

Following the idea of the previous example, we arrive at the statement: In a single roll of a fair die, we expect to get $1/6 = 0.1667$ aces, give or take 0.3727 aces or so.

How should we interpret this statement because we can *never* observe the expected value of 1/6 aces in a single roll? Well, let's suppose we repeated this experiment many times and record the results. We get a long list of 0's and 1's. If we take the average of this list, we get about $\mu = 1/6$.

Next we list the random sampling errors. This is a long list of the differences between what we got and the expected value. If a roll turns up *ace*, then our random sampling error is $1 - 1/6 = 5/6$; if not, then our random sampling error is $0 - 1/6 = -1/6$. This is the same as taking our list of 0's and 1's and subtracting 1/6 from every entry. The standard deviation of our list of random sampling errors is approximately the standard error, which is the standard deviation of all possible random sampling errors $\sigma = 0.3727$.

The values of parameters of populations such as μ and σ may be unknown, yet we assume that these concepts exist, just as we understand that every object in the physical universe has a mass even though we may not know what it is. Below are examples of such parameters, all of which we assume to exist, but many are not knowable.

Example 11.3 **Registration records.** Upon enrollment in college, every student supplies biographical information, including variables such as name, birth date, sex, local and permanent addresses, declared major, and so on. The Office of Registration and Records keeps such information. Here it is clear that parameters such as the percentage of female students and mean age can be found by examining the biographical pages of all students. Well, in theory, at least. In reality, there are many practical problems. There may be typographical and other kinds of transcription errors. There may be missing data. Also, for a variety of reasons, students may not be reporting correct information. Some students may have misunderstood questions, or they may be recalling it wrong, or they may be exaggerating some facts or even lying. In practice, *reported data* are prone to a variety of biases.

Example 11.4 **U.S. census records.** The U.S. census collects information about the entire U.S. population. Conceptually, this is similar to the biographical information collected by the college Office of Registrations and Records. Yet because of the size and complexity of the census, it is easier to see the possibility of errors in the information. First of all, some people such as the homeless are hard to reach. Also, some questions are easy to misunderstand. For example, people who have been unemployed for some time and have stopped looking for work, yet are still interested in working, may be put in the category "not in the work force" by some census takers and in the category "unemployed" by others.

Example 11.5 **Voter survey.** Consider the registered voters in your state and their responses to the question "Will you vote in the upcoming election?" The population consists of the responses. Unlike the university biographical information or the U.S. census, there is no register of responses. So a parameter such as the percentage of Yes responses is unknown. As a concept, this parameter exists, although for practical purposes it is not knowable until the election.

Example 11.6 **Destructive testing.** In considering the reliability of the U.S. stockpile of nuclear weapons, we cannot truly know whether a particular bomb will work until it is used. Conceptually, the proportion π of nuclear bombs that will successfully explode exists, but unless all are tested (and hence destroyed), the true value of π cannot be known.

Example 11.7 **Auto safety.** In determining the safety of automobiles, the situation is even more complex. In collisions of automobiles into a wall at speeds of 50 miles per hour, let π be the percentage of passenger lives who would be saved by air bags. The percentage can only be estimated by various means, but it can never be known. Yet the true percentage exists in theory.

Example 11.8 **Effectiveness of treatment.** In studying the effectiveness of drugs such as the Salk vaccine, one would like to know the true percentage of inoculated children who will still become ill with polio. Because we want to know the effectiveness of the vaccine for children who have not yet been inoculated, some not even yet born, we have a seemingly endless population. We can speak of a true percentage although we can never know it.

If the population box is finite—say, a box with N cards $x_1, x_2, \ldots, x_N$—and we compute the sample mean $\overline{x}$ of the *entire* box, it will turn out to be the same as the population mean μ. This is because the probability of getting any one card is $\frac{1}{N}$, which also appears in the computation of $\overline{x}$:

$$\mu = \overline{x} = \frac{1}{N} \sum_{k=1}^{N} x_k$$

On the other hand, if we compute the sample standard deviation s of the *entire* box, it differs from the population standard deviation σ because $\frac{1}{N-1}$ is used in the computation of

$$s = \sqrt{\frac{1}{N-1} \sum_{k=1}^{N} (x_k - \overline{x})^2}$$

rather than the probability $\frac{1}{N}$ of getting any one card, which is used in the computation of

$$\sigma = \sqrt{\frac{1}{N}\sum_{k=1}^{N}(x_k - \mu)^2}$$

We have the following relationship between the two:

$$\sigma = \sqrt{\frac{N-1}{N}}\, s$$

EXERCISES FOR SECTION 11.1

- **11.1** Model a single toss of a fair coin to count the number of heads.

    a. How many cards are in your population box?
    b. What are the numbers on the cards?
    c. In a single toss of a fair coin, we expect to get approximately _____ heads, give or take _____ heads or so.

  11.2 A freshmen class consists of 60% female students and 40% males. Select *one* student at random. Code the population as follows: 1 for female and 0 for male. Model this experiment.

    a. How many cards are in your population box?
    b. What are the numbers on the cards?
    c. In a random sample of one student from this freshman class, we expect to get approximately _____ females, give or take _____ females or so.

11.2 SAMPLES AND STATISTICS

A **simple random sample** from a population consists of n observations:

$$X_1, X_2, \ldots, X_n$$

where X_1 is the first observation, X_2 is the second, and so on. These are chosen in such a way that, at each stage, every element of the population has the same chance of being chosen. Depending upon the context, this may be *with* or *without* replacement.

Draws with replacement are independent, and thus computationally simpler.

Drawing such a sample is a random experiment. If the population is quantitative, then, as shown in the previous section, the expected value of the first element drawn random variable X_1 is the mean of the population $E(X_1) = \mu$, and the standard deviation is the standard deviation of the population $\text{SD}(X_1) = \sigma$.

Similarly, the expected value of the second element drawn X_2 is also the mean of the population $E(X_2) = \mu$, and the standard deviation is $\text{SD}(X_2) = \sigma$, and so on, for all n element of the sample. These random variables $X_1, X_2, \ldots, X_n$ are *identical but not equal*.

Example 11.9 **Identical but not equal.** Consider a population of heights of athletes X and a simple random sample $X_1, X_2, \ldots, X_n$. The random variable X_1 is obtained by measuring the height of the first randomly chosen athlete, X_2 is the height of the second chosen athlete, and so on. The random variables X_1 and X_2 are not equal; that is, the height of the first athlete measured will not always be the same as the height of the second athlete. But X_1 and X_2 have the same density function (regardless of whether draws are with or without replacement because we are dealing with unconditional probabilities). We say that the random variables are *identical*.

Because $X_1, X_2, \ldots, X_n$ are identical random variables, they all have the same mean and standard deviation:

$$E(X_1) = E(X_2) = E(X_3) = \cdots = E(X_n) = \mu_X \tag{11.1}$$

$$\text{SD}(X_1) = \text{SD}(X_2) = \text{SD}(X_3) = \cdots = \text{SD}(X_n) = \sigma_X \tag{11.2}$$

A real valued function $g(X_1, X_2, \ldots, X_n)$ of a random sample $X_1, X_2, \ldots, X_n$ is called a **statistic**.

Definition

A **statistic** *is a random variable of a sampling experiment. The term* **sampling distribution** *is commonly used to refer to a statistic as a random variable of a sampling experiment.*

We are already familiar with the sample count K as a statistic. The standard model is to draw a predetermined number of cards $X_1, X_2, \ldots, X_n$ with replacement from box of 0's and 1's and to count the 1's by adding up the n cards:

$$K = X_1 + X_2 + \cdots + X_n$$

For the binomial K, the draws are independent; for the hypergeometric K, the draws are made without replacement.

However, even if the population box is not dichotomous, we can add the numerical values of the sample and find the statistic

$$\text{SUM} = X_1 + X_2 + \cdots + X_n$$

Another statistic is the sample mean

$$\overline{X} = \frac{\text{SUM}}{n} = \frac{X_1 + X_2 + \cdots + X_n}{n}$$

This is also a random variable because the sample mean varies from sample to sample. Other statistics are the sample standard variance

$$S^2 = \frac{(X_1 - \overline{X})^2 + (X_2 - \overline{X})^2 + \cdots + (X_n - \overline{X})^2}{n - 1} \tag{11.3}$$

the sample median, the sample minimum, and so on.

This way of regarding a statistic is a leap. Parameters such as the population mean and population standard deviation are numbers. There is a natural inclination to think of the sample mean, the sample standard deviation, and any other statistic as numbers. Although it is true that for a given sample, the sample mean is a number, it varies from sample to sample. For the sampling experiment, the sample mean is a random variable, not a number. Recognizing statistics as random variables, we can talk about their expected values and standard deviations. Such observations give rise to various forms of the law of averages, which we will investigate in Sections 11.3 through 11.6.

It is important to be aware of whether the populations is dichotomous (in which case we consider the parameter of the population proportion π) or whether it is quantitative (in which case we consider the parameter of the population mean μ). Each population type has two forms of the law of averages.

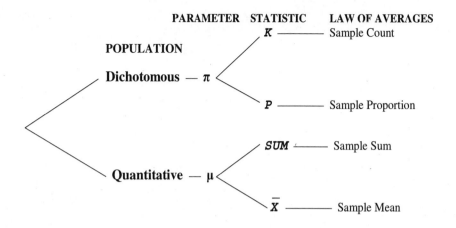

11.3 LAW OF AVERAGES FOR THE SAMPLE COUNT

The law of averages for the sample count is nothing new to us. It is just a restatement of the Central Limit Theorems for the Binomial Variable (Theorem 8.1), and the Poisson Limit Theorem (Theorem 9.7). Suppose that we have a Bernoulli population box with mean π and the standard deviation $\sigma = \sqrt{\pi(1-\pi)}$. We draw a random sample $X_1, X_2, \ldots, X_n$ of size n. The sample count

$$K = X_1 + X_2 + \cdots + X_n$$

is the number of successes in the sample. If draws are made with replacement, then K is a binomial random variable. If drawn without replacement, then K is a hypergeometric random variable. There are three parts to the law of averages for the statistic K.

First part The **first part** of the law of averages deals with the expected value of K, which we first proved in Exercise 5.34 on page 162 and again as Theorem 10.1 on page 280:

$$\mu_K = E(X_1) + E(X_2) + \cdots + E(X_n) = n\pi \tag{11.4}$$

Second part The **second part** of the law of averages deals with the random sampling error. Random sampling errors are deviations from the expected sum:

$$\text{Random sampling error} = K - \mu_K = K - n\pi$$

The standard deviation of all random sampling errors is the **standard error** of the sample count K, which is

$$\sigma_K = \text{SD(Random sampling error)} = \text{SD}(K - n\mu) = \text{SD}(K)$$

For sampling with replacement, we have by Theorem 10.8 on page 286

$$\sigma_K = \sqrt{n\pi(1-\pi)} \qquad (11.5)$$

For sampling without replacement from a population of size N, we multiply by the reduction factor to obtain

$$\sigma_K = \sqrt{\frac{N-n}{N-1}}\sqrt{n\pi(1-\pi)} \qquad (11.6)$$

Third part   The **third part** of the law of averages deals with the shape of the histogram of the random variable K. There are three cases:

3(a). If the sample size n is small, then we can generally compute the probabilities either by hand or with the help of technology. However, when $n \geq 100$, because of round-off error, these calculations may no longer be reliable.

3(b). If n is large [general rule: $n\pi \geq 5$ and $n(1-\pi) \geq 5$], then by the Central Limit Theorem for the binomial, the histogram of K follows the normal curve with mean $\mu_K = n\pi$ and standard deviation σ_K given in the second part (Equation 11.5 or 11.6).

3(c). If n is large and π is small [general rule: $n \geq 100$ and $n\pi < 5$][1], then by the Poisson Limit Theorem, the histogram follows the Poisson histogram with parameter $\lambda = \mu_K = n\pi$.

See Figure 11.1 for a comparison of the normal approximation of 3(b) and the Poisson approximation of 3(c).

---

[1] Or we could interchange success and failure and use $n > 100$ and $n(1-\pi) < 5$.

11.3 LAW OF AVERAGES FOR THE SAMPLE COUNT

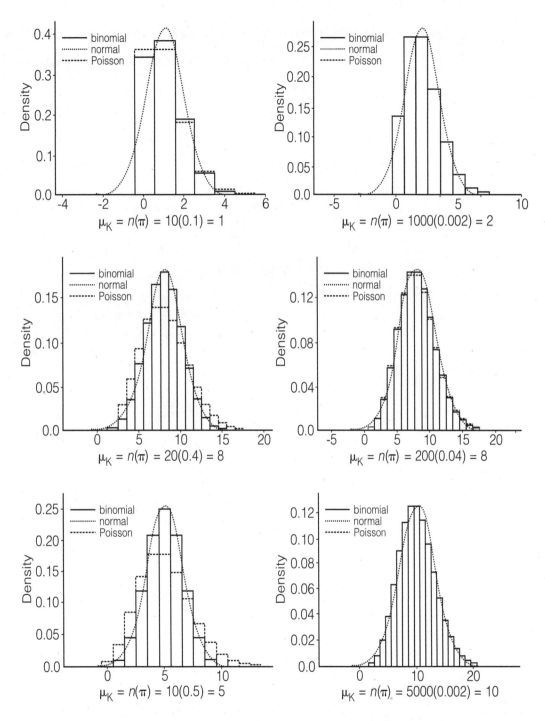

Figure 11.1 Normal and Poisson approximations of the sampling count K (sampling with replacement) for various values of π and n

Example 11.10 **Coin tosses.** Toss a fair coin 100 times and count the number of heads. Each sample of 100 tosses taken, from an infinite population of all possible tosses, can be represented as a string $X_1, X_2, X_3, \ldots, X_{100}$, where $X_j = 0$ if the jth toss comes up tails and $X_j = 1$ if the jth toss comes up heads. Here, $\pi = 0.5$ and $\sigma = 0.5$. By the law of averages, K follows a normal curve with $\mu(K) = 50$, and $\sigma_K = 5$. Let's find the following probabilities:

  a. $Pr(K = 50)$ b. $Pr(K > 63)$ c. $Pr(K \geq 57)$

Solution Because $n\pi = 100(0.50) = 50 > 5$ and $n(1-\pi) = 100(1-0.50) = 50 > 5$, we can use the normal approximation. The normal random variable is continuous but K is discrete, so we use a continuity correction:

a.
$$\begin{aligned}
Pr(K = 50) &= Pr(49.5 < K \leq 50.5) \\
&= Pr\left(\frac{49.5 - 50}{5} < \frac{K - n\pi}{\sqrt{n\pi(1-\pi)}} \leq \frac{50.5 - 50}{5}\right) \\
&\approx \Phi(0.1) - \Phi(-0.1) = 0.0797
\end{aligned}$$

b.
$$\begin{aligned}
Pr(K > 63) &= Pr(K > 63.5) \\
&= 1 - Pr(K \leq 63.5) \\
&\approx 1 - \Phi(\frac{63.5 - 50}{5}) = 1 - \Phi(2.7) = 0.0035
\end{aligned}$$

c.
$$\begin{aligned}
Pr(K \geq 57) &= Pr(K > 56.5) \\
&= 1 - Pr(K \leq 56.5) \\
&\approx 1 - \Phi\left(\frac{56.5 - 50}{5}\right) = 1 - \Phi(1.3) = 0.0968
\end{aligned}$$

Notice how easily these probabilities are computed. Carefully computing the probabilities directly from the binomial probability function, we find $Pr(K = 50) = 0.0796$, $Pr(K > 63) = 0.0035$, and $Pr(K \geq 57) = 0.0967$. So the normal approximation does not do badly here, but the relative error of this approximation increases in the extreme tails.

Notice that the relative error is larger when approximating the probability in part b of being 2.7 standard deviations above the mean than the probability in part c of being 1.3 standard deviation above the mean.

Example 11.11 **A hand of cards.** Draw a random hand of 13 cards from a 52-card deck. Find the probability of getting exactly 6 spades.

Solution This is sampling without replacement from a population box with

$$\pi = \frac{1}{4}, \text{ and } \sigma = \sqrt{\frac{1}{4} \cdot \frac{3}{4}}$$

So for $n = 13$ we have

$$\mu_K = n\pi = 13 \cdot \frac{1}{4} = 3.25$$

and

$$\sigma_K = \sqrt{\frac{52-13}{52-1}} \sqrt{13} \sqrt{\frac{1}{4} \cdot \frac{3}{4}} = \sqrt{1.8639706} = 1.365273$$

The exact answer is

$$Pr(K = 6) = \frac{\binom{13}{6}\binom{39}{7}}{\binom{52}{13}} = 0.041563$$

The sample size $n = 13$ is too small to meet either criterion 3(a) to use the normal approximation or criterion 3(c) to use the Poisson approximations. Out of curiosity, let's see what values they compute.

Using the normal approximation we have

$$\begin{aligned} Pr(K = 6) &= Pr(5.5 < K < 6.5) \\ &\approx Pr\left(\frac{5.5 - 3.25}{1.3625} < Z < \frac{6.5 - 3.25}{1.3625}\right) \\ &= \Phi(2.3805) - \Phi(1.6480) = 0.0410 \end{aligned}$$

Using the Poisson approximation we have

$$Pr(K = 6) \approx e^{-3.25} \frac{3.25^6}{6!} = 0.0635$$

In this example, the normal approximation fared better than anticipated, but the Poisson approximation did not. The Poisson approximation will do better when the mean and variance of the sample count K are nearly equal.

Example 11.12 ***Auditors.*** An auditor randomly selects 300 of 8000 vouchers and inspects them for deviations from controls. Suppose 60 of the 8000 vouchers exhibit deviations. What is the probability that no more than 1 voucher in the sample exhibits deviations?

Solution The expected number of deviations from controls is

$$\mu_K = n\pi = 300 \cdot \frac{60}{8000} = 2.25$$

and the approximate (Poisson) probability, with $\lambda = 2.25$ is

$$Pr(K \le 1) = f(0) + f(1) \approx e^{-2.25} + 2.25e^{-2.25} = 0.3425$$

Computation of the hypergeometric probability yields the exact value of 0.3270.

EXERCISES FOR SECTION 11.3

- **11.3** You are interested in counting the number of heads you get when tossing a fair coin 400 times. Fill in the blanks.

  a. In this experiment, you expect _____ heads, give or take _____ heads or so.
  b. Approximately 68% of the time you would get between _____ and _____ heads.

- **11.4** Suppose that 15% of all service contracts of a major appliance require service. Of 1000 customers, what is the probability that fewer than 120 require service? Use a normal approximation.

- **11.5** Suppose you know that 40% of the registered voters in a certain town are Democrats. A random-dial telephone sample is taken. You would like to draw a random sample from those who will actually vote in the election and then to know how they would actually vote. However, your phone sample is drawn from the population of nonvoters as well as voters. Moreover, not all voters will respond to unsolicited phone calls. Finally, some would-be voters are unsure how they will vote before the election or may change their minds before the election. Suppose you are able to devise a phone survey that overcomes most of these potential biases and you want to test it in this known population. You take a sample of 80.

  a. If your survey overcomes these biases, you expect _____ Democrats, give or take _____ Democrats or so, in your sample.
  b. What is the probability that your sample actually predicts a Republican victory; that is, what is the probability that your sample of 80 ends up with more Republicans than Democrats?

11.6 Suppose that past experience has show that 60% of all students in an introductory computer science course have programming errors on their first assignment. In a class of 30 student, find the following probabilities:

a. fewer than 15 students have errors

b. at least 22 students have errors

c. between 15 and 21, inclusive, have errors

11.4 LAW OF AVERAGES FOR THE SAMPLE SUM

The laws of averages of the previous two sections were for dichotomous populations. Consider a simple random sample $X_1, X_2, \ldots, X_n$ drawn from a *quantitative* population with mean μ and standard deviations σ. The statistic we will now consider is the sample sum:

$$\text{SUM} = X_1 + X_2 + \cdots + X_n$$

Notice that this is the same formula as for the sample count K, except that the population box need not be dichotomous. We will see that the three parts to the law of averages for the statistic SUM are similar to that of the sample count K.

First part

1. The **first part** deals with the expected value (Theorem 10.1 on page 280):

$$\mu_{\text{SUM}} = E(X_1) + E(X_2) + \cdots + E(X_n) = \mu + \mu + \cdots + \mu = n\mu$$

Second part

2. The **second part** of the law of averages deals with the size of the random sampling errors. The random sampling errors are deviations from the expected sum:

$$\text{Sampling error} = \text{SUM} - \mu_{\text{SUM}} = \text{SUM} - n\mu$$

And the standard error of SUM is the standard deviation of the sampling errors:

$$\sigma_{\text{SUM}} = \text{SD(Sampling error)} = \text{SD(SUM} - n\mu) = \text{SD(SUM)}$$

For sampling with replacement we have, by Theorem 10.8 on page 286,

$$\begin{aligned}\text{Var(SUM)} &= \text{Var}(X_1) + \text{Var}(X_2) + \cdots + \text{Var}(X_n) \\ &= \text{Var}(X) + \text{Var}(X) + \cdots + \text{Var}(X) \\ &= n\sigma^2\end{aligned}$$

Thus, we have
$$\sigma_{\text{SUM}} = \sqrt{\text{Var}(X)} = \sqrt{n}\sigma$$

For sampling without replacement from a population of size N, we multiply by the reduction factor to obtain

$$\sigma_{\text{SUM}} = \sqrt{\frac{N-n}{N-1}}\sqrt{n}\sigma$$

Third part 3. The **third part** of the law of averages deals with the density curve of the random variable SUM. There are two cases.

   3(a). If the population follows the normal curve, then the statistic SUM follows a normal curve as well, regardless of the sample size.

   3(b). Even if the population is not normal, then for large values of n, the random variable SUM still tends to follow a normal curve. What is considered large? It depends on the shape of the population. If the population is close to normal, then n need not be very large at all. Similarly, if the population density is defined on a bounded interval, and the density function is nearly symmetric about the mean, then n need not be very large. As a practical rule for most cases, $n \geq 40$ is considered sufficiently large, but $n \geq 50$ is better.

We show graphs in Figure 11.2 (page 307) of the densities of the statistic SUM for various sample sizes n when sampling with replacement from an exponential population. As the sample size increases, the center of SUM moves to the right. So, in order to better see the comparison to the normal, the horizontal scale is given in standard units. Notice that, even though the population is far from being normal, as the sample size increases from $n = 1, 2, 4, 16$ to 64, the density curves of the statistic SUM approach a normal curve. Formally, the third part of the law of averages is the Central Limit Theorem (Theorem 8.2 on page 234).

Example 11.13 *Dots on a fair die.* For the number of dots shown on a roll of a fair die, the probability mass function is uniform, as shown in Section 4.7 on page 138. The total number of dots shown when n dice are rolled is the variable SUM for a sample of size n.

11.4 LAW OF AVERAGES FOR THE SAMPLE SUM

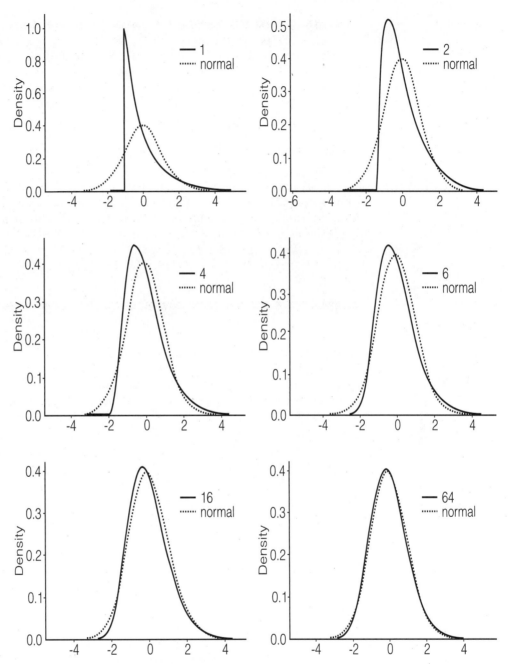

Figure 11.2 Normal approximation of the statistic SUM for sampling with replacement from an exponential population. The solid curves show the density curves of SUM for samples of size $n = 1$, 2, 4, 8, 16, and 64. Normal curves with the same means and standard deviations of SUM are shown dashed for comparison. The horizontal scale is in standard units.

Example 11.14 *Overloading an elevator.* The elevator in a campus building is rated for a maximum of 16 persons and 2500 pounds. Suppose that at class change time, passengers on this elevator, who are generally students loaded down with coats, backpacks, books, and so on, weigh an average of 150 lb with a standard deviation of 32 lb. (Does this estimate make sense based on the 68% and 95% common rules?) Find the chance that an elevator loaded with a random sample of 16 students will overload the elevator.

Solution Here we assume that the 16 students make up a small sample in comparison with the population of all students, so we can ignore the reduction factor. Also, we assume that weights of students follow roughly a normal curve. The relevant numbers: $n = 16$, $\mu = 150$, $\sigma = 32$, $E(\text{SUM}) = n\mu = 2400$, $\sigma_{\text{SUM}} = \sqrt{n}\sigma = 128$. So, $Pr(\text{Overload}) = Pr(\text{SUM} > 2500) = Pr(Z > 0.78125) = 0.2173$.

Example 11.15 *How much does a pound of butter weigh?* This question separates the statisticians from the mathematicians. Butter that you buy in a supermarket is generally sold in one pound cartons, each containing four separately wrapped sticks of butter. In the packaging plant, a machine squeezes the bulk butter into a continuous rectangular shape that is then cut into quarter-pound sticks. The sticks are then wrapped and packaged, four sticks to a carton. These cartons are further packed into boxes of 100 one-pound cartons each. The density of the butter that comes out of the machine varies, depending on many factors, including temperature, humidity, seasonal changes in dairy feed, and mechanical adjustments of the machinery.

The goal is to cut the butter into exactly quarter-pound (4-ounce) sticks. As with any continuous variable, there is a zero probability that a stick weighs exactly 4 ounces. This means that we can be almost certain that a stick does not weigh exactly 4 ounces, and a carton does not contain exactly 1 pound of butter. Furthermore, because the cartons are subject to government inspection, the machine is adjusted so that the sticks weigh, on the average, slightly more than 4 ounces—say, $\mu = 4.1$ ounces—with variability of $\sigma = 0.1$ ounce.

Assuming random variation in weight, the weight of a pound of butter is given by the variable SUM for $n = 4$. The expected value of SUM is $\mu_{\text{SUM}} = n\mu = 4 \times 4.1 = 16.4$ ounces. The *random error* of the statistic SUM is the difference between what we get and what we expect to get:

$$\text{SUM} - \mu_{\text{SUM}}$$

The *standard error* of SUM is the standard deviation of the random errors.

So,

$$\sigma_{\text{SUM}} = \text{SD}(\text{SUM} - \mu_{\text{SUM}}) = \text{SD}(\text{SUM}) = \sqrt{n}\sigma = 2 \times 0.1 = 0.2 \text{ ounces.}$$

This means that a pound carton of butter weighs about 16.4 ounces give or take 0.2 ounces or so. The probability that a pound of butter is underweight (that is, less than 16 ounces) is $Pr(W < 16) = Pr(z < -2) = \Phi(-2) = 2.28\%$.

Note that the standard error does not increase in proportion to the sample size n, but rather in proportion to $\sqrt{n}$.

A box of 100 pounds of butter contains 400 sticks, so its net weight will be SUM for $n = 400$. Such a box will contain about $400 \times 4.1 = 1640$ ounces (102.5 pounds) of butter, give or take $\sqrt{400} \times 0.1 = 2$ ounces or so. The 2.5 extra pounds, give or take 2 ounces or so, is part of the manufacturing loss that the packaging company incurs.

What is the probability that a 100-pound box of butter has a net weight more than 102 pounds?

Solution We may assume that SUM follows the normal curve with $\mu = 1640$ and $\sigma = 2$. We have $Pr(\text{SUM} > 1632) = Pr(Z > -4) = 0.99997$. This means that the company is almost certain to take a production loss of at least 2 pounds per 100-pound box.

Example 11.16 *Gambler's ruin.* Let's consider a game of tossing a fair coin where on heads you win one dollar and on tails you lose a dollar. This is called a fair game because on you have an equal chance of winning a dollar and losing a dollar. It is a common perception that, the more you play a fair game, the more likely you will come out even. In truth, the variability of your winnings and losses will increase with the number of games you play.

Here the population box model consists of one card marked +1 and the other −1, representing the winnings on each play, so $\mu = 0$ and $\sigma = 1$. The winnings after n plays of the game is the sample statistic SUM for the n draws from the box with replacement. By the law of averages, we have $\mu_{\text{SUM}} = n \cdot \mu = 0$ and $\sigma_{\text{SUM}} = \sqrt{n}\sigma = \sqrt{n}$. Furthermore, for a large number of plays of the game, the winnings follow a normal curve. We see that, as the number n of plays increases, although your expected winnings stay at 0, the standard deviation of your winnings $\sqrt{n}$ increases.

Now suppose that you have a dollars to play and your opponent has b dollars. As soon as your winnings reach $-a$ dollars, you are ruined and cannot continue. Similarly, as soon as your winnings reach $+b$ dollars, your opponent is ruined. The probability that your winnings stay between $-a$ and $+b$ dollars is

$$Pr(-a < \text{SUM} < +b) = Pr(-a/\sqrt{n} < z < +b/\sqrt{n})$$

which tends to 0 as n increases.

> *This means that the probability either you or your opponent is eventually ruined is 1.*

For example, after 25 games, your expected winning will be $0, give or take $\$\sqrt{25} = \$5$ or so. So, if you have $a = 5$ dollars and your opponent has more—say, $b = 100$ dollars—then by the 68% rule, there is roughly a 16% probability that your winnings will be $-\$5$ or less. Because you could have been ruined earlier than the 25th game, the probability that you are ruined by the 25th game is substantially more than 16% (whereas you opponent could not yet have been ruined). Actually, it can be shown that if you continue playing, the probability that you will be ruined before your opponent comes out to $100/105 = 95.2\%$. In general, if you have a dollars and your opponent has b dollars, then the probability that you will be ruined before your opponent is ruined is $b/(a+b)$.

> *The moral of the story: A casino with a big bank is almost certain to make money even if the games of chance are fair.*

EXERCISES FOR SECTION 11.4

- **11.7** A commuter plane will take 19 passengers. The safe payload for the passengers is 3800 pounds (that is, 200 pounds per passenger). Assuming passengers are randomly selected from a normal population with mean of 180 pounds and standard deviation of 40 pounds, how often will 19 passengers overload the plane?

- **11.8** A charter airliner packs customers in. It will take 80 passengers. The safe payload for the passengers is 16,000 pounds. On a full flight, this is an average of 200 pounds per passenger. Assuming passengers are selected at random from a population with mean of 180 pounds and standard deviation of 40 pounds, how often will 80 passengers overload the plane?

- **11.9** As a result of a US Airways commuter plane crash in January 2003, the National Transportation Safety Board has raised the assumed mean weight of passengers by 10 pounds. Repeat Exercise 11.8 assuming a mean weight of passengers to be 190 pounds, with standard deviation of 40 pounds.

11.10 Continuing with Exercise 11.9, winter adds another 5 pounds to the average weight of passengers. Repeat Exercise 11.8 assuming a mean weight of passengers to be 195 pounds, with standard deviation of 40 pounds.

- **11.11** The weights of lab mice follow a normal curve with mean 30 grams and standard deviation 5 grams.

    a. If a lab mouse is randomly selected, find the probability that it weighs over 35 grams.

    b. If four mice are randomly selected, find the probability that their total weight is over 140 grams.

    c. If four mice are randomly selected, find the probability that *all* of them weigh over 35 grams.

11.12 A warranty service shop takes an average of 40 minutes to repair a component with a standard deviation of 15 minutes.

    a. Find the mean and standard deviation time needed to sequentially repair 10 items.

    b. Assuming that repair times follow a normal curve, what is the probability that 10 items can be repaired within 6 hours?

- **11.13** A convenience store sells an average of 20 containers of yogurt each day with a standard deviation of 5 containers.

    a. What is the mean and standard deviation for the total sales for a week? Assume that daily sales follow a normal curve.

    b. If the store has an inventory of 150 containers, what is the probability that the supply will last for at least one week?

    c. What inventory is sufficient in order to be 99% certain the supply lasts one week?

11.5 LAW OF AVERAGES FOR THE SAMPLE PROPORTION

In a dichotomous population, coded 1 for success and 0 for failure, the mean is π, the probability of success on each trial, and the standard deviation is $\sigma = \sqrt{\pi(1-\pi)}$. The sample proportion P of successes in the sample is

$$P = \frac{K}{n}$$

The random variables K and P differ only by a change of scale. Hence, the law of averages for the random variable P can easily be derived from that of the sample count K. Here is a summary of **the law of averages for sample proportions P**.

First part   1. $\mu_P = \pi$.

Second part   2. For sampling with replacement,

$$\sigma_P = \sqrt{\frac{\pi(1-\pi)}{n}}$$

For sampling without replacement from a population of size N, also multiply by the reduction factor to obtain

$$\sigma_P = \sqrt{\frac{N-n}{N-1}}\sqrt{\frac{\pi(1-\pi)}{n}}$$

Third part   3(a). For small values of n, the probabilities of the statistic P can be computed directly with a calculator.

               3(b). For large n [general rule: $n\pi \geq 5$ and $n(1-\pi) \geq 5$], the statistic P follows the normal curve with mean π and standard deviation σ_P.

               3(c). For large n, where π is small [general rule: $n \geq 100$ and $n\pi < 5$], the probabilities of the statistic P can be obtained from the Poisson approximation of K as in Section 9.4 on page 255.

See Figure 11.3 on page 313 for an illustration of part 3(b).

11.5 LAW OF AVERAGES FOR THE SAMPLE PROPORTION

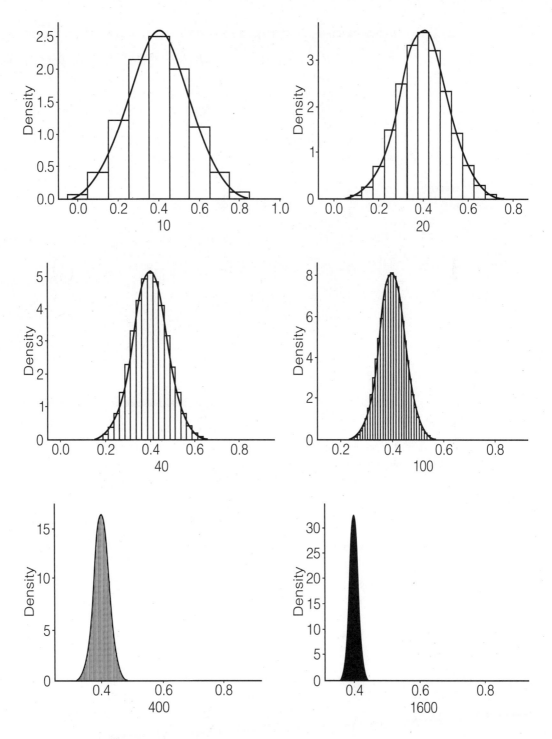

Figure 11.3 Normal approximation of the sample proportion P for $\pi = 0.40$ and sample sizes $n = 10, 20, 40, 100, 400$, and 1600

Example 11.17 *Polling voters.* In the 1964 presidential election, 60% voted Democratic and 40% voted Republican. Counting Republicans, the population mean is $\mu = \pi = 0.40$. For a random sample of size n, the proportion of Republicans in the sample is P. Because we have a large population, the draws are essentially independent, and we are able to treat the sample as if it were drawn with replacement. In a random random sample of size $n = 5$, let's find the probability that the majority of them ($P > 50\%$) consists of Republicans.

Solution For small n, computations of the binomial (and hypergeometric) probabilities can easily be calculated by computer or calculator. We have

$$\begin{aligned}Pr(P > 0.50) &= Pr(K > 2.5) = Pr(K=3) + Pr(K=4) + Pr(K=5) \\ &= \binom{5}{3}\pi^3(1-\pi)^2 + \binom{5}{4}\pi^4(1-\pi)^1 + \binom{5}{5}\pi^5(1-\pi)^0 \\ &= 0.317\end{aligned}$$

For such a small sample, there is a 31.7% chance of wrongly predicting a Republican victory even when the true percentage of Republicans is only $\pi = 0.40$.

For large n, such calculations would be tedious. Furthermore, because of round-off error in the calculations, they may no longer be reliable. However, in these cases, the law of averages permits us to use the normal curve in the case $n\pi \geq 5$ and $n(1-\pi) \geq 5$ or the Poisson probabilities in the case $n \geq 100$ and $n\pi < 5$, (or $n(1-\pi) < 5$, provided we interchange success and failure.

Example 11.18 *Polling voters.* Continuing Example 11.17, let's find the probability that a random sample of size 25 wrongly predicts a Republican victory.

Solution Here the sample proportion P follows the normal curve with mean $\mu_K = \pi = 0.40$ and standard error $\sigma_P = \sqrt{\frac{0.40 \cdot 0.60}{25}}$. In order to use the continuity correction we convert to the variable $K = nP$. We have $\mu_K = n\pi = 10$, and $\sigma_K = \sqrt{n\pi(1-\pi)} = \sqrt{6}$. Using the normal approximation, we have

$$Pr(P > 0.50) = Pr(K > 12.5) = Pr(Z > \frac{12.5 - 10}{\sqrt{6}}) \approx 0.154$$

Increasing the sample size from $n = 5$ to $n = 25$, the chance of erroneously predicting a Republican victory has gone down from 31.7% to 15.4%.

EXERCISES FOR SECTION 11.5

11.14 In Example 11.17 on page 314, as the sample gets larger, the chance of a wrong prediction gets smaller. Repeat Example 11.17 for samples of size:

    a. $n = 100$ **b.** $n = 400$ **c.** $n = 900$

• **11.15** Cook County has approximately four times as many registered voters as DuPage County. A simple random sample of registered voters is taken in each county to estimate the percentage of voters who would vote for a regional transportation revenue bond. Choose one of the following options. Other things being equal, a sample of 4000 in Cook County will be

     (a) four times as accurate   (b) twice as accurate   (c) as accurate

as a sample of 1000 in DuPage County. Explain why (c) is not the correct option even though the two samples take the same percentage of voters from each county.

11.16 Suppose 46% of all bass in a lake are male. In a random sample of 50 bass, find the mean and standard deviation for the *percentage* of bass in the sample. What is the probability that the sample will have between 42% and 50% males, inclusive?

• **11.17** Assume 48% of all live births are girls. In the month of September, Hospital A has 75 live births, and hospital B has 150 live births. Which hospital is more likely to be closer to the expected number of girls μ_K? Which one is more likely to be closer to the expected percentage of girls μ_P?

* * * * * * * *

C 11.18 In this class exercise, each student in class rolls a die 80 times. This can also be simulated by a programmable calculator. Each student should record both the number of aces and the proportion of aces in the 80 rolls.

    a. Answer the following questions. What is the population? What is the sample? What are μ and *sigma* of the population? What is the smallest possible value of the statistic K? What is the largest possible K? What is $E(K)$? What is the smallest possible random sampling error (in absolute value)? What is the largest possible random sampling error (in absolute value)? What is your random sampling error? What is σ_K?

    b. Answer the following questions. What is the smallest possible value of the statistic P? What is the largest possible P? What is $E(P)$? What is the smallest possible random sampling error (in absolute value)? What is the largest possible random sampling error (in absolute value)? What is your random sampling error? What is σ_P?

    c. After all students in class report their results, make histograms of the sample counts and sample proportions; compute the means and standard deviations. Compare the class results with the laws of averages for sample count and sample proportion.

11.6 LAW OF AVERAGES FOR THE SAMPLE MEAN

Recall that a random sample of size n consists of n random variables:

$$X_1, X_2, \ldots, X_n$$

The next statistic we will consider is the sample mean $\overline{X}$. Although, for a specific sample, the sample mean is a number

$$\overline{x} = \frac{x_1 + x_2 + \cdots x_n}{n}$$

for the sampling process, the sample mean is a random variable

$$\overline{X} = \frac{\text{SUM}}{n} = \frac{X_1 + X_2 + \cdots + X_n}{n}$$

This simple relationship between the sample sum and the sample mean

$$\overline{X} = \frac{\text{SUM}}{n}$$

leads to the mean of the statistic

$$E(\overline{X}) = E(\frac{\text{SUM}}{n}) = \frac{1}{n}E(\text{SUM}) = \frac{1}{n}n\mu = \mu \qquad (11.7)$$

In words, **the mean of the sample mean is the population mean.**

Similarly, for sampling with replacement, we have

$$\text{Var}(\overline{X}) = \text{Var}(\frac{1}{n}\text{SUM}) = (\frac{1}{n})^2\text{Var}(\text{SUM}) = (\frac{1}{n})^2 n\sigma^2 = \frac{\sigma^2}{n} \qquad (11.8)$$

and $\sigma_{\overline{X}} = \frac{\sigma}{\sqrt{n}}$.

The **law of averages for the random variable** $\overline{X}$ is thus an easy translation from the law of averages for SUM.

First part 1. $\mu_{\overline{X}} = \mu$.

Second part 2. For sampling with replacement,
$$\sigma_{\overline{X}} = \frac{\sigma}{\sqrt{n}}$$

For sampling without replacement from a population of size N, we must also multiply by the reduction factor to obtain
$$\sigma_{\overline{X}} = \sqrt{\frac{N-n}{N-1}}\frac{\sigma}{\sqrt{n}}$$

Third part 3(a). If the population follows the normal curve, then the statistic $\overline{X}$ follows a normal curve as well, regardless of the sample size.

3(b). Otherwise, even if the population is not normal, then for large values of n, the random variable $\overline{X}$ still tends to follow a normal curve. What is considered large? The rule for what constitutes a large sample is the same for the random variable $\overline{X}$ as for the random variable SUM. Namely, it depends on the shape of the population. If the population is close to normal, then n need not be very large at all. Similarly, if the population density is defined on a bounded interval, and the density function is nearly symmetric about the mean, then n need not be very large. As a practical rule, for most cases $n \geq 40$ is considered sufficiently large, but $n \geq 50$ is better.

Example 11.19 **Height of an athlete.** We return to the example of male athletes at the large university, where $\mu = 185$ cm and $\sigma = 6$ cm. Say that 9 of these athletes are waiting in another room for a regular physical examination. These athletes constitute a sample of size $n = 9$. If we had to guess the mean athlete's height of this sample, it is reasonable to use $\mu = 185$ cm. As with the estimate of the height of a single athlete, it is unlikely that your estimate of 185 cm would be the exact sample mean. The difference between μ and the sample mean will be approximately $\sigma_{\overline{X}} = \frac{\sigma}{\sqrt{9}} = \frac{6}{3} = 2$ cm. Recall from Example 11.1 that the estimate for a single athlete was accurate to within 6 cm, whereas for 9 athletes the accuracy improves (by a factor of $\sqrt{9}$) to within 2 cm. In this case, we assume that we know the population mean μ. However, if we knew the sample mean $\overline{x}$ but we did not know μ, then $\overline{x}$ can be used to estimate μ. The accuracy of the estimate is given by the standard error; in this case, 2 cm. It is clear from the second part of the law of averages that the accuracy of the sample mean $\bar{x}$ in estimating μ is related to the square root of the sample size n.

EXERCISES FOR SECTION 11.6

- **11.19** Consider a normal population with mean $\mu = 1$ and variance $\sigma^2 = 4$. Take a random sample of size 9 with replacement and compute the sample mean $\overline{X}$.

    a. Find $E(\overline{X})$.   **b.** Find $\text{Var}(\overline{X})$.
    c. Sketch the density curve of the statistic $\overline{X}$.
    d. Approximately 95% of the *population* lies between the values _____ and _____.
    e. There is a 95% chance that the *sample mean* $\overline{X}$ will be between _____ and _____. Caution: This answer is not the same as for part **d**.
    f. There is a 99% chance that the *sample mean* $\overline{X}$ will be between _____ and _____.

11.20 Consider a normal population with mean $\mu = 60$ and standard deviation $\sigma = 5$. Take a random sample of size 16 with replacement and compute the sample mean $\overline{X}$.

    a. Find $E(\overline{X})$.   **b.** Find $\text{SD}(\overline{X})$.
    c. Sketch the density curve of the statistic $\overline{X}$.
    d. Approximately 95% of the *population* lies between the values _____ and _____.
    e. There is a 95% chance that the *sample mean* $\overline{X}$ will be between _____ and _____. Caution: This answer is not the same as for part **d**.
    f. There is a 99% chance that the *sample mean* $\overline{X}$ will be between _____ and _____.

- **11.21** Fill in the blanks.

    If you roll a fair die 100 times, you can expect to get a mean of **a.** _____ dots per roll.

    The mean number of dots you actually get is denoted by the symbol **b.** _____.

    The difference between the mean number you expect and the mean number you actually get will be approximately **c.** _____ and is called **d.** _____.

11.22 A random sample of size $n = 100$ is taken from a population box with $\mu = 60$ and $\sigma = 10$.

a. Find the probability of the event $57 < \overline{X} < 63$.

b. Find the probability that $\overline{X}$ exceeds 62.

• **11.23** A random sample of size $n = 400$ is taken from a population box with $\mu = 100$ and $\sigma = 16$.

a. Find the probability of the event $98 < \overline{X} < 102$.

b. Find the probability that $\overline{X}$ exceeds 102.

11.24 I noticed that my car gets a mean of 22.8 miles per gallon (mpg) of gas in city driving. Suppose that with each tankful, the mileage varies by approximately 2 miles per gallon. I'm concerned that the mean mileage may have dropped after I changed to another type of tire; so I measure the mileage for 5 fill-ups.

a. If my mean mileage has *not* gone down, then I expect the average mileage of the 5 tankfuls to be _____ miles per gallon give or take _____ miles miles per gallon.

b. The average mileage of the 5 tankfuls turned out to be 21.9 mpg. Assuming that the mileage per tankful follows the normal curve with $\sigma = 2$ mpg, what is the probability that the sample average mileage is 21.9 mpg or less (that is, assuming my mean mileage has not changed)?

c. Am I justified in saying that the mean gas mileage as dropped? Explain.

• **11.25** A company that owns and services a fleet of cars for its sales force has found that the service lifetime of disc brake pads varies from car to car according to a normal curve with mean of 43,000 miles and standard deviation of 4,500 miles. A new and cheaper brand of brake pads is installed on 36 cars.

a. If the new brand of pads lasts as long as the previous type, you would expect the average life of brake pads on the 36 cars to be _____ miles give or take _____ miles.

b. The average life of the pads on these 36 cars turned out to be only 41,000 miles. What is the probability that the sample average life is 41,000 miles or less if the new pads last just as long as the old ones?

c. The company takes this probability as evidence that the average lifetime of the new brand of pads is less than that of the old ones. What would you say? Explain.

11.26 IQ's are normally distributed with a mean of 100 and a standard deviation of 15.

a. If a person is chosen at random, what is the chance that he or she will have a score over 115?

Two persons are chosen at random.

b. What is the chance that both will have a score over 115?

c. What is the chance that at least one will have a score over 115?

d. What is the chance that their total score will be over 230?
   e. What is the chance that their average score will be over 115?

• **11.27** Serum cholesterol levels in 17-year-olds follow the normal curve with mean 176 mg/dl and standard deviation of 30 mg/dl.

   a. For a randomly selected 17-year-old, what is the probability that he or she has serum cholesterol levels above 200?
   b. For a random sample of ten 17-year-olds, what is the probability that the mean serum cholesterol level is above 200?

11.7 THE Z STATISTIC

For a random sample $X_1, X_2, \ldots, X_n$, the **Z statistic** is the random variable

$$Z_{\overline{X}} = \frac{\overline{X} - \mu}{\sigma/\sqrt{n}}$$

This statistic is the standardized random variable $\overline{X}$, because $\mu_{\overline{X}} = \mu$ and $\sigma_{\overline{X}} = \sigma/\sqrt{n}$. Thus, the law of averages for the Z statistic is just a variant of the law of averages for the statistic $\overline{X}$.

The **law of averages for** $Z_{\overline{X}}$ is as follows:

First part   1. $\mu_{Z_{\overline{X}}} = 0$.

Second part 2. For sampling with replacement, we have $\sigma_{Z_{\overline{X}}} = 1$. For sampling without replacement, multiply by the reduction factor.

Third part   3(a). If the population is normal, then the statistic $Z_{\overline{X}}$ is standard normal.

   3(b). Otherwise, if n is large [general rule: $n \geq 40$], then the statistic $Z_{\overline{X}}$ follows the standard normal curve.

11.8 THE *T* STATISTIC

The T statistic is used in cases when the parameter σ is unknown and we use the sample standard deviation instead. For a random sample $X_1, X_2, \ldots, X_n$ of size n, with sample mean

$$\overline{X} = \frac{\sum_k X_k}{n}$$

and sample standard deviation

$$S = \sqrt{\frac{1}{n-1} \sum_k (X_k - \overline{X})^2}$$

we define the **T statistic** with $n-1$ *degrees of freedom* (we write $df = n-1$) as follows:

$$T_{n-1} = \frac{\overline{X} - \mu}{S/\sqrt{n}}$$

The difference between the T statistic and the Z statistic of the previous section is that the T statistic uses the sample standard deviation, whereas the Z statistic uses the population standard deviation σ.

> *The advantage of the **T** statistic over the **Z** statistic is that it can be computed without the knowledge of the parameter σ.*

If the sample size n is large, there is little difference between the two. However, for small n the **law of averages for the *T* statistic** is quite different.

First part   1. If the population is normal, then for each n greater than 2, we have

$$E(T_{n-1}) = 0$$

Second part   2. If the population is normal, then the statistic T_{n-1} has a density function given by the Student t_{n-1} curves described below.

Third part   3. Otherwise, if the population variance is finite and if n is large [general rule: $n \geq 40$], then the statistic T_{n-1} follows the standard normal curve.

The T statistic was first investigated by W. S. Gosset in 1908.[2] The density curves for the T_{n-1} statistics are called the Student t_{n-1} curves. If we are sampling from a normal population, then the t_{n-1} curves are bell shapes centered at 0, just as the standard normal curve z. The main difference is that the tails of the t_{n-1} curves are fatter due to added variability of the unknown population parameter σ. As the degrees of freedom $df = n - 1$ increase, the tails become narrower and the t_{n-1} curves rapidly approach the standard normal curve z. See Figure 11.4.

Areas under sections of the t curves can be found using tables or programmable calculators. Unfortunately, there is a different curve for each value of the sample size n, so tables (such as Table 4 on page 510 at the back of this book) give only summary information.

Although the results of W. S. Gosset assume that the underlying population is normal, empirical evidence shows that the t curves are rather insensitive to moderate departures from normality. However, in cases in which the population is distinctly nonnormal, such as when the population is Bernoulli, we must either use an exact method (such as in Chapter 6), take a larger sample so that we can use the Z statistic, or use some advanced (*non-parametric*) method that does not require normality.

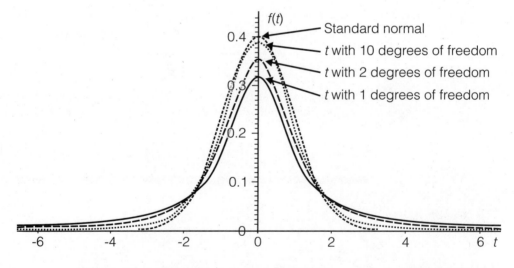

Figure 11.4 Student t curves for various degrees of freedom compared with the standard normal curve

---

[2]W. S. Gosset was a chemist for the Guinness Brewing Company in Dublin, Ireland. Because the company did not permit publication of "trade secrets" by employees, he used the pen name Student. A biography of W. S. Gosset can be found at the website www-gap.dcs.st-and.ac.uk/~history/Mathematicians/Gosset.html.

11.9 ESTIMATORS OF PARAMETERS; ACCURACY AND PRECISION (optional)

When we use a statistic, say W, to estimate a population parameter ω, the **random sampling error** of W is defined to be

$$\text{random sampling error} = W - \omega$$

The expected random sampling error or **bias** of the estimator is defined to be

$$\textbf{Bias:} \quad \beta_W = E(\text{Random sampling error}) = E(W - \omega) = \mu_W - \omega$$

If the bias is zero, $\beta_W = 0$, then we say that the statistic W is an **unbiased estimator of the parameter** ω.

The **efficiency** of an estimator is measured by the **mean squared sampling error**.

$$\textbf{Efficiency:} \quad \text{MSE} = E(W - \omega)^2$$

This looks similar to the variance.

$$\textbf{Variance:} \quad \sigma_W^2 = E(W - \mu_W)^2$$

except that the MSE is measured around the parameter of estimation ω instead of the mean of the statistic μ_W.

The term **accuracy** of an estimator refers to the square root of the MSE.

$$\textbf{Accuracy:} \quad \sqrt{\text{MSE}} = \sqrt{E(W - \omega)^2}$$

The term **precision** of an estimator refers to the square root of the variance.

$$\textbf{Precision:} \quad \sqrt{\text{Var}} = \sqrt{E(W - \mu_W)^2}$$

It turns out that

$$\text{MSE} = \sigma_W^2 + \beta_W^2 \tag{11.9}$$

So we have the relationship

$$\text{accuracy}^2 = \text{precision}^2 + \text{bias}^2$$

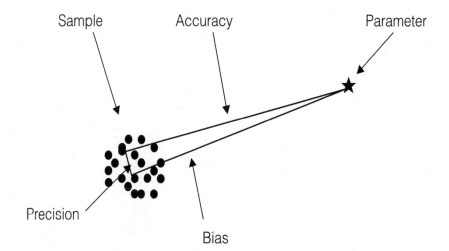

Note that, for an unbiased estimator, the accuracy is equal to the precision, and that the efficiency of the estimator is just its variance.

The estimator $\overline{X}$ When we use the sample mean $\overline{X}$ to estimate the population mean μ, the random sampling error is $\overline{X} - \mu$. The sample mean $\overline{X}$ is an unbiased estimator of μ because the bias is $\beta = E(\text{Sampling error}) = E(\overline{X} - \mu) = \mu - \mu = 0$. The efficiency of the sample mean is

$$\text{Var}(\overline{X}) = \text{Var}(\frac{1}{n}\text{SUM}) = (\frac{1}{n})^2 \text{Var}(\text{SUM}) = (\frac{1}{n})^2 n\sigma^2 = \frac{\sigma^2}{n}$$

The estimator S^2   Next we show that, for any random sample $\{X_1, X_2, \ldots, X_n\}$ of size n, the choice of the divisor $n-1$ for the sample variance

$$S^2 = \frac{1}{n-1} \sum_k (X_k - \overline{X})^2$$

makes it an unbiased estimator of the population variance σ^2.

11.9 ACCURACY AND PRECISION

THEOREM 11.1   *For sampling with replacement, the sample variance is an unbiased estimator of the population variance; that is,*

$$E(S^2) = \sigma^2$$

Proof   Because $S^2 = \frac{1}{n-1}\sum_i (X_i - \overline{X})^2$, by Equation 11.3 it is sufficient to show that

$$E(\sum_i (X_i - \overline{X})^2) = (n-1)\sigma$$

Recall from Equations 11.1, 11.2, 11.7, and 11.8 that

$$E(X_i) = \mu, \text{ for all } i \qquad (11.10)$$

$$\text{Var}(X_i) = E(X_i^2) - \mu^2 = \sigma^2, \text{ for all } i \qquad (11.11)$$

$$E(\overline{X}) = \mu \quad \text{and} \qquad (11.12)$$

$$\text{Var}(\overline{X}) = E(\overline{X}^2) - \mu^2 = \frac{\sigma}{n} \qquad (11.13)$$

Thus, we have

$$\begin{aligned}
E\left(\sum_i (X_i - \overline{X})^2\right) &= E\left(\sum (X_i^2 - 2\overline{X}X_i + \overline{X}^2)\right) \\
&= E\left(\sum X_i^2 - 2\overline{X}\sum X_i + \sum \overline{X}^2\right) \\
&= E\left(\sum X_i^2 - 2n\overline{X}^2 + n\overline{X}^2\right) \\
&= E\left(\sum X_i^2 - n\overline{X}^2\right) \\
&= \sum E(X_i^2) - nE(\overline{X}^2) \\
&= \sum (\sigma^2 + \mu^2) - n\left(\frac{\sigma^2}{n} + \mu^2\right) \text{ by (11.11) and (11.13)} \\
&= n\sigma^2 + n\mu^2 - \sigma^2 - n\mu^2 \\
&= (n-1)\sigma^2
\end{aligned}$$

Warning   Although S^2 is an unbiased estimator of σ^2, the statistic S is not an unbiased estimator of the parameter σ, except in the trivial case where S is constant $S = 0$ for all samples.

EXERCISES FOR SECTION 11.9

11.28 Four guns are tested by firing 12 shots each at a target.[3] Gun 1 was not clamped down well, so it wobbled, and it was aimed slightly low and to the right. Gun 2 was not clamped down well but was aimed correctly. Gun 3 was clamped down and aimed correctly. Gun 4 was clamped down, aimed very carefully, but the shots were affected by a slight breeze.

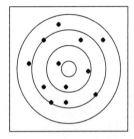 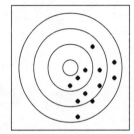 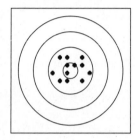 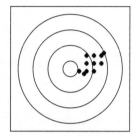

a. Which of the pattern of shots belong to which gun?
b. Which guns are biased? **c.** Unbiased? **d.** The most precise?
e. The most accurate? **f.** Smallest MSE?

11.29 It is known that 72% of the registered voters in Lake County, Indiana, are registered as Democrats. To test a new telephone sampling method, we call 500 Lake County voters and ask their party. We do this 5 times and obtain the following results for the percentage of Democrats. The results are

$$62.8\%, \quad 63.1\%, \quad 64.4\%, \quad 62.3\% \quad \text{and} \quad 65.2\%$$

a. Estimate the bias of this sampling method.
b. Estimate the precision of this sampling methods.
c. Estimate the accuracy of this sampling method.

11.30 Let $X_1, X_2,$ and X_3 be a random sample of waiting times from an exponential random variable X with a mean $\mu = 100$ hours. Consider the estimators $W = (X_1 + 2X_2 + X_3)/4$ and $\overline{X} = (X_1 + X_2 + X_3)/3$.

a. Show that W is an unbiased estimators of μ.
b. Show that $\overline{X}$ is an unbiased estimators of μ.
c. Find the efficiency of W.
d. Find the efficiency of $\overline{X}$.
e. Compare the efficiency of W with that of $\overline{X}$.

---

[3] This example is from T. Wonnacott & R. Wonnacott, *Introductory Statistics*, third edition, J. Wiley & Sons, 1977. This material is used by permission of John Wiley & Sons.

- **11.31** Show that for any random sample $\{X_1, X_2, \ldots, X_n\}$, a weighted mean

$$W = a_1 X_1 + a_2 X_2 + \cdots + a_n X_n, \text{ with } a_1 + a_2 + \cdots + a_n = 1$$

is always an unbiased estimator of μ.

11.32 Continuing with Exercise 11.31 above, show that $W = a_1 X_1 + a_2 X_2 + \cdots + a_n X_n$ in an unbiased estimator of μ only when $a_1 + a_2 + \cdots + a_n = 1$.

11.33 Show that the statistic

$$\frac{1}{n} \sum_k (X_k - \mu)^2$$

is an unbiased estimator of σ^2. You may follow the proof of Theorem 11.1.

11.34 Show that, for a dichotomous population, the statistic P is an unbiased estimator of the parameter π. What is the efficiency of P?

REVIEW EXERCISES FOR CHAPTER 11

- **11.35** Suppose you roll a fair die 100 times and count the number of aces.
    - **a.** What is the largest number of aces possible?
    - **b.** What is the smallest number of aces possible?
    - **c.** You expect _____ aces, give or take _____ aces or so.

11.36 Suppose you roll a fair die 100 times and count the total number of dots.
   - **a.** What is the largest number possible?
   - **b.** What is the smallest number possible?
   - **c.** You expect _____ dots, give or take _____ dots or so.

- **11.37** Suppose that 12% of all American adults are left-handed.
    - **a.** If you take a sample of 1000 American adults, you would expect _____% of them to be left-handed, give or take _____% or so.
    - **b.** What is the probability of getting a sample with less than 10% left-handers?

11.38 In a gubernatorial election in 1998, the incumbent won with 58% of the votes as opposed to 42% for her challenger. An exit poll would have correctly forecast the winner if the majority in the sample voted for the incumbent. Calculate the probability of a correct forecast if the poll had consisted of a random sample of size:

a. $n = 1$
b. $n = 5$
c. $n = 25$ (use normal approximation)

- **11.39** In the 1964 presidential election, 60% voted Democratic and 40% voted Republican. Calculate the probability that a random sample of voters would contain a majority of Democrats for a sample of size:

  a. $n = 1$
  b. $n = 3$
  c. $n = 100$ (use normal approximation)

11.40 A local airline flies aircraft with a load limit of 11,000 lb. It is estimated that passengers (with luggage) weigh an average of 200 lb with a standard deviation of 50 lb. What is the probability that 50 passengers will overload such an aircraft?

- **11.41** The vitamin B-2 content of a certain brand of vitamin pills follows the normal curve. The average content is claimed to be 30 mg with a standard deviation of 2 mg A quality-control inspector selects 25 pills for testing. What is the probability that the total vitamin B-2 content of these pills is less than 725 mg?

11.42 Suppose that a machine set for filling 1-lb boxes of sugar yields a weight W with $E(W) = 16.0$ ounces and $\sigma_W = 0.2$ ounces. A carton of sugar contains 48 boxes.

  a. Describe and sketch the density function of the weight T of the carton.
  b. Compute the probability that the total weight of the carton exceeds 48.2 lb (771.2 ounces).

- **11.43** Imagine a gambling casino consisting of 100 busy roulette tables. Suppose that each table brings in an average hourly profit of $50 with a standard deviation of $25.

  a. Fill in the blanks. The entire casino brings in an average hourly profit of _____ with a standard deviation of _____.
  b. Find the probability that the casino will make a profit less than $4500 on any given hour.

11.44 A group of 50,000 tax forms of people in a certain profession shows a mean gross income of $45,000, with a standard deviation of $25,000. Furthermore, 20% of the forms show a gross income over $60,000. A group of 400 forms is chosen at random for audit.

  a. Find the probability that less than 19% of the forms chosen for audit show gross incomes over $60,000.

b. Find the probability that the total gross income of the audited forms is over $19,000,000.

• **11.45** The life span of an electrical component follows the exponential curve with mean and standard deviation of 150 hours.

   a. What is the probability that the life span of such a component will exceed 200 hours?
   b. Suppose that a random sample of 50 of such components was examined. What is the probability that the mean of the sample will exceed 200 hours?

11.46 Consider the small population consisting of the set $\{2, 2, 4, 6, 10, 12\}$, which has mean $\mu = 6$ and $\sigma = \sqrt{44/3}$.

   a. List all 15 samples of size 2 without replacement.
   b. List all 15 sample means $\bar{x}$ of the above samples.
   c. Find $\mu_{\bar{x}}$ and $\sigma_{\bar{x}}$ of these sample means.
   d. For this population and $n = 2$, verify parts 1 and 2 of the law of averages for sampling without replacement. In other words, verify

$$\mu_{\bar{x}} = \mu \quad \text{and} \quad \sigma_{\bar{x}} = \sqrt{\frac{N-n}{N-1}} \frac{\sigma}{\sqrt{n}}$$

• **11.47** For a population with mean $\mu = 17$ and standard deviation $\sigma = 40$, what should the sample size n be in order to give you $\sigma_{\bar{X}}$ less than 5?

11.48 For a population with mean $\mu = 17$ and standard deviation $\sigma = 40$, what should the sample size n be in order to give you an $\sigma_{\bar{X}}$:

   a. less than 4?   b. less that 2?   c. less than 1?

• **11.49** In a dichotomous population with $\pi = 0.5$, what should the sample size be in order for the sample proportion P to have σ_P no more than:

   a. 0.03?   b. 0.02?   c. 0.01?

11.50 Show that for a fixed sample of size n, the largest possible standard error σ_K for the binomial variable K occurs when $\pi = 1 - \pi = 0.5$.

• **11.51** A poll is planned a few days before an election in a large city for a closely contested race for mayor. In light of the previous exercise, what size sample should be taken to guarantee an SE_P no larger than 0.01?

11.52 Illinois has about 20 times as many registered voters as Wyoming. Choose one of the following options and explain. Other things being equal, a random sample of 1800 registered voters taken in Illinois will give results of:

   (a) greater accuracy   (b) about the same accuracy   (c) less accuracy

than a random sample of 1800 registered voters taken in Wyoming.

- **11.53** In rush hour stop-and-go traffic on wet pavement, the delay between the time a car starts to move and the car behind it starts to move is an exponential random variable with a mean and standard deviation of $\theta = 2$ seconds. Suppose there are 16 cars ahead of you waiting at a red traffic light. When the light changes to green, find the mean and standard deviation for the waiting time before you are free to move (include a delay for the first car in line as well).

11.54 In a city, 30% of the work force would answer Yes if asked Q1: "Do you use public transportation to get to work?" Also, 60% would answer Yes if asked Q2: "Do you favor using gasoline taxes to repair roads and bridges?" In addition, 18% would answer Yes to both questions. An individual is to be selected at random from the city's work force and asked both questions.

Let A denote the event that the individual responds Yes to Q1.

Let B denote the event that the individual responds Yes to Q2.

Let A' and B' denote the complements of A and B, respectively.

a. Are A and B independent events? Explain.
b. Are A and *not* A independent events? Explain.
c. Show $Pr(A' \cap B) = (1 - Pr(A))Pr(B) = 0.42$.
d. Show $Pr(A' \cap B') = (1 - Pr(A))(1 - Pr(B)) = 0.28$.

* * * * * * * *

- **11.55** (Proof) This is a continuation of the previous exercise. Let X denote the random variable that equals 1 whenever A occurs and equals 0 otherwise. Similarly, let Y denote the random variable that equals 1 whenever B occurs and equals 0 otherwise.

    a. Find the mean and standard deviation of X.
    b. Find the mean and standard deviation of Y.
    c. Let $Z = X \cdot Y$. Find the mean and standard deviation of Z.
    d. Which of X, Y, and Z are Bernoulli random variables?
    e. Let $D = X - Y$. Find the mean and standard deviation of D.

11.56 (Proof) This is a continuation of the previous two exercises. Suppose we randomly select 100 individuals (with replacement) from the city's work force and ask each of them the two questions Q1 and Q2. Then, we record the jth individual's answers as (X_j, Y_j), where $X_j = 1$ if the answer to Q1 is Yes and $X_j = 0$ otherwise; and similarly, $Y_j = 1$ if the answer to Q2 is Yes and $Y_j = 0$ otherwise. These are 100 independent and identically distributed bivariate random variables. For each $j = 1, 2, \ldots, 100$, let $D_j = X_j - Y_j$ be the difference in responses to the two questions.

a. Show the expected value of SUM $= \sum_{j=1}^{100} D_j$ is -30.

b. Show the standard deviation of SUM $= \sum_{j=1}^{100} D_j$ is $\sqrt{45}$.

c. Use the Central Limit Theorem 8.2 to show that

$$Z_{SUM} = \frac{\sum_{j=1}^{100} D_j + 30}{\sqrt{45}}$$

is approximately standard normal.

d. Approximate $Pr\left(-0.43 \leq \frac{1}{100} \sum_{j=1}^{100} D_j \leq -0.17\right)$.

- **11.57** Prove the warning statement on page 325. In particular, show that, when sampling with replacement, the sample standard deviation S underestimates the population standard deviation σ. Use Theorem 7.2 on page 209 and the warning statement on page 135.

11.58 Let $V = 2Z - 1$ where Z be a random variable with probability density function $f(z) = \frac{1}{\sqrt{2\pi}} e^{-\frac{1}{2}z^2}$, $-\infty < z < \infty$.

a. Find the mean of V.
b. Find the standard deviation of V.
c. Find the probability density function of V.
d. Is V normally distributed?

C 11.59 Use software such as Minitab or Excel to simulate density curves of the random variable SUM such as in Figure 11.2 (on page 307) by taking random samples from an exponential distribution with mean and standard deviation equal to 3. Take 100 random sample of the following sizes, compute the sample sum for each of the samples, and then obtain histograms of the 100 sample sums:

a. $n = 1$ b. $n = 4$ c. $n = 9$ d. $n = 16$ e. $n = 25$

For each simulation, compare your results with the law of averages for the sample sum. Notice how the skew tends to disappear as n gets larger.

C 11.60 As in Exercise 11.59, take random samples from an exponential distribution with mean and standard deviation equal to 3. This time simulate the density curves of the random variable $\overline{X}$. Take 100 random sample of the following sizes, compute the sample mean for each of the samples, and then obtain histograms of the 100 sample means:

a. $n = 1$ b. $n = 4$ c. $n = 9$ d. $n = 16$ e. $n = 25$

For each simulation, compare your results with the law of averages for the sample mean. Notice how the skew tends to disappear as n gets larger.

Chapter 12

THE z AND t TESTS OF HYPOTHESES

12.1 THE z TEST

12.2 TWO-SIDED z TEST

12.3 BOOTSTRAPPING AND THE t TEST

12.4 WHICH IS THE NULL HYPOTHESIS?

In Chapter 6, we considered testing of hypotheses using the binomial and hypergeometric random variables. Now we will test hypotheses about parameters of a population by taking advantage of the law of averages described in Chapter 11. The z test uses the part of the law of averages (Central Limit Theorem), which states that for large samples, the statistics K, SUM, P, and $\overline{X}$ follow the normal curve. The t test deals with cases of small samples where the population σ is unknown. In Chapter 13, we will deal with situations in which we estimate a parameter of the population by examining a random sample without the benefit of a hypothesis. Then, in Chapter 14, we consider inferences involving comparisons of two populations.

12.1 THE z TEST

Example 12.1   *Making a better laser printer.*   In order to make laser printers more reliable, a manufacturer makes them with replaceable organic photoconductor (OPC) cartridges that contain many of the moving parts of the printer. Tests have shown that the OPC cartridges have a mean time to failure (MTF) of 30,000 pages with a standard deviation of 7500 pages. Engineers have come up with an improved design that they think will have a longer mean time to failure.

The new design is presented to management. Management wants proof that the new design has a longer life than the present model. Because in theory, any number of OPC cartridges can be manufactured, the population is, in effect, infinite. A decision on whether the new OPC cartridges are truly better has to be based on a finite sample. Hence, 100% certainty cannot be obtained. Suppose that management wants to be "99% confident" that the new OPC cartridges last longer.

The risks Here are the risks. If the new OPC cartridges are really no better than the present ones, but management decides to change production to the new design, then it is making a Type I error. This error will result in costs associated with change in production, advertising, and new inventory. On the other hand, if the new OPC cartridges have a better mean time to failure but management decides that the observed improvement is due to sample variability, then it is making a Type II error. The computer industry is very competitive. Such an error will result in loss of sales.

Management approves the manufacture and testing of a sample of 100 OPC cartridges with the new design. These 100 new cartridges are found to have a mean time to failure of 31,500 pages with a standard deviation of 7000 pages.

Let μ be the (unknown) mean time to failure of the OPC cartridges with the new design. The sample mean is $\bar{x} = 31,500$ pages. Does this prove that the improved OPC cartridges last longer? Or can the observed increase in mean time to failure be explained by sample variability?

The burden of proof lies in showing that the new OPC cartridges are indeed better. The null hypothesis is thus that they are no better. The alternative hypothesis is that the new OPC cartridges last longer:

$$H_0 : \mu = 30,000 \qquad H_a : \mu > 30,000$$

The null hypothesis is stated as an equality. We say that it is a **simple hypothesis**. The alternative is an inequality, a **compound hypothesis**.

> *The rules of statistical evidence require us to assume* $\mathbf{H_0}$
> *during the conduct of the test.*

Assuming H_0, the statistic $\overline{X}$ is normal with a mean of $E(\overline{X}) = \mu = 30{,}000$. What about the standard deviation? We have a **"historical"** population standard deviation of $\sigma = 7500$. On the other hand, the sample standard deviations is $s = 7000$. The decision of which one to use is part of the art of statistics. Here we go with $\sigma = 7500$ because if the new OPC cartridges are the same as the old, then the SD for the new OPC cartridges should be the same as for the old (however, the other choice is also defensible). This gives us a sampling standard error of $\sigma_{\overline{X}} = \frac{\sigma}{\sqrt{n}} = \frac{7500}{\sqrt{100}} = 750$ pages. Figure 12.1 is a graph of the statistic $\overline{X}$ under the assumption H_0.

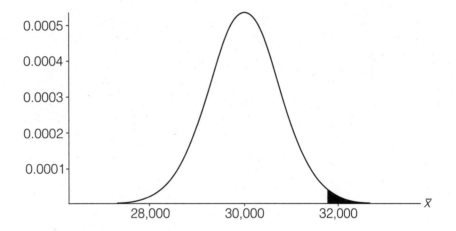

Figure 12.1 Sampling distribution of $\overline{X}$ for $n = 100$, $\mu = 30{,}000$, $\sigma = 7500$, $\mu_{\overline{X}} = 30{,}000$, and $\sigma_{\overline{X}} = 750$. The rejection region for $\alpha = 0.01$, consisting of the upper 1%, is shaded.

Solution — To meet the 99% criteria required by management, the sample mean has to fall in the upper 1% region. This is the **level of significance** $\alpha = 0.01$. The boundary between the shaded rejection region and the unshaded acceptance region is called the **critical value** $\overline{x}_{.99} = 31{,}743.75$, which is denoted $\overline{x}_c$. It is computed by first finding the critical value in standard units $z_c = z_{.99} = 2.326$. The **rejection region**, also called the **critical region**, is the interval $(\overline{x}_c, \infty) = (31{,}743.75, \infty)$, or in standard units $(z_c, \infty) = (2.326, \infty)$. This rejection region is shaded in Figure 12.1. Note that the observed $\overline{x} = 31{,}500$ does not fall in the rejection region. The decision must be to retain H_0. According to the decision rule of $\alpha = 0.01$, it has not been proven beyond a reasonable doubt that the new OPC cartridges have a longer MTF.

Critical value procedure

The **critical value procedure** is an alternative to the **P-value procedure** for testing hypotheses presented in Chapter 6. The critical value procedure works best in situations where the same test is being performed for different data sets, such as in quality control. The P-value procedure is preferred for reports and publications. Both have six steps. The table below illustrates the difference.

Step	Critical value procedure	P-value procedure
1	$H_0: \mu = 30,000$	$H_0: \mu = 30,000$
2	$H_a: \mu > 30,000$	$H_a: \mu > 30,000$
3	$\alpha = 0.01$	$\alpha = 0.01$
4	Critical value $$z_c = z_{1-\alpha} = 2.326$$ $$\bar{x}_c = \mu + z_c \frac{\sigma}{\sqrt{n}}$$	Test statistic $$z_s = \frac{\bar{x}_s - \mu}{\sigma/\sqrt{n}} = \frac{31500 - 30000}{750} = 2$$
5	Test statistic $$z_s = \frac{\bar{x}_s - \mu}{\sigma/\sqrt{n}} = 2$$ Compare z_s to z_c (or $\bar{x}_s$ to $\bar{x}_s$)	P-value $$= Pr(Z \geq z_s) = 0.02275$$ Compare P-value to α
6	In English: Insufficient evidence to conclude new OPC's have longer MTF.	In English: Insufficient evidence to conclude new OPC's have longer MTF.

Had management chosen the weaker level of significance $\alpha = 5\%$, the critical value would have been $\bar{x}_{.95} = 31,233.75$, the observed $\bar{x}$ would have fallen in the rejection region, and the null hypothesis would have been rejected.

> *In order to avoid* **data snooping,** *the level of significance, which determines the critical value, should be decided before the evidence is seen.*
>
> *The level of significance α of a test is the risk of rejecting the null hypothesis when the null hypothesis is actually true.*

Recall the terminology used earlier in Chapter 6. Evidence is said to be **significant** if it rejects the null hypothesis at the $\alpha = 0.05 = 5\%$ level of significance. Evidence is said to be **highly significant** if it rejects the null hypothesis at the $\alpha = 0.01 = 1\%$ level of significance.

In deciding the level of significance, one must take into consideration the risks of incorrectly rejecting a true null hypothesis as well as the risk of incorrectly retaining a false null hypothesis. There is no reason to stick with the 5% or 1% levels. Different situations and different readers require different standards for evidence. In scientific reports, it is not satisfactory to merely report *"the results were not significant"* or *"the results were highly significant."* Report the *P*-value as well.

Example 12.2 **More on the laser printer.** In the previous example, management had set the level of significance of $\alpha = 1\%$ in testing a sample of 100 of the redesigned OPC cartridges. The experiment was conducted and the evidence was not strong enough to lead to the rejection of the null hypothesis. How strong is the evidence? Suppose, for example, that in truth the new design increased the MTF by 1000 pages. How likely is it that a test of just 100 OPC cartridges will lead to the correct conclusion?

Solution Again, assuming the population standard deviation is 7500 and using the critical value $\overline{x}_{.99} = 31{,}744$ as the critical value for the test, we compute:

$$\begin{aligned} Pr(\text{Reject } H_0 \mid \mu = 31000, \sigma = 7500) &= Pr(\overline{X} \geq 31744 \mid \mu = 31000, \sigma = 7500) \\ &= Pr\left(\frac{\overline{X}-\mu}{\sigma_{\overline{X}}} \geq \frac{31744-31000}{7500/\sqrt{100}}\right) \\ &\approx Pr(Z \geq 0.992) \\ &= 0.1606 \end{aligned}$$

Power of the test We say that the **power of the test** $1 - \beta$ for a 1000-page improvement in MTF is 16.06%. This test fails to detect a 1000-page improvement in MTF for approximately $\beta = 84\%$ of samples of 100 redesigned OPC cartridges. If management thought that a 1000-page increase in MTF was of practical importance, then it should have tested more than 100 newly designed OPC cartridges, which would have made the test more powerful.

EXERCISES FOR SECTION 12.1

12.1 Under the same $\alpha = 1\%$ rule of the above test procedure, compute the power to detect a 1000-page improvement in MTF if 400 newly designed OPC cartridges are tested.

12.2 If management insisted on testing just 100 redesigned OPC cartridges and wanted a 50% chance of detecting a 1000-page improvement in MTF, what should the level of significance α be now?

- **12.3** Suppose that the mean income of 35-year-olds in the Unites States is $24,000. A random sample of 100 35-year-olds in North Dakota results in a sample mean of $23,500 and a sample standard deviation of $4000. Although we don't know the population standard deviation σ, the sample is large, so it is reasonable to use the sample standard deviation as an estimate of σ.

    a. At the 5% level of significance, should we conclude that 35-year-olds in North Dakota have lower incomes on average than the national average? State the null and alternative hypothesis, find the rejection region, and state your conclusion in common English.

    b. Suppose that the true mean income of 35-year-olds in North Dakota is actually $23,600. For the decision rule found in part a, find the probability β of committing a Type II error.

12.4 Suppose that the mean income of 35-year-olds in the United States is $24,000. A random sample of 100 35-year-olds in California results in a sample mean income of $24,600 and a sample standard deviation of $4000. Although we don't know the population standard deviation σ, the sample is large, so it is reasonable to use the sample standard deviation as an estimate of it.

    a. At the 5% level of significance, should we conclude that 35-year-olds in California have a higher average income than the national average? State the null and alternative hypotheses, find the rejection region, and state your conclusion in common English.

    b. Suppose that the true mean income of 35-year-olds in California is actually $24,500. For the decision rule found in part a, find the probability β of committing a Type II error.

- **12.5** In the preceding five years, entering students at a certain university had an average SAT verbal score of 612 points. A simple random sample of 100 students taken from this year's class showed the average SAT verbal score for students to be 630 with a standard deviation of 80 points. Does this show an improvement in verbal abilities (or can the observed difference be explained by chance variation)? Carry out the appropriate test at the $\alpha = 5\%$ level. Although we don't know the population standard deviation σ, the sample is large, so it is reasonable to use the sample standard deviation as an estimate of it.

12.6 National data show that, on the average, college freshmen work an average of 12.5 hours per week. One administrator does not believe that freshmen at her college spend so much time working. Her college has nearly 2000 freshmen. She takes a simple random sample of 50 and interviews them. On average, they work only 9.9 hours a week with a standard deviation of 12.3 hours. Is the difference between 9.9 and 12.5 real? Test at the 5% level of significance. Although we don't know the population standard deviation σ, the sample is large, so it is reasonable to use the sample standard deviation as an estimate of it.

12.2 TWO-SIDED z TEST

Example 12.3 *Testing people for extrasensory perception.* Let's set up a test to screen people for extrasensory perception (ESP). Three cards are face down on a table; one has a star, another has wavy lines, and the third has a large circle. The subject has to guess which of the three cards has the star. Evidence of ESP is obtained by counting successes in repeated trials—say, in $n = 150$ trials.

What would be needed to prove that someone has ESP? Weak evidence would not do it. Clearly, the null hypothesis must be that the person being tested has no ESP and the level of significance α must be very small. If we used the criterion $\alpha = 1\%$, then one out of every 100 people without ESP would pass the test. What would it mean if 1000 people were tested at the $\alpha = 1\%$ level and 10 of them passed? This should not be surprising; it is what we expect if *none* of them had ESP.

So, for a general screening, let's choose something smaller that 1%—say, $\alpha = 0.001 = 0.1\%$. Now what would be considered evidence of ESP? Results that cannot be explained by chance. If the null hypothesis is true, you expect the person to guess correctly $\pi = 1/3$rd of them. Actually you expect the sample count K to be approximately $E(K) = n\pi = 150(1/3) = 50$, give or take $\sigma_K = \sqrt{n\pi(1-\pi)} = \sqrt{(150)(1/3)(2/3)} = 5.7735$ or so.

Guessing *all* correct could not be explained by chance. Interestingly, guessing all wrong could not be explained by chance either. (Huh? Think of someone who actually has ESP, but who does not believe in it and is subconsciously trying to guess wrong.) The alternative hypothesis must be two-sided; that is, we reject H_0 whenever the sample count K is *too large* or *too small*.

Now we are ready to set up the test:

$$H_0 : \pi = 1/3 \qquad H_a : \pi \neq 1/3 \qquad \alpha = 0.001$$

Hypotheses say something about parameters of the population.

Decisions about the hypotheses and level of significance of a test must never consider the results of the experiment (no peeking!). Well, this is the ideal. Often you hear of the report of the results before you decide to test, so you have to ignore the data when you set up the hypotheses and decision rule. Certainly, the result of the test cannot be any part of the statement of the hypotheses.

The question of whether to have a one-sided alternative, such as in Example 12.1 on page 333, or a two-sided alternative, as in this example, can be controversial.

> *A decision of whether to have a one-sided alternative should be based on the goals of the test. It should never be based on the data used for the test; doing so would be a form of data snooping.*

If the subjects to be tested all *claimed* to have ESP, an observed value of K that is *too* small, such as $k = 36$, would have to be explained as merely random sampling error. It would then be reasonable to have a one-sided alternative $\pi > 1/3$. Because we are screening the general population, it is reasonable to have a two-sided alternative.

Because $n = 150$ is so large, the binomial exact test as described in Chapter 6 is not practical. We use the law of averages for the sample count. The statistic K follows the normal curve with mean 50 and standard deviation 5.77. See Figure 12.2.

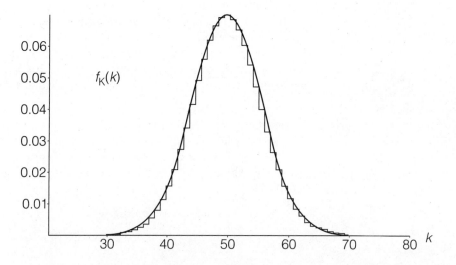

Figure 12.2 The probability histogram of the sample count K with a normal curve superimposed (here $n = 150$, $\mu_K = 50$, and $\sigma_K = 5.77$)

Solution The risk α of erroneously rejecting H_0 must be split between the two sides of the alternative. The rejection region is the tail to the left of $-z_{\frac{\alpha}{2}} = -z_{.0005} = -3.29053$ and the tail to the right of $z_{1-\frac{\alpha}{2}} = z_{.9995} = +3.29053$. If we do not use a continuity correction, then the critical values are:

$$k_c = 50 \pm 3.29053\sqrt{150 \cdot \tfrac{1}{3} \cdot \tfrac{2}{3}}$$

This results in the following decision rule (and rejection region):

Retain H_0 if $\quad 31.00212 \leq K \leq 68.99788$

Reject H_0 if $\quad K < 31.00212 \quad$ or $\quad K > 68.99788$

Of course, the sample count K has to be a whole number, so the decision rule is really:

Retain H_0 if $\quad 32 \leq K \leq 68$

Reject H_0 if $\quad K \leq 31 \quad$ or $\quad K \geq 69$

Effective significance level The *effective* significance level of the test is then the area of the left and right rejection regions, when they are adjusted for the discrete values of K. Using the normal approximation this computes to

$$\alpha = Pr(K \leq 31) + Pr(K \geq 69) = 0.000677 + 0.000677 = 0.00135$$

This normal approximation of a binomial probability compares favorably with the direct computation of the rejection region using the binomial probability mass function, which computes to

$$\alpha = 0.0004457 + 0.0008571 = 0.00130$$

Modified rejection region Because the effective α does not quite meet the $\alpha = 0.001$ level of significance we sought, let's reduce the size of the rejection region to

Retain H_0 if $\quad 31 \leq K \leq 69$

Reject H_0 if $\quad K \leq 30 \;$ or $\; K \geq 70$

We get $\alpha = 0.000366 + 0.000366 = 0.00073$ with a normal approximation (compared to a direct computation of $\alpha = 0.0004457 + 0.0004457 = 0.00089$).

With discrete random variables, it is often necessary to calculate the effective α and then modify the rejection region to meet the target level of significance

---

342   CHAPTER 12.   THE z AND t TESTS OF HYPOTHESES

Now let's say that someone guessed 64 correct out of the 150. This is the test statistic $k_s = 64$, which is 14 guesses better than the expected value of $n\pi = 50$, but it still falls in the acceptance region; no ESP proven for this person.

P-*value* What about the P-value for this person? If the alternative hypothesis $H_a : \pi > 1/3$ were one-sided, then the P-value would be the probability that someone without ESP would get 64 or more correct. Using the continuity correction, this computes to

$$\begin{aligned}
P\text{-value} &= Pr(K \geq k_s) = Pr(K \geq 64) = Pr(K > 63.5) \\
&\approx Pr\left(Z \geq \tfrac{63.5-50}{5.77350}\right) = Pr(Z \geq 2.338269) \\
&= 0.009687
\end{aligned}$$

However, here the alternative hypothesis is two-sided $H_a : \pi \neq 1/3$. Thus, the P-value is the probability that K differs from the expected value of $\mu_K = 50$ by at least 14. This computes to:

$$\begin{aligned}
P\text{-value} &= Pr(K \leq 36) + Pr(K \geq 64) \\
&= 2Pr(K \geq 64) = 2Pr(Z \geq 2.338269) \\
&= 0.019373
\end{aligned}$$

> *For a two-sided alternative hypothesis* $\mathbf{H_a}$, *the P-value is twice as large as for a one-sided alternative.*

See Figure 12.3 for a graph showing the P-value for a two-sided test.

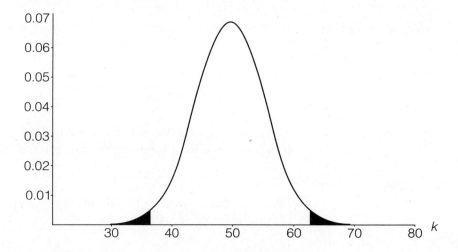

Figure 12.3 Shaded region indicates P-value for someone guessing $k_s = 64$ correct answers out of $n = 150$, when $H_0 : \pi = 1/3$, and the alternative hypothesis is two-sided $H_a : \pi \neq 1/3$

This example illustrates that reports of tests should explain the specific hypotheses (for example, one-sided or two-sided) as well as give the P-value.

For a discrete statistic, such as K, the effective α should be calculated, and reported. The effective α is the probability that the statistic K is in the rejection region when the rejection region is adjusted for whole value of K.

Below is a comparison of the critical value and P-value procedures for two-sided tests.

Step	Critical value procedure	P-value procedure
1	$H_0 : \pi = 1/3$	$H_0 : \pi = 1/3$
2	$H_a : \pi \neq 1/3$	$H_a : \pi \neq 1/3$
3	$\alpha = 0.001$	$\alpha = 0.001$
4	Critical values $$z_c = \pm z_{1-\alpha/2}$$ $$k_c = n\pi \pm z_c \sqrt{n\pi(1-\pi)}$$	Test statistic $$z_s = \frac{K_s - n\pi}{\sqrt{n\pi(1-\pi)}}$$
5	Test statistic $$z_s = \frac{K_s - n\pi}{\sqrt{n\pi(1-\pi)}}$$ Retain H_0 if z_s is between the two values of z_c (or k_s is between the two values of k_c); otherwise reject H_0.	P-value $$= 2Pr(Z \geq z_s)$$ Compare P-value to α.
6	In English: Insufficient evidence to conclude person has ESP.	In English: Insufficient evidence to conclude person has ESP.

EXERCISES FOR SECTION 12.2

- **12.7** A sample of size $n = 121$ is taken from a large dichotomous population. The sample proportion turns out to be $p = 0.64$. Perform a test of the hypothesis $\pi = 0.50$ against a two-sided alternative. Use $\alpha = 0.05$.

12.8 A psychic claims to have extrasensory perception (ESP). To test this claim, you have him guess which of three cards, lying face down on a table, is the ace of spades. He guesses correctly 28 times out of 72 attempts.

a. State the null hypothesis.

b. State the alternative hypothesis.

c. Compute the P-value.

d. Choose one option: Decide at the 5% level of significance whether he *does* or *does not* have ESP.

- **12.9** A professional basketball player used to make only 55% of his free throw shots. Now the player's agent claims he has improved and wants more money. The agent points out that the player has made 40 of 60 free throw attempts this season.

  a. Test whether the player has *improved* (this makes it a one-sided alternative) at the $\alpha = 0.05$ level. Be sure to state H_0 and H_a. Note that making 40 out of 60 free throws is the evidence. It cannot be used in the statement of the hypotheses. State your conclusion in plain English.

  b. Repeat the test at the $\alpha = 0.01$ level.

  c. Suppose, in fact, that the player has improved to the extent that he can make 70% of his free throws. Find the probability β that the procedure at the $\alpha = 0.01$ level would lead to the wrong conclusion.

12.10 A sample of size $n = 64$ is taken from a large population with a known standard deviation of $\sigma = 1.4$. The observed sample mean turns out to be $\bar{x} = 28.5$. Test the hypothesis that the population mean is $\mu = 30$ against a two-sided alternative. Use $\alpha = 0.05$.

- **12.11** A sample of size $n = 81$ is taken from a large population with an unknown standard deviation. The sample has a mean of $\bar{x} = 137$ and standard deviation of $s = 12$. Test the hypothesis that the population mean is $\mu = 140$ against a two-sided alternative. Use $\alpha = 0.05$.

12.12 A sample of size $n = 91$ is taken from a large population with an unknown standard deviation. The sample has a mean of $\bar{x} = 477$ and standard deviation of $s = 120$. Test the hypothesis that the population mean is $\mu = 500$ against a two-sided alternative. Use $\alpha = 0.05$.

- **12.13** Bolts used to assemble certain engine blocks should average 1.50 inches in length. A random sample of 400 such bolts taken from a single machine show an average of 1.504 inches and a standard deviation of 0.075 inches. Test with a two-sided alternative at the $\alpha = 0.05$ level of significance to determine whether this machine needs adjustment.

12.3 BOOTSTRAPPING AND THE t TEST

In statistics, **bootstrapping** refers to using information in the sample itself to infer something about the sampling variability of a statistic, rather than making perhaps unreasonable assumptions about the population from which the sample is drawn.[1] In this section, we are using the term **bootstrapping** to refer to the use of the sample standard deviation s in place of the population standard deviation σ. Later, in Chapter 17, we will consider other examples of bootstrapping.

Example 12.4 *Math aptitude test.* Suppose we want to estimate math scores of 6th-grade students in Illinois by using a California math aptitude test. The test has been normalized for California 6th-grade students to have a mean of $\mu_{CA} = 250$ and a standard deviation of $\sigma_{CA} = 15$. In order to test whether Illinois students are better than California students, we use a sample of, say $n = 100$ Illinois 6th graders, compute the sample mean $\overline{x}_{IL}$, and then use the population standard deviation σ_{IL} from Illinois to calculate the standard error for the mean. Because we do not know the parameter σ_{IL}, we have two alternatives. We could make the (perhaps unreasonable) assumption that the standard deviations for the two states Illinois and California are the same and thereby use the "historical" standard deviation σ_{CA}. Or we could estimate σ_{IL} by using the sample standard deviation s. If we use the sample s, we call it *bootstrapping*. We must then use the T statistic as described in Section 11.8 on page 321.

Example 12.5 *At-rest pulse.* We return to Exercise 6.9. A three-month program of physical fitness is conducted on five subjects. The at-rest pulse is measured before and after the program with the following results:

Subject	1	2	3	4	5
Before	73	77	68	62	72
After	68	72	64	62	71

In Exercise 6.9 on page 175, the sign test was used to test whether the average pulse has gone down. Because there is one tie (which is discarded), four (+)

---

[1] Bradley Efron used the term **bootstrap methods** in his article *Bootstrap methods: Another look at the jackknife*, Annals of Statistics **7** (1979), pp. 1–26, to refer explicitly to methods based on the idea of treating the sample as if it were the population from which that sample was drawn, and then studying the sampling distribution of the original statistic by sampling with replacement from this surrogate "population."

signs, and no $(-)$ signs, the P-value computes to P-value $= 1/16 = 0.0625$.
Now use the t test at the $\alpha = 0.05$ level of significance.

Solution Here $H_0 : \mu = 0$, $H_a : \mu > 0$, and $\alpha = 0.05$. The five subjects have an observed sample mean of $\overline{x} = 3$ and

$$s = \sqrt{\frac{2^2 + 2^2 + 1^2 + (-3)^2 + (-2)^2}{4}} = 2.3452$$

which makes the estimated standard error

$$SE_{\overline{x}} = \frac{s}{\sqrt{n}} = 1.0488$$

This gives us an observed

$$t_s = \frac{\overline{x} - \mu}{s/\sqrt{n}} = 2.8604$$

using $\mu = 0$ specified by the null hypothesis. On the TI-83, the P-value comes out to

$$P\text{-value} = Pr(T \geq t_s) = tcdf(2.8604, 99999999, 4) = 0.02296$$

If we use Table 4 on page 510, we must look in the row for four degrees of freedom $df = n - 1 = 4$. Then the observed $t_s = 2.8604$ falls between the upper tail probabilities 0.02 and 0.025. This means $0.02 < P$-value < 0.025.

Note that the sign test is not significant at the 0.05 level, whereas the t test is significant at the 0.05 level.

The t test has more power than the sign test. However, the t test assumes that the random sample comes from a population that is approximately normal. If such an assumption cannot be made, one cannot use the t test.

EXERCISES FOR SECTION 12.3

12.14 A sample of size $n = 12$ is taken from a large normal population with an unknown standard deviation. The sample has a mean of $\overline{x} = 7.38$ and standard deviation of $s = 0.74$. Test the hypothesis that the population mean is $\mu = 7$ against a two-sided alternative. Use $\alpha = 0.05$.

12.4 WHICH IS THE NULL HYPOTHESIS?

- **12.15** A sample of size $n = 18$ is taken from a large normal population with an unknown standard deviation. The sample has a mean of $\bar{x} = 16.3$ and standard deviation of $s = 3.32$. Test the one-sided alternative hypothesis that the population mean is greater than 15. Use $\alpha = 0.05$.

12.16 The drying times for paint under standard conditions follow the normal curve. A paint manufacturer claims that the mean drying time of its interior wall paint is 60 minutes. A consumer testing organization measures the drying time on 12 test areas. It observes that the 12 test areas have a mean drying time of 66.3 minutes, with a standard deviation of 8.4 minutes. Considering the results of the testing organization, should the manufacturer's claim be rejected at the $\alpha = 0.05$ level? Use Table 4 on page 510 to find bounds for the P-value.

12.4 WHICH IS THE NULL HYPOTHESIS?

If one of the hypotheses is simple and the other compound, then the decision of which is the null hypothesis and which is the alternative hypothesis is clear. The simple one is the null hypothesis. What if both hypotheses are compound? Let's look at this question in the context of an example.

Example 12.6   *Widgets.* Before a company is licensed to manufacture widgets, its widgets must be shown to satisfy a government standard. The government standard requires that fewer than a third break when dropped from a height of three feet. The company claims that its widgets meet or exceed this standard. A government official tests the company's claim. She takes a random sample of 10 widgets and tests them. Here π is the true proportion that break under the test.

These are the two hypotheses:

Widgets meet government standards ($\pi < 1/3$).

Widgets do not meet government standards ($\pi \geq 1/3$).

Note that both hypotheses are compound. However, in order to deny a license, the burden of proof lies with the government showing that the widgets do not meet standards, and this is the alternative hypothesis.

> *The alternative hypothesis is the hypothesis that you want to prove.*

This makes the null hypothesis the compound hypothesis $\pi < 1/3$. In order to make H_0 simple, we use the boundary value $\pi = 1/3$:

$$H_0 : \pi = \frac{1}{3} \qquad \text{versus} \qquad H_a : \pi > \frac{1}{3}$$

NOTES ON TECHNOLOGY

Test for mean

Minitab: Enter the data in column **C1**. Select **Stat→Basic Statistics**. Next select either **1-Sample Z** or **1-Sample t**. Select the column **C1**, check off the **Test mean** button, and enter the mean of H_0 and the inequality of the alternative (either **not equal, less than** or **greater than**. Hit **OK** and results will be displayed, including the P-value.

Excel: Enter your data in column **A**. Go to an empty cell and use the function **ZTEST**. The function can be found by hitting the icon f_x and choosing **Statistical→ZTEST**, and then hitting **OK**. Fill in the values. In the current version of Excel that we are using, the output is the one-sided P-value for a z test, even though it is stated to be the two-sided P-value.

The t test is a little more involved because we have to trick Excel to perform a two-sample test on a single sample. Enter your data of n values in column **A** and enter the mean to be tested in the corresponding entries of column **B**, as if it were a sample of the same size n, all of the values being the same. Go to an empty cell and use the function **TTEST**. The function can be found by hitting the icon f_x and choosing **Statistical→TTEST**, and then hitting **OK**. Enter the column A location of the sample as the first array, the column B location of the phantom sample as the second array, either 1 or 2 for one-sided or two-sided; and then enter **1** for Type. The output will be the P-value for a t test.

TI-83 Plus: Enter the data in list L1 or have the summary statistics available. Press **STAT→TESTS** and select either **Z-Test**, if knowledge of σ is included in the null hypothesis, or **T-Test**, if not. In the subsequent display, if your data are in list L1, select **Data**; otherwise select **Stats**. Press **ENTER** and then fill in the rest of the display. Select **CALCULATE** and hit **ENTER** again. The display will include the P-value.

REVIEW EXERCISES FOR CHAPTER 12

- **12.17** A claim is made that the average starting salary for a certain profession is $45,000. A survey is taken and a sample of 49 starting salaries is obtained. The sample average is $43,000 with an SD of $10,000. Test the hypothesis at the 0.05 level of significance that the mean is less than $45,000. State the null and alternative hypotheses. Find the rejection region and state your conclusion in common English.

12.18 A government agency monitors the temperature rise in the water 100 meters downstream from a water-cooled power station. Regulatory guidelines require that the rise not exceed 1.0°C. The agency collects 6 water samples and obtains a mean rise of 1.2°C with a standard deviation of 0.3°C. The agency believes that readings follow the normal curve.

    a. What decision rule should it use for $\alpha = 0.05$?
    b. What is its decision? State it in common English.

- **12.19** The vitamin B-2 content of a certain brand of vitamin pills follows the normal curve. The manufacturer claims that the average vitamin B-2 content is 30 mg with a standard deviation of 2 mg per pill. A quality-control inspector randomly selects 25 pills and measures their total vitamin B-2 content. It is not good to have either too much or too little vitamin; so the alternative hypothesis must be two-sided. Suppose that the inspector actually finds a total vitamin B-2 content of 725 mg in the 25 pills. Should the manufacturer's claim be rejected at the 5% level?

12.20 A random sample of 900 births consisted of 477 boys and 423 girls. Test the "equally likely births" hypothesis against the hypothesis that there are more boy births. Use a level of significance of 5%. State the null and alternative hypotheses, the P-value, and your decision in common English.

- **12.21** A student doing a survey at a large university found that her random sample of 160 students reported an average GPA of 3.5 with a standard deviation of 0.45. The registrar's office published the average GPA to be 2.85. Can the difference be explained by sampling variability? If not, how else can it be explained?

12.22 Some random variation in the thickness of plastic sheets produced by a machine is expected. To determine whether the thickness is within acceptable limits, 12 plastic sheets are randomly selected from a day's production and the thicknesses are measured in millimeters. The data:

    1.26 1.19 1.23 1.28 1.18 1.17 1.24 1.21 1.23 1.20 1.25 1.29

This sample mean is 1.2275 and standard deviation is 0.03864. Assume a normal population with unknown mean and variance. Test the hypothesis that the mean of the parent population is 1.2 mm vs. that it is not 1.2 mm. State H_0, H_a, use $\alpha = 0.05$, and state your conclusion in context of the problem.

- 12.23 Let K denote a binomial count based on $n = 210$ trials with unknown probability of success π. Use the Central Limit Theorem to construct a test of the hypothesis $H_0 : \pi = 0.9$ versus the alternative hypothesis $H_a : \pi < 0.9$ at the 7% level of significance. For what values of the sample count K should H_0 be rejected?

12.24 A basketball player has hit 60% of his shots from the floor. In the next 100 shots, he makes 70 baskets. Has his shooting improved? Test with a one-sided alternative at the $\alpha = 0.05$ level of significance.

- 12.25 Test the hypothesis of Exercise 6.33 on page 192 using the z test instead of the binomial exact test. If the judge is fair (H_0) in choosing 100 potential jurors from the panel of 102 women and 248 men, then the expected proportion of women chosen is $\pi = \frac{102}{102+248} = 0.291$. The observed proportion is $p = \frac{9}{100} = 0.09$. Use $\alpha = 0.01$.

	Chosen	Not chosen	Row total
Women	9	93	102
Men	91	157	248
Column total	100	250	350

12.26 Test the hypothesis of Exercise 6.31 on page 192 using the z test instead of the binomial exact test. A certain kind of computer chip has a failure rate of 8%. During quality-control inspection of a batch of 100, 15% are found to be defective. Is this within historical limits or is the failure rate out of control? Use $\alpha = 0.05$.

- 12.27 Light bulbs of a certain type have a mean lifetime of 750 hours with a standard deviation of 50 hours. Engineers have modified a manufacturing process and believe that their newly designed bulbs should have greater mean life with no change in variability. They plan to put 100 light bulbs on life test and reject the null hypothesis in favor of the alternative hypothesis if their average life time exceeds 761.63 hours.

   a. State the null and alternative hypotheses.
   b. Sketch the power curve for mean life times between 750 and 775.
   c. Using this procedure, what is the probability of retaining the null hypothesis if the mean life time is 750 hours?
   d. Using this procedure, what is the probability of correctly rejecting the null hypothesis if the mean life time is 760 hours? 770 hours?

Chapter 13

ESTIMATION WITH CONFIDENCE

13.1 DIFFERENCE BETWEEN CONFIDENCE AND PROBABILITY

13.2 TWO-SIDED CONFIDENCE INTERVALS

13.3 ONE-SIDED CONFIDENCE INTERVALS

13.4 BOOTSTRAPPING AND THE t CURVES

13.5 MARGIN OF ERROR AND SAMPLE SIZE

13.6 INTERVAL ESTIMATE OF PROPORTION π

13.7 SMALL SAMPLE INTERVAL ESTIMATES OF PROPORTIONS: BINOMIAL EXACT INTERVALS

In hypothesis testing, we start with a statement about a parameter, and then we examine a random sample to determine whether its data are consistent with that hypothesis. In standard interval estimation, we have no hypothesis about the parameter. A sample is examined to arrive at an estimate of a parameter. In spite of having less to work with, the technical aspects of interval estimation are simpler than that of testing hypotheses.

13.1 DIFFERENCE BETWEEN CONFIDENCE AND PROBABILITY

Consider your professor walking around the front of the classroom. Say that whenever she is lecturing, she spends 80% of the time within easy reach—say, within 3 feet—of the chalkboard. She moves around. Her location relative to the chalkboard is variable. The location of the chalkboard is fixed (the parameter). If you know where the chalkboard is, you can say that there is an 80% probability that she is within 3 feet of the chalkboard.

Note the obvious: that whenever the professor is within 3 feet of the chalkboard, the chalkboard is within 3 feet of the professor.

Now suppose that the room turns dark because of a power failure and you lose your orientation. But you can still locate your professor because her watch emits some light. You can be 80% *confident* that the chalkboard can be found within 3 feet of her. Because the chalkboard does not move, its location is not variable. We do not use the term *probability* when referring to the location of the chalkboard. We use the term *confidence* instead. Probability refers to the variable location of the professor; and confidence refers to the fixed location of the chalkboard.

So that's the difference between probability and confidence.

13.2 TWO-SIDED CONFIDENCE INTERVALS

Example 13.1 *Filling soft drink bottles.* Now let's consider a production line of bottles of a soft drink. The same production machinery is used to fill different size bottles, depending on the production run. In other words, when production of a particular size bottle fills the allotted warehouse space, production switches to one of 12-ounce, 16-ounce, 1-liter, or 2-liter containers. Each time, the machinery is adjusted for the new containers. You may have noticed in a supermarket that each 2-liter bottle is filled with a slightly different amount of soda. The slightly different heights of soft drink you see in the bottle necks are due to small differences in containers and fill. This is random variability. Say, for this particular machinery, the fill is known to vary according to the normal curve with $\sigma = 0.01$ liters. The value of $\sigma = 0.01$ represents the precision of the machine; after years of production, this value would be well-established. If production is switched to 2-liter bottles, the machine is adjusted so that the average fill is slightly *over* 2 liters in order to guard against the occasional underfill, yet management wants to overfill by as little as possible in order to reduce production loss. Because the fill is adjustable, we set it and test the fill of the bottles coming out. Afterward, we periodically take a random sample of, say, $n = 6$ bottles, and pour the contents into a calibrated beaker in order to make sure that the machine continues to operate within limits.

13.2 TWO-SIDED CONFIDENCE INTERVALS

Sampling theory

First we consider the situation from the point of view of sampling theory: Suppose that we know the value of the parameter μ, the mean fill for the machine. Say $\mu = 2.03$. Then any random sample of $n = 6$ will have an average fill $\overline{X}$ of 2.03 liters give or take $\sigma_{\overline{X}} = \frac{\sigma}{\sqrt{n}} = \frac{0.01}{\sqrt{6}} = 0.0041$ liters or so. Because the fill follows the normal curve, we know that 95% of the time the sample average fill will be within $z_{.975}\sigma_{\overline{X}} = 1.96\sigma_{\overline{X}} = 1.96(0.0041) = 0.008$ liters of $\mu = 2.03$ liters, as shown in Figure 13.1. We write

$$Pr(|\overline{X} - 2.03| \leq 0.008) = Pr(2.022 \leq \overline{X} \leq 2.038) = 0.95$$

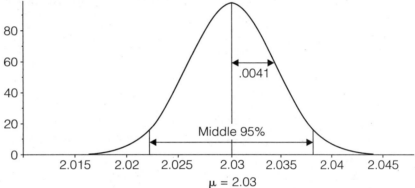

Figure 13.1 **Sampling from a known population with $n = 6$, $\mu = 2.03$, and $\sigma_{\overline{x}} = .0041$**

Similarly, 99% of the time the sample average will be within $z_{.995}\sigma_{\overline{X}} = 2.576(0.0041) = 0.0105$ liters of $\mu = 2.03$ liters:

$$Pr(|\overline{X} - 2.03| \leq 0.0105) = Pr(2.0195 \leq \overline{X} \leq 2.0405) = 0.99$$

Estimation theory

Now we take the point of view of estimation: Suppose that we observe the value of the sample average $\overline{x}$, but we do not know the parameter μ. Say a random sample of 6 bottles, when poured into the calibrated beaker, shows a sample sum of 12.3 liters. This makes the sample mean fill $\overline{x} = \frac{12.3}{6} = 2.05$ liters. The variability of the machine for *each* bottle is known to be $\sigma = 0.01$ liters. For 6 bottles, the variability of $\overline{X}$ is the standard error $\sigma_{\overline{X}} = \frac{\sigma}{\sqrt{n}} = \frac{0.01}{\sqrt{6}} = 0.0041$ liters; that is, the observed $\overline{x} = 2.05$ is within approximately

0.0041 liters of the unknown μ (see Figure 13.2). This is like using the light of your professor's watch in a dark room to guess the position of the blackboard.

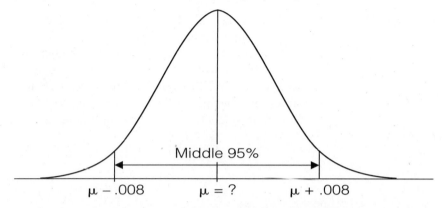

Figure 13.2 Sampling from a population with unknown μ

Because $\overline{X}$ follows the normal curve, we know that there is a 95% chance that $\overline{X}$ is within $1.96\sigma_{\overline{X}} = 1.96(0.0041) = 0.008$ liters of the unknown μ. We observed a value of the variable $\overline{X}$ to be $\overline{x} = 2.05$ liters, so we say that we are 95% *confident* that μ is within 0.008 liters of 2.05 liters, or

$$Conf(|2.05 - \mu| \leq 0.008) = Conf(2.042 \leq \mu \leq 2.058) = 0.95$$

This is the 95% confidence interval.

$$95\% \text{ CI:} \quad \mu = 2.05 \pm 0.008$$

Similarly, the 99% confidence interval is

$$99\% \text{ CI:} \quad \mu = \overline{x} \pm z_{.995}\sigma_{\overline{X}} = 2.05 \pm 2.576(0.0041) = 2.05 \pm 0.0105 \text{ liters}$$

or

$$99\% \text{ CI:} \quad 2.0395 \leq \mu \leq 2.0605$$

Simulation Figure 13.3 shows 20 random samples from the same normal population with $\sigma = 0.01$. For each of the 20 samples, the 95% confidence intervals for μ is drawn. Note that the true mean of this population is $\mu = 2.03$, which is fixed. What varies are the 20 confidence intervals. Also, note that in roughly 1 in

20 cases (that is, $100\% - 95\% = 5\%$ of the time) the 95% confidence interval does not contain the target μ.

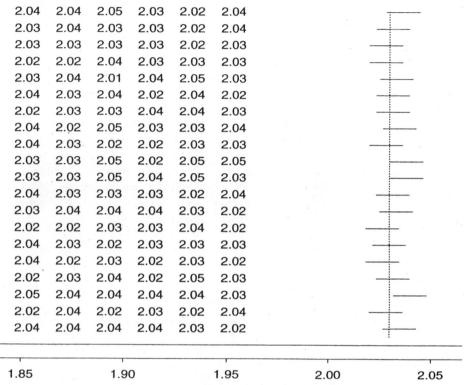

Figure 13.3 Each of the 20 rows is a random sample of size $n = 6$ from a normal population with $\mu = 2.03$ and $\sigma = 0.01$. For each sample, the 95% confidence interval for the mean is shown next to it. Approximately 5% of the confidence intervals do not cover the population mean.

EXERCISES FOR SECTION 13.2

- **13.1** A simple random sample of size $n = 9$ is taken from a large normal population with an unknown mean μ and a known standard deviation $\sigma = 8$. If $\bar{x} = 155.4$, find a:

  a. 95% confidence interval for μ

  b. 99% confidence interval for μ

13.2 A simple random sample of size $n = 25$ is taken from a large normal population with an unknown mean μ and a known standard deviation $\sigma = 0.9$. If $\bar{x} = 6.5$, find a:

a. 95% confidence interval for μ

b. 90% confidence interval for μ

- **13.3** An airline flies small aircraft with a load limit of 10,900 lb. The weights of a random sample of 100 passengers (with luggage) follow the normal curve with mean of $\bar{x} = 200$ lb. Assume the standard deviation is $\sigma = 40$ lb.

    a. Approximately 95% of the passengers weigh between ____ and ____ lb.

    b. Find the 95% confidence interval for the population mean weight μ of passengers on this airline.

    c. Estimate the probability that 50 passengers will overload such an aircraft.

13.4 A local airline wishes to estimate the weight of hand luggage taken on board by passengers. The weights of a random sample of 225 pieces of hand luggage follow a normal curve with a mean of $\bar{x} = 20$ lb. Assume a standard deviation of $\sigma = 5$ lb.

    a. It is estimated that approximately 95% of the hand luggage weighs between ____ and _____.

    b. Find the 95% confidence interval for the population mean weight μ of hand luggage on this airline.

    c. Find the 99% confidence interval for the population mean weight μ of hand luggage on this airline.

13.3 ONE-SIDED CONFIDENCE INTERVALS

Returning to the 99% confidence interval for the soft drink example above:

$$99\% \text{ CI:} \quad 2.0395 \leq \mu \leq 2.0605$$

we can conclude, from one side, that we are better than 99% confident that the true mean fill μ is at least 2.0395 liters *and at the same time* that μ is at most 2.0605 liters.

The city Department of Weights and Measures is not concerned about the upper limit 2.0605. The department wants to make sure that there is enough soft drink. For this purpose, a one-sided estimate is in order.

Given the data above, one can conclude that 99% of the time a 6-bottle sample mean $\overline{X}$ would be no more than $z_{.99} = 2.326$ standard errors above the population mean μ.

With probability 0.99, we have

$$\overline{X} \leq \mu + z_{.99}\sigma_{\overline{X}} = \mu + 2.326 \cdot \frac{(0.01)}{\sqrt{6}} = \mu + 0.0095$$

For the observed value $\overline{x} = 2.05$ of the variable $\overline{X}$, we have the 99% confidence statement

$$99\% \text{ CI:} \quad \mu \geq 2.05 - 0.0095 = 2.0405$$

Similarly, for a directional confidence interval going the other way, we can write

$$99\% \text{ CI:} \quad \mu \leq 2.05 + 0.0095 = 2.0595$$

Note that we can be 99% confident in each of the two statements separately. Each of the one-sided 99% confidence intervals carries a 1% risk that corresponds to a 1% tail of the normal curve. Because the tails do not intersect, there is 2% risk for the combined statement. Joining both 99% confidence statements produces a 98% confidence statement:

$$98\% \text{ CI:} \quad 2.0405 \leq \mu \leq 2.0595$$

EXERCISES FOR SECTION 13.3

- **13.5** A simple random sample of size $n = 16$ is taken from a large normal population with an unknown mean μ and a known standard deviation $\sigma = 1.5$. If $\overline{x} = 18.3$, find a:

  a. 95% confidence interval of the form $\mu \geq \cdots$
  b. 99% confidence interval of the form $\mu \geq \cdots$

  13.6 A simple random sample of size $n = 12$ is taken from a large normal population with an unknown mean μ and a known standard deviation $\sigma = 0.75$. If $\overline{x} = 5.3$, find a:

  a. 95% confidence interval of the form $\mu \leq \cdots$
  b. 90% confidence interval of the form $\mu \leq \cdots$

- **13.7** An insurance agent wants to estimate the mean age of residents in a community. He takes a simple random sample of 100 and finds a mean age of 28.2 years and a standard deviation of 9.1 years.

  a. Find a two-sided 95% confidence interval for the mean age in this community.

b. Find a one-sided 95% confidence interval of the form $\mu \leq \cdots$.

13.8 A real estate developer want to estimate the mean selling price of homes in a community. A sample of $n = 50$ has a mean of $\bar{x} = \$195,230$ with a standard deviation of $s = \$63,125$.

    a. Find a 95% two-sided confidence interval for the mean selling price of homes in this community.

    b. Find a one-sided 95% confidence interval of the $\mu \geq \cdots$.

13.4 BOOTSTRAPPING AND THE t CURVES

To make a point estimate of μ, the observed $\bar{x}$ is a good estimate. What is lacking in a point estimate is its accuracy. In order to make an interval estimate of μ, we need the **standard error of the mean** $\sigma_{\bar{X}} = \frac{\sigma}{\sqrt{n}}$. In the above example, we knew the value of σ but not that of μ. In most cases of estimation, we do not know either parameter. One option is to use the known sample standard deviation s instead of the population standard deviation σ. Using the sample standard deviation s to obtain an **estimated standard error** $\text{SE}(\bar{x}) = \frac{s}{\sqrt{n}}$ is an example of bootstrapping. In general, bootstrapping is the use of the sample data to estimate the accuracy of an estimate based on that data.

> *In cases where the sample size is large [general rule: $n \geq 40$], using the sample value s instead of σ to find the standard error makes little practical difference.*

Computer simulation

A computer simulation was performed of 1000 random samples of size 40 drawn with replacement from a box containing the 6 tickets $\{1, 2, 3, 4, 5, 6\}$. This is the uniform population box of the number of dots obtained on a roll of a fair die. The tickets in this box have mean $\mu = 3.5$ and standard deviation $\sigma = \sqrt{\frac{35}{12}} = 1.7078$. However, instead of using σ to compute the standard error, we use s. For sample sizes $n = 40$, the standardized sample means

$$t = \frac{\bar{x} - \mu}{s/\sqrt{n}}$$

were computed for each of these 1000 samples. The results are shown in Figure 13.4.

13.4 BOOTSTRAPPING AND THE t CURVES

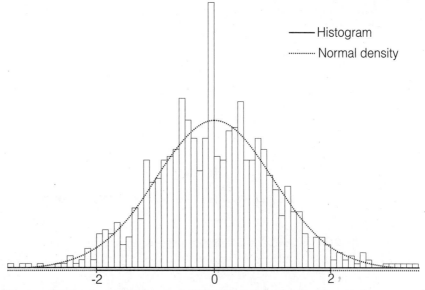

Figure 13.4 t statistics for 1000 random samples of size $n = 40$ from a population box of cards labeled 1 through 6

Notice that the t statistics follow the standard normal curve rather closely. (With more samples and narrower classes, the histogram would have followed the normal curve even better.)

Now we form one-sided and two-sided confidence intervals for each sample and for various confidence levels ranging from 80% to 99%. The results are summarized in the following table.

Percentage of confidence intervals that include the population mean based on 1000 random samples of size 40 from the set $\{1, 2, 3, 4, 5, 6\}$

Confidence level	One-sided right-bounded	One-sided left-bounded	Two-sided
85	86	86	86
90	90	91	90
95	95	95	95
97	97	97	97
98	98	98	97
99	98	99	99

Here we used the normal curve percentiles to form the confidence intervals. For example, the 1000 different 99% right-bounded confidence intervals were each of the form

$$\mu \leq \bar{x} + z_{.99}\frac{s}{\sqrt{n}} = \bar{x} + 2.326\frac{s}{\sqrt{40}}$$

for 1000 values of $\bar{x}$ and s. The last line of the table indicated that 98% of the right-bounded intervals included the population mean (actually, 984 of the 1000 intervals included $\mu = 3.5$). Of course, in this case, we knew the population mean was 3.5 so there was no need to estimate it. What this example does show is that, even though the population is discrete and uniformly distributed, the confidence intervals based on sample means, estimated standard errors, and the normal curve percentiles exhibit the appropriate coverages.

What happens if the population is skewed?

Another computer simulation was performed taking 1000 samples of size 40 with replacement from a box population very skewed to the right consisting of 3% 1's, 9% 2's, 16% 3's, 19% 4's, 23% 5's, and 30% 6's. For samples of size 40, the sampling distribution of the statistic $\overline{X}$ is only *slightly* skewed. So confidence intervals based on the normal approximation are only slightly off the mark. Results of a computer simulation are shown in Figure 13.5.

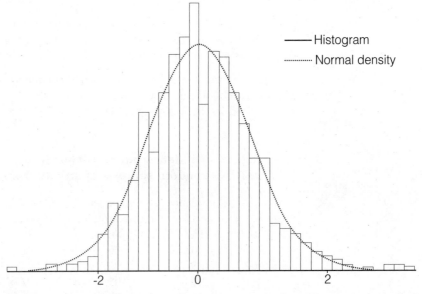

Figure 13.5 t statistics from 1000 samples of size $n = 40$ drawn from a box with cards distributed 3% 1's, 9% 2's, 16% 3's, 19% 4's, 23% 5's, and 30% 6's

The following table compares these results with various confidence intervals.

13.4 BOOTSTRAPPING AND THE t CURVES

Percentage of confidence intervals that include the population mean based on 1000 random samples of size 40 from a set containing 3% 1's, 9% 2's, 16% 3's, 19% 4's, 23% 5's, 30% 6's

Confidence level	One-sided right-bounded	One-sided left-bounded	Two-sided
85	85	85	85
90	90	91	90
95	95	96	95
97	96	98	97
98	97	99	98
99	98	100	98

Moral of the story

For samples of size 40 or larger, we can reasonably use the z curve (Table 3 in the back of the book) to compute confidence intervals even though we substitute the estimated standard error $\text{SE}_{\overline{x}} = \frac{s}{\sqrt{n}}$ for $\sigma_{\overline{x}} = \frac{\sigma}{\sqrt{n}}$. The t curves, which were designed to use the estimated SE, are based on the assumption that the parent population is normal, but this condition has been found not to be so strict. So, if we know that the parent population is approximately normal, the t table (Table 4 on page 510) is preferred. If the population is dichotomous (hence, not normal), the t table is never used. Bootstrapping when estimating a dichotomous parameter π has its own rules; these are discussed in Sections 13.6 and 13.7.

Example 13.2

Timing a computer program. A computer program, which is intended to run on a busy network, is being tested. The run times are measured $n = 400$ times. The results follow the normal curve. The 400 timings average $\overline{x} = 3.80$ seconds with a standard deviation of $s = 0.40$ second.

This means that

99% of timings

> 99% of the time the program completes within
> $\overline{x} + z_{.99}s = 3.80 + 2.326(0.40) = 4.730$ *seconds.*

Note that the standard deviation $s = 0.40$ is used here, not the SE, because the statement is concerned with the population of timings, which follow the normal curve. This does not involve an estimate of μ (and we could have done this in Chapter 8).

On the other hand, a 99% confidence interval is concerned with accuracy of the sample mean as an estimate of the true mean timing μ. It therefore uses $\text{SE}_{\overline{x}} = \frac{s}{\sqrt{n}} = \frac{0.40}{\sqrt{400}} = 0.02$.

99% confidence for μ

> **With 99% confidence we can say that μ is no more than $\bar{x} + z_{.99}\text{SE}_{\bar{x}} = 3.80 + 2.3264(0.02) = 3.847$ seconds.**

Notice that we used bootstrapping with the z curve, because the sample size $n = 400$ is so large. Table 4 on page 510 does not have a row for $df = 399$. For a more conservative estimate, we can use the t curve with $df = 100$, which has a percentile $t_{.99} = 2.3642$. For $df = \infty$, the percentiles are the ones for the z curve; here $z_{.99} = 2.3264$. However, each gives us the same confidence interval to within *three* decimal places.

Example 13.3 *Estimating car mileage.* A consumer testing organization uses three identical cars to determine gas mileage of a particular model. It is known that gas mileage depends upon many factors, causing the mileage to follow a normal curve. These are the results:

Car	Mpg
A	23
B	25
C	24

Let's find a 95% confidence interval for the true gas mileage for this car model.

We know that the t statistic has $df = n - 1 = 2$ degrees of freedom and follows the t_2 curve. Using Table 4 on page 510, row $df = 2$, we see $t_{2,0.975} = 4.303$. So the two-sided 95% confidence interval is

$$-4.303 \leq \frac{\bar{x} - \mu}{s/\sqrt{n}} \leq 4.303$$

Because the sample of three cars has $\bar{x} = 24$ and $s = 1$, this results in

$$\text{95\% CI:} \quad \mu = \bar{x} \pm 4.303 \frac{s}{\sqrt{n}} = 24 \pm 2.484 \quad \text{or}$$

$$\text{95\% CI:} \quad 21.52 \leq \mu \leq 26.48$$

13.4 BOOTSTRAPPING AND THE t CURVES

Summary: Confidence Intervals for μ

1. **Known σ.** If a random sample is taken from a population with unknown mean μ but known standard deviation σ, then a two-sided $1-2\alpha$ confidence interval for μ can be computed from the equation

$$z_\alpha \leq \frac{\bar{x}-\mu}{\sigma/\sqrt{n}} \leq z_{1-\alpha} \tag{13.1}$$

In particular, a 95% confidence interval has the form

$$95\% \text{ CI:} \quad \bar{x}-(1.96)\frac{\sigma}{\sqrt{n}} \leq \mu \leq \bar{x}+(1.96)\frac{\sigma}{\sqrt{n}} \quad \text{or}$$

$$95\% \text{ CI:} \quad \mu = \bar{x} \pm (1.96)\frac{\sigma}{\sqrt{n}} \tag{13.2}$$

The conditions for this confidence interval of Formula 13.2 are as follows.

> a. *The sample must be randomly drawn.*
>
> b. *The draws must be independent of each other (if draws are made without replacement from a small population, then the reduction factor $\sqrt{\frac{N-n}{N-1}}$, as described in Part 2 of the law of averages for the statistic $\overline{X}$ on page 317, must be used).*
>
> c. *The sample size n is large according to Part 3 of the law of averages for the statistic $\overline{X}$ as given on page 317; that is, either the population itself is normal, or if not, then $n \geq 40$ or so.*

2. **Large sample.** If a *large* sample is taken from a population with unknown mean μ and unknown standard deviation σ, then a two-sided $1-2\alpha$ confidence interval for μ can be computed as in Part 1 above, except that the population standard deviation σ is replaced by the sample standard deviation s. The criterion of a large sample size n is that $n \geq 30$ or so.

3. **Sample from normal population, unknown $mathbf\sigma$.** If a simple random sample is taken from an approximately normal population with unknown mean μ and unknown standard deviation σ, then a two-sided $1-2\alpha$ confidence interval for μ can be computed from the equation

$$t_{df,\alpha} \leq \frac{\bar{x}-\mu}{s/\sqrt{n}} \leq t_{df,1-\alpha} \tag{13.3}$$

In particular, for a sample of size $n = 5$, we have $df = 4$ and $t_{4,0.975} = 2.7765$. So a 95% confidence interval has the form

$$95\% \text{ CI:} \quad \bar{x} - (2.7765)\frac{s}{\sqrt{n}} \leq \mu \leq \bar{x} + (2.7765)\frac{s}{\sqrt{n}} \quad \text{or}$$

$$95\% \text{ CI:} \quad \mu = \bar{x} \pm (2.7765)\frac{s}{\sqrt{n}} \tag{13.4}$$

The conditions for this confidence interval of Formula 13.4 are as follows.

> a. *The sample must be randomly drawn.*
>
> b. *The draws must be independent of each other (if draws are made without replacement from a small population, then the reduction factor $\sqrt{\frac{N-n}{N-1}}$, as described in Part 2 of the law of averages for the statistic $\overline{X}$ on page 317, must be used).*
>
> c. *The population must be approximately normal. The sample need not be large.*

The good news about the t curves is that the normality condition for the parent population is not strict. Even for populations that are quite far from normal, experience has shown that the t curves provide good interval estimates.

NOTES ON TECHNOLOGY

Confidence interval for mean

Minitab: Enter the data in column **C1**. Select **Stat→Basic Statistics**. Next select either **1-Sample Z** or **1-Sample t**. Select the column **C1**, be sure that the button **Confidence interval** is checked, and select the confidence level. The 1-Sample Z interval also requires a value for σ. Hit **OK** and the confidence interval will be displayed.

Excel: Enter your data in column **A**. To obtain a confidence interval based on a t curve, use one of the options of **Tools → Data Analysis → Descriptive Statistics**. Check off **Confidence Level for Mean** as well as **Summary Statistics**. The last item of the output is the margin of error, called Confidence Level, which has to be added and subtracted from the sample mean $\bar{x}$ to obtain the confidence interval based on a t curve.

To obtain a confidence interval based on the z curve, use the function **Confidence**, which can be found by hitting the icon $\mathbf{f_x}$ and choosing

Statistical→**CONFIDENCE**, and then hitting **OK**. Fill in α, σ, and n. For example, $\alpha = 0.05$ for $1 - \alpha = 95\%$ confidence. The output will be the margin of error that has to be added and subtracted from the sample mean $\bar{x}$ to obtain the confidence interval.

TI-83 Plus: Enter the data in list L1 or have the summary statistics available. Press **STAT**→**TESTS**, and select either **ZInterval**, if you know σ, or select **TInterval** if not. In the subsequent display, select **Data**, if your data are in list L1, or **Stats** otherwise. Press **ENTER**, and then fill in the rest of the display. For example, you may choose to change the confidence level. Then select **CALCULATE**, and hit **ENTER** again.

EXERCISES FOR SECTION 13.4

- **13.9** A simple random sample of size $n = 70$ has a mean $\bar{x} = 87$ and a sample standard deviation $s = 0.5$.

    a. Find a 95% confidence interval.

    b. Find a 98% confidence interval for the population mean.

    c. How important is the fact that we do not know the shape of the population?

13.10 A simple random sample of size $n = 100$ has a mean $\bar{x} = 1023$ and a sample standard deviation $s = 5.3$.

    a. Find a 95% confidence interval.

    b. Find a 99% confidence interval for the population mean.

    c. How important is the fact that we do not know the shape of the population?

- **13.11** A simple random sample of size $n = 4$ has a mean $\bar{x} = 85$ and a sample standard deviation $s = 7.7$. Assuming the population follows the a normal curve, find a:

    a. 95% confidence interval

    b. 99% confidence interval for the population mean

13.12 A simple random sample of size $n = 12$ has a mean $\bar{x} = -2.4$ and a sample standard deviation $s = 3.0$. Assuming the population follows the normal curve, find a:

    a. 95% confidence interval

    b. a 99% confidence interval for the population mean

- **13.13** Suppose 12 subjects are selected at random from a large population. The serum cholesterol level of each of these subjects is measured. The sample mean turns out to be 150 mg/dl, and the sample standard deviation is 25 mg/dl. Construct a 90% confidence interval for the population mean.

13.14 A consumer testing organization used 5 identical cars to determine gas mileage for a certain model. It is known that gas mileage follows the normal curve. These were the results (in miles per gallon):

$$24 \quad 22 \quad 27 \quad 23 \quad 24$$

   a. Construct a 90% confidence interval for the true gas mileage for this model.

   b. Construct a 95% one-sided confidence interval of the form $\mu \geq \cdots$.

- **13.15** A consumer testing organization measures the burning time of 10 fluorescent tubes. The results (in hours):

$$974 \quad 794 \quad 1093 \quad 839 \quad 824 \quad 1157 \quad 879 \quad 747 \quad 1044 \quad 897$$

   a. Find the mean and standard deviation of the data.

   b. Find the 95% one-sided confidence interval for the mean burning time for all fluorescent tubes of this kind of the form $\mu \leq \cdots$.

   c. The manufacturer claims that the mean burning time is at least 1000 hours. Would you accept this claim at the 5% level of significance?

13.16 In 1879, A. A. Michelson made 100 determinations of the velocity of light in air using a modification of a method proposed by the French physicist Foucault.[1] The numbers are in kilometers per second and have had 299,000 subtracted from them:

850	740	900	1070	930	850	950
980	980	880	1000	980	930	650
760	810	1000	1000	960	960	960
940	960	940	880	800	850	880
900	840	830	790	810	880	880
830	800	790	760	800	880	880
880	860	720	720	620	860	970
950	880	910	850	870	840	840
850	840	840	840	890	810	810
820	800	770	760	740	750	760
910	920	890	860	880	720	840
850	850	780	890	840	780	810
760	810	790	810	820	850	870
870	810	740	810	940	950	800
810	870					

[1] The data given here are as reported by Stephen M. Stigler in *Do Robust Estimators Work with Real Data?*, The Annals of Statistics **5** (1977), p. 1075; they can be downloaded from the Web at http://lib.stat.cmu.edu/DASL/.

Assume this is a random sample from a hypothetical population of measurements. Construct a 99% two-sided confidence interval for the mean of the population of all possible measurements.

⋆ **13.17** Construct a box plot for the above data. Are there any outlying values? Are the data skewed? The time-ordered data are listed left to right, top to bottom. Stigler grouped these in blocks of 20. Plot the data over time. For these data, does it appear reasonable to estimate the speed of light by adding 299,000 to the estimated population mean? Stigler applied the corrections used by Michelson to adjust for speed of light in a vacuum and suggests that 734.5 is the appropriate true value for comparison with these measurements.

13.5 MARGIN OF ERROR AND SAMPLE SIZE

In laboratory experiments, the margin of error refers to the standard error of the estimator of a parameter. Let's say that a parameter ω is being estimated by a statistic W. A sample is drawn, the value w of the statistic W and an estimated standard error SE_W are computed from the sample, and then we write

$$\omega = \text{w} \pm \text{SE}_W$$

Typically, when n measurements $x_1, x_2, \ldots, x_n$ are taken to estimate a population mean μ, the mean of the measurements $\overline{X}$ is used as an estimator; and the estimated standard error is $\text{SE}_{\overline{X}} = \frac{s}{\sqrt{n}}$. In such cases, an expression such as

$$\mu = 132 \pm 4.5$$

should be read as follows: The value 132 denotes the mean of the measurements $\overline{x}$, and 4.5 denotes the estimated standard error. For large n, this would also be the 68% confidence interval. For small n, the t_{n-1} curves can be used to obtain a confidence interval.

Example 13.4 *Determining the appropriate sample size for estimating* μ. When a statistical study is begun, an important question concerns the size of the sample needed. Consider the mean burning times μ of fluorescent tubes of the type mentioned in Exercise 13.15. The data for the 10 tubes

| 974 | 794 | 1093 | 839 | 824 | 1157 | 879 | 747 | 1044 | 897 |

yield sample mean $\bar{x} = 924.8$ hours, standard deviation of $s = 136.6$ hours, and standard error $\sigma/\sqrt{n}$ estimated by $SE_{\bar{x}} = s/\sqrt{n} = 136.6/\sqrt{10} = 43.2$ hours. So we can report

$$\mu = 924.8 \pm 43.2 \text{ hours}$$

Let's say that we wanted to estimate the mean burning time to within a margin of error of no more than 15 hours. The equation to solve is

$$\frac{\sigma}{\sqrt{n}} \leq 15$$

Solving for n, we obtain

$$n \geq \left(\frac{\sigma}{15}\right)^2$$

To estimate the size of a sample needed for a study we mujst estimate σ. If no estimate is available, a preliminary sample such as the 10 tubes of Exercise 13.15 may be needed.

In this preliminary sample, $s = 136.6$. Using this estimate, we need

$$n \geq \left(\frac{136.6}{15}\right)^2 = 82.94$$

tubes. This value should be rounded up. This means that we need to test at least $n = 83$ fluorescent tubes to obtain a a margin of error no more than 15 hours.

If we wanted a 95% confidence interval with a margin of error E

$$95\% \text{ CI:} \quad \mu = \bar{x} \pm E$$

of no more than 25 hours, the equation to solve would be

$$E = z_{.975} \frac{\sigma}{\sqrt{n}} \leq 25$$

Solving for n, we obtain

$$n = \left(\frac{z_{.975}\sigma}{E}\right)^2 \geq \left(\frac{1.96\sigma}{25}\right)^2$$

Using the same preliminary $s = 136.6$, it turns out we need

$$n \geq \left(\frac{1.96 \times 136.6}{25}\right)^2 = 114.69$$

tubes. This means that we need to test at least $n = 115$ fluorescent tubes to obtain a 95% confidence interval with a margin of error no more than 25 hours.

Notice that for a given level of confidence, the margin of error is inversely related to the square of the sample size. So, in order to obtain half the margin of error, we need a sample four times as large.

Example 13.5 in the next section shows how to determine the appropriate sample size to estimate a population proportion π.

EXERCISES FOR SECTION 13.5

13.18 Thirty-four males from the U.S. Navy diving and special warfare community who had chronic knee and lower back pain were part of a study to test a treatment with over-the-counter drugs.[2] The study reported a mean weight of 87 ± 2 kilograms. Explain this in your own words. What is $\bar{x}$? What is s? Find the 95% confidence interval of the mean weight.

• 13.19 Suppose we want to estimate the mean age of all people in the civilian work force in the Unites States to obtain a 99% confidence interval with a margin of error of no more than one year. Let's use 50 years as an approximation of the range of 95% of this population. Dividing by 4, we approximate the population standard deviation to be 12.5 years. Why do we use the divisor 4? Using this estimate, how large a sample is needed to obtain a 99% confidence interval with a margin of error of no more than one year?

13.20 We want to estimate the mean number of hours that students work to within a 95% margin of error of no more that 2 hours. If we estimate the standard deviation to be 9 hours, how large a sample is needed?

• 13.21 How large a random sample is required to obtain an estimate of the mean IQ of college students to within a 99% margin of error of 2 points? Use $\sigma = 15$.

13.22 How large a random sample is required to obtain an estimate of the mean IQ of college students to within a 95% margin of error of 1 points? Use $\sigma = 15$.

13.6 INTERVAL ESTIMATE OF PROPORTION π

A dichotomous population, coded 1 for success and 0 for failure, has the mean π, the probability of success on each trial, and it has the standard deviation

---

[2]Lt. Christopher T. Leffler et al., *Glucosamine, chondroitin, and manganese ascorbate for degenerative joint disease of the knee and low back: A double-blind, placebo-controlled study*, Military Medicine **164** (1999), pp. 85–91.

$$\sigma = \sqrt{\pi(1-\pi)}$$

In random sampling from such a population, we can use the sample proportion P to estimate the population proportion π because P is an unbiased estimator of π, that is, $E(P) = \pi$. For *independent* Bernoulli trials, the standard error is

$$\sigma_P = \sqrt{\frac{\pi(1-\pi)}{n}}$$

If the sample size n is large, the statistic P follows the normal curve. Thus, a two-sided $1 - 2\alpha$ confidence interval has the form

$$z_\alpha \leq \frac{p - \pi}{\sqrt{\pi(1-\pi)/n}} \leq z_{1-\alpha} \tag{13.5}$$

In particular, a 95% confidence interval has the form

$$\text{95\% CI:} \quad p - (1.96)\sigma_P \leq \pi \leq p + (1.96)\sigma_P \quad \text{or}$$

$$\text{95\% CI:} \quad \pi = p \pm (1.96)\sigma_P \tag{13.6}$$

Bootstrapping approach

To avoid a quadratic solution of Equation 13.5 for the variable π, if the sample size n is large, we can bootstrap using the

Estimated standard error $\mathbf{SE}_P = \sqrt{\frac{p(1-p)}{n}}$

instead of $\sigma_P = \sqrt{\frac{\pi(1-\pi)}{n}}$. This results in a two-sided $1 - 2\alpha$ confidence interval

$$\textbf{Bootstrap } (1 - 2\alpha) \textbf{ CI:} \quad \pi = p \pm z_\alpha \mathbf{SE}_P = p \pm z_\alpha \sqrt{\frac{p(1-p)}{n}} \tag{13.7}$$

> *A sample from a dichotomous population is generally considered to be large if* $n\pi \geq 5$ *and* $n(1 - \pi) \geq 5$. *This means that for confidence intervals, a random sample of size n should be expected to contain at least 5 successes and at least 5 failures for* all *values of π in the confidence interval.*

13.6 INTERVAL ESTIMATE OF PROPORTION

Conservative approach

In estimation of π, unlike estimation of μ of the soft drink Example 13.1 on page 352, we cannot know the value of $\sigma = \sqrt{\pi(1-\pi)}$ without knowing the parameter π itself. For a graph of σ as a function of π, see Figure 13.6.

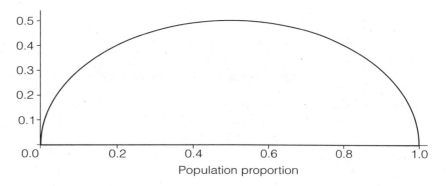

Figure 13.6 The value of $\sigma = \sqrt{\pi(1-\pi)}$ as a function of the parameter π

Note that the graph has a horizontal tangent at $\pi = 0.50$. The maximum value of σ occurs there, so it is always true that $\sigma \leq \sqrt{0.50(1-0.50)} = 0.50$. Using the maximum instead of σ gives us more conservative (that is, larger) confidence intervals. Because $\sigma_P = \frac{\sigma}{\sqrt{n}}$, this gives:

***A conservative estimate of the standard error for* P:**
$$\max\{\sigma_\mathbf{P}\} = \frac{0.50}{\sqrt{\mathbf{n}}} = \frac{1}{2\sqrt{\mathbf{n}}}$$

For large n, the statistic P follows the normal curve. This means that the margin of error for the 68% confidence interval is no more than $\frac{1}{2\sqrt{n}}$ and the margin of error for 95% confidence is no more than $\frac{1.96}{2\sqrt{n}}$.

This results in a two-sided 95% confidence interval

$$95\% \text{ CI:} \quad \pi = p \pm \frac{1.96}{2\sqrt{n}}$$

Example 13.5 *Determining appropriate sample size for opinion polls.* In opinion polls, what size of a simple random sample is needed for a margin of error of no more than 4%? In the language of opinion polling, unlike laboratory experimentation, the margin or error refers to the one for the 95% confidence interval. Solving $0.04 \geq \frac{1.96}{2\sqrt{n}}$, for n, we obtain $n \geq 600.25$. Many polls take samples of size about 625 in order to comfortably attain this margin of error. Similarly, $n \geq 1068$ is needed for a 3% margin of error. Recall that the Harris Poll discussed in Example 1.1 on page 1 used a sample of $n = 1256$, and it claimed a 3% margin of error.

Example 13.6 *Percentage of bass in a lake.* A random sample of 144 fish caught in Lake Wobegon included 40 bass. Estimate the percentage of catchable fish in the lake that are bass.

Solution The sample proportion is $p = 40/144 = 0.2778$. A conservative 95% confidence interval is

$$\pi = \frac{40}{144} \pm 1.96 \cdot \frac{0.5}{\sqrt{144}} = 0.2778 \pm 0.0817$$

or

Conservative 95% CI: $19.6\% \leq \pi \leq 35.9\%$

The sample is sufficiently large to use bootstrapping. This results in a 95% confidence interval as follows:

$$\begin{aligned}\pi &= p \pm 1.96 \cdot \frac{\sqrt{p(1-p)}}{\sqrt{144}} \\ &= \frac{40}{144} \pm 1.96 \sqrt{\frac{(40/144)(104/144)}{144}} \\ &= 0.2778 \pm 0.0732\end{aligned}$$

which gives us the 95% interval

Bootstrap 95% CI: $20.5\% \leq \pi \leq 35.1\%$

This is an improvement over the conservative estimate without bootstrapping.

13.6 INTERVAL ESTIMATE OF PROPORTION

It may be unrealistic to believe that this is an unbiased estimate of the proportion of bass in the lake. We are assuming that every fish in the lake had the same chance of being caught, which may be unreasonable. If there are a lot of bass fishers and they know their stuff, bass may be disproportionately selected for capture. It could be that the fishers are choosing larger hooks, making it less likely that they catch smaller fish. Or perhaps the smart fish are less likely to be caught, leaving a bias for less experienced (smaller) bass. In summary, it is better to view this as a confidence interval for the long-run proportion of bass *caught* in the lake, not the *true* proportion of bass in the lake.

> *It is not appropriate to use the t curves (Table 4 on page 510) when bootstrapping an interval estimate of a population proportion π.*
>
> *The t curves are to be used only when bootstrapping a population variable that is*
>
>   1. *quantitative and*
>   2. *follows approximately a normal curve.*
>
> *When estimating a proportion π, the population variable is dichotomous (Bernoulli), not quantitative. Although the law of averages shows that the statistic P may follow a normal curve, a Bernoulli variable, as shown in Figure 5.1 on page 145, does not follow a normal curve.*

EXERCISES FOR SECTION 13.6

- **13.23** A sample of size $n = 150$ is taken from a large dichotomous population. There are $k = 66$ successes in the sample. Find a 95% confidence interval of the form $\pi \geq \cdots$:

    a. using the conservative method

    b. using bootstrapping

13.24 A sample of size $n = 60$ is taken from a large dichotomous population. There are $k = 44$ successes in the sample. Find a (two-sided) 95% confidence interval for π:

  a. using the conservative method

  b. using bootstrapping

- **13.25** In the early 1900s, Karl Pearson carried out a coin-tossing experiment in which a single coin was tossed 24,000 times, resulting in 12,012 heads. Construct a (two-sided) 99% confidence interval for the long-run percentage of heads.

13.26 You plan to take a simple random sample of your campus to find out what proportion of students are employed for 15 hours per week while classes are in session. You plan to present your results as a 95% confidence interval, and you want the width of the interval to be no more than 0.20; that is, the margin of error should be no more than 0.10.

   a. What is the smallest sample size that you would need to be sure your confidence interval would be this narrow?

   b. If you are willing to guess that this proportion is no more than 0.25, what is the smallest sample size that you would need to be sure your confidence interval would be this narrow?

13.27 The state Motor Vehicle Department plans to estimate the proportion of drivers who have not received any tickets for moving violations during the past three years. How large a random sample of its records of current drivers should it take to be within 0.02 of the true proportion with

   a. 68% confidence?   b. 99% confidence?

13.28 A random sample of 200 car owners is selected, and 140 are found to have air-conditioned cars. Find a two-sided 98% confidence interval for the percentage of car owners who have air-conditioned cars. Find a one-sided 98% confidence interval of the form $\pi \geq \cdots$.

13.29 Continuing with Exercise 13.18, 34 males from the U.S. Navy diving and special warfare community who had chronic knee and lower back pain alternated between treatment and placebo in the double-blind study. Fourteen of them had improvement of symptoms under the treatment versus the placebo.

   a. Use the conservative estimate of standard error to find a one-sided 95% confidence interval of the form $\pi \geq \cdots$.

   b. Find the P-value for the one-sided alternative hypothesis $\pi > 0.50$.

13.7 SMALL SAMPLE INTERVAL ESTIMATES OF PROPORTIONS: BINOMIAL EXACT INTERVALS

For small samples, a direct use of the binomial probabilities can be made, instead of the normal approximation, as was done for the binomial exact test of Chapter 6. The calculations can be tedious, so graphs simplify the computation of confidence intervals.

13.7 BINOMIAL EXACT INTERVALS

Example 13.7 *Exact confidence interval for π.* Suppose that a dichotomous experiment is replicated 10 times and that it turns out to be successful 3 of those times. We want to construct a 90% confidence interval for the long-run proportion of success. What values of π should be included in the interval? It should include only those π_0 that would be retained in testing the null hypothesis

$$H_0 : \pi = \pi_0 \quad \text{versus} \quad H_a : \pi \neq \pi_0$$

To this end, we plot the following functions of π_0 in Figure 13.7:

$$Pr(K \leq 3) = \sum_{k=0}^{3} \binom{10}{k} \pi_0^k (1-\pi_0)^{10-k}$$

and

$$Pr(K \geq 3) = 1 - \sum_{k=0}^{2} \binom{10}{k} \pi_0^k (1-\pi_0)^{10-k}$$

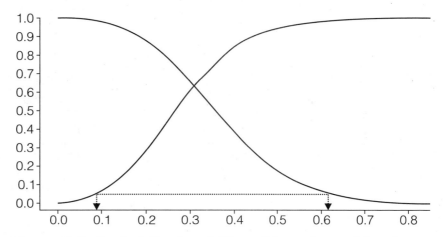

Figure 13.7 Dashed line is a 90% confidence interval for the population proportion $0.087 < \pi < 0.607$ based on a random sample of size $n = 10$ with $k_S = 3$ successes. The vertical scale corresponds to the one-sided P-values.

The probability curves trace out P-values and the interval where both curves are above the line P-value $= 0.05$ includes all values of π_0 for which the null hypothesis would be retained. This is the 90% confidence interval $0.087 \leq \pi \leq 0.607$, as indicated by the dashed interval. Note that this interval is not symmetric about the point estimate 3/10.

This method is also the basis of the graphs in Table 5 on pages 511–512 at the back of this book, which can be used to find 95% and 99% confidence intervals. Here the confidence intervals are read off the vertical scale between the two curves representing the sample size n. For example, for a random sample of size $n = 20$, if the sample statistic is $p = 0.35$, then we can read the 95% confidence interval from the vertical axis of Table 5 on page 511, looking above the point $p = 0.35$ on the horizontal axis and reading the confidence interval off the vertical axis between the two $n = 20$ curves. We get $0.154 \leq \pi \leq 0.592$. Again, notice that this interval is not symmetric about the point 0.35. Actually, some 55% of the interval exceeds 0.35.

Also, if the observed success rate is $p = 0.35$, then the observed failure rate is $q = 1 - p = 0.65$, so the 95% confidence interval for the long run proportion of failures is $1 - 0.592 \leq 1 - \pi \leq 1 - 0.154$, or $0.408 \leq 1 - \pi \leq 0.846$.

The graphs of the two-sided 95% confidence intervals in Table 5 illustrate the following:

- The narrowing of the confidence interval with increasing sample size n.

- The widening of the confidence intervals for fixed sample size n, as the sample proportion p approaches 0.50.

- The more nearly perfect symmetry of the confidence intervals about p, as p approaches 0.50.

- The larger range of sample proportions for which the corresponding confidence interval appears symmetric about p, as the sample size n increases.

How well do these confidence intervals work? For each population proportion π and sample size n, let us define the coverage $c(\pi)$ function[3] as the probability that the confidence interval computed from a random sample of size n will actually include π.

---

[3] Suppose that for each sample count K, we compute the sample proportion P and the 95% confidence interval $L_K \leq \pi \leq R_K$ or $[L_K, U_K]$. For each interval $[L_K, U_K]$, let the indicator function $I_K(\pi)$ be the function that has the value 1 on the interval $[L_K, U_K]$ and zero elsewhere. Then

$$c(\pi) = \sum_{k}^{n} I_k(\pi) \binom{n}{k} \pi^k (1-\pi)^{n-k}$$

Figure 13.8 shows the **coverage functions** for the small-sample 95% confidence intervals and the large sample confidence intervals with bootstrapping. The coverage functions $c(\pi)$ look jagged and are discontinuous because a small change in π may cause π to be included or excluded from a 95% confidence interval [and hence $Pr(K = k)$ is either included or excluded from the $c(\pi)$ sum].

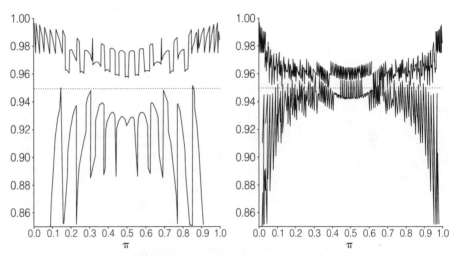

Figure 13.8 Coverage curves of 95% confidence interval when sample size is $n = 20$ on the left and $n = 100$ on the right. In each case, the upper coverage curve is for the small-sample method using Table 5 on page 511 and the lower coverage curve is for the large sample method using Equation 13.6 of Section 13.6.

Note that the coverage is always at least 0.95 for the small sample method 95% confidence intervals. However, because the binomial random variable is discrete, the coverage is not exactly 95%. Not surprisingly, the large sample method does not work well for small samples (the coverage is often less than 0.95). However, its performance improves with increasing sample size, especially if π is not too far from 0.50.

NOTES ON TECHNOLOGY

Confidence Interval for Proportion

We find a 90% confidence interval for π for Example 13.7 with $k = 3$ successes out of $n = 10$ trials.

Minitab: Minitab uses a binomial exact interval estimate, which is equivalent to using Table 5 at the back of this book. Select **Stat→Basic**

Statistics. Next select **1 Proportion.** Check the **Summarized Data** button and enter the number of trials **10** and number of successes **3**. Go to the **Options** submenu to enter the confidence level of **90%**. The confidence interval displayed will be (0.087264, 0.606624) as we obtained in Example 13.7.

Excel: Excel does not have a reasonably easy way of finding confidence intervals for proportions as part of its standard tools.

TI-83 Plus: The standard interval given by the TI-83 Plus is a confidence interval based on the normal curve, so it should be used only for large samples. Press **STAT→TESTS** and select either **1-PropZInt**. Complete the display by entering **3** for the variable x and **10** for the variable n and change C-Level to **.90**. Then select **Calculate**. The resulting interval is (.06164,.53836), which, in this case, is not close to the exact confidence interval because the sample is small.

To compute the exact confidence interval for small samples, we find the interval shown in Figure 13.7 using the **Solver** function of the TI-83 Plus. **Solver** is a subitem of **MATH**. Complete **eqn:** $0 = 0.05 - binomcdf(10, X, 3)$ (you may have to bring the cursor up). Then bring the cursor down to the prompt **X =** and hit **SOLVE**, which is **ALPHA-ENTER**. After a few seconds, the display should show $X = .6066241610$, which is the upper bound of the 90% confidence interval. For the lower bound, bring the cursor up and edit the equation solver to read eqn: $0 = 0.95 - binomcdf(10, X, 2)$. Then scroll down to the line **X =** and hit **SOLVE**. Now the display should show the lower bound $X = .08726443391$. The resulting interval (.08726,.60662) agrees with the confidence interval obtained in Example 13.7. The numbers entered for this $1 - \alpha = 90\%$ confidence interval are $\alpha/2 = 0.05$, $1 - \alpha/2 = 0.95$, $n = 10$, $k = 3$, and $k - 1 = 2$. It may take several seconds to compute each bound. You can speed things up a little by making a reasonable guess for the bounds, such as $k/n = 0.30$, and entering 0.30 at the **X =** prompt before hitting **Solve**.

EXERCISES FOR SECTION 13.7

13.30 A sample of 44 men showed that 3 are red–green color blind. Use Table 5 on page 511 or appropriate technology to find a 95% confidence interval for the percentage who are red–green color blind.

• **13.31** A sample of 100 women showed that 7 were shorter than 5 feet. Use Table 5 on page 512 or appropriate technology to find a 99% confidence interval for the percentage who are shorter than 5 feet.

13.7 BINOMIAL EXACT INTERVALS

13.32 In a clinical trial of a drug to treat hypertension, 2% of 200 subjects developed a dry cough. Find a 99% confidence interval for the percentage of users of this drug who develop a dry cough. Notice that the sample is too small for the method of bootstrapping as given on 370 of Section 13.6.

  a. Do this exercise using Table 5 on page 512 or appropriate technology.

  b. Do this exercise the wrong way using bootstrapping.

• **13.33** To study whether cell phones interfere with the operation of heart pacemakers, some interference was detected by an electrocardiogram with a certain model cell phone in 4 out of 40 pacemaker users. Find a 95% confidence interval. Notice that the sample is too small for the method of bootstrapping as given on page 370 of Section 13.6.

  a. Do this exercise using Table 5 on page 511 or appropriate technology.

  b. Do this exercise the wrong way using bootstrapping.

13.34 In Exercise 13.28, a random sample of 200 car owners found 140 with air-conditioning. Use Table 5 on page 512 or appropriate technology to find a small-sample 99% confidence interval for the percentage of car owners who have air-conditioned cars.

• **13.35** Continuing with Exercise 13.29, 34 males from the U.S. Navy diving and special warfare community who had chronic knee and lower back pain alternated between treatment and placebo in the double-blind study. Fourteen of them had improvement of symptoms under the treatment versus the placebo. Use Table 5 on page 511 or appropriate technology to find a small-sample two-sided 95% confidence interval.

13.36 Repeat Example 13.6 on page 372 using Table 5 on page 511 or appropriate technology to find a small-sample 95% confidence interval. A random sample of 144 fish caught in Lake Wobegon included 40 bass.

13.37 It might seem reasonable that a 95% confidence interval for a long run proportion π constructed with direct use of the binomial random variable would have a coverage of 95%. Suppose the sample size is $n = 20$. There are 21 possible confidence intervals depending on how many successes k are observed.

  a. Suppose $\pi = 0.80$. Show that the probability 0.80 will be included in the small-sample 95% confidence interval is 0.978. This indicates that the method is a bit conservative in this case.

  b. Repeat part **a.** using large-sample methods and the conservative estimate of the standard error for P.

REVIEW EXERCISES FOR CHAPTER 13

13.38 A college is interested in the finding the average number of credit hours students have upon graduation. The minimum needed for graduation is 128 hours. The college used anecdotal evidence that 160 hours is the maximum of any recent graduate.

   a. Use one of the common rules mentioned in Chapter 2 to estimate the standard deviation.

   b. How large a sample size should be chosen to obtain a 95% interval with a margin of error of two credit hours?

• **13.39** Suppose 49 samples of a solution were analyzed for copper concentration in grams per liter (g/l). These averaged out to 12.37 g/l, with an SD of 0.21 g/l. Fill in the blanks.

   a. Each reading was approximately _____ g/l give or take _____ g/l or so.

   b. The true concentration of copper is estimated as _____ g/l; this estimate is likely to be off by _____ g/l or so.

   c. If another reading were to be taken it would be approximately _____ g/l give or take _____ g/l or so.

   d. Find the 95% confidence interval for the true unknown concentration.

13.40 A Republican and a Democratic candidate were running for a congressional seat. On election day, upon leaving polling places, a simple random sample of voters were asked how they voted. Of 900 such voters, 486 said they voted for the Republican candidate.

   a. Assuming they told the truth, find a 95% (one-sided) confidence interval for the minimum percentage who voted Republican.

   b. Can you predict a Republican victory with 95% confidence?

• **13.41** A random sample of 100 bass caught at Lake Wobegon has a mean length of 35.5 centimeters with a standard deviation of 5 centimeters. Find a 98% confidence interval for the mean length of all catchable bass in Lake Wobegon

13.42 An inspector tested the life of 144 batteries by keeping flashlights on until the light deteriorated by 50%. The results followed a normal curve. The average life of batteries tested was 20 hours with a standard deviation of 3 hours.

   a. Fill in the blanks. Approximately 90% of the batteries of this type would last between _____ and _____ hours.

   b. Find the 99% confidence interval for the true average life of batteries of this type.

REVIEW EXERCISES FOR CHAPTER 13

- **13.43** For a certain scale used in a chemistry laboratory, let X denote the measured weight and μ the true weight of the object being weighed. We will assume that X is a normally distributed random variable and that the scale is perfectly calibrated so that the expected value of X is μ. The standard deviation of a single weighing on this scale is $\sigma = 50$ micrograms. (A microgram is 10^{-6} grams.)

    a. Suppose that nine independent weighings of an object will be made and the sample mean $\overline{X}$ will be calculated. What is the standard deviation of the random variable $\overline{X}$?

    b. The sample mean in part **a.** turned out to be $\overline{x} = 0.100000$ grams. Construct 68% confidence limits for the true weight of the object.

    c. A laboratory assistant does not know that $\sigma = 50$ micrograms. However, for the assistant's sample of 9 weighings of the object he finds the sample mean to be 0.1000 grams and the sample standard deviation to be 100 micrograms. How should the assistant construct a 95% confidence interval for μ?

    d. A weighing is considered a success if the difference between the measured weight X and the true weight μ is less than 64 micrograms in absolute value (that is, a success occurs if $|x - \mu| < 64$ micrograms). What is the probability that a randomly selected weighing will result in a success? Recall that $\sigma = 50$ micrograms.

    e. Let π denote the probability of success. Suppose three independent weighings are to be made. What is the probability that they will all result in success? State your answer in terms of π.

    f. A laboratory assistant does not know the value of π. He decides to test the hypothesis H_0 that $\pi = 0.90$ versus the alternative hypothesis H_a that $\pi < 0.90$. He plans to take $n = 10$ independent weighings of a standard object whose weight is known to be 1 gram. He lets K equal the total number of successes in these $n = 10$ independent trials. Is K a binomial random variable? If H_0 were true, what would be the mean and variance of K?

    g. The laboratory assistant plans to reject H_0 as false if he observes $K \leq 7$. What is the probability of making a Type I error? If H_a were true and $\pi = 0.80$, what is the probability of making a Type II error?

13.44 The body mass index (BMI) is computed as $BMI = weight[kg]/(height[m])^2$. Consider the percentile rank BMI for 34 white, non-Hispanic girls from the fall 1998 kindergarten class as given in Exercise 6.29 on page 192. If the white, non-Hispanic girls in the Early Childhood Longitudinal Study ECLS-K study mirrored the children in the reference population for the CDC Growth Charts, then we would expect the percentile ranks to be uniformly distributed with a mean of $\mu = 50$ and a standard deviation of $\sigma = 100/\sqrt{12}$.

    a. Construct a 95% two-sided confidence interval for the mean percentile rank for all the white, non-Hispanic girls in the ECLS-K study.

b. Use this confidence interval to test the hypothesis that the mean is not 50 at a 5% level of significance.

13.45 Repeat the test of the above exercise but this time for Hispanic boys as given in Exercise 6.28 on page 191.

13.46 Fly-By-Nite airlines weighs hand luggage of a sample of passengers flying from Chicago to Florida. The following data are obtained (weight in pounds):

$$\begin{array}{cccccccc} 30 & 31 & 30 & 34 & 15 & 28 & 29 \\ 16 & 16 & 7 & 23 & 22 & 6 & 38 \\ 12 & 15 & 26 & 19 & 11 & 25 & 24 \end{array}$$

a. Make a table of the data with classes of width 10.
b. Sketch a histogram.
c. Find the mean.
d. Find the standard deviation.
e. Using the above results and assuming the population is normal, find the 95% confidence interval for the true mean weight of hand luggage on such flights to Florida.

• **13.47** The body mass index (BMI) is computed as $BMI = weight[kg]/(height[m])^2$. The following data are for the change in BMI for 34 white, non-Hispanic girls from the fall of kindergarten to the spring of first grade (roughly 18 months). For girls of this age, BMI normally increases slightly during this time period. These girls are a random sample from the 4854 white, non-Hispanic girls in the Early Childhood Longitudinal Study ECLS-K who had their height, weight, and age measured in the fall of 1998 and again in the spring of 2000. Construct a 90% confidence interval for the mean change in BMI.

$$\begin{array}{cccccccccc} 4.04 & 1.27 & -0.16 & -0.06 & 3.48 & 0.43 & 0.41 & 0.72 & 0.41 & 0.51 \\ -0.25 & 1.15 & -0.14 & 0.29 & 1.77 & -0.94 & 1.16 & 0.75 & 1.36 & 0.45 \\ 0.43 & 0.40 & 2.21 & 0.83 & 1.43 & 0.16 & 1.34 & 2.70 & 2.12 & 0.70 \\ -0.73 & 0.03 & -0.94 & 0.44 \end{array}$$

13.48 Construct a 90% confidence interval as in the preceding exercise but this time for a sample of 59 Hispanic boys. The following data are for the change in BMI for 59 Hispanic boys from the fall of kindergarten to the spring of first grade (roughly 18 months). The sample mean is $\bar{x} = 1.020$, and the standard deviation is $s = 1.944$.

$$\begin{array}{cccccccccc} 0.95 & 0.00 & 0.73 & 1.84 & 0.45 & 2.14 & -1.26 & 1.68 & -0.11 & 10.29 \\ 1.65 & 0.11 & -0.56 & 1.17 & 0.48 & -1.72 & 1.31 & -0.20 & -0.08 & 2.39 \\ 2.27 & 2.03 & 3.95 & 0.82 & 0.29 & -0.48 & 0.89 & -0.17 & -0.60 & -0.13 \\ -0.18 & 1.52 & 4.48 & 5.34 & -0.26 & 0.17 & 1.15 & 1.24 & 0.49 & -0.40 \\ -1.12 & -0.53 & -0.20 & 1.83 & 3.45 & 0.16 & 3.33 & 3.98 & -1.99 & 0.78 \\ 0.60 & -0.49 & 1.25 & 2.25 & 0.63 & 2.42 & -1.01 & -0.05 & 1.21 \end{array}$$

REVIEW EXERCISES FOR CHAPTER 13

- **13.49** Use the random sample of 59 Hispanic boys in Exercise 6.28 on page 191 to construct a one-sided upper 99% confidence interval of the form $\pi \geq \cdots$ for the percentage of all Hispanic boys in the ECLS-K study whose BMI is at or above the 95th percentile on the CDC Growth Charts.

13.50 The 1990 census data showed that in Cook County, Illinois, there were 2,369,624 workers 16 years and older and that their commute to work averaged 32 minutes with a standard deviation of 22 minutes. The 43,588 people who worked at home were each assigned a commuting time of 0 minutes. The data also showed that 1,733,985 of these workers commuted by car (alone or in a car pool).

   a. Can we conclude that approximately 68% of these workers had commute times between 10 and 54 minutes? Explain!

   b. Can we conclude that a 95% confidence interval for the percentage of workers in Cook County age 16 and over who commuted by car in 1990 is $73.175533\% \pm 0.000036\%$? Explain.

- **13.51** In 1995, a researcher takes a survey in Cook County, Illinois (1.46 million households), to determine how far the head of household travels to work. A simple random sample of 2500 households is chosen and the occupants are interviewed. It is found that the mean commuting time for the head of households is 35 minutes and the standard deviation is 25 minutes. If possible, find an approximate 95% confidence interval for the mean commuting time of all heads of households in Cook County. If this is not possible, explain why.

13.52 We continue with Exercise 13.51 above. The researcher interviewed all persons age 16 and over in the sample households: There were 4070 such persons. The mean commute of these persons was 33 minutes, and the standard deviation was 24 minutes. In 1800 of the sample households the head of household commuted by car.

   a. If possible, find an approximate 95% confidence interval for the mean commute time of all people age 16 and over in Cook County. If this is not possible, explain why.

   b. If possible, find an approximate 95% confidence interval for the percentage of all households in Cook County where the head of household commutes by car. If this is not possible, explain why.

- **13.53** A new method was used to measure the acceleration due to gravity at Little Rock, Arkansas. The results of 225 measurements had a mean of 978.721 cm/sec^2, with a standard deviation of 0.045 cm/sec^2. Assume no bias in the measurements. Fill in the blanks in part **a.** and answer the rest True or False.

   a. The true acceleration due to gravity is estimated as _____; this estimate is likely to be off by approximately _____ or so.

b. 979.721 ± 0.006 cm/sec^2 is an approximate 95% confidence interval for the mean of the 225 readings.

c. 979.721 ± 0.006 cm/sec^2 is an approximate 95% confidence interval for the true speed of light.

d. There is a 95% probability or so that the next reading will be in the range 979.721 ± 0.006 cm/sec^2.

e. Approximately 95% of the 225 readings were in the range 979.721 ± 0.006 cm/sec^2.

f. If another 225 readings are made, there is about a 95% probability that their average will be in the range 979.721 ± 0.006 cm/sec^2.

* * * * * * * * *

13.54 Solve the inequality 13.5 for π to obtain a $1 - 2\alpha$ confidence interval:

$$z_\alpha \leq \frac{p - \pi}{\sqrt{\pi(1-\pi)/n}} \leq z_{1-\alpha} \qquad (13.5)$$

We can do this by first observing that the symmetry of the normal curve results in $z_\alpha = -z_{1-\alpha}$. Thus, we can convert the inequality to:

$$\left| \frac{p - \pi}{\sqrt{\pi(1-\pi)/n}} \right| \leq z_{1-\alpha} \qquad (13.8)$$

Finally, we can square both sides of the inequality 13.8 and use the quadratic equation. The solution is known as the $1 - 2\alpha$ **score confidence interval** for π. It is an improvement of Formula 13.7 on page 370 but more complicated.

Chapter 14

TWO-SAMPLE INFERENCE

14.1 MATCHED PAIR SAMPLES

14.2 INDEPENDENT SAMPLES

14.3 WELCH'S FORMULA

14.4 INDEPENDENT SAMPLES WITH EQUAL VARIANCES

So far we have tested hypotheses and made estimates about a single population mean or proportion. However, we often need to make comparisons between two populations based on samples from each population. This is the subject of the current chapter. The simplest situation occurs when the samples consist of matched pairs as we will consider in Section 14.1. When matched pairs are not practical, comparison may be based on independent samples drawn from the populations. Cases where we have more than two populations will be studied in Chapter 16.

14.1 MATCHED PAIR SAMPLES

In double-blind experiments, the best results are obtained when the same subject can be used for both treatment groups. For example, in testing a new drug treatment for osteoporosis, each subject may alternate between a placebo and an experimental drug every three months. Bone density is then measured after each treatment. The treatments may even be randomized to further ensure double blindness. Using the same subjects for both treatments reduces the variability between subjects and thus increases the confidence that any difference observed can be attributed to the treatment.

When the same subject cannot be used, sometimes twins are compared. Or subjects may be matched in other ways. Even then, matched pairs are not always possible. For example, there is no practical way to compare different surgical procedures for open heart surgery by the use of matched pairs.

Comparing the difference between two populations with matched pair samples is no more difficult than making inferences about single populations. We have done this before. Example 12.5 on page 345 is an example. For each matched pair, we compute the difference of the variable and treat this difference as a single variable. The example below is typical.

Example 14.1 *Repeating an SAT.* Below are scores of the mathematical part of the Scholastic Aptitude Test (SAT) for five randomly selected students. Each of them took the test twice. Let's find a 95% confidence interval for the increase in score.

	Student				
	1	2	3	4	5
First attempt	540	610	480	580	440
Second attempt	560	600	530	590	470
Improvement	20	−10	50	10	30

Solution Let the random variable W be the improvement. For these five values of W, we have $\bar{w} = 20$ and

$$s^2 = \frac{(20-20)^2 + (-10-20)^2 + (50-20)^2 + (10-20)^2 + (30-20)^2}{4} = 500$$

Assuming that W follows the normal curve, the 95% confidence interval for the improvement is

$$95\% \text{ CI:} \quad \mu = 20 \pm t_{4, 0.975} \frac{\sqrt{500}}{\sqrt{5}} = 20 \pm 27.76 \quad \text{or}$$

$$95\% \text{ CI:} \quad -7.76 \leq \mu \leq 47.76$$

Note that the 95% confidence interval includes the value 0. This means that a hypothesis of no change ($H_0 : \mu_W = 0$) is consistent with the data.

The hypothesis that the mean test scores have changed ($H_a : \mu_W \neq 0$) has not been proven at the $\alpha = 0.05$ level of significance. The null hypothesis H_0 of no change must be retained. It is important to understand that we have not proven the null hypothesis. The burden of proof is to disprove H_0, which is difficult to do with a sample so small.

***

Example 14.2 *Battleship paint.* Suppose that the U.S. Navy wants to test two barnacle-resistant paints on eight battleships. One could randomly assign the battleships to the two paints, four to paint A and four to paint B. However, there could be a great deal of variability in resistance to barnacles because battleships sail in different waters, spend different times at sea, sail at different speeds, and perform different duties. Such a test would lack power because of the high variability and the very small samples.

A better way to test the barnacle-resistant paints is to do a paired sample comparison. Paint one side of each battleship with paint A and the other with paint B. The sides should be randomly chosen—say, by a toss of a coin. Because, where one side of the battleship goes, so does the other, variability due to different waters, times, speeds, and so on, are eliminated.

After the ships returned to port, the following data were recorded of the number of barnacles per square meter found clinging to the ships. Test for a difference between the paints at the $\alpha = .05$ level.

	\multicolumn{8}{c}{Battleship}		
	1	2	3	4	5	6	7	8	Mean	SD
Paint A	65	45	38	88	30	51	64	38	52.38	19.03
Paint B	48	36	45	74	31	39	58	31	45.25	14.77
Difference	17	9	−7	14	−1	12	6	7	7.13	7.92

Solution Let the random variable D be the difference in the number of barnacles. For this test we use $\bar{d} = 7.13$, $s_d = 7.92$ and $n = 8$. Because we cannot peek at the data, the alternative hypothesis must be two-sided:

$$H_0 : \mu_D = 0 \quad \text{and} \quad H_a : \mu_D \neq 0$$

Assuming that D follows the normal curve, we have

$$t_s = \frac{\bar{d} - \mu_D}{s_d/\sqrt{n}} = \frac{7.13 - 0}{7.92/\sqrt{8}} = 2.546$$

With $df = 7$, we obtain P-value $= .038$; so we reject H_0. We conclude at the 5% level that there is a difference in the paints. In fact, we can conclude that paint B collects fewer barnacles.

EXERCISES FOR SECTION 14.1

- **14.1** Consider the following matched-pair data set. Assume normality of the parent populations. Find a 95% confidence interval for the difference in means.

$$x: \quad 7 \quad 9 \quad 9 \quad 6 \quad 8 \quad 8$$
$$y: \quad 4 \quad 4 \quad 3 \quad 4 \quad 9 \quad 5$$

14.2 Consider the following matched-pair data set. Test the hypothesis that the population means are equal against a two-sided alternative. Assume normality of the parent populations and use $\alpha = 0.10$.

$$x: \quad 38 \quad 28 \quad 26 \quad 28 \quad 28$$
$$y: \quad 36 \quad 32 \quad 38 \quad 38 \quad 34$$

- **14.3** In a double-blind test, 16 patients were given an experimental sleeping pill one night and a placebo on another night. On the average, the patients slept longer under the effect of the sleeping pill than the placebo. The 16 patients slept an average of 1.58 hours longer, with an SD of 1.23 hours, using the sleeping pill instead of the placebo. Find the 99% confidence interval for the average increase in sleeping time for patients using the sleeping pill.

14.4 A study was conducted to determine the effect of a training program of geometric shapes on 5-year-old children's ability to complete a puzzle. The time in seconds required to complete the puzzle before and after the training program appear below. Find a 99% confidence interval for the mean difference in times.

	Subject					
	1	2	3	4	5	6
Before	63	73	49	57	78	58
After	57	72	40	45	79	49

NOTES ON TECHNOLOGY

Paired Samples

Minitab: Enter the data in columns **C1** and **C2**. Select **Stat→Basic Statistics→paired t** and then the columns **C1** and **C2**. Be sure to check the settings under the Options button.

Excel: Enter your data in columns **A** and **B**. Go to an empty cell and use the function **TTEST**. The function can be found by hitting the icon f_x and choosing **Statistical→TTEST**, and then hitting **OK**. Enter the column A location of the sample as the first array, the column B location as the second array, either 1 or 2 for one-sided or two-sided, and **1** for Type. The output will be the *P*-value for a paired *t* test. A similar procedure can be found under **Tools→Data Analysis→t-Test: Paired Two Sample for Means**.

TI-83 Plus: Enter the two data sets in lists **L1** and **L2**. Then enter the differences in **L3** by clearing the screen and entering **L1 - L2, STO→, L3**. You can test the difference and find a confidence for the difference following the instructions in the Notes on Technology on pages 348 and 364, respectively. Be sure to change the default list to **L3** in the input options of these TI-83 Plus procedures.

14.2 INDEPENDENT SAMPLES

In comparing two treatments, sometimes matched pair comparisons are not possible. Let's consider situations where independent samples are assigned to two treatments.

Example 14.3   **Rat chow.**   Lab rats are randomly assigned to several groups and given special diets for three weeks. Ten rats are randomly chosen for each diet, but 20 are given the standard diet. When several alternative treatments are being compared to a standard treatment, it is important to have reliable values for the standard treatment. Thus, the size of the sample that receives the standard treatment is often larger than any of the other samples.

Here let's compare just one of the special diets—say, diet X—with the standard diet Y. Suppose that the mean weight gain for the $n_X = 10$ rats on diet X was $\bar{x} = 5.20$ grams and the sample standard deviation for this group was $s_x = 0.80$ grams. The mean and standard deviation for the weight gains of the $n_Y = 20$ rats on diet Y were $\bar{y} = 4.80$ grams and $s_y = 0.60$ grams, respectively.

Diet	n	mean	SD
X	10	5.20	0.80
Y	20	4.80	0.60

We want to know whether the difference in weight gains $\bar{x} - \bar{y}$ under the two diets is "real"; that is, if the experiment were to be repeated many times, do we consistently find that diet X produces different weight gains than the standard one? The burden of proof is on the results of this one experiment to show that there is a difference in weight gains for the two parent *populations*. Obviously, there is a difference in mean weight gains of the two *samples* of lab rats. We would not expect the two *sample* means to be the same even if the all rats were fed the same diet. (Consider drawing two samples from the same population box; it's unlikely that the sample means are exactly the same.)

We test the null hypothesis that there is no difference in mean weight gains against the alternative that there is a difference.

Solution In terms of the population means, the null and alternative hypotheses are

$$H_0 : \mu_X - \mu_Y = 0 \text{ and}$$
$$H_a : \mu_X - \mu_Y \neq 0$$

Notice that the hypotheses must be statements about the population means μ_X and μ_Y, not the observed sample means $\bar{x}$ and $\bar{y}$.

We use the difference in the sample means $\overline{X} - \overline{Y}$ to test for the difference in population means. Because the samples are independent, so are their sample means. According to Theorems 10.6 and 10.8, and the law of averages for the sample mean in Chapter 11, we have

$$\mu_{(\overline{X}-\overline{Y})} = E(\overline{X} - \overline{Y}) = E(\overline{X}) - E(\overline{Y}) = \mu_X - \mu_Y$$

$$\sigma_{(\overline{X}-\overline{Y})} = \sqrt{\text{Var}(\overline{X} - \overline{Y})} = \sqrt{\text{Var}(\overline{X}) + \text{Var}(\overline{Y})} = \sqrt{\frac{\sigma_X^2}{n_X} + \frac{\sigma_Y^2}{n_Y}}$$

Because we do not know σ_X and σ_Y, we bootstrap and use the estimated standard error

$$\text{SE}_{(\bar{x}-\bar{y})} = \sqrt{\frac{s_x^2}{n_X} + \frac{s_y^2}{n_Y}}$$

The t statistic is then

$$t_s = \frac{(\bar{x} - \bar{y}) - \mu_{\overline{X}-\overline{Y}}}{\sqrt{\frac{s_x^2}{n_X} + \frac{s_y^2}{n_Y}}}$$

If the sample sizes were large, we could comfortably use the normal table of Table 3 on page 508. However, $n_X = 10$ is not large, so we must use one of the t curves of Table 4 on page 510.

What are the degrees of freedom? Surprisingly, statisticians do not agree on the correct degrees of freedom. There are three approaches that are common. In this section, we consider the conservative approach. In Section 14.3, we present the approach of Welch's formula. Finally, in Section 14.4 we consider a method that can be used when the parent populations have equal variances.

Conservative approach

> A conservative way to compute the degrees of freedom is to use the minimum
>
> $$\mathbf{df} = \min(\mathbf{n_X} - 1, \mathbf{n_Y} - 1) \qquad (14.1)$$

Conservative solution

For our case, this is $df = 9$. So for our rats-on-a-diet data, the observed t statistic is

$$t_s = \frac{5.20 - 4.80}{\sqrt{\frac{0.80^2}{10} + \frac{0.60^2}{20}}} = 1.3969$$

Using $\alpha = 0.05$ as our level of significance and a two-sided alternative, we find the 2.5 and 97.5 percentiles on the t_9 curve, which turn out to be ± 2.262. We state the decision rule as follows:

$$\text{Retain } H_0 \text{ if } |t| \leq 2.262 \qquad \text{Reject } H_0 \text{ if } |t| > 2.262$$

Because the absolute value of the observed $t = 1.3969$ does not exceed 2.262, the t statistic fails to fall into the rejection region and the null hypothesis is retained.

To compute the P-value, we use Table 4 on page 510 and note that $t_s = 1.3969$ falls between $t_{9,0.95} = 1.833$ and $t_{9,0.90} = 1.383$.[1] Recall that for two-sided alternatives, we double the upper tail probabilities of 0.05 and 0.10, so

$$0.10 < P\text{-value} < 0.20$$

We conclude that the data are consistent with the null hypothesis. The difference in average weight gains can be explained by sample variability.

Equivalently, H_0 is retained at the $\alpha = 5\%$ level because $\mu_X - \mu_Y = 0$ is contained in the 95% confidence interval

$$\begin{aligned}
95\% \text{ CI:} \quad \mu_X - \mu_Y &= (\bar{x} - \bar{y}) \pm t_{9,0.975} \text{SE}_{(\bar{x}-\bar{y})} \\
&= (5.80 - 4.80) \pm 2.262\sqrt{\frac{0.80^2}{10} + \frac{0.60^2}{20}} \\
&= 0.4 \pm 0.6477
\end{aligned}$$

---

[1] Or we can use technology, such as the *tcdf* function on the TI-83 Plus, to obtain $Pr(T \geq 1.3969) = 0.09797$. Double this to obtain the P-value $= 0.1959$ for the two-sided alternative.

EXERCISES FOR SECTION 14.2

- **14.5** Suppose independent random samples from two normal populations gave the following results:

  $n_X = 8 \quad \bar{x} = 35 \quad s_x = 4$
  $n_Y = 12 \quad \bar{y} = 20 \quad s_y = 3$

  a. Compute the standard error.

  b. Find a conservative 95% confidence interval for the difference $\mu_X - \mu_Y$.

14.6 Suppose independent random samples from two normal populations gave the following results:

  $n_X = 25 \quad \bar{x} = 16 \quad s_x = 4.7$
  $n_Y = 45 \quad \bar{y} = 20 \quad s_y = 2.3$

  a. Compute the standard error.

  b. Find a conservative 95% confidence interval for the difference $\mu_X - \mu_Y$.

- **14.7** Suppose independent random samples from two normal populations gave the following results:

  $n_X = 10 \quad \bar{x} = 48 \quad s_x = 6.4$
  $n_Y = 25 \quad \bar{y} = 44 \quad s_y = 5.0$

  a. Compute the standard error.

  b. Test the hypothesis $\mu_X - \mu_Y > 0$ at the 5% level using a conservative df.

14.8 Suppose independent random samples from two normal populations gave the following results:

  $n_X = 6 \quad \bar{x} = 12 \quad s_x = 1.8$
  $n_Y = 15 \quad \bar{y} = 9 \quad s_y = 1.5$

  a. Compute the standard error.

  b. Test the hypothesis $\mu_X - \mu_Y \neq 0$ at the 5% level using a conservative df.

- **14.9** A study was conducted to test for the difference in salaries of union and nonunion workers performing the same task at two different companies. A random sample of 100 union workers showed an average salary of $26,240 with a standard deviation of $1000, while a random sample of 81 nonunion workers showed an average salary of $25,930 with a standard deviation of $1200. Test for a difference in salary between union and nonunion workers at the 5% level using a conservative df. State:

  a. H_0 **b.** H_a **c.** the decision rule

  d. Choose one option.
  Decision: There (is a) (is no) significant difference between the average salary of union and nonunion workers performing these tasks.

14.10 An experiment was done to see whether open-book tests make a difference. A calculus class of 48 students agreed to be randomly assigned by the draw of cards to take a quiz either open-book or closed-book. The quiz consisted of 30 integration problems of various difficulty. Students were to do as many as possible in 20 minutes. The 24 students taking the test closed-book got an average of 15 problems correct with a standard deviation of 2.5. The open-book group got an average of 12.5 correct with a standard deviation of 3.5. What do you say? Use a two-sided test with $\alpha = 0.05$ and a conservative df.

• **14.11** Does the use of candy increase restaurant tipping?[2] Ninety-two dining parties and two seasoned waiters were part of an experiment in which, just prior to delivering the check to the table, the waiters were randomly given either a red or a black card. If the card was red, each person in the dining party was given a fancy foil-wrapped piece of chocolate when the check was delivered. If the card was black, the check was delivered without chocolate. For the 46 dining parties who received the chocolate, the mean tip was 17.8% and the standard deviation was 3.06%. For the 46 who received no chocolate, the mean tip was 15.1% and the standard deviation was 1.89%. Test whether chocolate increased the percentage of the tip at the $\alpha = 0.01$ level using a conservative df.

14.12 Spelling tests were given to a random sample of 5th graders in a large school system. There were 40 boys and 50 girls in the sample. The boys had a mean score of 75.1 and a variance of 103.9. The girls had a mean of 78.1 and a variance of 90.6. Is the average score for girls higher than for boys? Test at the 0.03 level of significance using a conservative df. Be sure to state the null and alternative hypotheses and the P-value. State your decision in plain English in the context of the problem.

14.3 WELCH'S FORMULA

We can strengthen the above test for a difference of means $\mu_X - \mu_Y$ by using the same t statistic

$$t_s = \frac{\overline{x}_1 - \overline{x}_2}{\text{SE}_{\overline{x}_1 - \overline{x}_2}}$$

but calculating the degrees of freedom as follows.

---

[2] David Strohmetz, Bruce Rind, Reed Fisher, & Michael Lynn, *Sweetening the till: The use of candy to increase restaurant tipping*, Journal of Applied Social Psychology **32(2)** (2002), pp. 300–309.

Welch's formula

$$df = \frac{\left(\frac{s_x^2}{n_X} + \frac{s_y^2}{n_Y}\right)^2}{\left(\frac{s_x^2}{n_X}\right)^2/(n_X - 1) + \left(\frac{s_y^2}{n_Y}\right)^2/(n_Y - 1)} \quad (14.2)$$

The computed value should be rounded down (truncated) to an integer.

This formula makes the calculations much more complicated than the conservative way (Equation 14.1), but it raises the degrees of freedom.

Example 14.4 *Rat chow again.* Let's now use Welch's formula to test for a difference in weight gain.

Diet	n	mean	SD
X	10	5.20	0.80
Y	20	4.80	0.60

Welch's solution Welch's formula yields

$$df = \frac{\left(\frac{0.80^2}{10} + \frac{0.60^2}{20}\right)^2}{\left(\frac{0.80^2}{10}\right)^2/8 + \left(\frac{0.60^2}{20}\right)^2/19} = \frac{0.006724}{0.000472164} = 14.24$$

We truncate to an integer; in this case, $df = 14$. Previously, we used $df = 9$. Using the same $\alpha = 0.05$, we compare the previously calculated t statistic $t = 1.3969$ with the 97.5 percentile on the t_{14}-curve, which is 2.145. Because the absolute value of the t statistic does not exceed 2.145, the t statistic still fails to fall in the rejection region.

We can compute the new P-value $= 0.1842$. Although this is still much larger than $\alpha = 0.05$, this stronger test results in a smaller P-value than the previous P-value $= 0.1959$.

Having a larger df strengthens the test; that is, the test is more likely to reject the null hypothesis, when it is false.

EXERCISES FOR SECTION 14.3

• **14.13** Suppose independent random samples from two normal populations gave the following results:

$$n_X = 25 \quad \bar{x} = 16 \quad s_x = 4.7$$
$$n_Y = 45 \quad \bar{y} = 20 \quad s_y = 2.3$$

a. Evaluate Welch's formula.
b. Find a 95% confidence interval for $\mu_X - \mu_Y$.

14.14 Suppose independent random samples from two normal populations gave the following results:

$$n_X = 8 \quad \bar{x} = 35 \quad s_x = 4$$
$$n_Y = 12 \quad \bar{y} = 20 \quad s_y = 3$$

a. Evaluate Welch's formula.
b. Find a 95% confidence interval for $\mu_X - \mu_Y$.

• **14.15** Suppose independent random samples from two normal populations gave the following results:

$$n_X = 10 \quad \bar{x} = 48 \quad s_x = 6.4$$
$$n_Y = 25 \quad \bar{y} = 44 \quad s_y = 5.0$$

a. Evaluate Welch's formula.
b. Test the hypothesis $\mu_X - \mu_Y > 0$ at the 5% level using Welch's df.

14.16 Suppose independent random samples from two normal populations gave the following results:

$$n_X = 6 \quad \bar{x} = 12 \quad s_x = 1.8$$
$$n_Y = 15 \quad \bar{y} = 9 \quad s_y = 1.5$$

a. Find the standard error.
b. Test the hypothesis $\mu_X - \mu_Y \neq 0$ at the 5% level using Welch's df.

• **14.17** A comparison of the number of calories in 28 vegetarian and 28 nonvegetarian entrees served in a college dining hall are summarized in the table below.

Group	n	mean	SD
Vegetarian	28	351	119
Nonvegetarian	28	322	87

For both the vegetarian and nonvegetarian groups, the number of calories follow the normal curve.

    a. Use Welch's formula to find a 95% confidence interval for the difference in calories for the two groups.

    b. Use a conservative df to find a 95% confidence interval.

14.18 A random sample of 25 families with income over $100,000 had an average of 2.2 children with a standard deviation of 1.2 children. A random sample of 20 families with income under $100,000 had an average of 2.8 children with a standard deviation of 1.8. Test whether there is a difference in the average number of children for the two income groups. Use $\alpha = 0.01$ and Welch's formula.

• **14.19** To determine the effectiveness of a certain vitamin supplement, the following data were obtained:

Weight increase for two groups of mice

Control group	Treatment group
12	18
19	16
14	23
	20
	23

    a. Use Welch's formula to construct a 90% confidence interval for the difference of the means. Assume the populations are normal.

    b. Test at the $\alpha = 0.05$ level whether the vitamins treatment results in a higher weight gain.

14.4 INDEPENDENT SAMPLES WITH EQUAL VARIANCES

If two populations with unknown means μ_X and μ_Y have unknown but equal variances, we can estimate the common variance σ by taking a weighted average of sample variances. This estimate, using sample variances s_X^2 and s_Y^2, is called the **pooled variance**

Pooled variance

$$s_p^2 = \frac{\sum(x-\bar{x})^2 + \sum(y-\bar{y})^2}{(n_X-1)+(n_Y-1)} = \frac{(n_X-1)s_x^2 + (n_Y-1)s_y^2}{(n_X-1)+(n_Y-1)}$$

14.4 INDEPENDENT SAMPLES WITH EQUAL VARIANCES

The estimated standard error of the difference of means is then

Pooled SE

$$\text{SE}_{(\bar{x}-\bar{y})} = \sqrt{\frac{s_p^2}{n_X} + \frac{s_p^2}{n_Y}} = s_p\sqrt{\frac{1}{n_X} + \frac{1}{n_Y}}$$

In pooling, we gain power by being able to *add* the degrees of freedom

Pooled df

$$df = (n_X - 1) + (n_Y - 1) = n_X + n_Y - 2 \qquad (14.3)$$

instead of taking the minimum (for the conservative way) or using Welch's formula.

Example 14.5 *More rat chow.* For the rats-on-a-diet example, if we assume the null hypothesis that both rat diets have essentially the same effect, then not only are the means equal, but so are the standard deviations.

Diet	n	mean	SD
X	10	5.20	0.80
Y	20	4.80	0.60

Pooled solution An assumption of equal variances changes the hypotheses as follows.

$$H_0 : \mu_X - \mu_Y = 0 \text{ and } \sigma_X - \sigma_Y = 0$$
$$H_a : \mu_X - \mu_Y \neq 0$$

Here we assume that the difference in standard deviations

$$s_x = 0.80, \quad s_y = 0.60$$

is due to sample variability. Thus,

$$s_p^2 = \frac{9(0.80)^2 + 19(0.60)^2}{9 + 19} = 0.45 \quad \text{with} \quad df = 9 + 19 = 28$$

Recalculating the t statistic by replacing both s_x^2 and s_y^2 by s_p^2, we obtain

$$t_s = \frac{\overline{x} - \overline{y} - \mu_{\overline{X}-\overline{Y}}}{\sqrt{\frac{s_p^2}{n_X} + \frac{s_p^2}{n_Y}}} = \frac{5.20 - 4.80}{\sqrt{0.45 \left(\frac{1}{10} + \frac{1}{20}\right)}} = 1.5396$$

The 97.5 percentile of t_{28} is 2.048. Because the absolute value of our t statistic does not exceed this value, we again retain the null hypothesis.

We calculate P-value $= 0.13488$ for the two-sided alternative. The data are again in agreement with the null hypothesis, so the difference in average weight gains under the two diets can be explained by chance. However, notice that the P-value under this assumption of equal standard deviations is smaller than before. This reflects the greater power of this test when, in fact, the population standard deviations are equal.

Below we compare *proportions* based on independent random samples from two dichotomous populations

---

Example 14.6 **Widgets.** A quality-control engineer suspects that the proportion of defective widgets differs significantly between two production shift. Of $n_1 = 400$ widgets sampled from the first shift's production, $k_1 = 60$ had defects, while of $n_2 = 300$ widgets sampled from the second shift only $k_2 = 30$ had defects.

Shift	n	k
1	400	60
2	300	30

Let's test the hypotheses below at the $\alpha = 5\%$ level of significance.

$$H_0: \pi_1 - \pi_2 = 0 \qquad H_A: \pi_1 - \pi_2 \neq 0$$

Pooled solution Clearly, if $\pi_1 = \pi_2$, then the population variances are equal. So we use a **pooled sample proportion** of

$$p = \frac{k_1 + k_2}{n_1 + n_2} = \frac{90}{700}$$

to obtain a **pooled standard deviation** of

$$s_p = \sqrt{p(1-p)} = \sqrt{\left(\frac{90}{700}\right)\left(\frac{610}{700}\right)}$$

The estimated standard error for $p_1 - p_2$ is

14.4 INDEPENDENT SAMPLES WITH EQUAL VARIANCES

$$\begin{aligned}
\text{SE}_{(p_1-p_2)} &= \sqrt{\frac{s_p^2}{n_1} + \frac{s_p^2}{n_2}} \\
&= s_p\sqrt{\frac{1}{400} + \frac{1}{300}} \\
&= \sqrt{\left(\frac{90}{700}\right)\left(\frac{610}{700}\right)}\sqrt{\frac{1}{400} + \frac{1}{300}} \\
&= 0.02557
\end{aligned}$$

Because the parent populations are dichotomous and thus not normal, it is not appropriate to use t curves. The test statistic z_s computes as follows:

$$z_s = \frac{(p_1 - p_2) - (\pi_1 - \pi_2)}{\text{SE}_{(p_1-p_2)}} = \frac{0.05}{0.02557} = 1.955$$

The rejection rule for the z test with a level of significance of α:

$$\text{Reject } H_0 \text{ if } |z| > z_{(1-\frac{\alpha}{2})}$$

For $\alpha = 0.05$, we have $z_{.975} = 1.96$, and for $\alpha = 0.055$, we have $z_{.9725} = 1.9189$. So the null hypothesis would be retained at the 5% significance level but not at the 5.5% significance level.

EXERCISES FOR SECTION 14.4

14.20 Suppose independent random samples from two normal populations gave the following results. Assume the parent populations have equal variances.

$n_X = 8 \quad \bar{x} = 35 \quad s_x = 4$

$n_Y = 12 \quad \bar{y} = 20 \quad s_y = 3$

a. Find the pooled standard deviation.

b. Find a 95% confidence interval for $\mu_X - \mu_Y$.

• **14.21** Suppose independent random samples from two normal populations gave the following results. Assume the parent populations have equal variances.

$n_X = 6 \quad \bar{x} = 16 \quad s_x = 2.7$

$n_Y = 8 \quad \bar{y} = 20 \quad s_y = 2.3$

a. Find the pooled standard deviation.

b. Find a 95% confidence interval for $\mu_X - \mu_Y$.

14.22 Suppose independent random samples from two normal populations gave the following results. Assume the parent populations have equal variances.

$n_X = 6 \quad \bar{x} = 12 \quad s_x = 1.8$

$n_Y = 15 \quad \bar{y} = 9 \quad s_y = 1.5$

a. Find the pooled standard deviation.

b. Test the hypothesis $\mu_X - \mu_Y \neq 0$ at the 5% level.

• **14.23** Suppose independent random samples from two normal populations gave the following results. Assume the parent populations have equal variances.

$n_X = 10 \quad \bar{x} = 48 \quad s_x = 5.4$

$n_Y = 25 \quad \bar{y} = 44 \quad s_y = 5.0$

a. Find the pooled standard deviation.

b. Test the hypothesis $\mu_X - \mu_Y > 0$ at the 5% level.

14.24 Repeat Exercise 14.19 assuming that the standard deviations of the populations are the same.

Weight increase for two groups of mice

Control group	Treatment group
12	18
19	16
14	23
	20
	23

• **14.25** Suppose the systolic blood pressure was measured for a random sample of 25 men selected from a large group of men ages 25–30 who drink at least 16 ounces of carbonated soft drinks per day. The sample has a mean of 129.87 mmHg with standard deviation 15.32 mmHg. Suppose a second random sample of size 30 is drawn from a group men ages 25–30 who do not drink carbonated soft drinks. This sample has a mean systolic blood pressure of 119.71 mmHg with standard deviation 18.23 mmHg. Test the hypothesis that there is a significant difference in the mean systolic blood pressure of these groups. Assume the population variances are equal and use the $\alpha = 1\%$ level of significance.

Group	n	mean	SD
Soft drink	25	129.87	15.32
Control	30	119.71	18.23

14.26 Using the data of Exercise 14.25 above, find the 95% confidence interval of the difference in systolic blood pressure between the two groups. Again, assume the population variances are equal.

• **14.27** A retailer is trying to determine the difference between the mean number of screws of a particular size in prepackaged containers supplied by two competing companies. A random sample of 10 containers from Company A produces

14.4 INDEPENDENT SAMPLES WITH EQUAL VARIANCES

a mean of 35.3 with a variance of 5.8, while a random sample of 9 containers from Company B produces a mean of 31.7 with a variance of 11.6. Find a 95% confidence interval of the true difference in means assuming the populations to be normal with equal variances.

14.28 Using the data of Exercise 14.27 above, test whether there is a significant difference in the mean number of screws for the two companies. Assume equal variances and use $\alpha = 0.05$.

• **14.29** To test the idea that the game is not over until the last quarter, a sample of football games and a sample of basketball games were collected from the 1990 and 1990–91 seasons, respectively. For the 93 football games sampled, 72 of teams leading after 3 quarters went on to win (the other 21 lost). For the 189 basketball games, the team leading after 3 quarters won 150 times. Test the hypothesis that the proportion of games won by the late-in-the-game leader is the same for the two sports.[3] Use $\alpha = 0.05$.

14.30 A test was given to 500 eighth graders to see whether calculators help them in solving word problems.[4] Half of them were permitted to use calculators, and the other half could use only paper and pencil. A certain question was answered correctly by 18 students in the calculator group and 59 students in the paper-and-pencil group. Use a pooled sample proportion to test whether calculators made a difference in answering this question. Use $\alpha = 0.01$.

• **14.31** Test the hypothesis of Exercise 6.34 on page 193 using the z test as in Example 14.6 on page 398 instead of Fisher's exact test.

	Success	Failure	Row total
New drug	38	4	42
Standard treament	14	7	21
Column total	52	11	63

14.32 Test the hypothesis of Exercise 6.36 on page 193 using the z test as in Example 14.6 on page 398 instead of Fisher's exact test.

	Cold	No cold	Row total
Vitamin C	17	123	140
Placebo	31	108	139
Column total	48	231	279

---

[3] These data are from Table 1 of the article by Harris Cooper, Kristina M. DeNeve, and Frederick Mosteller, *Predicting Professional Sports Game Outcomes from Intermediate Game Scores*, Chance **5** (1992), No. 3–4, pp. 18–22.

[4] *The Third National Mathematics Assessment: Results, Trends and Issues*, National Assessment of Educational Progress, Princeton, 1983. This result was cited in David Freedman, Robert Pisani, & Roger Purves, *Statistics*, third edition, W. W. Norton & Co, 1998. The question was: "One hundered students are going on a field trip. Each bus carries 30 students. How many buses are needed?" The correct answer is 4. The most common incorrect answer was 3.33.

NOTES ON TECHNOLOGY

Difference of Means

Minitab: Enter the data in columns **C1** and **C2**. Select **Stat→Basic Statistics**. Then choose **2-Sample t** or **2 Proportions**. Check the **Samples in different columns** button and enter **C1** and **C2**. Welch's formula for *df* is used in the 2-Sample *t* test. Summarized data can also be used for the 2 Proportion test.

Excel: Enter your data in columns **A** and **B**. Go to an empty cell and use the function **TTEST**. The function can be found by hitting the icon f_x and choosing **Statistical→TTEST**, and then hitting **OK**. Enter the column A location of the sample as the first array, the column B location as the second array, either 1 or 2 for one-sided or two-sided, and Type **2** for equal variances or Type **3** for unequal variances. The output will be the *P*-value. A similar procedure can be found under **Tools→Data Analysis→t-Test: Two Sample Assuming Unequal Variances** and **Tools→Data Analysis→t-Test: Two Sample Assuming Equal Variances**

TI-83 Plus: Enter the two data sets in lists **L1** and **L2** or have the summary statistics ready. Press **STAT→TESTS** and select either **2-SampZ-Test** or **2-SampT-Test** or **2-PropZTest**. In the subsequent display, select **Data** if your data are in lists L1 and L2, or select **Stats** otherwise. Press **ENTER**, and then fill in the rest of the display. Select **CALCULATE** and hit **ENTER** again. The display will include the *P*-value.

REVIEW EXERCISES FOR CHAPTER 14

- **14.33** A random sample of size 5 taken from a normal population results in a mean of 70 and standard deviation of 10. A second random sample of size 12 taken from another normal population results in a mean of 60 and a standard deviation of 8. Find the 90% confidence interval for the difference of the population means.

    a. Use the conservative degrees of freedom.

    b. Use Welch's formula.

14.34 In an experiment conducted in 1972, each of 48 male bank supervisors was given a copy of a personnel file and asked to judge whether the person should be promoted to a routine branch manager job or whether the person's file should be held and other applicants interviewed.[5] The 48 personnel files were identical except that 24 indicated the applicant was male and the other 24 indicated the applicant was female. Of the 24 "male" files, 21 were recommended for promotion. Of the 24 "female" files, 14 were recommended for promotion. Assume this experiment randomly assigned the 48 bank supervisors to one of two treatment groups. One group of 24 received the female personnel files, and the other group received the male files. Observe whether the applicant is approved for promotion. For each group, compute the proportion of applicants who were recommended for promotion. Test the hypothesis that there is no difference in promotion-recommendation rates for these two treatments. Should there be a one-sided or two-sided alternative? Compute a P-value. Is it reasonable to attribute the observed difference in promotion rates to chance? Does the data provide convincing evidence that the bank supervisors discriminated against female applicants?

- **14.35** Test the hypothesis of Exercise 6.37 on page 193 using the z test as in Example 14.6 on page 398 instead of Fisher's exact test.

	Flu	No flu	Row total
Shot	3	12	15
No shot	18	15	33
Column total	21	27	48

14.36 An experiment was conducted in 1997 using the drug infliximab to treat Crohn's disease. In the study, 37 patients were given the drug and 36 were given a placebo. After four weeks, 24 of the treated patients showed a clinical improvement while 6 of the placebo patients did so. Test at the 5% level to see whether the improvement rate is significantly greater in the treatment group. Assume equal variances as in Example 14.6 on page 398.

- **14.37** Compare the Department of Defense Test for Espionage and Sabotage (TES) with the Counterintelligence Scope Polygraph (CSP) as described in Exercises 3.72 and 3.73. If we exclude the inconclusive ratings, the CSP correctly identified the 56 programmed guilty (PG) subjects 34 times, and the TES identified the 30 PG subjects 25 times. Test for a difference in the rate of identifying programmed guilty subjects. Let $\alpha = 0.05$.

14.38 This is a continuation of the above exercise. If we exclude the inconclusive ratings, the CSP correctly identified 126 programmed innocent (PI) subjects 120 times; and the TES identified 54 PI subjects 48 times. Test for a difference in the rate of identifying programmed innocent subjects at the 5% level.

- **14.39** Do Example 14.1 on page 386 about repeating an SAT the *wrong* way by assuming independent samples instead of paired samples. Use Welch's formula, which yields $df = 7.41$.

---

[5] From the article by B. Rosen and T. Jerdee, *Influence of sex role stereotypes on personnel decisions*, Journal of Applied Psychology **59** (1974), pp. 9–14.

14.40 Do the battleship paint Example 14.2 on page 387 the *wrong* way by assuming independent samples instead of paired samples. Use Welch's formula, which yields $df = 13.19$. Explain why the conclusion is different.

* * * * * * * * *

14.41 Suppose $n = 100$ adults are selected at random (with replacement) from a large population and asked two Yes or No questions. Let K_1 denote the number of Yes answers to the first question, and let π_1 denote the probability of a Yes responses to this question.

    a. Use the Central Limit Theorem to show that

$$Z_1 = \frac{K_1 - 100\pi_1}{\sqrt{100\pi_1(1-\pi_1)}}$$

    follows the standard normal curve.

    b. To test the hypothesis $H_0 : \pi_1 = \frac{1}{2}$ versus $H_a : \pi_1 \neq \frac{1}{2}$, the following rejection region is proposed. What is the significance level for this statistical test?

$$C = \left\{ K_1 : \left| \frac{K_1 - 50}{5} \right| > 1.96 \right\}$$

    c. What is the power of this test when $\pi_1 = 0.30$?

    d. Let K_2 denote the number of Yes answers to the second question, and let π_2 denote the probability of a Yes response to this question. Let $p_1 = \frac{K_1}{100}$, and $p_2 = \frac{K_2}{100}$. Also, let p_{12} be the proportion of those sampled who respond Yes to both questions. Show Z_D follows the standard normal curve:

$$Z_D = \frac{(K_1 - K_2) - 100(\pi_1 - \pi_2)}{\sqrt{100(p_1(1-p_1) + p_2(1-p_2) - 2(p_{12} - p_1 p_2))}}$$

    e. How would you construct an interval estimate for $\pi_1 - \pi_2$, the difference in the proportion of Yes responses to the two questions?

    f. Suppose you decide to test the null hypothesis

$$H_0 : \pi_1 - \pi_2 = 0 \quad \text{versus} \quad H_a : \pi_1 - \pi_2 \neq 0$$

    What critical region should you use at the $\alpha = 0.05$ level of significance?

    g. Suppose in your random sample of 100 adults 20 answered Yes to both questions, 30 answered Yes to the first question, and 60 answered Yes to the second question. Construct an approximate 95% confidence interval for the difference in the proportion of Yes responses to the two questions in the adult population.

    h. For the data in part **g**, would you reject the null hypothesis using your test in part **f**? Informally describe your conclusions.

14.42 A double-blind experiment[6] was designed to compare the effectiveness of a high-dose regime of bromocriptine compared with a low-dose regime in the treatment of Parkinson's disease. Patients were randomly allocated to one of the two treatment groups and given increasing amounts of the drug according to a prescribed schedule and then given a clinical rating by a trained neurologist during the first six months of treatment. The incremental increases in dosage continued until the patients showed a 33% improvement on their clinical rating scale over their average pretreatment score. An important question was how quickly the patients achieved 33% improvement under the two treatment regimes. Here are the summary data for the patients who showed this level of improvement during the six months of the study:

Treatment group	Number of patients	Mean	Standard deviation
Slow	37	11.64	7.31
Fast	28	10.00	5.43

Use these data to test for a difference of means:

a. using a conservative df. **b.** using Welch's df.

c. assuming the two population variances are equal.

Everitt computed $t = 1.00$, $df = 62$, and P-value $= 0.3$.

d. Which t test did Everitt compute?

e. Was it a one-sided or two-sided alternative?

Remark about Exercise 14.42

Everitt warns his readers that it is not appropriate to use the t test for these data. Although this is a randomized double-blind experiment, there is a serious problem. This analysis involves the selective use of data. Only patients who showed a 33% improvement were compared. Those patients who did not show this much improvement in six months were not included in the analysis. Also, dropouts of the trial were excluded. Everitt presents the following table:

Treatment groups after 6 months	Slow treatment (Percent of total)	Fast treatment (Percent of total)
Reaching improvement criterion	37 (46%)	28 (52%)
Not reaching improvement criterion	12 (15%)	5 (9%)
Dropouts due to severe side effects	21 (26%)	19 (35%)
Dropouts due to noneffects	10 (13%)	2 (4%)
Total number of patients	80 (100%)	54 (100%)

The following remarks go well beyond what we have discussed up to this point and are included only to remind you that sometimes there

---

[6]From B. S. Everitt, *Statistics in Psychiatry*, Statistical Science **2** (1987), pp. 107–15.

are more advanced statistical techniques that can shed light on difficult circumstances. Everitt points out that a regression approach (specifically, D. R. Cox's proportional-hazards model) is more appropriate and that the data for the patients who were dropouts or who did not reach the improvement criteria should be treated as *censured observations*. These data are censured because we do not know how long it would take for these patients to reach the improvement criteria because they withdrew from the study before reaching the improvement criteria. In effect, we only have a lower bound for the time it would have taken to reach the improvement criteria, and that is the time until they withdrew from the study. Whenever patients drop out of a study or their conditions are no longer followed T time units after their treatments began, data on these patients are censured at time T. Without going into detail, we point out that this method leads to much longer mean time to reaching improvement criterion than the sample means given in the initial table. Also, we see that the fast treatment group had slightly higher proportion of patients reaching the improvement criteria within the course of the experiment and lower proportion of patients not reaching the improvement criteria, or dropping out due to noneffects. This suggests that the difference in mean time to reach the improvement criteria is somewhat larger than would be anticipated from the original analysis.

- **14.43** Refer to Exercises 9.46 and 9.47 on page 266. Let Z be a test statistic and suppose the null hypothesis is rejected for sufficiently large values of Z. So the P-value is $Pr(Z \geq z_s \mid H_0) = 1 - F_0(z_s)$. Of course before the sample data is collected, z_s is unknown. Define a random variable $Q = 1 - F_0(Z_s)$. So Q is the random P-value whose value depends on the actual random sample.

    a. Suppose that, when H_0 is true, Z is has a standard normal distribution. Show that Q has a uniform distribution on the unit interval $[0, 1]$.

    b. Suppose that, when H_a is true, Z has a normal distribution with mean $\mu_Z = 1$ and standard deviation $\sigma_Z = 1$. Find the density function of Q.

14.44 Imagine a large population of two types, normal and abnormal. Suppose 90% of the population is normal. Suppose that there is a test and when tested the normal population test scores are standard normal but the abnormal population scores are normal with mean and standard deviation both 1. Devise a statistical test to detect an abnormal individual. State the null and alternative hypotheses. Suppose an individual's test score is $z = 1.5$. Compute the P-value. What proportion of those individuals whose P-values are about 5% are normal?

Chapter 15

CORRELATION AND REGRESSION

15.1 INTRODUCTION

15.2 SCATTER PLOTS

15.3 THE CORRELATION COEFFICIENT

15.4 FITTING A SCATTER PLOT BY EYE

15.5 THE REGRESSION LINE

15.6 ESTIMATION WITH REGRESSION

15.7 THE REGRESSION PARADOX

15.8 TESTING FOR CORRELATION

15.9 CORRELATION IS NOT CAUSATION

15.1 INTRODUCTION

So far we have mostly dealt with description and inference involving a single random variable. However, in the exact tests of Chapter 6 and the two-sample inference of Chapter 14, we made comparisons of a "treated" group with a "control" group. We can take the view that this is a comparison of a dichotomous variable X (which has value 1 for "treatment" and value 0 for "control") with a response random variable Y. For example, if we are testing a drug therapy for high blood pressure, we can compare the blood pressure Y of a control group ($X = 0$) with that of a treatment group ($X = 1$) to determine whether the drug is effective.

In this chapter, instead of a dichotomous input, we consider the effect of various dosages of the treatment. The **independent variable** (also called the **stimulus** or **predictor**) is the dosage X. The **dependent variable** (also called the **response variable** or **outcome**) is the blood pressure Y.

There are two questions we want to consider:

Correlation   How strong is the relationship between the dosage and blood pressure? This is called the **correlation** between X and Y.

Regression   How do we use this information to predict the response for a given dosage? This is **regression** theory.

Although it is often desirable to consider the simultaneous effect of several drug combinations, or of other variables, such as sex and age, here we restrict ourselves to a *single* independent variable (typically denoted by X) and a response variable Y; this is called **simple regression** as opposed to **multiple regression**. Also, we will only consider cases where the response to treatment follows a straight line, called **linear regression**, as opposed to following some other curve.

Cases where the stimulus has a curved response or where several stimuli are considered can be studied in subsequent courses on regression theory.

15.2 SCATTER PLOTS

The independent variable may be set by the experimenter, as would be the case in a study of dosage of a drug versus blood pressure. Or both the independent variable and response may be observed variables. This would be the case in a study of the relationship between age X and height Y of children obtained from a random sample.

Example 15.1   *Age and heights of children.* Consider the following table of ages (in months) and heights (in centimeters) of eight children.

Table 15.1. Age x in months versus height y in centimeters

Child ID	x	y
1	24	87
2	48	101
3	60	120
4	96	159
5	63	135
6	39	104
7	63	126
8	39	96

We consider age to be the stimulus and height the response because we want to use age to predict height.[1] We plot the eight pairs of data points (x, y) using the Cartesian coordinate system. This is called the **scatter plot** of the data (see Figure 15.1).

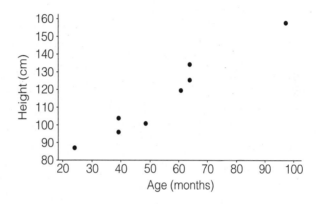

Figure 15.1 Scatter plot of height versus age for eight children

Positive correlation

The upward trend agrees with common sense: As children age, they grow. We say there is a positive correlation between age and height. The points *follow* a straight line with positive slope; however, the relationship is statistical, not deterministic. For example, children 5 and 7 are the same age but they have different heights, and the same for children 6 and 8.

Negative correlation

The following weights of thirty-eight 1978–79 model cars[2] (in 1000s of pounds) and their fuel economy (in miles per gallon) show a negative correlation because as weight increases, the fuel economy declines. Notice, however, in Figure 15.2, that the plot is slightly curved.

---

[1] It is reasonable to consider the reverse. In trying to guess a person's age, we are doing informal regression by taking into consideration many factors, including height. Then height is an independent variable and age is the dependent variable.

[2] H. V. Henderson & P. F. Velleman, *Building multiple regression models interactively*, Biometrics **37** (1981), pp. 391–411.

Table 15.2. Weight x in pounds versus fuel economy y in mpg

ID	x	y	ID	x	y	ID	x	y	ID	x	y
01	4360	16.9	02	4054	15.5	03	3605	19.2	04	3940	30.0
05	2155	30.0	06	2560	27.5	07	2300	27.2	08	2230	30.9
09	2830	20.3	10	3140	17.0	11	2795	21.6	12	3410	16.2
13	3380	20.6	14	3070	20.8	15	3620	18.6	16	3410	18.1
17	3840	17.0	18	3725	17.6	19	3955	16.5	20	3830	18.2
21	2585	26.5	22	2910	21.9	23	1975	34.1	24	1915	35.1
25	2670	27.4	26	1990	31.5	27	2135	29.5	28	2679	28.4
29	2595	28.8	30	2700	26.8	31	2556	33.5	32	2200	34.2
33	2020	31.8	34	2130	37.3	35	2190	30.5	36	2815	22.0
37	2600	21.5	38	1925	31.9						

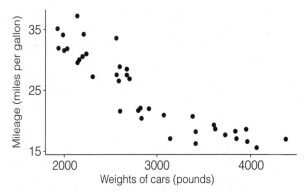

Figure 15.2 Scatter plot of mileage versus weight of 38 cars

Zero correlation   Figure 15.3 shows a computer-generated scatter plot of 250 points with correlation close to zero. Here $\bar{x} = 0.04, s_x = 2.0, \bar{y} = 0.13, s_y = 1.0$. Often such a plot is oval in shape with no upward or downward trend from left to right.

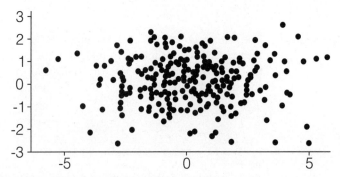

Figure 15.3 Scatter plot with correlation close to zero
(here $\bar{x} = 0.4, s_x = 2.0, \bar{y} = 0.13, s_y = 1.0$)

15.3 THE CORRELATION COEFFICIENT

Five numbers are generally used to summarize a scatter plot of data. Four of the numbers are the means and standard deviations of the two variables separately: $\bar{x}, \bar{y}, s_x$, and s_y.

> *The fifth number, the **correlation coefficient** r, is a measure of the strength of the linear relationship between* X *and* Y.

In order to motivate the procedure to compute the correlation coefficient, we create standardized axes z_X and z_Y, which are centered at the **point of averages** or **centroid** or **joint mean** $(\bar{x}, \bar{y})$, and whose units are in standard deviations s_x and s_y, respectively. Given a data point (x, y) in the original coordinate system, the new coordinates are (z_x, z_y), where $z_x = \frac{x-\bar{x}}{s_x}$ and $z_y = \frac{y-\bar{y}}{s_y}$ are the z-scores of x and y, respectively (see Figure 15.4).

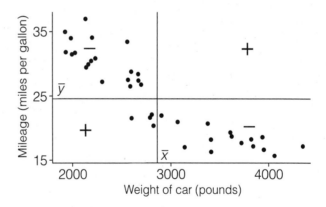

Figure 15.4 **Scatter plot with signs of the products** $z_x \cdot z_y$ **in four quadrants**

The standardized coordinate system divides the scatter plot into four quadrants. In the first quadrant, the z-scores z_x and z_y are both positive. In the third quadrant, the z-scores z_x and z_y are both negative. In both cases, the products of the z-scores $z_x \cdot z_y$ are all positive there. For data points in the second quadrant, we have $z_x < 0$ and $z_y > 0$, making the products $z_x \cdot z_y < 0$. Similarly, in the fourth quadrant, we have $z_x > 0$ and $z_y < 0$, also making the products $z_x \cdot z_y < 0$:

For a sample of size n, the average of these n products is the correlation coefficient r. Here the average uses the divisor $n-1$, the same as the degrees of freedom of s_x and s_y.

$$r = \sum_{\text{all } x,y} \frac{z_x \cdot z_y}{n-1} = \frac{\sum_{i=1}^{n}(x_i - \overline{x})(y_i - \overline{y})}{(n-1)s_x s_y} = \frac{\sum_{i=1}^{n}(x_i - \overline{x})(y_i - \overline{y})}{\sqrt{\sum_i (x_i - \overline{x})^2 \sum_i (y_i - \overline{y})^2}} \quad (15.1)$$

A scatter plot has a positive correlation if the data points are predominantly in the first and third quadrants, which is what makes the average of the products $z_x \cdot z_y$ positive. Conversely, a scatter plot with a negative correlation has most data points in the second and fourth quadrants, which makes r negative.

Table 15.3. Computation of r for age versus height of eight children

Child ID	x	y	$(x-\overline{x})^2$	$(y-\overline{y})^2$	z_x	z_y	$z_x \cdot z_y$
1	24	87	900	841	−1.3718	−1.2230	+1.6776
2	48	101	36	225	−0.2744	−0.6326	+0.1735
3	60	120	36	16	+0.2744	+0.1687	+0.0463
4	96	159	1764	1849	+1.9205	+1.8134	+3.4825
5	63	135	81	361	+0.4115	+0.8013	+0.3297
6	39	104	225	144	−0.6859	−0.5061	+0.3471
7	63	126	81	100	+0.4115	+0.4217	+0.1735
8	39	96	225	400	−0.6858	−0.8434	+0.5785
Totals	432	928	3348	3936	0 ✓	0 ✓	+6.8089

The five-number summary of the scatter plot is

$$\overline{x} = \frac{432}{8} = 54 \qquad \overline{y} = \frac{928}{8} = 116$$

$$s_x = \sqrt{\frac{3348}{7}} = 21.87 \qquad s_y = \sqrt{\frac{3936}{7}} = 23.71 \qquad r = \frac{6.8089}{7} = 0.9727$$

One must be careful with the signs of the z_x, z_y, and $z_x \cdot z_y$ entries. All of the entries in the $z_x \cdot z_y$ column turned out to be positive in this example, but this is not typical.

15.3 THE CORRELATION COEFFICIENT

> *The correlation coefficient is symmetric with respect to the variables* X *and* Y; *that is, it's the same regardless of which is the independent variable and which is the dependent one. And there are no units of measure for the coefficient* r.

Figure 15.5 shows some computer-generated scatter plots along with their correlation coefficients.

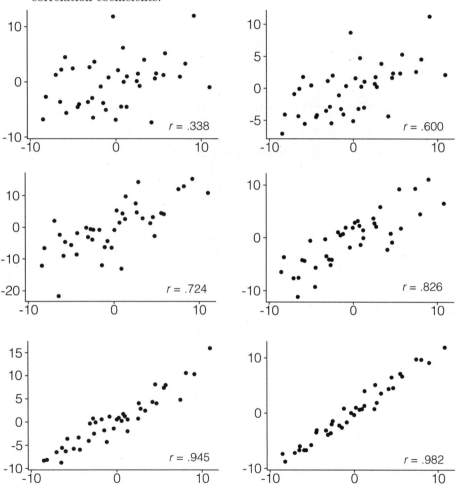

Figure 15.5 Some computer-generated scatter plots, each of 40 points. Correlation coefficients are indicated on each plot.

> *The correlation coefficient is always between* -1 *and* $+1$
>
> $$-1 \leq r \leq +1$$

Population correlation coefficient

Recall from Section 10.3 that the correlation coefficient between random variables X and Y is

$$\rho = E(Z_X \cdot Z_Y)$$

For bivariate data $\{(x_1, y_1), (x_2, y_2), \ldots, (x_n, y_n)\}$, obtained from a sample of n observations of the random variables X and Y, the correlation coefficient r is *approximately* an unbiased estimator of the parameter ρ. In other words,

$$\rho \approx E(r)$$

and the approximation improves as the sample size n increases.

The major axis

Scatter plots form constellations of data points. Often we can see an elliptical shape. If $r > 0$, then the major axis of the scatter plot goes through the point of averages and has slope 1 with respect to the standardized coordinate system, or slope s_y/s_x in original units. The major axis has the equation

Major axis for positive r

$$z_y = z_x \quad \text{or} \quad \frac{y - \bar{y}}{s_y} = \frac{x - \bar{x}}{s_x} \quad \text{or} \quad y - \bar{y} = \frac{s_y}{s_x}(x - \bar{x})$$

If $r < 0$, then the major axis has a negative slope with equation

Major axis for negative r

$$z_y = -z_x \quad \text{or} \quad \frac{y - \bar{y}}{s_y} = -\frac{x - \bar{x}}{s_x} \quad \text{or} \quad y - \bar{y} = -\frac{s_y}{s_x}(x - \bar{x})$$

And if $r = 0$, then we take the major axis to be the horizontal line through $(\bar{x}, \bar{y})$.

Major axis for zero r

$$z_y = 0, \quad \text{or} \quad y - \bar{y} = 0$$

15.3 THE CORRELATION COEFFICIENT

The correlation coefficient measures the strength of the *linear* relationship of two variables. Strongly related variables can have a correlation coefficient that is zero or close to zero. Figure 15.6 shows a scatter plot of 100 points with a curved trend, which clearly has a strong association between X and Y but has $r = 0.0$. (Also, $\bar{x} = 0.7$, $\bar{y} = 75.5$, $s_x = 60.9$, $s_y = 25.8$.)

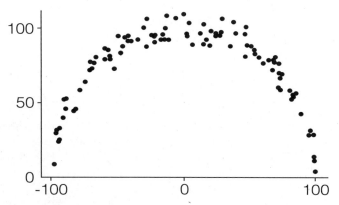

Figure 15.6 Strongly associated variables with zero correlation

The following theorem gives us two more formulas for r.

THEOREM 15.1 *Let $\overline{xy}$ denote the mean of products $x \cdot y$ for all data points (x,y). Then*

$$r = \frac{n}{n-1} \frac{\overline{xy} - \bar{x} \cdot \bar{y}}{s_x \cdot s_y} = \frac{\sum xy - (\frac{1}{n}) \sum x \sum y}{\sqrt{\sum(x-\bar{x})^2 \sum(y-\bar{y})^2}}$$

Proof

$$\begin{aligned}
r &= \sum \frac{z_x \cdot z_y}{n-1} \\
&= \frac{1}{n-1} \sum \frac{(x-\bar{x}) \cdot (y-\bar{y})}{s_x \cdot s_y} \\
&= \frac{1}{n-1} \frac{\sum xy - \bar{y}\sum x - \bar{x}\sum y + n\bar{x} \cdot \bar{y}}{s_x \cdot s_y} \\
&= \frac{1}{n-1} \frac{\sum xy - n\bar{y} \cdot \bar{x} - n\bar{x} \cdot \bar{y} + n\bar{x} \cdot \bar{y}}{s_x \cdot s_y} \\
&= \frac{1}{n-1} \frac{\sum xy - n\bar{x} \cdot \bar{y}}{s_x \cdot s_y} \\
&= \frac{n}{n-1} \frac{\overline{xy} - \bar{x} \cdot \bar{y}}{s_x \cdot s_y}
\end{aligned}$$

This is the first formula of the theorem. The second one follows from an expansion of the formulas for s_x and s_y. Details are left as Exercise 15.7 on page 418.

Taking advantage of these formulas, we can recalculate the correlation coefficient for ages versus heights using fewer fractions.

Table 15.4. Computation of r for ages versus heights

Child ID	x	y	xy	$(x-\overline{x})^2$	$(y-\overline{y})^2$
1	24	87	2088	900	841
2	48	101	4848	36	225
3	60	120	7200	36	16
4	96	159	15264	1764	1849
5	63	135	8505	81	361
6	39	104	4056	225	144
7	63	126	7938	81	100
8	39	96	3744	225	400
Totals	432	928	53643	3348	3936

The correlation coefficient computes to

$$r = \frac{\sum xy - (\frac{1}{n})\sum x \sum y}{\sqrt{\sum(x-\overline{x})^2 \sum(y-\overline{y})^2}} = \frac{53643 - 432 \cdot 928/8}{\sqrt{3348 \cdot 3936}} = 0.9727$$

NOTES ON TECHNOLOGY

Correlation Coefficient and Scatter Plot

Minitab: Enter the paired data in columns **C1** and **C2**. Select **Stat→Basic Statistics** and then choose **Correlation**. Enter columns **C1** and **C2**.

To view the scatter plot, select **Graph→Plot**, enter **C1** and **C2** for X and Y, and then hit **OK**.

Excel: Enter the paired data in columns **A** and **B**. Go to an empty cell and use the function **CORREL**. The function can be found by hitting the icon f_x and choosing **Statistical→CORREL**, and then hitting **OK**.

To view the scatter plot, click the Chart Wizard and select the chart identified as **XY(Scatter)** and hit **Next**. Insert the data range (for example, write A1:B21 if you have two columns of 21 values) and click **Finish**.

15.3 THE CORRELATION COEFFICIENT

TI-83 Plus: Enter the paired data sets in lists **L1** and **L2**. Press **STAT→TESTS→LinRegTTest**. Select **CALCULATE** and hit **ENTER**. The display will include the correlation coefficient r.

To view the scatter plot, hit the **STAT PLOTS** button (this is **2nd→Y=**), select **Plot1**, and select **On** and the picture that resembles a scatter plot (this is the first picture) for **Type**. Then click **ZOOM→ZoomStat**.

EXERCISES FOR SECTION 15.3

- **15.1** Consider the following scores of seven students in a statistics course, where x denotes the score on the first test and y that of the second.

	\multicolumn{7}{c}{Student ID}
	1	2	3	4	5	6	7
x:	40	50	70	80	80	100	70
y:	80	70	85	65	85	95	80

  a. Draw a scatter plot.
  b. Find an equation for the major axis and superimpose it on the scatter plot.
  c. Compute the correlation coefficient.

  15.2 Consider the following scores of six students in a statistics course, where x denotes the score on the first test and y that of the second.

	\multicolumn{6}{c}{Student ID}
	1	2	3	4	5	6
x:	30	50	60	70	60	90
y:	65	70	60	65	50	80

  a. Draw a scatter plot.
  b. Find an equation for the major axis and superimpose it on the scatter plot.
  c. Compute the correlation coefficient.

- **15.3** Consider the data set $\{(0,2), (2,1), (4,3)\}$. Find the correlation coefficient using the method of Table 15.3.

  15.4 Consider the data set $\{(5,4), (6,3), (6,0), (7,1)\}$. Find the correlation coefficient using the method of Table 15.4.

- **15.5** For the following data set, compute the five numbers used to summarize a scatter plot.

$$\begin{array}{c|cccccc} x: & 10 & 20 & 25 & 25 & 30 & 40 \\ y: & 7 & 5 & 4 & 3 & 4 & 1 \end{array}$$

- **15.6** For the following data set, compute the five numbers used to summarize a scatter plot.

$$\begin{array}{c|cccccc} x: & 12 & 8 & 6 & 4 & 6 & 0 \\ y: & 27 & 27 & 22 & 12 & 32 & 42 \end{array}$$

- **15.7** (Proof) Complete the proof of Theorem 15.1 on page 415 by deriving the second formula.

15.4 FITTING A SCATTER PLOT BY EYE

The famous statistician Sir Francis Galton (1822–1911) studied the relationship between heights of 1078 sons and their fathers. Figure 15.7 shows a scatter plot of his results.[3]

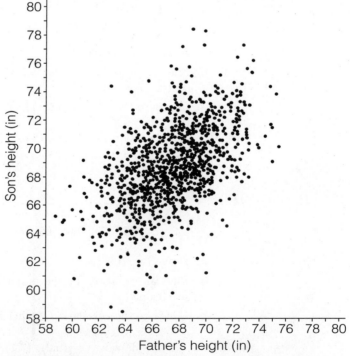

Figure 15.7 Heights of 1078 fathers x and their sons y

---

[3] The three scatter plots of this section are by David Freedman, Robert Pisani, & Roger Purves, *Statistics*, third edition, W. W. Norton & Co, 1998. Used by permission of W. W. Norton & Company, Inc.

It is fairly easy to estimate by eye the mean height of the 1078 fathers, measured along the x-axis, to be approximately 68 inches. Also, the mean height of the 1078 sons, measured along the y-axis, can be estimated to be approximately 69 inches.

Estimates of means: $\quad \bar{x} \approx 68 \quad \bar{y} \approx 69$

Next we estimate s_x and s_y by applying the 68%, 95%, and 99.7% rules. Although different people may come to slightly different results, it's not hard to see that the intervals $x = 68 \pm 2.5$, $x = 68 \pm 5.0$, and $x = 68 \pm 7.5$ contain approximately 68%, 95% and 99.7% of the fathers' heights, respectively. The same holds for the intervals $y = 69 \pm 2.5$, $y = 69 \pm 5.0$, and $y = 69 \pm 7.5$ of the sons' heights. We obtain

Estimates of SDs: $\quad s_x \approx 2.5 \quad s_y \approx 2.5$

This is enough to sketch the major axis, which is shown by the dashed line superimposed on the scatter plot in Figure 15.8. The equation of the major axis (as obtained from the formulas on page 414) is $y - 69 = \left(\frac{2.5}{2.5}\right)(x - 68)$ or

Major axis: $\quad y = x + 1$

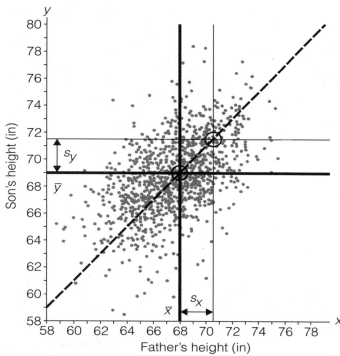

Figure 15.8 Scatter plot with major axis sketched as a dashed line

Now suppose we want to guess the height of a randomly selected son. If we do not know his father's height, our best guess is the mean of all sons $\overline{y} = 69$ inches. This guess will be accurate to $s_y = 2.5$ inches or so.

But now suppose that we know his father's height—say, $x = 72$ inches—which is $z_x = 1.6$ standard deviations above average height of all fathers. Because of the positive correlation, we expect the son's height also to be above the average height of all sons. Take a vertical slice of the scatter plot about $x = 72$ as shown in Figure 15.9 by dashed vertical lines. The points in this slice consists of all data points for which the fathers have heights of 72, rounded to the nearest inch. Now we estimate by eye the average height of those sons in this strip and mark it with an X; it appears to be the point $(72, 71)$. This is the conditional mean, denoted by $\hat{y}$, of sons' heights under the condition $x = 72$. We write:

For $x = 72$, the predicted y is $\hat{y} = 71$.

Note that this predicted point falls *below* the major axis but *above* the horizontal line $\overline{y} = 69$.

Repeat this for a father's height of $x = 64$ inches. These data points are between the solid vertical lines in Figure 15.9. Here $z_x = -1.6$ and we mark with another X the estimated conditional mean:

For x, the predicted y is $\hat{y} = 67$.

Note that this predicted point is *above* the major axis but *below* $\overline{y} = 69$.

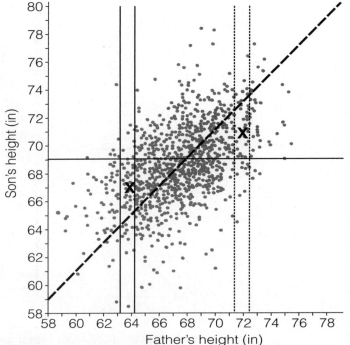

Figure 15.9 Scatter plot with two conditional means estimated

If we continue marking the conditional means, we get the X's on the graph in Figure 15.10.

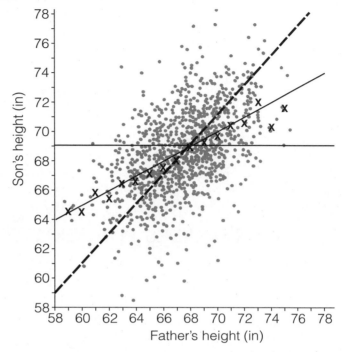

Figure 15.10 The X's show the mean height of the sons for each value of the father's height. The solid line is the best fit of a regression line. The diagonal dashed line is the major axis.

The curve joining the conditional means is called the **regression curve** $\hat{y} = f(x)$. If the regression curve is a straight line, it is called the **regression line** $\hat{y} = a + bx$. Although it is unlikely that, for a given data set, the marked conditional means will line up to form a perfect straight line, it makes sense here to say that the regression curve *follows* a straight line.

> Note that the regression line is a better fit of the X's within the central body of the scatter plot; the X's deviate more from the fitted line as the scatter plot thins out near the edges.

Note that the regression line goes through the point of averages. Also, note that the regression line is not as steep as the major axis. Actually, the regression line in this example is 50% as steep as the major axis. It is no coincidence that the correlation coefficient for heights of fathers versus sons is $r = 0.50$. This relationship between r and the slope of the regression line is proved in Theorem 15.2 below. We obtain the following equation of the regression line.

Regression line
$$\hat{y} - 69 = 0.50 \left(\frac{2.5}{2.5}\right)(x - 68) \quad \text{or} \quad \hat{y} = 0.50x + 35$$

> 1. **Both the regression line and the major axis intersect at the point of averages $(\overline{x}, \overline{y})$.**
> 2. **Whereas, the slope of the major axis is s_y/s_x, the regression line is r times as steep with slope rs_y/s_x.**

It is important to develop a sense of eyeing a rough estimate in a scatter plot. In fitting a regression line by eye, just a couple of X's are sufficient for a sketch. Our estimates come quite close to the calculated values of $\overline{x} = 67.7$, $\overline{y} = 68.7$, $s_x = 2.74$, $s_y = 2.81$, $r = 0.51$. These calculated values lead to a regression equation of $\hat{y} = 0.52x + 33.29$.

EXERCISES FOR SECTION 15.4

15.8 We can estimate temperature by listening to crickets. Below is a scatter plot of number of chirps per minute versus air temperature in degrees Fahrenheit measured 20 times.

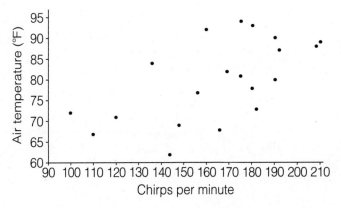

Choose the best option.

a. $\overline{x}$ is closest to: 120   135   150   165   180   195
b. $\overline{y}$ is closest to: 65   70   75   80   85   90
c. s_x is closest to: 10   20   30   40   50   60
d. s_y is closest to: 5   10   15   20   25   30
e. Using your choices, find an equation of the major axis.
f. Fit the regression line by eye.
g. Comparing the slopes of the major axis and the regression line, it seems that the correlation coefficient is closest to:
     −0.25   0.00   0.10   0.30   0.60   0.80

15.5 THE REGRESSION LINE

Each time we mark an X in a slice of the scatter plot, we are finding the mean of the dots in that slice. For each data point (x_i, y_i), the difference between the predicted value $\hat{y}_i = mx_i + b$ and the observed value y_i is called the **residual** value $e_i = y_i - \hat{y}_i$. Recall from Exercise 2.22 that the mean of a set is the number that minimizes the RMS deviation from that number. To obtain the line of best fit for *all* n data points, it makes sense to formally define the regression line as follows.

Residual standard deviation s_e

> *The regression line is that line that minimizes the RMS of the residuals*
> $$s_e = \sqrt{\frac{\sum_i e_i^2}{n-2}}$$

Here we divide by $df = n - 2$ degrees of freedom because the sample regression line involves estimating two unknown parameters, the slope and the y-intercept. Recall that when a single parameter is being estimated, the sample standard deviation s has $df = n - 1$ degrees of freedom. In general, when k parameters are being estimated, then there are $df = n - k$ degrees of freedom.

We use the symbol s_e for the RMS deviation to indicate that this is the "standard deviation" about the regression line, whereas s_y is the "standard deviation" about the mean $\bar{y}$. The residual standard deviation has the same units of measurement as s_y.

THEOREM 15.2 *The regression line of bivariate data goes through the point of averages $(\bar{x}, \bar{y})$ and has slope $r\frac{s_y}{s_x}$.*

Proof The regression line is the line
$$\hat{y} = a + bx \qquad (15.2)$$
which, for all possible values of a and b, has the smallest possible RMS of the residuals
$$s_e = \sqrt{\frac{\sum_i e_i^2}{n-2}} = \sqrt{\frac{\sum_i (y_i - \hat{y})^2}{n-2}}$$
This is equivalent to finding a and b that minimizes SSE, the *Sum of Square of the Residuals*:
$$\text{SSE} = \sum_i (y_i - \hat{y})^2 = \sum_i (y_i - (a + bx_i))^2 \qquad (15.3)$$

First, we find the intercept a. Taking a derivative of SSE with respect to a, we obtain

$$\frac{\partial}{\partial a}\text{SSE} = -2\sum_i (y_i - (a + bx_i))$$

This derivative is equal to zero when

$$\sum_i (y_i - (a + bx_i)) = 0 \quad \text{or}$$

$$\sum_i y_i - na - b\sum_i x_i = 0 \quad \text{or}$$

$$n\bar{y} - na - bn\bar{x} = 0$$

Solving for a we obtain the equation

$$a = \bar{y} - b\bar{x}$$

Plugging this solution of a into the regression equation $\hat{y} = a + bx$, we obtain

$$\hat{y} = b(x - \bar{x}) + \bar{y} \tag{15.4}$$

Because this equation goes through the point of averages $(\bar{x}, \bar{y})$, the first statement of the theorem is proved.

To find the value of the slope b that minimizes SSE, we first substitute Equation (15.4) into (15.3) to obtain

$$\text{SSE} = \sum_i (y_i - \hat{y})^2 = \sum_i ((y_i - \bar{y}) - b(x_i - \bar{x}))^2 \tag{15.5}$$

Taking a derivative with respect to b, we obtain

$$\begin{aligned}(\text{SSE})' &= -2\sum_i ((y_i - \bar{y}) - b(x_i - \bar{x}))(x_i - \bar{x}) \\ &= -2\left(\sum_i (y_i - \bar{y})(x_i - \bar{x}) - b\sum_i (x_i - \bar{x})^2\right)\end{aligned}$$

This derivative is equal to zero when

$$b = \frac{\sum_i (y_i - \bar{y})(x_i - \bar{x})}{\sum_i (x_i - \bar{x})^2} \tag{15.6}$$

Observing that $s_x^2 = \frac{\sum_i (x_i - \bar{x})^2}{n-1}$, we have

$$b = \frac{\sum_i (y_i - \bar{y})(x_i - \bar{x})}{(n-1)s_x^2} = \left(\frac{\sum_i (y_i - \bar{y})(x_i - \bar{x})}{(n-1)s_x s_y}\right) \frac{s_y}{s_x}$$

which reduces to $b = r\frac{s_y}{s_x}$ by Equation 15.1 on page 412. The second statement is proved.

Regression equations

The theorem results in three forms of the equation of the regression line:

$$\hat{y} - \bar{y} = r\frac{s_y}{s_x}(x - \bar{x}) \quad \text{or} \quad \frac{\hat{y} - \bar{y}}{s_{\hat{y}}} = r\frac{x - \bar{x}}{s_x} \quad \text{or} \quad z_{\hat{y}} = r z_x \qquad (15.7)$$

Example 15.2 *Age and heights of children.* Using the first equation $\hat{y} - \bar{y} = r\frac{s_y}{s_x}(x - \bar{x})$ and Table 15.4, we get the regression equation for age versus height of children:

$$\hat{y} - 116 = 0.9727 \sqrt{\frac{3936}{3348}} (x - 54)$$

or

$$\hat{y} = 1.055x + 59.048 \qquad (15.8)$$

Using this equation, we obtain the table of residuals below. The table shows that the sum of the residuals is 0 (this is equivalent to checking that the fitted line goes through the point of averages) and SSE = 212.00 (for any other choice of a and b, the sum of squares would be larger).

Table 15.5. Ages x in months versus heights y in centimeters

Child ID	x	y	$\hat{y}$	$e = y - \hat{y}$	e^2
1	24	87	84.36	+2.64	6.97
2	48	101	109.67	−8.67	75.20
3	60	120	122.33	−2.33	5.42
4	96	159	160.30	−1.20	1.68
5	63	135	125.49	+9.51	90.40
6	39	104	100.18	−3.82	14.59
7	63	126	125.49	+0.51	0.26
8	39	96	100.18	−4.18	17.47
Totals	432	928	928 ✓	0 ✓	212.00

Figure 15.11 shows a graph of the regression line superimposed on the scatter plot.

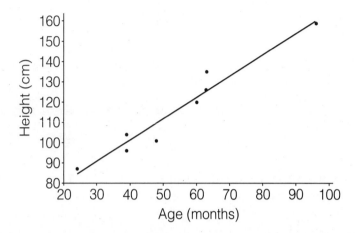

Figure 15.11 Scatter plot of age versus height of children, with regression line superimposed

NOTES ON TECHNOLOGY

Equation of Regression Line

Minitab: Enter the paired data in columns **C1** (for x) and **C2** (for y). Select **Stat→Regression→Regression**. Enter columns **C2** for Response and **C1** for Predictors. To obtain the scatter plot with the regression line superimposed, select **Stat→Regression→Fitted Line Plot**. Enter columns **C2** for Response(Y) and **C1** for Predictor(X).

Excel: Enter the paired data in columns **A** and **B**. Select **Tools→Data Analysis→Regression**. Enter the Y range, such as **B1:B21**, and the X range, such as **A1:A21**. To obtain the scatter plot with the regression line superimposed, check the **Line Fit Plots** box.

TI-83 Plus: Enter the paired data sets in lists **L1** and **L2**. Press **STAT→TESTS→LinRegTTest**. Select **CALCULATE**, hit **ENTER**. The display will include the y-intercept a and slope b.

EXERCISES FOR SECTION 15.5

- **15.9** Continuing with Exercise 15.1, on page 417:

	Student ID						
	1	2	3	4	5	6	7
x:	40	50	70	80	80	100	70
y:	80	70	85	65	85	95	80

Find an equation of the regression line.

15.10 Continuing with Exercise 15.2, on page 417:

	Student ID					
	1	2	3	4	5	6
x:	30	50	60	70	60	90
y:	65	70	60	65	50	80

Find an equation of the regression line.

- **15.11** Continue with Exercise 15.3, on page 417, for the data set

$$\{(0,2), (2,1), (4,3)\}$$

   a. Find an equation of the regression line.

   b. Compute SSE using the method of Table 15.5 on page 425.

15.12 Continue with Exercise 15.4, on page 417, for the data set

$$\{(5,4), (6,3), (6,0), (7,1)\}$$

   a. Find an equation of the regression line.

   b. Compute SSE using the method of Table 15.5 on page 425.

- **15.13** Continuing with Exercise 15.5 on page 418:

x:	10	20	25	25	30	40
y:	7	5	4	3	4	1

   a. Find an equation of the regression line.

   b. Compute s_e using a table like Table 15.5 on page 425.

15.14 Continuing with Exercise 15.6 on page 418:

x:	12	8	6	4	6	0
y:	27	27	22	12	32	42

   a. Find an equation of the regression line.

   b. Compute s_e using a table like Table 15.5 on page 425.

15.6 ESTIMATION WITH REGRESSION

Suppose we know that a child is 60 months old and comes from the same population as the 8 children of our age–height data set. We want to predict the height of the given child. One of the children in the data set was actually 60 months old and was 120 centimeters tall. We do not want to use this single data point to estimate the height of the given child. Actually, the (60,120) point lies below the regression line, so 120 centimeters would be an underestimate.

Recall the summary of the scatter plot for the 8 children

$$\overline{x} = 54, \ \overline{y} = 116, \ s_x = 21.87, \ s_y = 23.71, \ r = 0.9727$$

We can use the regression line $z_{\hat{y}} = rz_x$ to find a prediction in three steps.

Three-step regression procedure

We are given $x = 60$.

1. Standardize x: $z_x = \frac{x - \overline{x}}{s_x} = \frac{60 - 54}{21.87} = 0.2744$.

2. Regression: $z_{\hat{y}} = rz_x = 0.9727(0.2744) = 0.2669$.

3. Compute $\hat{y}$ from $z_{\hat{y}}$:

   $\hat{y}$ is 0.2669 SD's above average, so

   $\hat{y} = 116 + 0.2669(23.71) = 122.33$ centimeters.

Accuracy of regression estimate

If a child's age is *unknown*, then the best estimate of the child's height is to guess the average of 116 centimeters. This estimate will be accurate to $s_y = 23.71$ centimeters or so.

However, if the age is *known*—say, $x = 60$ months—then the best estimate is given by the regression line

$$\hat{y} = 122.33 \text{ centimeters}$$

The accuracy of the estimate depends on the variability of the data *about the regression line*, and that is measured by the residual standard deviation[4] s_e.

---

[4]Here we are assuming the variability about the regression line does not depend on the value of x. This is called **homoscedasticity**. We can check for homoscedasticity by looking at a plot of the residuals. The size of the residuals should not have a trend or pattern when we look at them from left to right.

Recalling from Table 15.5 that SSE = 212.00, we calculate the residual standard deviation:

$$s_e = \sqrt{\frac{SSE}{n-2}} = \sqrt{\frac{212.00}{6}} = 5.944 \text{ centimeters} \qquad (15.9)$$

The information $x = 60$ improved our prediction for the height from 116 ± 23.71 to 122.33 ± 5.944. This improvement is due to the strong correlation between age and height, which makes the residual standard deviation small. In Theorem 15.3, we will show that s_e is almost $s_y\sqrt{1-r^2}$. So, if the correlation coefficient had been zero, there would have been no improvement.

If the variable of height follows the normal curve, we can say that 95% of all children who are 60 months old have height in the range $\hat{y} \pm z_{.975}s_e = 122.33 \pm 1.96(5.944) = 122.33 \pm 11.65$ centimeters.

The computation of SSE used in Table 15.5 is tedious. The following lemma gives a nicer formula. Then the following Theorem 15.3 gives a formula in terms of r.

Lemma $\quad SSE = (1-r^2)\sum(y_i - \bar{y})^2$

Proof $\quad$ Let's expand Equation (15.5):

$$\begin{aligned} SSE &= \sum_i (y_i - \hat{y})^2 = \sum_i ((y_i - \bar{y}) - b(x_i - \bar{x}))^2 \\ &= \sum \{(y_i - \bar{y})^2 - 2b(y_i - \bar{y})(x_i - \bar{x}) + b^2(x_i - \bar{x})^2\} \\ &= \sum (y_i - \bar{y})^2 - 2b\sum(y_i - \bar{y})(x_i - \bar{x}) + b^2 \sum(x_i - \bar{x})^2 \end{aligned}$$

By Equation (15.1), we have

$$\sum(y_i - \bar{y})(x_i - \bar{x}) = r\sqrt{\sum(x_i - \bar{x})^2 \sum(y_i - \bar{y})^2}$$

Substituting this into the middle term of the last expression, we obtain

$$SSE = \sum(y_i - \bar{y})^2 - 2br\sqrt{\sum(x_i - \bar{x})^2 \sum(y_i - \bar{y})^2} + b^2 \sum(x_i - \bar{x})^2$$

Now we substitute $b = r\frac{s_y}{s_x} = r\frac{\sqrt{\sum(y_i - \bar{y})^2}}{\sqrt{\sum(x_i - \bar{x})^2}}$ from Theorem 15.2, to obtain

$$SSE = \sum(y_i - \bar{y})^2 - 2r^2 \sum(y_i - \bar{y})^2 + r^2 \sum(y_i - \bar{y})^2 = (1-r^2)\sum(y_i - \bar{y})^2$$

THEOREM 15.3 $s_e = s_y\sqrt{1-r^2}\sqrt{\frac{n-1}{n-2}}$

Proof We show that
$$\frac{s_e^2}{s_y^2} = \left(\frac{n-1}{n-2}\right)(1-r^2)$$

Because
$$s_e^2 = \frac{\sum(y_i - \hat{y}_i)^2}{n-2}$$

and
$$s_y^2 = \frac{\sum(y_i - \bar{y})^2}{n-1}$$

then
$$\frac{s_e^2}{s_y^2} = \left(\frac{n-1}{n-2}\right)\frac{\sum(y_i - \hat{y}_i)^2}{\sum(y_i - \bar{y})^2} = \left(\frac{n-1}{n-2}\right)\frac{\text{SSE}}{\sum(y_i - \bar{y})^2}$$

The theorem now follows from the lemma above.

━━━━━━━━━━━━━━━━━━━━━━━━━━━━━━━━━━━━━━

The factor $\sqrt{\frac{n-1}{n-2}}$ is the adjustment for the difference of degrees of freedom between s_y and s_e. For $n \geq 10$, this factor is close to 1 and can often be ignored.

> **The residual standard deviations s_e is almost $s_y\sqrt{1-r^2}$ and approaches equality as the sample size n gets large.**

If we recalculate s_e for the age–height data using Theorem 15.3, we get the same value as before in Formula (15.9):

$$s_e = s_y\sqrt{1-r^2}\sqrt{\frac{n-1}{n-2}} = 23.71\sqrt{1-(0.9727)^2}\sqrt{\frac{7}{6}} = 5.944$$

EXERCISES FOR SECTION 15.6

- **15.15** A large study[5] of over 1000 men found that the relationship between height x, in inches, and length of forearm y, in inches, had the five-number summary:

$$\bar{x} = 69 \qquad \bar{y} = 18 \qquad s_x = 2.5 \qquad s_y = 1.0 \qquad r = 0.80$$

---

[5] Based on a study of K. Pearson & A. Lee, *On the laws of inheritance in man*, Biometrika **2** (1903), pp. 357–462.

Use the three-step procedure to predict the forearm length of a man whose height is:

    a. unknown   **b.** 69 inches   **c.** 73 inches   **d.** 64 inches

15.16 A large study of over 1000 married couples found the following relationship between heights of husbands x and wives y, in inches:

$$\bar{x} = 68 \qquad \bar{y} = 64 \qquad s_x = 2.7 \qquad s_y = 2.5 \qquad r = 0.35$$

Use the three-step procedure to predict the height of a wife whose husband has height:

    a. unknown   **b.** 69 inches,   **c.** 73 inches   **d.** 64 inches

• **15.17** Continuing with Exercise 15.15, assume forearm lengths follow the normal curve.

    a. Find the percentage of men whose forearm length is over 19 inches.

    b. Of men who are 73 inches tall, find the percentage whose forearm length is over 19 inches.

15.18 Continuing with Exercise 15.16, assume heights of wives follow the normal curve.

    a. Find the percentage of wives who are shorter than 60 inches.

    b. Of husbands who are 64 inches tall, find the percentage whose wives are shorter than 60 inches.

15.7 THE REGRESSION PARADOX

Let's return to Galton's data of the heights of fathers versus sons. The fathers who were 72 inches tall were 4 inches above average ($z_x = 1.6$), but their sons averaged $\hat{y} = 71$ inches, which is only 2 inches above average ($z_y = 0.8$). Also, fathers who were 64 inches tall were 4 inches below average ($z_x = -1.6$), but their sons averaged $\hat{y} = 67$ inches, which is only 2 inches below average ($z_{\hat{y}} = -0.8$).

In summary, the sons of tall fathers were taller than average, but not as much above average as their fathers. The sons of short fathers were shorter than average, but not as much below average as their fathers. This phenomenon is called **regression toward the mean** or the **regression effect**.

The trend toward the mean will continue in the next generation of sons. Galton had noticed that various hereditary traits tend toward to the mean in successive generations. High values of a trait are followed by lower ones, and low values are followed by higher ones.

Now we reverse the role of the independent and dependent variable, and attempt to predict the height of a man's father. Say that a man is $y = 73$ inches tall. The predicted height of his father is denoted $\hat{x}$ (see Figure 15.12). Here we take a horizontal strip about $y = 73$ and find the mean height of the fathers in that strip.

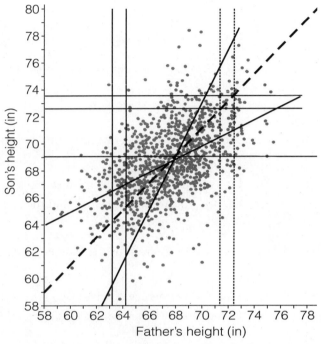

Figure 15.12 The two regression lines and the major axis

The correlation coefficient is symmetric with respect to x and y, so it's still $r = 0.50$. Because the son is $z_y = 1.60$ standard deviations above average, we predict his father to be $r = 0.50$ times as tall, that is, $z_{\hat{x}} = 0.80$ standard deviations above average. This makes the prediction of the height of the father $\hat{x} = 0.80(2.5) + 68 = 70$ inches. Regression toward the mean happens in regression from x to y and from y to x.

> *The regression line of* **x** *to* **y** *is not the same as the regression line of* **y** *to* **x**. *They both go through the point of averages, but they are symmetric to each other about the major axis of the scatter plot:*
>
> $$\hat{y} - \overline{y} = r\frac{s_y}{s_x}(x - \overline{x})$$
>
> $$\hat{x} - \overline{x} = r\frac{s_x}{s_y}(y - \overline{y})$$

In pretest–posttest situations, subjects who do very well on the first test generally do better than average on the second test, but not as much above average as on their first test. Those who do poorly on the first test generally do below average on the second test, but not as much below average as on their first test. When this happens, this is nothing other than the regression effect.

> *The regression effect is a consequence of the shape of a scatter plot for a linear association, and it holds for all scatter plots. Falling into a incorrect conclusion about the cause of this phenomenon is called the* **regression fallacy.**

Air force pilots have to take two test landings as part of their training. They are rated on both. It was observed[6] that when pilots are praised for their good first landing, they generally do worse on their second. And conversely, when they are criticized for the poor landings, they do better on their next one. The conclusion that "praise spoils, criticism improves" is not warranted by the facts. This is yet another example of the regression fallacy.

The **placebo effect** is the effect on subjects who have been given a placebo treatment when they think that it's real medicine. This effect is particularly strong in illnesses that involve pain, depression, and certain heart and digestive conditions. There is a component of regression toward the mean in the placebo effect. The course of an illness does not progress at a constant rate. Very bad periods are often followed by periods in which the patient feels better. If one enters a double-blind experiment because of a worsening of a condition, the experiment is in a pretest–posttest situation. Regression theory says that, on the average, subjects will do better in the posttest. Some researchers[7] even claim that there is no placebo effect other than regression toward the mean.

EXERCISES FOR SECTION 15.7

- **15.19** Use the information of Exercise 15.16 on page 431

$$\overline{x} = 68 \qquad \overline{y} = 64 \qquad s_x = 2.7 \qquad s_y = 2.5 \qquad r = 0.35$$

to estimate the height of a man whose wife is 68 inches tall. Also give a 95% prediction interval for his height.

---

[6]D. Kahneman and A. Tversky, *On the Psychology of Prediction*, Psychology Review **80** (1973), pp. 237–52. This example is also discussed in K. McKean *Decisions*, Discovery, (1985) pp. 22–31.

[7]Asbjorn Hrobjartsson and Peter C. Gotzsche, *Is the Placebo Powerless? An Analysis of Clinical Trials Comparing Placebo with No Treatment*, The New England Journal of Medicine **344**, No. 21 (May 24, 2001).

15.20 **(Forensics.)** Use the information of Exercise 15.15 on page 430

$$\bar{x} = 69 \qquad \bar{y} = 18 \qquad s_x = 2.5 \qquad s_y = 1.0 \qquad r = 0.80$$

to estimate the height of a man whose forearm length is 17 inches. Also give a 99% prediction interval for his height.

• **15.21** The IQ scores of men and women follow the normal curve with mean 100 and standard deviation 15, and the correlation coefficient between IQ scores of husbands and wives is approximately $r = 0.50$. Thus, for wives with IQ score of 130, their husbands have an average IQ of 115. Find the average IQ of the wives of those husbands with an IQ of 115. Why is the answer not equal to 130?

15.22 Assume that heights and weights of men between 20 and 30 years old follow normal curves and have correlation coefficient $r = 0.47$. If a male is in the 90th percentile in height, what is his predicted percentile in weight?

15.23 Four-year-old children in a gifted program have their IQ scores tested each year. It is noted that at the end of the first year in the program, the average IQ scores has gone down. Explain the regression effect in this context and the misleading regression fallacy.

15.8 TESTING FOR CORRELATION

Just because there is a sample correlation ($r \neq 0$) does not prove that there is a true correlation ($\rho \neq 0$) between variables X and Y. Testing whether there is a correlation is a test for

$$H_0 : \rho = 0$$

against a two-sided alternative

$$H_a : \rho \neq 0$$

or against a one-sided alternative

$$H_a : \rho > 0 \quad \text{or} \quad H_a : \rho < 0$$

The sample regression line

$$\hat{y} = a + bx$$

is based on a sample. So it's an estimate of the true regression line

$$E(Y) = \alpha + \beta x$$

15.8 TESTING FOR CORRELATION

> *Because the slope of the true regression line is $\beta = \rho \frac{\sigma_Y}{\sigma_X}$, a test of $\mathbf{H_0}: \rho = 0$ is equivalent to a test of $\mathbf{H_0}: \beta = 0$.*

In this section, we will find the expected value and standard error of the slope b of the regression line.

Here we take the view that the values $\{x_1, x_2, \ldots, x_n\}$ are determined by the experimenter. Hence, they will not be variables, but constants. The response variables for each stimulus x_i will be Y_i, with $E(Y_i) = \alpha + \beta x_i$.

We assume the following in this analysis:

1. Independence of the random variables.

2. Homoscedasticity, that is, $\text{Var}(Y_i) = \sigma_e^2$ for all $i = 1, 2, 3, \ldots, n$.

The slope of the sample regression line is

$$b = \frac{\sum_i (x_i - \overline{x})(y_i - \overline{y})}{\sum_i (x_i - \overline{x})^2} = \sum_i \left[\frac{(x_i - \overline{x})}{\sum_i (x_i - \overline{x})^2} \right] (y_i - \overline{y}) = \sum_i \left[\frac{(x_i - \overline{x})}{\sum_i (x_i - \overline{x})^2} \right] y_i$$

The last equality follows from the fact that

$$\sum_i [(x_i - \overline{x})\overline{y}] = \overline{y} \sum_i (x_i - \overline{x}) = 0$$

The slope b is an observation of the random variable

$$B = \sum_i \left[\frac{(x_i - \overline{x})}{\sum_i (x_i - \overline{x})^2} \right] Y_i$$

The following theorem shows that the slope of the sample regression line is an unbiased estimator of β, the slope of the true regression line.

THEOREM 15.4 $E(B) = \beta$

Proof
$$E(B) = \sum_i \left[\frac{(x_i - \overline{x})}{\sum_i (x_i - \overline{x})^2} \right] E(Y_i) = \sum_i \left[\frac{(x_i - \overline{x})}{\sum_i (x_i - \overline{x})^2} \right] (\alpha + \beta x_i)$$
$$= \frac{\alpha \sum_i (x_i - \overline{x}) + \beta \sum_i (x_i - \overline{x})(x_i)}{\sum_i (x_i - \overline{x})^2}$$

Because $\sum_i (x_i - \overline{x}) = 0$ we have

$$\sum_i (x_i - \overline{x}) x_i = \sum_i (x_i - \overline{x}) x_i - \overline{x} \sum_i (x_i - \overline{x})$$
$$= \sum_i (x_i - \overline{x}) x_i - \sum_i (x_i - \overline{x}) \overline{x}$$
$$= \sum_i (x_i - \overline{x})^2$$

This last expression leads to

$$E(B) = \frac{\alpha \cdot 0 + \beta \sum_i (x_i - \overline{x})^2}{\sum_i (x_i - \overline{x})^2}$$
$$= \beta \frac{\sum_i (x_i - \overline{x})^2}{\sum_i (x_i - \overline{x})^2} = \beta$$

THEOREM 15.5 $\text{Var}(B) = \dfrac{\sigma_e^2}{\sum_i (x_i - \overline{x})^2}$

Proof As was shown in Theorem 10.8, variances add for independent random variables and coefficients are squared. Thus,

$$\text{Var}(B) = \text{Var}\left(\sum_i \left[\frac{(x_i - \overline{x})}{\sum_i (x_i - \overline{x})^2} \right] Y_i \right)$$
$$= \sum \left[\frac{(x_i - \overline{x})}{\sum_i (x_i - \overline{x})^2} \right]^2 \text{Var}(Y_i)$$
$$= \sum \left[\frac{(x_i - \overline{x})}{\sum_i (x_i - \overline{x})^2} \right]^2 \sigma_e^2$$
$$= \frac{\sigma_e^2}{\sum_i (x_i - \overline{x})^2}$$

If we use the formula of Theorem 15.3 on page 430, we can obtain the estimated standard error of β:

$$\text{SE}_b = \frac{s_e}{\sqrt{\sum(x_i - \overline{x})^2}} = \frac{s_e}{s_x\sqrt{n-1}} = \frac{s_y}{s_x}\sqrt{\frac{1-r^2}{n-2}} = \frac{b}{r}\sqrt{\frac{1-r^2}{n-2}} \quad (15.10)$$

If the population of Y follows a normal curve and we have homoscedasticity, we may find confidence intervals β using t curves with $df = n - 2$ degrees of freedom.

Example 15.3 *Age and height of children.* Recall that for the age–height data, the regression equation (15.8) was

$$\hat{y} = 59.048 + 1.055x$$

and by Table 15.4 and 15.5, we had $s_e = 5.944$ and $\sum(x_i - \overline{x})^2 = 3348$. The 95% confidence interval for β is thus

Confidence Interval

$$95\% \quad \text{CI:} \quad \beta = b \pm t_{6, 0.975} \text{SE}_b = 1.055 \pm 2.447 \frac{5.944}{\sqrt{3348}} = 1.055 \pm 0.251$$

In order to test for a positive correlation $\rho > 0$ between age and height of children, we test for a positive slope of the regression line $\beta > 0$:

$$H_0 : \beta = 0 \qquad H_a : \beta > 0$$

$$P\text{-value} = Pr(b \geq 1.055 \mid \beta = 0) = Pr(t_6 \geq \frac{1.055 - 0}{5.94/\sqrt{3348}}) = Pr(t_6 \geq 10.2696)$$

Using the t table in the back of the book, we see that P-value < 0.0005. So we can say beyond a reasonable doubt (P-value < 0.0005) that there is a positive correlation between age and height of children. Although we gave a statistical proof of the obvious, this example shows that a random sample of $n = 8$ children is sufficient to prove it.

EXERCISES FOR SECTION 15.8

15.24 Given a scatter plot of 12 data points with $\bar{x} = 5, \bar{y} = 100, s_x = 1.5, s_y = 9.0$, and $r = -0.20$, assume normality and homoscedasticity.

  a. Find a 99% confidence interval for β.

  b. Test for a negative correlation $\rho < 0$.

• **15.25** Given a scatter plot of 7 data points with $\bar{x} = 50, \bar{y} = 8, s_x = 15, s_y = 2$, and $r = 0.30$, assume normality and homoscedasticity.

  a. Find a 95% confidence interval for β.

  b. Test for a positive correlation $\rho > 0$.

15.26 Continue with Exercises 15.1 on page 417 and 15.9 on page 427. Assume normality and homoscedasticity.

  a. Find a 90% confidence interval for the slope β of the true regression line.

  b. Also test for a positive correlation $\rho > 0$.

• **15.27** Continue with Exercises 15.2 on page 417 and 15.10 on page 427. Assume normality and homoscedasticity.

  a. Find a 90% confidence interval for the slope β of the true regression line.

  b. Also test for a positive correlation $\rho > 0$.

15.28 Continue with Exercises 15.3 on page 417 and 15.11 on page 427. Assume normality and homoscedasticity.

  a. Find a 95% confidence interval for the slope β of the true regression line.

  b. Also test for a correlation $\rho \neq 0$.

15.29 Continue with Exercises 15.4 on page 417 and 15.12 on page 427. Assume normality and homoscedasticity.

  a. Find a 95% confidence interval for the slope β of the true regression line.

  b. Also test for a positive correlation $\rho > 0$.

15.30 Continue with Exercises 15.5 on page 418 and 15.13 on page 427. Assume normality and homoscedasticity.

  a. Find a 99% confidence interval for the slope β of the true regression line.

  b. Also test for a negative correlation $\rho < 0$.

15.31 Continue with Exercises 15.6 on page 418 and 15.14 on page 427. Assume normality and homoscedasticity.

  a. Find a 99% confidence interval for the slope β of the true regression line.

  b. Also test for a negative correlation $\rho < 0$.

15.9 CORRELATION IS NOT CAUSATION

We should observe extreme caution in interpreting correlation and regression as **causation**. An observed relationship between two variables may, of course, be purely coincidental, as it may be with any statistical test. However, interpreting correlation has additional dangers.

Common response

A recent newspaper article headline reads "Kids who regularly eat breakfast get better grades." Of course, it is possible that eating breakfast regularly could contribute to better grades, but there are other explanations. Kids who regularly eat breakfast generally have better home environments. They are generally better supervised, including supervision of homework assignments, sleep, hygiene, medical care, and so on. If, as a part of an experiment, the children with better home environments were to be deprived of breakfast, they would probably be getting better grades anyway. In short, a better home environment contributes to both better grades and regularly eating breakfast. This is called a **common response**. Good grades and good breakfast habits are common responses to a better home environment.

Colleges use SAT or ACT scores as predictors of success in college. But this does not mean that high scores *cause* success. It is more likely that both high scores and success are a common response to good study habits (and other factors such as good health and intelligence).

Confounding factors.

It had long been argued by the tobacco industry that the higher rate of cancer and heart disease experienced by smokers is not due to smoking but to other factors. Among them are that smokers are more likely to drink alcoholic beverages, have aggressive personalities, and in general have more "sloppy" lifestyles than nonsmokers. Because these factors are associated with more cancer and heart disease, and these factors tend to be more prevalent among smokers than nonsmokers, it was argued that these are **confounding factors** of the association.

In the United States, people who drink coffee with caffeine do not have more heart disease than those who do not drink coffee. However, those who drink decaffeinated coffee have significantly more heart disease. When such a finding was first announced, many people in the media wondered whether the process of removing caffeine added some harmful chemical. Yet none of the three common methods of removing caffeine adds anything that does not evaporate at the brewing temperatures of coffee. The obvious answer is the fact that elderly people and people with heart problems often switch to decaffeinated coffee. Both factors are confounded with coffee drinking.

Extrapolation Few relationships between variables are linear for all values of x and y. If a plot of the residuals shows that they are randomly distributed around 0, then it is safe to **interpolate** (that is, to use predicted values $\hat{y}$ for values of x that are in the range of the data). But making predictions outside this range (**extrapolation**) must be done with caution. It is a matter of judgment how far one can go beyond the range of the independent variables $\{x_1, x_2, \ldots, x_n\}$ to predict the response.

If we compute the correlation coefficient and the regression line of weight versus mileage of cars for the 38 cars of Table 15.2 on page 410, we get

$$\bar{x} = 2868, \quad \bar{y} = 24.76, \quad s_x = 707, \quad s_y = 6.54, \quad r = -0.9033 \quad \text{and}$$

Regression equation: $\quad \hat{y} = 48.6935 - 0.008361x \quad\quad (15.11)$

For example, $x = 6000$ is outside the range of the data. Using this regression equation for $x = 6000$ pounds, we get a *negative* value for the mileage, $\hat{y} = -1.47$.

If one compares the car weight versus fuel economy data with the fitted regression line in Figure 15.13, we can see that the relationship is curved. A plot of the residuals shows a trend from negative values to positive and then back to negative. Even a linear regression *within* the range of the data is questionable. Extrapolation beyond this range will result in predictions that deviate significantly from the truth.

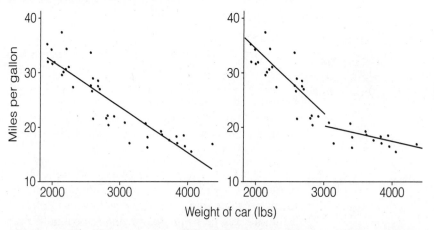

Figure 15.13 Scatter plot of weight versus fuel consumption from Table 15.2 on page 410. At left, regression line superimposed on scatter plot. At right, scatter plot split into light cars (below 3000 pounds) and heavy cars (over 3000 pounds) with two regression lines.

In cases of curved association, one can split the scatter plot into smaller intervals where the associations are linear. This will make for better interpolation. It will also improve close extrapolation, but it will not remove the danger of far extrapolation.

A good way to correct for nonlinearity is to transform either x or y. The transformation $y \to 1/y$ will straighten the car mileage scatter plot.

REVIEW EXERCISES FOR CHAPTER 15

15.32 Given $n = 10$, $\bar{x} = 5$, $\bar{y} = 25$, $s_x = 3$, $s_y = 10$, $r = 0.45$, assume normality and homoscedasticity.

  a. Find an equation of the regression line.

  b. If $x = 6$, find the prediction $\hat{y}$.

  c. Find a 95% confidence interval for the slope of the true regression line.

  d. Test for a positive correlation at the $\alpha = 0.05$ level.

• **15.33** Given $n = 8$, $\bar{x} = 75$, $\bar{y} = 85$, $s_x = 12$, $s_y = 16$, $r = 0.45$, assume normality and homoscedasticity.

  a. Find an equation of the regression line.

  b. If $x = 63$, find the prediction $\hat{y}$.

  c. Find a 95% confidence interval for the slope of the true regression line.

  d. Test for a positive correlation at the $\alpha = 0.05$ level.

15.34 Consider the percentage of U.S. adults and adolescent with AIDS who survived more than 24 months in the years from 1993 to 1997.[8] Assume normality and homoscedasticity.

Year:	1993	1994	1995	1996	1997
Percent surviving ≥ 24 months:	60	64	74	83	87

  a. Find an equation of the regression line.

  b. Extrapolate to predict the survival rate for 1998.

  c. Give a 95% confidence interval for β.

  d. Test for a positive slope of the true regression line.

---

[8] Lisa M. Lee et al., *Survival After AIDS Diagnosis in Adolescents and Adults During the Treatment Era, United States, 1984–1997*, Journal American Medical Assoc. **285** (2001), pp. 1308–15.

- **15.35** Measurement of bones of five Neanderthal fossils found the following measurements of arm bones (humerus) and leg bones (femur), in millimeters.[9] Assume normality and homoscedasticity.

	\multicolumn{5}{c}{Specimen}				
	1	2	3	4	5
Humerus:	312	335	286	312	305
Femur:	430	458	407	440	422

  a. Find the correlation coefficient.

  b. Find an equation of the major axis.

  c. Find an equation of the regression line.

  d. Give a 95% confidence interval for β.

  e. Estimate the length of a femur when a humerus is 450 mm long.

---

[9] Prof. Erik Trinkaus, quoted on page 534 in Andrew F. Siegel, Charles J. Morgan, *Statistics and Data Analysis, An Introduction*, second edition, John Wiley & Sons, 1996.

Chapter 16

INFERENCE WITH CATEGORICAL DATA

16.1 INTRODUCTION

16.2 COMMENTS ON THE DEFINITION OF χ^2

16.3 TESTING GOODNESS OF FIT

16.4 CONTINGENCY TABLE TESTS

16.5 ONE-SIDED CHI-SQUARE TESTS
FOR THE 2×2 CONTINGENCY TABLE

16.6 COMPOUND HYPOTHESES (optional)

16.1 INTRODUCTION

Recall that a variable is categorical if it assigns each observation to one of several categories. Blood type, as considered in Exercise 2.1 on page 25, is an example. The categories are the blood Types O, A, B, and AB. In this chapter, we discuss methods for analyzing categorical data. We will concentrate on a technique based on the chi-square (χ^2) statistic.

We will investigate two important statistical problems.

Goodness of fit   1. How reasonable is it to maintain that observed data from a population could have been generated by a proposed probability model? For example, in Exercise 2.1 on page 25 we compared the blood type of 40 children with that of the population of the United States. The question is whether the blood types of these 40 children differ from the general U.S. population?

Contingency table analysis

2. Here we consider a more complicated situation. Based on observed data from a population, how reasonable is to maintain that two categorical variables are independent? For example, is marital status (with categories never-married, married, divorced, separated, and widowed) independent of employment status (with categories employed, unemployed, and not in labor force)? Or, on the contrary, is at least one of the marital states related to at least one of the employment states? Because marital status and employment status are *categorical variables*, it makes no sense to find means and test for a difference of means as we can with *quantitative variables*.

Some goodness-of-fit examples

- **Fairness of a die.** Suppose we roll a die many times and record which face turns up each time. We can use these data to decide if your die rolling technique is fair. The probability model to test is whether all sides are equally likely.

- **Winning lottery numbers.** Consider the Illinois Lottery Pick Four game. In this game, each player selects four digits. When the winning numbers are drawn, the lottery officials use machines to randomly pick a winning four-digit number. Looking at the data of 2684 draws in the year 1996 given in Example 3.20 on page 86, we can test whether every number has the same chance of winning. The model to test is whether the winning digits are uniformly distributed.

- **Birth order.** Can the gender of children in large families be described by the same probability model used to describe a sequence of tosses of a fair coin? This question was first posed in Section 10.1 on page 274.

- **Geiger counter data.** Rutherford and Geiger collected data on the number of alpha particles emitted by radioactive iodine. In Exercise 9.43 on page 265, we asked whether it is reasonable to expect that their data follow a Poisson random variable. In this chapter, we can formally test this goodness of fit.

Some contingency table examples

Interests of boys and girls. A survey was administered to 478 grade school students in Michigan who were asked, What would you most like to do at school: Make good grades, be good at sports, or be popular?[1]

Below are some contingency table problems.

---

[1] Data for this kind of problem can be found online at http://lib.stat.cmu.edu/DASL/Stories/Students'Goals.html.

- Do fourth graders have the same goals as sixth graders?
- Do boys and girls have the same goals?
- Do students in urban school districts have the same goals as students in suburban districts?

The questions here are whether attributes of one categorical variable are independent of that of another. The alternative hypothesis is that *at least one* attribute (e.g., boy) is statistically related to another (e.g., sports).

***

Example 16.1   *Fairness of a die.*   Suppose we roll a die 900 times and tabulate the various outcomes.

Face	Observed count
1	167
2	156
3	151
4	128
5	132
6	166
Total	900

The probability model is that all faces are equally likely; so we expect $900/6 = 150$ outcomes of each face. Comparing what we got to what we expected results in the following table.

Face j	Observed count O_j	Expected count E_j	Observed − Expected $O_j - E_j$
1	167	150	+17
2	156	150	+6
3	151	150	+1
4	128	150	−22
5	132	150	−18
6	166	150	+16
Totals	900	900	0

Notice that the face 4 deviated the most from the expected; we observed a sample count of only 128 when we expected 150. This is a deficit of 22. *If we were to single out this face*, we could use the methods of Chapters 6 or 12. The two-sided z test gives us

$$P\text{-value} = Pr(|K - 150| \geq 22) = 2Pr\left(z \leq \frac{128 - 150}{\sqrt{900(1/6)(5/6)}}\right) = 0.0491$$

This is (barely) significant at the $\alpha = 0.05$ level.

To single out the category that deviates the most from the expected, is an example of **data snooping**. *We should not look at the data to decide on what to test.*

> *If we have many categories, it's not unusual to have at least one of them deviate significantly from the expected value even if the parent population fits the probability model in question.*

To test the fairness of the die, we want a test that takes into consideration *all* of the categories. Notice that the column of deviations $O_j - E_j$ always adds up to 0. To measure the discrepancy between what we observe and what we expect, we want to do something similar to taking the RMS of these deviations, as we did in the definition of the standard deviation. This time, however, we will not take the square root, so the statistic we define is more analogous to the grouped variance

$$s^2 = \sum_j \frac{f_j(x_j - \bar{x})^2}{n-1}$$

Formally, we define the **chi-square statistic** as follows:

$$\chi^2 = \sum_j \frac{(O_j - E_j)^2}{E_j}$$

This statistic is a weighted average of the terms $(O_j - E_j)^2$, just as the grouped variance is a weighted average of the terms $(x_j - \bar{x})^2$. It makes sense that the χ^2 statistic give weights inversely proportional to the expected counts E_j. A deviation of $O_j - E_j = 10$ from an expected count of $E_j = 10$ should have greater weight than a deviation of $O_j - E_j = 10$ from an expected count of $E_j = 100$ because the relative error is greater in the former than the latter.

> *Remember that χ^2 statistics are always computed from count data, not proportions or percentages. Also, with the exception of the situation of Section 16.5 with $df = 1$, the χ^2 tests always have two-sided alternatives.*

For our rolling-a-die example, the chi-square statistics is the sum of six terms:

Face	Observed count	Expected count	Observed − Expected	χ^2 terms
j	O_j	E_j	$O_j - E_j$	$(O_j - E_j)^2/E_j$
1	167	150	+17	289/150
2	156	150	+6	36/150
3	151	150	+1	1/150
4	128	150	−22	484/150
5	132	150	−18	324/150
6	166	150	+16	256/150
Totals	900	900	0	1390/150

We have $\chi^2 = 1390/150 = 9.267$.

How do we interpret this statistic? Theorem 16.1 below shows that the expected value of this χ^2 statistic is actually 5. It's clear that the greater the discrepancy between the observed counts and the expected counts, the greater the chi-square statistic. The question can be rephrased. Is the value of the chi-square statistic $\chi^2 = 9.267$ so large that the fairness of the die is statistically rejected?

16.2 COMMENTS ON THE DEFINITION OF χ^2

Formally, the chi-square statistic is based on a **multinomial random variable**. A multinomial random variable is like a binomial random variable except that, instead of counting successes, each observation is classified as being in exactly one of c categories. For a random sample of n items, let K_j denote the count of items in category j, where $j = 1, 2, \ldots, c$. The probability mass function is

$$f(k_1, k_2, \ldots, k_c) = \frac{n!}{k_1! k_2! \cdots k_c!} \pi_1^{k_1} \pi_2^{k_2} \cdots \pi_c^{k_c} \tag{16.1}$$

where $k_j \geq 0$, $\pi_j \geq 0$, $k_1 + k_2 + \cdots + k_c = n$ and $\pi_1 + \pi_2 + \cdots + \pi_c = 1$.

The multinomial coefficients $\dfrac{n!}{k_1! k_2! \cdots k_c!}$ are extensions of the binomial coefficients $\dfrac{n!}{k!(n-k)!}$. They count the number of combinations of n things chosen $k_1, k_2, \ldots, k_c$ at a time.[2]

Note that for $c = 2$ this agrees with the binomial probability mass function

$$f(k) = \binom{n}{k} \pi^k (1-\pi)^{n-k}$$

---

[2] Recalling footnote 2 of Chapter 5 on page 150, the multinomial coefficients are the number of n-letter words one can make with k_1 A's, k_2 B's, k_3 C's, and so on.

The rules for a binomial count K of Section 5.4 (on page 147) must be extended to allow for more than two categories,

Rule 1. The number of trials n is predetermined.

Rule 2. For each j, the random variable K_j is the number of times category j occurs in the n trials.

Rule 3. The trials are independent.

Rule 4. The probability of getting category j on a single trial is π_j, and it is the same for every trial.

THEOREM 16.1 *The sampling distribution of the χ^2 statistic has mean $E(\chi^2) = c - 1$*

Proof For n independent trials, each of the counting random variables K_j is a binomial random variable with probability of success π_j. Using Formula 5.6 on page 151, we have $E(K_j - n\pi_j)^2 = \text{Var}(K_j) = n\pi_j(1 - \pi_j)$. Then using Theorem 10.1, we have

$$E(\chi^2) = \sum_{j=1}^{c} \frac{E(K_j - n\pi_j)^2}{n\pi_j} = \sum_{j=1}^{c} \frac{n\pi_j(1 - \pi_j)}{n\pi_j} = c - 1$$

Remark For $c = 2$ categories, a multinomial random variable is binomial. The two categories are 0 (failure) and 1 (success). In the notation of the binomial random variable, the count of successes is denoted $K_1 = K$, so the count for the failures is then $K_0 = n - K$.

Category j	Observed count O_j	Expected count E_j	Observed − Expected $O_j - E_j$
0 (failure)	$n - K$	$n(1 - \pi)$	$-K + n\pi$
1 (success)	K	$n\pi$	$K - n\pi$
Totals	n	n	0

It turns out that for $c = 2$ the chi-square statistic is the square of the standardized statistic $Z = \frac{K - n\pi}{\sqrt{n\pi(1-\pi)}}$ for the binomial random variable K:

16.2 COMMENTS ON THE DEFINITION OF χ^2

$$\chi^2 = \frac{(-K+n\pi)^2}{n(1-\pi)} + \frac{(n\pi-K)^2}{n\pi}$$

$$= \frac{\pi(-K+n\pi)^2 + (1-\pi)(n\pi-K)^2}{n\pi(1-\pi)}$$

$$= \frac{(K-n\pi)^2}{n\pi(1-\pi)} = Z^2$$

Formally, the χ^2 density curves are defined as follows:

$$f_{\chi^2}(x) = \begin{cases} c^{-1} x^{\frac{\nu}{2}-1} e^{-\frac{x}{2}} & \text{for } x > 0 \\ 0 & \text{otherwise} \end{cases} \quad (16.2)$$

where ν = degrees of freedom and the constant

$$c = \int_0^\infty x^{\frac{\nu}{2}-1} e^{-\frac{x}{2}} \, dx = \Gamma(\frac{\nu}{2}) 2^{\frac{\nu}{2}}$$

It turns out that when $\nu = 1$ the χ^2 density function is the same as the density function of Z^2 where Z is standard normal. Also, when $\nu = 2$, the χ^2 density function is the exponential density function with $\theta = 2$. Figure 16.1 shows some graphs of χ^2 densities for various degrees of freedom. The densities are all skewed to the right but tend to look more like normal density curves as the degrees of freedom increase.

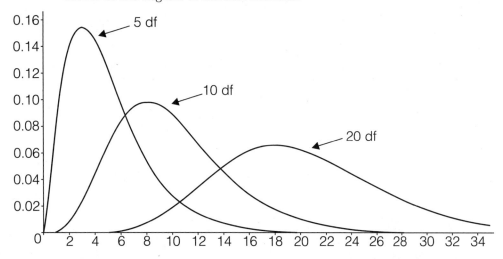

Figure 16.1 Chi-square curves with 5, 10, and 20 degrees of freedom

EXERCISES FOR SECTION 16.2

- **16.1** (Proof) Prove that if the random variable T has a χ^2 density with ν degrees of freedom, then $E(T) = \nu$. Use Formula (16.2).

16.2 (Proof) (Proof) Prove that if the random variable T has a χ^2 density with ν degrees of freedom, then $\text{Var}(T) = 2\nu$. Use Formula (16.2).

- **16.3** (Proof) Let K_j $(j = 1, \ldots, c)$ be random variables with probability mass function given by Equation 16.1. Show that K_1 is a binomial random variable with parameters n and π_1.

16.4 (Proof) Let K_j $(j = 1, \ldots, c)$ be random variables with probability mass function given by Equation 16.1 with $c > 3$. Show that K_1, K_2 and a new $K_3' = n - \{K_1 + K_2\}$ are multinomial random variables with parameters $\pi_1, \pi_2, \pi_3' = 1 - \pi_1 - \pi_2$ and n.

- **16.5** (Proof) Let K_j $(j = 1, \ldots, c)$ be random variables with probability mass function given by Equation 16.1 with $c > 2$. Show that, when two categories are collapsed into one, say $K = K_{c-1} + K_c$, then the random variables $K_j, j = 1, \ldots, c-2$ and K are multinomial random variables with probability mass function

$$f(k_1, \ldots, k_{c-2}, k) = \frac{n!}{k_1! \cdots k_{c-2}! k!} \pi_1^{k_1} \cdots \pi_{c-2}^{k_{c-2}} \pi^k \quad \text{where} \quad \pi = \pi_{c-1} + \pi_c$$

16.6 (Proof) Let K_j $(j = 1, \ldots, c)$ be random variables with probability mass function given by Equation 16.1. Show $\text{Cov}(K_1, K_2) = -n\pi_1\pi_2$.

* * * * * * * * *

- **16.7** (Proof) Use the theorems in Section 10.4 to prove that, for the sampling distribution of the χ^2 statistic, we have

$$\text{Var}(\chi^2) = 2(c-1) + \frac{\sum_{j=1}^{c} \frac{1}{\pi_j} - (c+1)^2 + 3}{n}$$

Thus, for large n, we have $\text{Var}(\chi^2) \approx 2(c-1)$.

16.3 TESTING GOODNESS OF FIT

The chi-square statistic compares sample categorical data with what would be expected if the data were generated from a sampling distribution. Large values of the test statistic are taken to be evidence against the null hypothesis. The good news is that, for large samples, the sampling distribution of the chi-square statistic under the assumption of the null hypothesis has a definite distribution (given in Table 6 on page 513 at the back of this book) that depends only on its degrees of freedom.[3] The degrees of freedom are determined by the number of categories c. For goodness-of-fit problems, we have $df = c - 1$. A random sample is generally considered to be large if all of the expected counts E_j are at least 5.

> As a general rule, if the expected counts satisfy $E_j \geq 5$, for all j, then percentiles of χ^2 can be found in Table 6 on page 513 at the back of this book.

The rule that all of the expected counts be at least 5 is consistent with the General Rule (page 231) for the normal approximation of binomial counts. However, it is quite conservative. A more liberal rule is that almost all of the expected counts be at least 5, but one or two can be less than 5. For small samples, the exact P-value can be computed as we will do in Example 16.3. Also, some statistical packages compute exact P-values.

Example 16.2 *Rolls of a die.* Figure 16.2 shows a histogram of the sampling distribution of χ^2 statistics for the experiment of rolling $n = 8$ fair dice. The chi-square density curve with $df = c - 1 = 5$ is also shown. We see that, when testing for the goodness of fit of a uniform probability model at the $\alpha = 0.05$ level, the chi-square distribution is a reasonable approximation, even though all of the expected counts $E_1 = E_2 = E_3 = E_4 = E_5 = E_6 = 8/6 = 1.333$ are less than 5.

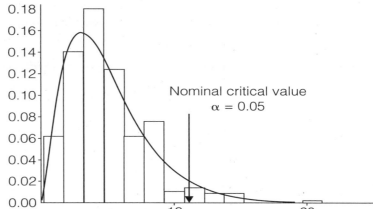

Figure 16.2 A histogram of the χ^2 statistic for the rolls of 8 fair dice and the χ^2 density curve with $df = 5$ superimposed

---

[3] Remember how, for large samples, we approximated binomial random variables by normal random variables? Similarly, multinomial random variables can be approximate by multivariate normal random variables. A chi-square statistic is a sum of c squares of multinomial random variables. For large samples, the chi-square statistic is thus approximately a sum of squares of multivariate normal random variables. This sum is a chi-square random variable with $c - 1$ degrees of freedom.

Now to answer the question of the fairness of the die. We calculated $\chi^2 = 9.267$. There are $c = 6$ categories, so $df = 5$. All of the expected counts are $E_j = 150 \geq 5$. We have P-value $= Pr(\chi^2 \geq 9.267)$, which is the area of the shaded region in Figure 16.3. Using Table 6 on page 513, we see that 90th percentile of the χ^2 curve with $df = 5$ is 9.2363 and the 95th percentile is 11.0705. Thus, we have $0.05 < P$-value < 0.10. Actually, using technology we obtain P-value $= 0.099$. Our conclusion is that at the 5% level of significance, the value of χ^2 is not large enough to reject the fairness of the die.

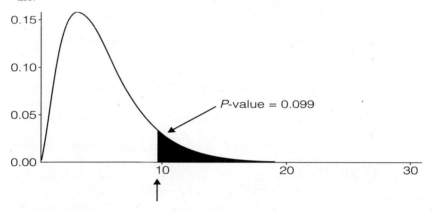

Figure 16.3 The χ^2 curve with 5 degrees of freedom (underneath arrow points to $\chi^2 = 9.267$)

In the example below, we show how to compute the exact P-value for a small sample.

Example 16.3 **Rolling a die five times.** Suppose we roll a die $n = 5$ times and get three 3's and two 6's. There are $c = 6$ categories, for the 6 faces of a die, and the probability of any outcome can be found by using the multinomial probability mass function (16.1) with $\pi_1 = \pi_2 = \cdots = \pi_6 = 1/6$. So, the probability of rolling three 3's and two 6's is

$$\frac{5!}{0!0!3!0!0!2!}\left(\frac{1}{6}\right)^0\left(\frac{1}{6}\right)^0\left(\frac{1}{6}\right)^3\left(\frac{1}{6}\right)^0\left(\frac{1}{6}\right)^0\left(\frac{1}{6}\right)^2 = \frac{10}{6^5} \approx 0.0013$$

Of course, if we roll a fair die $n = 5$ times, we expect to get $n\pi_1 = 5/6$ 1's, and $n\pi_2 = 5/6$ 2's, and $n\pi_3 = 5/6$ 3's, and so forth. Now, if we got three 3's and two 6's when we rolled 5 dice, we can make a table as we did on page 447.

Face	Observed count	Expected count	Observed − Expected	χ^2 terms
j	O_j	E_j	$O_j - E_j$	$(O_j - E_j)^2/E_j$
1	0	5/6	−5/6	5/6
2	0	5/6	−5/6	5/6
3	3	5/6	+13/6	169/30
4	0	5/6	−5/6	5/6
5	0	5/6	−5/6	5/6
6	2	5/6	+7/6	49/30
Totals	5	5	0	$\chi^2 = 318/30 = 10.6$

Notice that the observed value of the chi-square statistic is quite a bit more than the expected value of $c - 1 = 5$. Is it unusual to observe a chi-square of 10.6 or more when you were expecting it to be 5?

Solution We cannot use Table 6 on page 513 because the sample is too small. Recall that the expected counts should be at least 5. However, we can calculate the P-value exactly. In this dice-roll example, there are $6^5 = 7776$ equally likely possible outcomes. We compute the χ^2 statistic in each case and thus derive the sampling distribution. The following table illustrates the computations.

Sampling distribution of χ^2 statistics[4]

Split	χ^2	Probability
5-0	25.0	$\frac{6!}{1!5!} f(5,0,0,0,0,0) = 6/6^5$
4-1	15.4	$\frac{6!}{1!1!4!} f(4,1,0,0,0,0) = 150/6^5$
3-2	10.6	$\frac{6!}{1!1!4!} f(3,2,0,0,0,0) = 300/6^5$
3-1-1	8.2	$\frac{6!}{1!2!3!} f(3,1,1,0,0,0) = 1200/6^5$
2-2-1	5.8	$\frac{6!}{2!1!3!} f(2,2,1,0,0,0) = 1800/6^5$
2-1-1-1	3.4	$\frac{6!}{1!3!2!} f(2,1,1,1,0,0) = 3600/6^5$
1-1-1-1-1	1.0	$\frac{6!}{5!1!} f(1,1,1,1,1,0) = 720/6^5$

As we see from this table, the chance of getting a χ^2 statistic of 10.6 or more is $\frac{6}{6^5} + \frac{150}{6^5} + \frac{300}{6^5} = \frac{456}{7776} = 0.0586$. This is the P-value.

If the dice are fair, we anticipate rolling the 5 dice $\frac{7776}{456} \approx 17$ times or so to observe a chi-square statistic as big as 10.6.

---

[4]The split 3–2, for example, refers to getting three of one kind and two of another, such as three aces and two 2's. Any particular 3–2 split has the same chance $\frac{10}{6^5}$ using the multinomial formula. There are 30 possible 3–2 splits. We can solve this counting problem with a multinomial coefficient by choosing one of the 6 possible faces to be tripled, 1 to be doubled, and 4 to be omitted. This gives us the probability $30 \times \frac{10}{6^5} = \frac{300}{6^5}$ of a 3–2 split.

Example 16.4   **Illinois Lottery Pick Four game.** Data for the first $n = 2684$ draws were presented in Example 3.20 on page 86. The null hypothesis is that this is a random sample of 2684 digits. There are $c = 10$ categories, one for each of the possible digits $0, 1, 2, \ldots, 9$. Formally, if π_j is the probability of the digit j, then

$$H_0: \pi_j = \frac{1}{10} \quad \text{for all } j \qquad H_a: \pi_j \neq \frac{1}{10} \quad \text{for at least one } j$$

Test this hypothesis.

Solution   Under the null hypothesis, the expected number of 0's is $n\pi_0 = 268.4$. Similarly, the expected number of every other digit is 268.4. We can compute the chi-square statistic from the following table.

Digit j	Observed count O_j	Expected count E_j	χ^2 Tterms $(O_j - E_j)^2/E_j$
0	265	268.4	0.04307
1	255	268.4	0.669
2	261	268.4	0.204
3	296	268.4	2.838
4	263	268.4	0.1086
5	262	268.4	0.1526
6	249	268.4	1.402
7	290	268.4	1.738
8	284	268.4	0.9067
9	259	268.4	0.3292
Totals	2684	2684.0	$\chi^2 = 8.39195$

So the computed value of the chi-square statistic is 8.39195. If the null hypothesis is true, then the sampling distribution of the chi-square statistic is the chi-square distribution with $c - 1 = 9$ degrees of freedom. Using Table 6 on page 513 we see that the 75th percentile of the χ^2 distribution is 11.3887 and the 10th percentile is 4.1682. Thus, $0.25 < P\text{-value} < 0.90$. Because the P-value is so large, we retain the null hypothesis. We cannot conclude that any one digit predominated. The empirical probabilities calculated in Example 3.20 on page 86 are consistent with the theoretical ones given by a uniform random variable.

Example 16.5   **Birth order.** Let's return to the data presented in Example 10.2 on page 272 from 15,162 families of 5 (or more) children. Let K denote the number of girls among the first 5 children. Can 5 tosses of a fair coin as given in Example 10.1 on page 268 be used to model the number of girls in families of 5 children?

Solution Formally, we assume the $n = 15{,}162$ families to be a random sample. The null hypothesis is that the number of girls is like the number of heads in a random sample of 15,162 tosses of 5 fair coins.

$$H_0 \ : \ Pr(K = k) = \binom{5}{k}(0.5)^k(1-0.5)^{5-k} \quad \text{for all } k$$

$$H_a \ : \ Pr(K = k) \neq \binom{5}{k}(0.5)^k(1-0.5)^{5-k} \quad \text{for at least one } k$$

The follow table illustrates the computation of the chi-square statistic.

k	O	E	O − E	$(O-E)^2/E$
0	549	473.8	+75.2	11.93
1	2476	2369.1	+106.9	4.827
2	4753	4738.1	+14.9	0.047
3	4621	4738.1	-117.1	2.895
4	2245	2369.1	-124.1	6.497
5	518	473.8	+44.2	4.121
Totals	15162	15162.0	0.0	$\chi^2 = 30.318$

The expected counts are computed using

$$E_k = n\binom{5}{k}(0.5)^k(1-0.5)^{5-k} = \binom{5}{k}\frac{15162}{32}$$

If the null hypothesis is true, then the sampling distribution of the chi-square statistic is the chi-square distribution with $c - 1 = 5$ degrees of freedom and $E(\chi^2) = 5$. Using Table 6 on page 513, we see that P-value $= Pr(\chi^2 > 30.318) < 0.001$. Because the P-value is very small, we reject the null hypothesis.

The data do not fit the proposed probability model. Looking at the above table, it appears that there were too many families with no or one girl and too few with four girls.

EXERCISES FOR SECTION 16.3

16.8 Suppose that data fall into 4 classes with the frequencies 45, 55, 63, 47. Test whether the data come from a population with a uniform distribution.

• **16.9** Suppose that data fall into 4 classes with the frequencies 47, 12, 18, 3. Perform a goodness-of-fit test to determine whether the data come from a population with a 9:3:3:1 ratio.

16.10 Recall Gregor Mendel's work with garden pea plants as described in Example 5.4 on page 146. According to his genetic theory, if pea plants with yellow and green seeds are crossed and the first-generation seeds are all yellow, then the second-generation plants have either yellow or green seeds. These second-generation plants should come in proportions 3 yellow to 1 green. Do a goodness-of-fit test to see if his 8023 second-generation plants, of which 6022 were yellow and 2001 were green, is consistent with his theory. Use $\alpha = 0.01$.

• **16.11** Test the hypothesis of Exercise 6.31 on page 192 using the χ^2 goodness-of-fit test instead of the binomial exact test. A certain kind of computer chip has a failure rate of 8%. During quality-control inspection of a batch of 100, 15% are found to be defective. Is this within historical limits or is the failure rate out of control? Use $\alpha = 0.05$.

16.12 Suppose we roll a die 5 times and get a 4, a 2, and three 1's. Compute, as was done in Example 16.3, the χ^2 statistic and the corresponding P-value for testing whether the die is fair.

• **16.13** Suppose we roll a die five times and get a 1, a 2, a 4, and two 5's. Compute, as was done in Example 16.3, the χ^2 statistic and the corresponding P-value for testing whether the die is fair.

16.14 Let's take another look at the data in Table 10.5 on page 274 of 15,162 families of 5 children.

Table 10.5: Pairs in families of 5 children

Pair	f	$f\%$
F F	15,109	24.913%
F M	14,906	24.578%
M F	14,886	24.545%
M M	15,747	25.965%
$\sum = 60,648$		$\sum = 100.001\%$

Is there enough evidence to show that the gender of the next child is not independent of the previous?

    a. State the null and alternative hypotheses.
    b. Create a table to compute the chi-square statistic
    c. Compute the P-value.
    d. State your conclusion about the adequacy of the fair coin toss model in this case.

16.15 Let's take another look at the data in Example 10.2 on page 272. Let y denote the number of changes in sequence and k denote the number of girls. Can the five tosses of a fair coin model (Example 10.1) with y denoting the number of changes in sequence and k the number of heads be used to model the families of five children? The null hypothesis probabilities are given in Table 10.1 on page 271

a. Create a table to compute the chi-square statistic and show that the chi-square statistic is 39.33.
b. Explain why the sampling distribution of the chi-square statistic is chi-square with 13 degrees of freedom.
c. Compute the P-value.
d. State your conclusion about the adequacy of the fair coin toss model in this case.

16.4 CONTINGENCY TABLE TESTS

Sometimes two or more attributes are used to describe people or things. In Example 2.2 on page 22, bachelor's degree recipients were classified by race and field. In Example 10.4 on page 277, residents of Illinois were classified by gender and age (two of many conceivable attributes). In Example 10.6 on page 283, a fair coin-tossing experiment was described by two attributes: the number of heads on the first toss and the number of heads on the second toss. Two attributes are independent if the relative frequency of subjects in each category of one attribute does not depend on the value of the other attribute. Let (X, Y) denote the values of the two attributes of a randomly selected subject from the target population. Recall from Section 10.2 on page 275 that discrete random variables X and Y are independent if and only if the joint probability table is a multiplication table of the marginal probabilities. Thus, race and field were dependent attributes for science and engineering bachelor's degree recipients in 1993 in Example 2.2. Also gender and race are dependent attributes for residents of Illinois according to the census data in Example 10.4. In general, we will say that two attributes are independent if and only if the corresponding random variables X and Y are independent.

Suppose a simple random sample is drawn from a population with two or more attributes. How could you use these data to test the hypothesis that the attributes are independent?

Example 16.6 **Interests of boys and girls.** Consider a survey that included the following question:

What would you most like to do at school?

A. Make good grades.

B. Be popular.

C. Be good at sports.

Randomly selected sixth graders were asked this question, and the answers were combined with demographic data as summarized in the table below.

Data count table

Gender	Grades	Popular	Sports	Total
Boys	45	18	24	87
Girls	51	37	8	96
Total	96	55	32	183

(Goal)

Dividing by $n = 183$, we obtain the joint *relative* frequency table, which we can compare to the multiplication table of its margins:

Relative frequency table

Gender	Grades	Popular	Sports	Total
Boys	0.2459	0.0984	0.1311	0.4754
Girls	0.2787	0.2022	0.0437	0.5246
Total	0.5246	0.3005	0.1749	1.0000

(Goal)

Multiplication table of its margins

Gender	Grades	Popular	Sports	Total
Boys	0.2494	0.1429	0.0831	0.4754
Girls	0.2752	0.1577	0.0917	0.5246
Total	0.5246	0.3005	0.1749	1.0000

(Goal)

Now it is easy to see that 52.46% of the 183 children in the sample are girls and 30.05% of the children wanted most to be popular at school. However, 20.22% of the children are girls who wanted most to be popular at school, not the $0.5246 \times 0.3005 = 15.77\%$ obtained from the multiplication table of the margins. We note similar discrepancies when comparing other cells in these tables.

Certainly, these attributes are not independent for the data in the sample, but can the same be said for the entire population of sixth graders from which this sample is drawn? We can test the hypothesis of independence using a chi-square statistic.

Solution There are 2 levels of gender and 3 levels of interest, so $2 \times 3 = 6$ cells in all. Under the assumption of independence (the null hypothesis) we can estimate the chance of a randomly selected sixth grader being in a particular cell by the corresponding entry in the multiplication table of the margins. Then the expected cell count is just this estimated probability times the sample size:

**Estimated expected counts
(assuming independence of attributes)**

	Goal			
Gender	Grades	Popular	Sports	Total
Boys	45.64	26.15	15.21	87.00
Girls	50.36	28.85	16.79	96.00
Total	96.00	55.00	32.00	183.00

Using this table and the data count table, we now compute the chi-square statistic in the usual way:

$$\chi^2 = \sum \frac{(O-E)^2}{E}$$
$$= \frac{(45-45.64)^2}{45.64} + \frac{(18-26.15)^2}{26.15} + \frac{(24-15.21)^2}{15.21} + \frac{(51-50.36)^2}{50.36} + \frac{(37-28.85)^2}{28.85} + \frac{(8-16.79)^2}{16.79}$$
$$= 14.53$$

We want to use a χ^2 distribution as given in Table 6 on page 513 at the back of this book—but which one? The general rule for computing degrees of freedom is:

> **Degrees of freedom =
> number of classes − number of estimated parameters − 1**

Here there are 6 classes and 3 estimated parameters,[5] so $df = 6-3-1 = 2$. For a contingency table with c columns and r rows, this comes out to

$$df = (c-1)(r-1)$$

Thus, the P-value for the chi-square test of independence of these attributes is $Pr(\chi^2 \geq 14.53)$ with $df = 2$. Using Table 6 on page 513, we see that

---

[5] We estimated the expected cell counts by estimating the marginal probability mass function for each attribute. There are three levels of goals and two levels of gender, so we estimated five marginal probabilities. However, the marginal probabilities have to add to 1, and we can use the total probability rule twice (once for the row and once for the column). This means that there are only three probabilities to estimate. The other two can be found by subtraction. For example, if we knew the probability that a randomly selected sixth grader was a boy, we could easily compute the probability of a girl because these two probabilities sum to 1.

P-value < 0.001. We conclude that sixth-grade boys and girls have different opinions on what they most like to do at school.

Let's now summarize the chi-square test of independence of two attributes, A and B, based on a random sample of size n. Given the hypotheses

$$H_0 : A \text{ and } B \text{ are independent attributes}$$
$$H_a : A \text{ and } B \text{ are dependent attributes}$$

we will summarize a test of independence at the α level of significance.

Notation: Suppose the attribute A has r categories $A_1, A_2, \ldots, A_r$ and B has c categories $B_1, B_2, \ldots, B_c$, with observed data table:

Contingency table

A	B_1	B_2	$\cdots$	B_c	$Totals$
A_1	O_{11}	O_{12}	$\cdots$	O_{1c}	R_1
A_2	O_{21}	O_{22}	$\cdots$	O_{2c}	R_2
$\cdots$	$\cdots$	$\cdots$	$\cdots$	$\cdots$	$\cdots$
A_r	O_{r1}	O_{r2}	$\cdots$	O_{rc}	R_r
Totals	C_1	C_2	$\cdots$	C_c	n

The row totals above $R_1, R_2, \ldots, R_r$ and the column totals $C_1, C_2, \ldots, C_c$ are the marginal counts. It is important to remember that these are counts, *not the relative frequencies*. The table of expected values:

Expected count table

A	B_1	B_2	$\cdots$	B_c	$Totals$
A_1	E_{11}	E_{12}	$\cdots$	E_{1c}	R_1
A_2	E_{21}	E_{22}	$\cdots$	E_{2c}	R_2
$\cdots$	$\cdots$	$\cdots$	$\cdots$	$\cdots$	$\cdots$
A_r	E_{r1}	E_{r2}	$\cdots$	E_{rc}	R_r
Totals	C_1	C_2	$\cdots$	C_c	n

The marginal counts $R_1, R_2, \ldots, R_r$ and $C_1, C_2, \ldots, C_c$ are the same for both tables and are obtained from the data table. The expected counts E_{jk} are computed with the formula

$$E_{jk} = \frac{R_j C_k}{n}$$

Then we compute the χ^2 statistic as follows:

$$\chi^2 = \sum_{j=1}^{r}\sum_{k=1}^{c} \frac{(O_{jk} - E_{jk})^2}{E_{jk}}$$

The decision rule is

Retain H_0 if $\chi^2 \leq \chi^2_{df,1-\alpha}$ Reject H_0 if $\chi^2 > \chi^2_{df,1-\alpha}$

where $\chi^2_{df,1-\alpha}$ denotes the $1 - \alpha$ percentile of the χ^2 distribution with $df = (c-1) \times (r-1)$ degrees of freedom.

Here, as with goodness-of-fit tests, it is best if all of the expected counts are at least five, but this is a conservative rule.

> As a general rule, if the expected counts satisfy $E_{jk} \geq 5$, for all j and k, then percentiles of χ^2 can be found in Table 6 on page 513 at the back of this book.

Example 16.7 *Scanner overcharges.*[6] State and county officials routinely check out scanners at California supermarkets and retail stores. In 1998, they randomly inspected 499 such stores. At each store, 30 items were selected and scanned. The number of overcharges among the 30 scanned items was recorded for each store. Three or more overcharges was considered a significant violation. Does it appear that the number of overcharges is independent of the type of retail store?

Contingency table

Type of store	No overcharges	One or two overcharges	Three or more overcharges	Total
Food	104	50	11	165
Drug	50	37	12	99
Other	148	73	14	235
Total	302	160	37	499

Here we have a random sample of 499 purchases from a large population. A formal test of independence of the two attributes is left as an exercise (Exercise 16.17 on page 463).

---

[6]The data are reported in the California Department of Food and Agriculture News Release CDFA99-046 (July 8, 1999), www.cdfa.ca.gov.

Example 16.8 ***Test for a trend.*** The 1998 survey was a follow-up of a similar survey involving 300 randomly selected stores in 1992. The question we want to consider is whether there had been a change in the pattern of overcharges.

Contingency table

Year	No overcharges	One or two overcharges	Three or more overcharges	Total
1998	302	160	37	499
1992	165	107	28	300
Total	467	267	65	799

The sampling method is different than in the previous examples. We have independent samples from two multinomial distributions, and we want to test the hypothesis that these distributions are the same. Under the null hypothesis that these distributions are the same, we can combine the samples into one large random sample of size 799. Because the marginal distributions of a multinomial distribution are binomial, it seems natural to estimate each cell probability by the corresponding relative frequency. The good news is that the chi-square statistic for this table turns out to be algebraically equivalent to the one used to test for independence of attributes. This is left as Exercise 16.19 on page 463.

NOTES ON TECHNOLOGY

Contingency Table Tests

This is how to use technology to perform contingency table tests.

Minitab: Enter the observed counts in columns. Select **Stat→Tables→Chi-Square Test**. Enter the columns containing the table in the square provided. Then hit **OK**.

Excel: Enter the observed counts in columns. In another location, calculate the expected counts. Go to an empty cell and use the function **CHITEST**. The function can be found by hitting the icon f_x and choosing **Statistical→CHITEST** and then hitting **OK**. Enter the Actual_range (for example A1:B3) for the place the observed counts are located and then enter the Expected_range. Then hit **OK**.

TI-83 Plus: Press **MATRIX** and select **EDIT** and hit **ENTER**. Enter the dimensions of the matrix, row by column, and then enter each of the values of the contingency table. Press **STAT→TESTS** and select χ^2-**Test**. The observed table is matrix A and the expected table will be computed as matrix B. Select **CALCULATE** and hit **ENTER**.

EXERCISES FOR SECTION 16.4

16.16 Perform a test for independence of the row and column attributes at the $\alpha = 0.05$ level.

	1	2	3	Total
x	53	63	47	163
y	65	60	67	192
Total	118	123	114	355

• **16.17** Use the 1998 scanner data given in Example 16.7 on page 461 to test whether the number of overcharges is independent of the type of retail store. Use $\alpha = 0.05$.

16.18 To investigate whether calorie intake has an effect on the incidence of chronic disease, researchers divided 120 rhesus monkeys into two equal groups; over the span of more than 10 years, one group ate without limit and the other group consumed only 70% as many calories as the first group.[7] In the well-fed group, 25 animals experience some form of chronic disease, compared to 13 in the low-calorie group. Test for independence of calorie intake and chronic disease at the $\alpha = 0.05$ level.

	Disease	No disease	Total
Low-cal	13	47	60
Well-fed	25	35	60
Total	38	82	120

• **16.19** Compute the χ^2 statistic for the data in Example 16.8 on page 462 and test whether there has been a change in pattern of overcharges. Use $\alpha = 0.05$.

16.20 Test the hypothesis for independence of the attributes of "Sex" and "Chosen by Judge" of Exercise 6.33 on page 192. This time use a χ^2 contingency table test with $\alpha = 0.01$.

	Chosen	Not chosen	Row total
Women	9	93	102
Men	91	157	248
Column total	100	250	350

• **16.21** Referring to the acupuncture therapy for cocaine addiction as described in Exercise 6.40 on page 194, an acupuncture treatment was compared to a sham placebo treatment and relaxation therapy. Below is the resulting table for the subjects that completed the eight-week program. Test for independence of the treatments at the $\alpha = 5\%$ level. Explain why the P-value obtained may

---

[7] J. Travis, *Low-cal diet may reduce cancer in monkeys*, Science News **158** (2000), p. 341.

not be reliable.

	Success	Failure	Row total
Acupuncture	7	6	13
Placebo	4	13	17
Relaxation	2	20	22
Column total	13	39	52

16.22 Recall Gregor Mendel's work with garden pea plants as described in Example 5.4 on page 146. This is a summary of his results.

Trait	Dominant	Recessive	Totals
Seed color	6022	2001	8023
Seed texture	5474	1850	7324
Pod color	428	152	580
Stem length	787	277	1064
Flower type	651	207	858
Pod type	882	299	1181
Flower color	705	224	929
Totals	14949	5010	19959

Notice that for each trait the dominant and recessive characteristics come proportional roughly 3 to 1. Test the hypothesis that dominant and recessive characteristics come in equal proportions. Follow the procedure of Example 16.8 and Exercise 16.19. Use $\alpha = 0.01$.

16.5 ONE-SIDED CHI-SQUARE TESTS FOR THE 2 × 2 CONTINGENCY TABLE

Chi-square tests for independence of several attributes have two-sided alternatives. In the 2 × 2 case, however, the chi-square table test is equivalent to a z test. As was shown in the remark on page 448, in the 2 × 2 case we have $df = 1$, and the χ^2 density curve is the square of the normal.

This means that we can use the chi-square contingency table test as a replacement of the binomial exact test and Fisher's exact test. By dividing the P-value obtained from the two-sided chi-square test by two, we can test hypotheses with one-sided alternatives. It must be emphasized that this works only for the 2 × 2 case, which has $df = 1$.

16.5 ONE-SIDED CHI-SQUARE TESTS FOR 2 × 2 TABLE

Example 16.9   *Flu shots and the flu.* Let's reconsider Exercise 6.37 on page 193. We compare the incidence of the flu between 15 college students who received the flu shot and 33 who had not. Here is the 2 × 2 table from Exercise 6.37:

Contingency table

	Flu	No flu	Total
Shot	3	12	15
No shot	18	15	33
Total	21	27	48

Let π_1 be the probability that a college student who received a flu shot gets the flu. And let π_2 be the probability that a college student without a flu shot gets the flu. The observed proportions are $p_1 = \frac{3}{15} = 0.200$ and $p_2 = \frac{18}{33} = 0.545$. We can state the test as

$$H_0 : \pi_1 = \pi_2 \qquad H_a : \pi_1 < \pi_2$$

Solution   Under the assumption of the null hypothesis, the expected values as given by the 2 × 2 table:

Expected counts

	Flu	No flu	Total
Shot	6.5625	8.4375	15
No shot	14.4375	18.5625	33
Total	21	27	48

Notice that, although there were only 3 students *observed* in the Shot/Flu cell, the *expected* count was 6.5625. Indeed, all of the expected counts are at least 5. The table satisfies the $E_{jk} \geq 5$ rule. The computation of the chi-square statistic:

$$\begin{aligned}
\chi^2 &= \sum \frac{(O-E)^2}{E} \\
&= \frac{(3-6.5625)^2}{6.5625} + \frac{(12-8.4375)^2}{8.4375} + \frac{(18-14.4375)^2}{14.4375} + \frac{(15-18.5625)^2}{18.5625} \\
&= 5.001
\end{aligned}$$

According to Table 6 on page 513, the χ^2 percentile is between 0.025 and 0.05. Because the alternative is one-sided, the P-value is half that for a two-sided alternative. We have $0.025/2 < P\text{-value} < 0.05/2$ or $0.0125 < P\text{-value} < 0.025$. Using technology, we obtain a χ^2 percentile of 0.02533 and $P\text{-value} = 0.02533/2 = 0.0127$.

The evidence is significant at $\alpha = 0.05$ that the risk of the flu is lower among college students who get the shot.

EXERCISES FOR SECTION 16.5

- **16.23** Test the hypothesis of Exercise 6.32 on page 192 using a one-sided χ^2 contingency table test.

	Breast cancer	No breast cancer	Row total
Raloxifen	22	5107	5129
Placebo	39	2537	2576
Column total	61	7644	7705

16.24 Test the hypothesis of Exercise 6.34 on page 193 using $\alpha = 0.05$ and a one-sided χ^2 contingency table test instead of Fisher's exact.

	Success	Failure	Row total
New drug	38	4	42
Standard treament	14	7	21
Column total	52	11	63

- **16.25** Repeat Exercise 6.40 on page 194 but this time use a one-sided χ^2 test at the $\alpha = 0.05$ level.

	Success	Failure	Row total
Acupuncture	7	6	13
Placebo	4	13	17
Column total	11	19	30

16.26 In a study comparing the severity of strokes with the use of aspirin, the following data were obtained by researcher Janet L. Wilterdink, M.D., at Brown University. Patients who had strokes were asked whether they had taken aspirin in the last week. A National Institute of Health scale was used to rate the severity of the stokes. Does aspirin reduce the risk of *severe* strokes? Do a one-sided χ^2 test at the 1% level of significance. Combine the Mild stroke and Moderate stroke columns in order to obtain a 2×2 table and $df = 1$.

	Mild stroke	Moderate stroke	Severe stroke	Total
Aspirin	256	204	49	509
No aspirin	329	324	113	766
Total	585	528	162	1275

- **16.27** Repeat the one-sided test of Exercise 16.26 above but this time test whether a higher percentage of the strokes are mild among the aspirin users.

16.6 COMPOUND HYPOTHESES (optional)

In Example 16.5 on page 454, we saw that the birth order of children could not be modeled by a fair coin toss. In retrospect, this was not surprising because, as was shown by John Arbuthnot in 1710 (see Example 6.5), male births are more frequent than female births, whereas for a fair coin the chances of heads and tails are equal. So we should have guessed that a better model would have been to model a sequence of births as a sequence of Bernoulli trials with π representing the probability of a female child. The value of π should be some value a little less than 0.50, but which one? It seems reasonable to use the data to estimate π and then to use this value to test for a goodness of fit. This is an example of a **compound hypothesis**.

If the null hypothesis is compound, then the expected cell counts must be estimated from the data. The chi-square statistic is computed as before but with the expected cell counts replaced by estimated cell counts.

Some care must be taken in estimating the expected cell counts. However, if we use the sample mean as an estimator of the population mean, and if the null hypothesis is true, then the sampling distribution of the chi-square statistic is chi-square with the degrees of freedom computed by the following rule.

> *Degrees of freedom =*
> *number of classes − number of estimated parameters − 1*

Example 16.10   *Birth order continued.* In Example 16.5 on page 454, we decided that the number of girls in families of 5 children cannot be modeled by the number of heads in 5 tosses of a fair coin; that is, with $\pi = 0.50$. Let us now consider whether a binomial distribution with $n = 5$ and some other value of the parameters π will fit the data. We estimate π by the proportion of girls in the 15,162 five-child families of Example 10.2 on page 272. There are 37,415 girls among the $5 \cdot 15162 = 75,810$ children. This gives us the us the estimate $\pi = 37415/75810 = 0.493526$. Let K denote the number of girls among the first 5 children. Using the data of Example 10.2, the follow table illustrates the computation of the chi-square statistic:

k	O	E	$(O-E)^2/E$
0	549	505.24	3.790
1	2476	2461.72	0.083
2	4753	4797.77	0.418
3	4621	4675.31	0.631
4	2245	2277.99	0.478
5	518	443.97	12.344
Totals	15162	15162	$\chi^2 = 17.744$

Solution The estimated expected values are computed as $15162 \times f_K(k)$ using the binomial probability mass function as given by Equation (5.1) on page 149 with $n = 5$ and $\pi = 0.493536$. Under the null hypothesis, the sampling distribution of the chi-square statistic is chi-square with $df = 6 - 1 - 1 = 4$ degrees of freedom. The P-value is $Pr(\chi^2 \geq 17.744)$. Using Table 6 on page 513, we see that $0.001 < P$-value < 0.01. Using technology, the actual value is P-value $= 0.00138464$. So again we would reject the null hypothesis. Under the null hypothesis, the expected value of the chi-square statistic is 4. Looking at the components of the chi-square statistic, it appears that there are too many 5-girl families and 5-boy families for this set to be described by a binomial probabilities. Perhaps the data set includes enough identical twins to make the independent trial assumption of the binomial probabilities untenable.

Example 16.11 *More on birth order.* In Example 16.10, we decided that the number of girls in families of 5 children cannot be modeled by a binomial distribution. Still there appears to be a certain symmetry to the birth-order data set. Given that there is one girl among the 5 children, the number of births seems to be uniformly distributed over the 5 possible permutations. For the 2476 one-girl families of Example 10.2 on page 272, the computation of the chi-square statistic for the null hypothesis that the 5 permutations are equally probable is illustrated in the following table:

Sequence	O	E	$(O-E)^2/E$
MMMMF	516	495.2	0.874
MMMFM	478	495.2	0.597
MMFMM	524	495.2	1.670
MFMMM	472	495.2	1.090
FMMMM	486	495.2	0.171
Totals	2476	2476	$\chi^2 = 4.404$

Now we test whether the 5 permutations are equally probable.

Solution Under the null hypothesis, the sampling distribution of this chi-square statistic follows the chi-square distribution of Table 6 on page 513 with 4 degrees of freedom. We have P-value $= Pr(\chi^2 \geq 4.404)$. By Table 6, P-value > 0.250, so the null hypothesis is retained.

Example 16.12 *Even more on birth order.* Continuing with the above example, given that there are 2 girls among the 5 children, let's test whether the number of births is uniformly distributed over the 10 possible permutations.

Solution For the 4753 two-girl families of Example 10.2 on page 272, we can construct a chi-square test that the number of births seems to be uniformly distributed over the 10 possible permutations. The computed value of the statistic is $\chi^2 = 8.862$ and P-value $= Pr(\chi^2 \geq 8.862) = 0.45$. So the null is retained.

Example 16.13 ***And more.*** To test the null hypothesis that, given the number of girls, all the possible birth order sequences are equally likely, we can combine these tests of uniformity. Adding the 6 individual chi-square statistics creates the combined chi-square statistic $0 + 4.40388 + 8.86198 + 11.8911 + 5.85746 + 0 = 31.0145$. Adding the degrees of freedom for these 6 individual chi-square statistics gives the degrees of freedom for the combined statistic, $df = 0+4+9+9+4+0 = 26$; so P-value $= Pr(\chi^2 \geq 31.0145) = 0.23$. The null hypothesis is retained. Given the number of girls among the first 5 children, the data are consistent with the hypothesis that all possible permutations of birth order are equally likely.

REVIEW EXERCISES FOR CHAPTER 16

16.28 Consider the blood types of the 40 children given in Exercise 2.1 on page 25. Are these blood types consistent with that of a random sample from the general U.S. population? Do a goodness-of-fit test at the $\alpha = 0.05$ level.

• **16.29** Consider the percentile rank BMI for 34 white, non-Hispanic girls from the fall of 1998 kindergarten class as given in Exercise 6.29. Divide the interval [0,100] into sixths. How many of the 34 scores do you expect to be in each of these intervals? Construct a chi-square goodness-of-fit test. State the null hypothesis, alternative hypothesis, P-value, and conclusion.

16.30 Consider the Salk vaccine trial date as given in Table 1.1 on page 12. Use the attribute *Treatment*, consisting of the two categories of Vaccine and Placebo, and the attribute *Response* consisting of the three categories of Paralytic & fatal polio, Nonparalytic polio, and No polio. Test for independence of attributes at the $\alpha = 0.01$ level. State your results in plain English.

• **16.31** Consider the ranking of facial wrinkles in smokers and nonsmokers as describes in Exercise 2.38 on page 67. Do a contingency table test of the data at the $\alpha = 0.05$ level of significance. Because there are so few scores of 5 or 6, combine these two into a single category. Be sure to state what you are testing and state your conclusion in plain English.

16.32 Repeat Exercise 16.22 on page 464 but test the hypothesis that the proportions are all 3 to 1. Construct 7 separate χ^2 statistics and combine them following the procedure of Example 16.11.

• **16.33** Repeat Exercise 6.14 on page 184 using a χ^2 test instead of the binomial exact test.

16.34 Repeat the analysis for a trend as done in Example 16.8 but include three years as given by the table below. Use $\alpha = 0.05$.

	No overcharges	One or two overcharges	Three or more overcharges	Total
1998	302	160	37	499
1996	185	83	32	300
1992	165	107	28	300
Total	652	350	97	1099

• **16.35** Use the data in Exercise 16.26 on page 466 and perform a two-sided contingency table test for an association between aspirin use and severity of strokes.

16.36 Consider the Rutherford and Geiger data as given in Exercise 9.43 on page 265. Perform a goodness-of-fit test to determine whether the numbers of alpha particles emitted follow the Poisson probabilities. Use a 5% level.

Chapter 17

RESAMPLING METHODS

17.1 OVERVIEW

17.2 PARAMETRIC BOOTSTRAPPING

17.3 NONPARAMETERIC BOOTSTRAPPING

17.4 PERMUTATION TESTS

17.5 COMPUTER-ASSISTED PARAMETRIC RESAMPLING

17.6 COMPUTER-ASSISTED NONPARAMETRIC RESAMPLING

17.1 OVERVIEW

Here we extend the bootstrapping idea further than in Section 13.4.

Suppose we have a random sample of size n from a population and wish to use it to estimate a parameter ω. The parameter ω might be the mean μ, the standard deviation σ, the proportion π, the correlation coefficient ρ, the median, or some other quantile of the population. From the random sample, we calculate a statistic W that is used to estimate the parameter ω. The statistic W may be the sample mean $\overline{X}$, sample standard deviations S, and so on. It is important to assess the quality of the estimator W by

- approximating the bias of our estimation procedure,
- assigning a standard error to our estimate, and
- finding a confidence interval for our parameter.

To do so, we need to know something about the sampling distribution of our estimator W. Instead of relying directly on normal z curves or Student's t curves to describe sampling distributions, the **bootstrap methods** that we discuss in this chapter, use computer simulations to approximate the distribution of W. They work as follows.

Suppose that the unknown population has a cumulative distribution F. The bootstrap method replaces the unknown population distribution F with a known **bootstrap distribution** F^* derived from the sample data.

There are three bootstrap methods we will consider: parametric, nonparametric, and permutation.

1. The **parametric bootstrap** method assumes that the distribution F is known, except for some unknown parameters.

   For example, it could be known that a population is exponential with a cumulative distribution function

   $$F(x) = \begin{cases} 1 - e^{-\frac{x}{\theta}} & : \quad \text{for } x > 0 \\ 0 & : \quad \text{for } x < 0 \end{cases}$$

   but the population mean θ is unknown. We can estimate the parameter θ by the sample mean $\bar{x}$ of the random sample of size n. Then the bootstrap distribution F^* would be the exponential distribution F with the unknown parameter replaced by its estimate $\theta^* = \bar{x}$. The sampling distribution of $\bar{X}$ can be approximated by computer simulations based on many samples of size n drawn from the (bootstrap) exponential population with mean $\theta^* = \bar{x}$. These samples are called **resamples**, and this procedure is called **parametric resampling**. Typically, the number of resamples R is 1000 or more. We could then use these resamples to estimate a 90% confidence interval for θ by examining the 5th and 95th percentiles of the $\bar{x}$'s obtained from these R resamples.

2. The **nonparametric bootstrap** method does not assume knowledge of the distribution F. Here, the distribution of F is replaced by F^*, the empirical distribution function of the sample data. So the resamples are obtained by drawing many samples of size n, with replacement, from the sample itself! This is called **nonparametric resampling**. A nonparametric bootstrap 90% confidence interval for the parameter ω could be obtained by taking the 5th and 95th percentiles of the statistic W among these resamples.

3. The **permutation bootstrap** method is a nonparametric bootstrap method applied to tests for differences of treatments. The experimenter randomly assigns the sample units to the various treatments. In **permutation resampling**, the treatment groups are pooled and resampling is done without replacement.

In resampling, we use computer simulations instead of looking up percentiles in tables. In certain situations, such as when estimating μ of a normal population, the parametric resampling method will give us essentially the same results as we would have obtain from the t table (Table 4 on page 510). This was called bootstrapping in Section 13.4 on page 358. In other situations, resampling methods will give us results that cannot be obtained otherwise.

17.2 PARAMETRIC BOOTSTRAPPING

Example 17.1 *Estimating a population mean.* It is known that explosives used in mining leave a crater that is circular in shape with a diameter that follows an exponential distribution:

$$F(x) = \begin{cases} 1 - e^{-\frac{x}{\theta}} & : \quad \text{for } x > 0 \\ 0 & : \quad \text{for } x < 0 \end{cases}$$

Recall from Chapter 9 that for an exponential random variable the mean is $\mu = \theta$. Suppose a new form of explosive is tested. The sample crater diameters (in centimeters) are as follows:

| 121 | 847 | 591 | 510 | 440 | 205 | 3110 | 142 | 65 | 1062 |
| 211 | 269 | 115 | 586 | 983 | 115 | 162 | 70 | 565 | 114 |

The sample mean is $\overline{x} = 514.15$, and the standard deviation is $s = 685.60$.

The 90% confidence interval for μ using the t curve with 19 degrees of freedom as described in Section 13.4 on page 363 is

$$90\% \text{ CI:} \quad \mu = \overline{x} \pm t_{.95} \frac{s}{\sqrt{n}} = 514.15 \pm 1.7291 \frac{685.60}{\sqrt{20}} \quad \text{or}$$

$$90\% \text{ CI:} \quad 249.07 \leq \mu \leq 779.23$$

It is inappropriate to use this confidence interval because it is based on the assumption that the parent population is normal. Examining the box plot of this sample (Exercise 17.1) shows that this assumption is in doubt. Thus, we have no reason to believe that the coverage for this method is truly 90%.

The parametric resampling method replaces the exponential population distribution F with unknown mean μ by the known exponential distribution F^* with mean $\mu^* = \overline{x} = 514.15$. Then resamples of size $n = 20$ are drawn from this surrogate population.

We took $R = 1000$ such samples or size $n = 20$ and computed the sample mean of each of the 1000 samples. Figure 17.1 shows a histogram of these results.

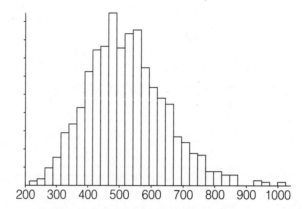

Figure 17.1 Histogram of the values of $\bar{x}$ for $R = 1000$ resamples of the 20 crater diameters of Example 17.1

Trimming off 5% from each tail gives us our bootstrap confidence interval. The 50th smallest (5th percentile) resample mean was 332.508, and the 951st one (95th percentile) was 726.450. This gives us our 90% confidence interval.

Parametric resampling 90% CI: $332.51 \leq \mu \leq 726.45$

Notice that the histogram is skewed to the right. This means that the confidence interval is not centered at the sample mean $\bar{x} = 514.15$.

As another example illustrating parametric bootstrap methods, suppose we use the sample standard deviation s to estimate the standard deviation σ of a normal population when both μ and σ are unknown. Because μ and σ are unknown, we cannot draw samples from this normal population, but we can easily estimate these parameters from our sample data and draw samples from a surrogate normal population with mean $\mu^* = \bar{x}$ and standard deviation $\sigma^* = s$.

Example 17.2 *Estimating a population standard deviation.* Consider a random sample of 20 scores:

107.853	109.640	93.927	104.021	118.092
67.698	111.109	101.384	124.980	102.511
118.291	96.477	117.134	110.180	110.413
80.494	109.089	98.480	96.813	113.998

The sample mean and standard deviation are $\bar{x} = 104.6292$ and $s = 13.4351$, respectively.

Draw a box plot for these data (Exercise 17.3). What do you find?

If we think the population follows the normal curve but do not know the population mean or standard deviation, we resample by replacing the unknown population distribution by a normal distribution with mean $\mu^* = 104.6292$ and standard deviation $\sigma^* = 13.4351$. We then draw random samples of size 20 with replacement from this surrogate population.

We could study the resampling distribution S^* (just as we studied the resamples $\overline{X}^*$ in the previous example). However, here we choose to use the ratio s^*/σ^* because the distribution of $W = S/\sigma$ does not depend on the unknown σ (See Exercise 17.4).

Compute $w^* = s^*/13.4351$, the ratio of the sample standard deviation s^* to the estimated population standard deviation $\sigma^* = 13.4351$. The difference between this ratio and 1 is the **relative sampling error** for our simulation experiment.

We should replicate this simulation experiment many times. In this way, we hope to learn about the sampling distribution of $W = S/\sigma$.

The results of replicating this sampling distribution $R = 5000$ times are shown in Figure 17.2.

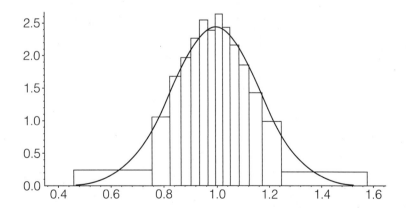

Figure 17.2 The values of $W^* = s^*/13.4351$ for $R = 5000$ resamples of the 20 scores of Example 17.2 with normal curve superimposed. All of the rectangles have the same areas because the classes were chosen to have approximately the same number of resamples.

Parametric resampling confidence interval

We can construct an approximate 90% confidence interval for σ by sorting the $R = 5000$ ratios and choosing the 5th percentile (the 250th smallest of the simulated sampling errors is 0.7306) and the 95th percentile (the 250th largest of the simulated sampling errors is 1.2512). So

$$Pr(0.7306 \leq \frac{S}{\sigma} \leq 1.2512) \approx 0.9$$

Solve these inequalities for σ to find

$$Pr(\frac{S}{1.2512} \leq \sigma \leq \frac{S}{0.7306}) \approx 0.9$$

Replacing S by the observed standard deviation $s = 13.4351$ gives the lower bound $13.4351/1.2512 = 10.74$ and the upper bound $13.4351/.7306 = 18.39$ of our 90% parametric bootstrap confidence interval for σ.[1]

Parametric resampling 90% CI: $10.74 \leq \sigma \leq 18.39$

Relative sampling error

The mean of these $R = 5000$ values turned out to be $\overline{W^*} = 0.98496$. Consequently, $0.98496 - 1 = -0.015$ is the approximate mean of the relative sampling error.

Bias estimate

Recall that the sample standard deviation s underestimates the population standard deviation σ on average (see Exercise 11.57). As discussed in Section 11.9, the bias of the estimator S is

$$\beta_S = E(S - \sigma) = E(S) - \sigma$$

---

[1] We did not actually have to draw samples from a normal population in this case. It can be shown that the sampling statistic

$$\frac{(n-1)S^2}{\sigma^2}$$

has a chi-square distribution with $n - 1$ degrees of freedom. So a 90% confidence interval for σ can be computed using Table 6 on page 513 and the formula

$$\sqrt{\frac{n-1}{\chi^2_{.95}}} (S) \leq \sigma \leq \sqrt{\frac{n-1}{\chi^2_{.05}}} (S)$$

So $\sqrt{\frac{19}{30.1435}} (13.4351) \leq \sigma \leq \sqrt{\frac{19}{10.1170}} (13.4351)$

which computes to $[10.67, 18.41]$. This is only slightly different from our approximate interval $[10.74, 18.39]$.

17.2 PARAMETRIC BOOTSTRAPPING

To make a bootstrap approximation of this bias, we resample from our surrogate normal population and compute S^* the sample standard deviation. The bootstrap random sampling error is

$$S^* - \sigma^*$$

So the bootstrap estimate of the bias is

$$\beta_{S^*} = E(S^* - \sigma^*) = E(S^*) - \sigma^*$$

With many resamples, the approximate bootstrap estimate of the bias is the average of the bootstrap random sampling errors.

Estimate of bias   Because the mean of our $R = 5000$ values turned out to be $\overline{W^*} = 0.98496$, our estimate of the bias is -0.202. The calculations follow:

$$E(S^* - \sigma^*) \approx \overline{S^*} - \sigma^* = (\overline{W^*} - 1)\sigma^* = (0.98496 - 1)(13.4351) = -0.202$$

So we have evidence that, on average, the sample standard deviation slightly underestimates the population standard deviation. The standard deviation of these $R = 5000$ ratios turned out to be 0.1574. Thus, our estimate of the standard deviation of the sampling distribution of S is

$$\text{SD} = (0.1574)(13.4351) = 2.114$$

We can approximate the root mean squared error of our estimate (using Formula 11.9 from page 323 in Section 11.9) by

$$\sqrt{\text{MSE}} = \sqrt{bias^2 + \text{SD}^2} = \sqrt{(0.202)^2 + (2.114)^2} = 2.12$$

> *So as an estimate of the population σ, we expect our sample s to be off the mark by approximately 2 or so.*

EXERCISES FOR SECTION 17.2

Comment Some of these exercises require the use of a computer. You may find Section 17.5 on Computer-Assisted Resampling useful for these exercises.

17.1 Make a box plot of the data of Example 17.1. Describe the skew and outliers.

17.2 Repeat Example 17.1 for the data below. Assume that the data came from a sample of size $n = 20$ drawn from an exponential population with an unknown mean. Find a parametric bootstrap 90% confidence interval for the population mean. The sample mean is $\bar{x} = 108$ and the standard deviation is $s = 110.6$.

| 33 | 107 | 329 | 4 | 13 | 26 | 374 | 118 | 147 | 30 |
| 30 | 196 | 141 | 56 | 168 | 2 | 12 | 39 | 69 | 266 |

• **17.3** Find the box plot for the data in Example 17.2 on page 474. Discuss skewness and outlying values.

17.4 Recall $s = \sqrt{\dfrac{\sum_{i=1}^{n}(X_i - \overline{X})^2}{n-1}}$.

Show $\dfrac{s}{\sigma} = \sqrt{\sum_{i=1}^{n} \dfrac{(Y_i - \bar{Y})^2}{n-1}}$ where $Y_i = \dfrac{X_i - \mu_X}{\sigma}$

For example, if X is a normal random variable with mean μ and standard deviation σ, then Y is the standard normal random variable. Because the statistic Y does not depend on either μ or σ, neither does the statistic S/σ, which is a function of the Y's.

• **17.5** The following data were collected by Charles Darwin (1809–1882)[2]. Pairs of plants, one cross-fertilized and one self-fertilized at the same time, and whose parents were grown from the same seed, were planted and grown in the same pot. The differences in height (measured in eighths of an inch) of 15 pairs of plants (cross-fertilized minus self-fertilized *Zea mays*) raised by Charles Darwin follow:

49 −67 8 16 6 23 28 41 14 29 56 24 75 60 −48

This is a paired-difference experiment. Assume the differences are normally distributed. Use the method described in Section 13.4 to construct a 98% confidence interval for the expected difference.

17.6 Use the Darwin data of Exercise 17.5 above. Construct a 90% parametric bootstrap confidence interval for the standard deviation. Assume the differences follow a normal curve.

---

[2] A description of the data set can be found at the website
http://lib.stat.cmu.edu/DASL/Stories/student.html.

17.3 NONPARAMETERIC BOOTSTRAPPING

The above procedure is an example of parametric bootstrapping. If we did not want to assume that we were sampling from a normal population or some other population with a specified shape, then we must extract all the information about the population from the sample itself. We bootstrap a sampling distribution for our estimator by drawing samples with replacement from our original data. This is called **nonparametric bootstrapping**.

Example 17.3   *Estimating a population mean.*   Again consider the random sample of 20 scores of Example 17.2:

107.853	109.640	93.927	104.021	118.092
67.698	111.109	101.384	124.980	102.511
118.291	96.477	117.134	110.180	110.413
80.494	109.089	98.480	96.813	113.998

Recall that the sample mean and standard deviation were $\bar{x} = 104.6292$ and $s = 13.4351$, respectively. The box plot of this data (Exercise 17.3) is skewed to the left. For this reason, we choose here not to make the normality assumption of Example 17.2.

We suppose this is a random sample from a population with unknown distribution, and we want to estimate the population mean μ by using the sample mean $\bar{x}$. In order to know something about the quality of our estimate, we need to know something about the sampling distribution of the sample mean. However, we do not know the population distribution, so we use the sample to estimate it.

> *Because our data consist of* $n = 20$ *observations, we estimate the sampling distribution of the sample mean by the sampling distribution of the sample mean based on random samples (with replacement) of size* $n = 20$ *from the data itself!*

Of course, it is very tedious to compute this sampling distribution directly, so we approximate it by drawing many random samples of size $n = 20$ from the data.

To estimate the sampling error, we compute the difference between the resample mean $\overline{x^*}$ and the mean of the data $\mu^* = \bar{x}$.

Bootstrap-t confidence interval

In order to compute confidence intervals for the population mean, we approximate the sampling distribution of a T statistic, as defined in Section 11.8, by taking many random samples with replacement of size $n = 20$ from the data and computing a t statistic for each of them:

$$T = \sqrt{n}\left(\frac{\overline{X} - \mu}{S}\right) = \sqrt{20}\left(\frac{\overline{X} - 104.6292}{S}\right)$$

A confidence interval based on such a T statistic we call a **bootstrap-t confidence interval**.

The result of $R = 5000$ such samples is shown in Figure 17.3.

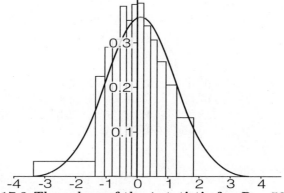

Figure 17.3 The values of the t statistic for $R = 5000$ resamples of the 20 scores of Example 17.2

These t statistics ranged in value from -3.34 to 6.86 with a mean of 0.1353 and a standard deviation of 1.1120. For comparison, a normal curve with the same mean and standard deviation is also shown. Notice that the t statistics are slightly skewed to the right. To find an approximate two-sided 90% confidence interval for the population mean, we follow the same procedure as described in Sections 13.2 and 13.4, except we use quantiles from the $R = 5000$ sampled t statistics in place of the corresponding quantiles of the t curve with 19 degrees of freedom:

$$\begin{aligned}
0.90 &= Pr\left(t_{.05} \leq T \leq t_{.95}\right) \\
&= Pr\left(t_{.05} \leq \sqrt{n}\left(\frac{\overline{X} - \mu}{S}\right) \leq t_{.95}\right) \\
&= Pr\left(\overline{X} - t_{.95}\left(\frac{S}{\sqrt{n}}\right) \leq \mu \leq \overline{X} - t_{.05}\left(\frac{S}{\sqrt{n}}\right)\right)
\end{aligned}$$

Because the 5th and 95th percentiles of the $R = 5000$ sampled T statistics are -1.477 and 2.1022, respectively, the approximate nonparametric bootstrap 90% two-sided confidence interval for μ is

$$104.6292 - 2.102237 \left(\frac{13.4351}{\sqrt{20}}\right) \leq \mu \leq 104.6292 - (-1.456744) \left(\frac{13.4351}{\sqrt{20}}\right)$$

or

$$\text{Nonparametric boostrap-}t\ 90\%\ \text{CI: } 98.31 \leq \mu \leq 109.07$$

If we had believed the sample was from a normal population and used the parametric bootstrap (that is, if we had resampled from a normal population with mean $\mu^* = 104.6292$ and standard deviation $\sigma^* = 13.4351$), then our resulting confidence interval would have been approximately the bootstrapping confidence interval we found in Chapter 13 using the Student's t curve with 19 degrees of freedom. Table 4 on page 510 gives us the 5th and 95th percentiles to be -1.7291 and 1.7291, respectively. So the parametric bootstrap 90% two-sided confidence interval for μ is

$$\text{Parametric bootstrap-}t\ 90\%\ \text{CI: } 99.44 \leq \mu \leq 109.82$$

just as would have been obtained using the procedure described in Section 13.4 on page 358.[3]

What is the bias?

We took $R = 5000$ random samples and each time computed this difference $\overline{x}^* - \overline{x}$. We found the average of these differences to be 0.021. This is our resampling approximation. Specifically, it is an approximation of the nonparametric bootstrap estimate of the bias of the sample mean as an estimate of the population mean. It is close to zero. Actually, we know from Section 11.9 that

$$E(\overline{X}) = \mu$$

and hence the bias is exactly zero in this case. The slight error in our approximation was caused because we looked at $R = 5000$ resamples instead of all possible random samples of size $n = 20$ from this data set.

What is the SD of $\overline{X}$?

We found the standard deviation of the 5000 differences to be 2.931. Again we know from Section 11.9 that

$$\text{Var}(\overline{X}) = \frac{\sigma^2}{n}$$

so the standard deviation of these sample means is exactly

$$\frac{13.0949}{\sqrt{20}} = 2.928$$

---

[3]You have already seen some simulation results for t statistics in Chapter 11 and are probably not too surprised that the nonparametric and parametric confidence interval methods produced similar results, even though the sample size is smaller here.

Example 17.4 *Estimating a population standard deviation.* This is a continuation of Example 17.2. Again we plan to estimate the population standard deviation σ by the sample standard deviation s, but this time we are unwilling to assume the population is normally distributed. We use the nonparametric bootstrap approach introduced in Example 17.3. The information we use about the population is contained in our data. It is a random sample of $n = 20$ observations.

> We estimate the sampling distribution of S/σ by the sampling distribution of the S^*/σ^* based on random samples (with replacement) of size $n = 20$ from the data itself!

Solution Here S^* denotes the sample standard deviation of a random sample from the data, and σ^* is the population standard deviation of the $n = 20$ observations. Recall the formula

$$\sigma = \sqrt{\frac{1}{N} \sum_{i=1}^{N} (x_i - \mu)^2}$$

from page 47 in Section 2.11. For the present data set, we have $\sigma^* = 13.0949$. The results for $R = 5000$ such samples are shown in Figure 17.4.

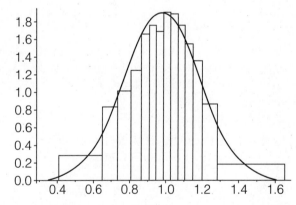

Figure 17.4 The values of s^*/σ^* for $R = 5000$ resamples of the 20 scores of Example 17.2 with normal curve superimposed

These $s^*/\sigma^* = s^*/13.0949$ statistics ranged in value from 0.4066 to 1.6509 with a mean of 0.979 and a standard deviation of 0.210. For comparison, a normal curve with the same mean and standard deviation is also shown in the figure above.

The mean relative error is $0.97887 - 1 = -0.021$, and our approximation of the bias is $(0.97887 - 1)(13.0949) = -0.277$. This means that we have evidence that, on average, the sample standard deviation slightly underestimates the population standard deviation. The standard deviation of these $R = 5000$ ratios turned out to be 0.21029. So $(0.21029)(13.0949) = 2.753$ is our estimate of the standard deviation of the sampling distribution of s. We can approximate the root mean squared error of our estimate (using Formula 11.9 on page 323 in Section 11.9) by

$$\sqrt{\text{MSE}} = \sqrt{bias^2 + \text{SD}^2} = \sqrt{(-0.2767)^2 + (2.7537)^2} = 2.7676$$

As an estimate of the population σ, we expect our sample s to be off the mark by approximatley 2.8 or so.

We can also find a 90% confidence interval as was done in Example 17.2. The 5th and 95th percentile of the $R = 5000$ s^*/σ^* ratios are 0.61583 and 1.31841, respectively. Inverting this interval gives

$$[13.4351/1.3184, 13.4351/0.61583] \text{ or}$$

Nonparametric resampling 90% CI: $10.19 \leq \sigma \leq 21.82$

How well do these bootstrap methods work?

The sample given in Example 17.2 was actually a random sample of 20 drawn from a normal population with mean 100 and standard deviation 15. Using methods beyond the scope of this book it can be shown that in this situation the bias of the sample standard deviation is -0.1960. So both bootstrap approaches worked well here.

To better illustrate how well these methods work, we drew 1000 samples from a normal population with mean $\mu = 100$ and standard deviation $\sigma = 15$. We estimated the population standard deviation by the sample standard deviation and then computed bootstrap estimates of the bias and standard error (that is, the standard deviation of the sampling distribution) of this estimate.

The results are shown in the following tables. We also constructed bootstrap 90% confidence intervals. The coverage for the parametric bootstrap confidence intervals was 88.9%, but for the nonparametric bootstrap it was only 82.0%. The nonparametric bootstrap would have performed better if the sample size had been much larger than 20. And the parametric bootstrap could behave much more poorly if the population did not closely follow the normal curve!

Descriptive statistics for parametric and nonparametric bootstrap

Parametric bootstrap

Variable	Mean	Median	Standard deviation
Lower bound 90	11.023	10.940	1.762
Upper bound 90	18.904	18.817	3.016
Bias	−0.191	−0.187	−0.190
SE	2.397	2.390	0.384

Coverage for parametric bootstrap = 0.889

Nonparametric bootstrap

Variable	Mean	Median	Standard deviation
Lower bound 90	11.956	11.919	2.031
Upper bound 90	19.150	19.095	3.342
Bias	−0.550	−0.535	0.132
SE	2.191	2.107	0.601

Coverage for nonparametric bootstrap = 0.820

Example 17.5   *Estimating a population correlation coefficient.* The correlation coefficient ρ is defined in Section 10.3. It measures how closely two variables are to being linearly related. Suppose (X_1, Y_1), (X_2, Y_2), ...,(X_n, Y_n) is a simple random sample of n observations from a bivariate population. The sample correlation coefficient r is defined to be the correlation coefficient ρ of X and Y when equal probability is assigned to each of the n points in the sample. (This definition is consistent with the computational formula for r presented later in Theorem 15.1 .)

We can estimate the population's correlation coefficient by the sample correlation coefficient. We can approximate the bias of r as an estimate of ρ by bootstrapping.

We will compare two dimensions of flowers of the iris species *I. Virginica*, the petal length X and the sepal length Y. It is anticipated that larger blossoms are larger in all dimensions, and so X and Y are positively correlated. A sample of 50 observations is shown below. These 50 data points are part of a larger data set collected by R. A. Fisher (and available online as a DASL[4] data set called Fisher's Irises).

---

[4] http://lib.stat.cmu.edu/DASL/Stories/Fisher'sIrises.html

x	y	x	y	x	y	x	y
45	49	51	63	56	63	60	63
48	60	51	65	56	63	60	72
48	62	51	69	56	64	61	72
49	56	52	65	56	64	61	74
49	61	52	67	56	67	61	77
49	63	53	64	57	67	63	73
50	57	53	64	57	67	64	79
50	60	54	62	57	69	66	76
50	63	54	69	58	65	67	77
51	58	55	64	58	67	67	77
51	58	55	65	58	72	69	77
51	58	55	68	59	68		
51	59	56	61	59	71		

The scatter plot of these data is shown in Figure 17.5.

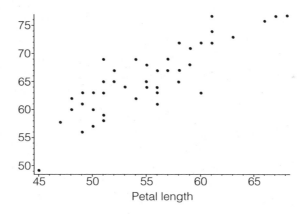

Figure 17.5 Scatter plot of the above data

The sample correlation coefficient is $r = 0.8642$. The histogram in Figure 17.6 shows the distribution of r^* based on $R = 2000$ resamples. It has a mean of 0.8613. Notice that it is skewed to the left. The approximate bias of the sample correlation coefficient as an estimate of the population's correlation coefficient is $0.8613 - 0.8642 = -0.0029$.

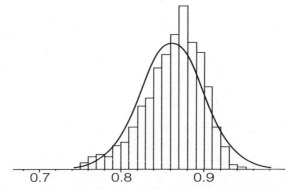

Figure 17.6 Histogram of r^* with normal curve superimposed

Solution We would like to construct a 90% confidence interval for ρ. We will base the interval on the sampling distribution of

$$T = \sqrt{n}\frac{r-\rho}{1-r^2}$$

instead of the sampling distribution of r because the sampling distribution of T is less dependent on ρ. The standard deviation of T will be close to 1 for large samples from populations that are approximately normal.

A bar graph of our $R = 2000$ resamples of

$$t^* = \sqrt{n}\frac{r^*-r}{1-r^{*2}}$$

is shown in Figure 17.7. Notice that it is skewed to the right.

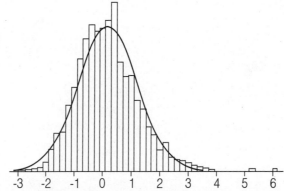

Figure 17.7 Histogram of t^* with normal curve superimposed

We approximate the 5th and 95th percentile of T by the 100th smallest and 100th largest of these T^* values. Then we invert the interval

$$t_{.05} \leq \sqrt{n}\frac{r-\rho}{1-r^2} \leq t_{.95}$$

to obtain the approximate 90% confidence interval

$$r - t^*_{.95}\frac{1-r^2}{\sqrt{n}} \leq \rho \leq r - t^*_{.05}\frac{1-r^2}{\sqrt{n}}$$

The 90% confidence interval is

Nonparametric resampling-t 90% CI: $0.79 \leq \rho \leq 0.91$

EXERCISES FOR SECTION 17.3

Comment   Some of these exercises require the use of a computer. You may find Section 17.5 on Computer-Assisted Resampling useful for these exercises.

- **17.7** Use the data in Example 17.2 on page 474. Construct an approximate 98% nonparametric bootstrap-t confidence interval for the population's median score. Base your estimate on your bootstrap approximation to the sampling distribution of the statistic

$$T = \frac{\text{med}_{\text{sample}} - \text{med}_{\text{population}}}{\text{MAD}_{\text{population}}}$$

Refer to Section 2.11 for a description of MAD, the mean absolute deviation from the median:

$$\text{MAD} = \frac{1}{n}\sum_{i=1}^{n} |x_i - \text{med}|$$

17.8 Fisher's iris data[5] includes measurements of petal width x and sepal width y for the species *I. Virginica*. Construct an approximate 90% nonparametric bootstrap confidence interval for the correlation coefficient ρ between petal width and sepal width.

---

[5]This data set can be found at the website http://lib.stat.cmu.edu/DASL/Stories/Fisher'sIrises.html

x	y	x	y	x	y	x	y
17	25	15	28	18	29	25	33
18	30	20	32	24	34	18	32
18	28	23	31	21	28	25	36
20	28	20	30	22	28	19	28
18	30	13	30	24	31	23	30
18	27	19	27	21	33	18	29
20	25	23	32	25	33	20	38
15	22	23	34	23	32	21	30
19	25	21	31	22	30	20	28
19	27	18	31	18	25	22	38
19	27	18	30	16	30	23	26
24	28	21	30	23	32		
18	30	14	26	21	30		

17.4 PERMUTATION TESTS

Finally, we describe a simple procedure for comparing treatments in randomized controlled experiments when the sample sizes are very small and you are not sure you are sampling from normal data.

Example 17.6   *Gopain dosage* The following discussion is based on an example by the well-known biostatistician D. J. Finney described in *Statistics for Biologists*, Chapman & Hall Ltd. (1980). He imagines that 12 laboratory mice are randomly assigned to two groups of size 7 and 5, respectively. The first group receives the standard dosage of the drug Gopain while the second group receives double dosage. It is conjectured that the higher dosage will reduce the concentration of salt in the blood. After 12 hours, the concentration of salt in the blood is measured with the following results:

Dosage	Data	Sum	Mean
Standard	14 8 3 20 18 9 12	84	$84/7 = 12$
Double	4 10 7 1 8	30	$30/5 = 6$

Although double dosage of Gopain reduced the concentration of salt in the blood in the samples, when investigating a scientific phenomena we should remain careful in making inferences about the populations. Would similar results be obtained if the experiment were done again under identical conditions? Or can the observed differences between the two groups be attributed solely to the variability in response of individual mice?

Solution

We formulate a null hypothesis

H_0: The difference in dosages of Gopain does not effect the concentration of salt in the blood of laboratory mice.

against a one-sided alternative

H_a: Double dosage of Gopain reduces effect the concentration of salt in the blood of laboratory mice.

If the null hypothesis is true, then the 12 levels of concentration of salt in the blood do not depend on the treatment the mouse received, and the splitting of these 12 measurements into groups of size 7 and 5 occurred by an act of randomization; that is, each of the 792 ways of assigning the 12 measurements to the two groups is equally likely.

What is unusual about the actual experimental data is that the average for the second group is much smaller than the average for the first. Only 33 of these 792 ways of dividing up the 12 measurements into 2 groups gives a sum for the second group of 30 or less (and so an average of 6 or less; hence, a difference in averages of 6 or more). Thus, when the null hypothesis is true, the probability of getting a result at least as extreme as the observed result is P-value $= 33/792 = 0.042$. A sufficiently small P-value would justify the rejection of the null hypothesis (in favor of an alternative hypothesis that the increased dosage reduces the concentration of salt in the blood).

In fact, it is tedious to go through the list of 792 possible groupings of the 12 measurements. Standard statistical packages allow us to easily extract random samples from this list and compute the necessary statistics. We can use this data *resampling* method to estimate the P-value.

One simulation run of $R = 1000$ randomly selected samples of 5 of these 12 measurements, without replacement, resulted in 42 that had sums no greater than 30. In this case, the P-value would be estimated by $42/1000 = 0.042$. Of course, using methods developed in Section 13.6, the estimated standard error of this estimate is

$$\sqrt{(0.042)(0.958)/1000} = 0.0063$$

Thus, a 95% confidence interval for the P-value is

$$95\% \text{ CI:} \quad P\text{-value} = 0.042 \pm 1.96 \times 0.0063 = 0.0420 \pm 0.0124$$

so it is not surprising that the estimated P-value is close to the permutation test P-value.

490   CHAPTER 17.   RESAMPLING METHODS

A stem plot of these 1000 sums shows that the simulated sampling distribution is bell-shaped. This suggests that the normal theory test might be appropriate to approximate the P-value.

The usual two-sample t test (assuming normally distributed populations with equal population variances) results in an observed test statistic $t_s = 2.01$, which results in a one-sided P-value $= 0.036$ based on 10 degrees of freedom.

> *A resampling approach to hypothesis testing is appropriate when:*
>
> - *randomization was used in the assignment of subjects to treatment groups;*
>
> - *the sample sizes are small; and*
>
> - *the assumption that the populations are approximately normal may not be warranted.*

EXERCISES FOR SECTION 17.4

Comment   Some of these exercises require the use of a computer. You may find Section 17.5 on Computer-Assisted Resampling useful for these exercises.

- **17.9**   Assume the mice in the experiment described in Exercise 14.19 had been randomly assigned to the two groups. Use resampling methods to construct a P-value of a test to determine whether the vitamin treatment results in higher weight gain.

Weight increase for two groups of mice

Control group	Treatment group
12	18
19	16
14	23
	20
	23

17.10   In order to test the endurance of two exterior latex paints, a consumer group divided a cedar board into eight parts and randomly assigned four parts to each paint. They checked the painted board each month and recorded how long each part lasted. These are the data (in months):

Paint A	4	45	59	50	16
Paint B	38	19	1	19	11

Use permutation methods to test whether the paints had different lasting times.

17.5 COMPUTER-ASSISTED PARAMETRIC RESAMPLING

Here we describe how to use Minitab, Excel, and Maple, to draw parametric bootstrap resamples from the sample of 20 scores of Example 17.2 on page 474:

107.853	109.640	93.927	104.021	118.092
67.698	111.109	101.384	124.980	102.511
118.291	96.477	117.134	110.180	110.413
80.494	109.089	98.480	96.813	113.998

- We assume the sample of 20 scores is drawn from a normal population with unknown mean and variance.

- We estimate the mean and standard deviation by the sample mean and standard deviation.

- Then we generate random samples from a normal population with this mean and variance.

- These resamples are used to study the sampling distribution of the sample standard deviation.

MINITAB Comments below are set off to the right with the symbol #.

1. Place the 20 scores of Example 17.2 in column **C1**.

2. If you have the Student Version of Minitab, you are limited to 5000 entries in you worksheet. We show how to generate 400 samples of size 20 from column **C1**, 200 samples at a time. On the other hand, if you have the full version of Minitab, all 400 samples can be generated at the same time.

    Choose **Calc** → **Calculator**. Enter the constant **K1** in the "Store result in variable:" box, and select the expression **MEAN(C1)**.

    Similarly, store the expression **STDEV(C1)** in the constant **K2** .

    Choose **Calc** → **Random Data** → **Normal** and fill in the menu with
    ```
    Generate 200 rows of data     # 400 for full version
    Store in column(s) C3-C22
    Mean K1
    Standard deviation K2
    ```

3. In order to compute all 200 (or 400 for full version) sample standard deviations and store them in **C23** (or **C2** for full version), choose **Calc** → **Row Statistics** and fill in the menu with

> Statistic standard deviation
> Input variables: C3-C22
> Store results in C23   # C2 for full version

4. For the Student Version, generate another 200 samples of size 20 and store all 400 sample standard deviations in **C2** as follows.

    Choose **Manip** → **Copy Columns...** and fill in the menu with
    > Copy from C23
    > To Columns C2

    Repeat steps **2** and **3** above.

    Then choose **Manip** → **Stack Columns** and fill in the menu with
    > Stack from the following columns: C23 C2
    > Store the stacked data in: C2

5. To save space on your worksheet, you may choose to erase unneeded columns. Choose **Manip** → **Erase Variables...** and fill in the menu with
    > Columns and constants to erase: C3-C23

6. After you have generated the sample standard deviations for many random samples, you can use them as a proxy for the sampling distribution of the sample standard deviation.

Macro You can automate and speed up this procedure by using a Minitab macro to generate many random samples from a normal population and store their standard deviations in a column on the Minitab worksheet.

Suppose you have stored the 20 scores in C1 and want to store the standard deviations of 4000 resamples in C2:

Choose **Edit** → **Command line editor** and enter the line
> %bootsSDp.MAC C1 20 4000 C2

Here **bootsSDp.MAC** is the name of a file containing the macro that you have stored (as a text file with line breaks) in your Minitab Macro folder. The file begins with MACRO and ends with ENDMACRO as shown below. This macro is for the parametric bootstrapping of sample standard deviations. It assumes that parent population is normal. As usual, comments are set off to the right with the symbol #.

17.5 COMPUTER-ASSISTED PARAMETRIC RESAMPLING

```
MACRO
BootsSDp OBSERVED T R SD
                    # BootsSDp is the name of the macro
                    # OBSERVED is the column containing the original data
                    # T = sample size
                    # R =number of random bootstrap samples desired.
                    # SD is the column that will contain the SDs of resamples
MCONSTANT T R k.1-k.2
MCOLUMN OBSERVED SD C.1-C.T
Let k.1 = MEAN(OBSERVED)
Let k.2 = STDEV(OBSERVED)
Random R C.1-C.T;
Normal k.1 k.2.
RStDev C.1-C.T SD.
ENDMACRO
```

EXCEL Be sure you have loaded the Analysis Tool Pak as described in the Notes on Technology on page 41.

1. Place the 20 scores into cells **A3..A22**.

2. Compute the sample mean and standard deviation of these 20 scores by placing the formulas =**AVERAGE(A3:A22)** in cell **B1** and =**STDEV(A3:A22)** in cell **B2**.

3. Generate a random sample from a normal population with this mean and variance by placing the following formula in cell **D2**:
    =**NORMINV(RAND(),$B$1,$B$2)**

    This is the key step—placing a randomly generate normal variate into cell **D2**. It represents the first value in the first sample.

4. Select **D2..W2** and fill right.

    This completes the first sample of size 20 from the original 20 data points.

5. Select **D2..W2** through **D1001..W1001** and fill down.

    This completes the generation of 1000 samples of size 20 from the normal population with the same mean and standard deviation as the original 20 scores.

6. Copy **D2..W2** through **D1001..W1001** and paste special-values only.

    This prevents the 1000 resamples of size 20 from being recalculated.

7. Place the function =**STDEV(D2:W2)** in cell **C2**.

    This computes the standard deviation of the first sample.

8. Select **C2..C1001** and fill down to compute the standard deviations of all 1000 resamples.

9. To compute the S^*/σ^* values into cells **X2..X1001**, put the function =C2/$B$2 in cell **X2**, then select **X2..X1001** and fill down.

10. Finally, copy **X2..X1001** and paste special-values only and then sort **X2..X1001**.

Bootstrap confidence interval

To compute the lower and upper bounds for an approximate 90% parametric bootstrap confidence interval for the population standard deviation σ, compute the following functions:

=STDEV(A3:A22)/X952

and

=STDEV(A3:A22)/X50

Normal theory confidence interval

To compute the lower and upper bounds for a 90% normal theory confidence interval for the population standard deviation σ, compute the following functions.

=B2/SQRT(CHIINV(0.05,19)/19)

and

=B2/SQRT(CHIINV(0.95,19)/19)

Bias

To approximate the bias, compute the difference between the mean of the 1000 SDs and the mean of the scores:

= AVERAGE(C2:C1001) − AVERAGE(A3:A22)

MAPLE

This Maple session below shows how to generate a random sample from a normal population with the same mean and standard deviation as the original data. This session generates the standard deviations for many such samples. These are used to construct confidence intervals.

- The computer cursor is indicated by the symbol >

- Lines that are comments, not read by the software, are indicated by the symbol #

- Maple output is indented.

```
>   with(stats):with(stats[statplots]):
      SampleData := [107.853, 109.640, 93.927, 104.021, 118.092, 67.698, 111.109, 101.384,
        124.980, 102.511, 118.291, 96.477, 117.134, 110.180, 110.413, 80.494, 109.089,
        98.480, 96.813, 113.998]:
>   SampleMean := describe[mean](SampleData);
      SampleMean := 104.6292000
>   SampleSD :=describe[standarddeviation[1]](SampleData);
      SampleSD := 13.43511525
>   stats[random,normald[SampleMean,SampleSD]](20);
      90.13007422, 104.3334307, 69.89568723, 98.67379926, 91.14999768, 104.2547668,
      125.1344756, 96.49933470, 106.8331130, 113.4026626, 97.36007327, 133.3135182,
      107.1069561, 96.34471253, 98.60089601, 115.6973711, 108.0166363, 107.2064598,
      115.7532588, 81.22164029
>   describe[standarddeviation[1]]([stats[random,normald[SampleMean,SampleSD]](20)]);
      11.88409943
>     # the SDs for 1000 samples are easily computed
>   SampleSDs := seq(describe[standarddeviation[1]]
              ([stats[random,normald[SampleMean,SampleSD]](20)]), j = 1 .. 1000):
>     # the Z* = sigma*/s* statistics are computed and sorted. The 5th and 95Th
      percentiles are computed and compared to the corresponding normal theory
      percentiles.
>   SortedZstars := sort([SampleSDs])/SampleSD:[SortedZstars[50],SortedZstars[951]];
      [.7265369364, 1.266641748]
>   [sqrt(stats[statevalf,icdf,chisquare[19]](.05)/19),
              sqrt(stats[statevalf,icdf,chisquare[19]](.95)/19)];
      [.7297084155, 1.259563985]
>     # parametric bootstrap 90% confidence interval for sigma
>   [SampleSD/SortedZstars[951],SampleSD/SortedZstars[50]];
      [10.60687860, 18.49199205]
>     # Normal theory 90% confidence interval for sigma;
>   [SampleSD/sqrt(stats[statevalf,icdf,chisquare[19]](.95)/19),
              SampleSD/sqrt(stats[statevalf,icdf,chisquare[19]](.05)/19)];
      [10.66648095, 18.41162163]
```

17.6 COMPUTER-ASSISTED NONPARAMETRIC RESAMPLING

Now we describe how to use Minitab, Excel, and Maple to draw nonparametric bootstrap resamples from the same sample of 20 scores of Example 17.2 on page 474 to estimate the population standard deviation σ:

107.853	109.640	93.927	104.021	118.092
67.698	111.109	101.384	124.980	102.511
118.291	96.477	117.134	110.180	110.413
80.494	109.089	98.480	96.813	113.998

Instead of using a surrogate normal population for resampling, here the surrogate population is the sample itself. Resamples of size 20 are drawn, with replacement, from this sample. These resamples are then used to study the sampling distribution of the sample standard deviation.

MINITAB

The key step in nonparametric bootstrapping is the repeated sampling of the original data with replacement. Suppose the 20 scores of Example 17.1 are stored in column C1. To generate 500 samples of size 20 from a column C1 and store them in columns C2-C21:

1. Choose **Calc** → **Random Data** → **Sample From Columns** and fill in the menu with
   ```
   Store 500 rows from Column(s) C1
   Store samples in:  Columns C2
   Be sure to check the box Sample With Replacement
   ```

2. Choose **Edit** → **Edit Last Dialog** and replace **C2** with **C3**.

3. Continue until you have generated data in all columns **C2** through **C21**.

   Now the first row of columns **C2-C21** represents your first resample, the second row represents your second resample, and so on.

4. To compute the sample SD for each of your resamples and store them in a column labeled **SD**: Choose **Calc** → **Row Statistics...** and fill in the menu with
   ```
   Indicating standard deviation
   Input variables:  C2-C21
   Store result in:  SD
   ```

Macro

If you are up to a little more programming, it is possible to automate this procedure by creating a Minitab macro and storing it in a file in the Macro folder of your Minitab folder (probably called Mtbwin or Mtbwinst if you are using a Windows operating system).

Here is a Minitab macro for nonparametric bootstrap confidence intervals for the population SD.

This macro illustrates the use of a DO loop to fill T columns with R resamples. The columns C.1–C.12 and the constants k.1–k.12 are locally defined variables and will not be on your worksheet once the macro has been run. However, the column containing the standard deviations of each of the 4000 resamples will be on your worksheet and available for further analysis.

17.6 COMPUTER-ASSIST. NONPARAMETRIC RESAMPLING

```
MACRO
BootsSDnp OBSERVED T R alpha SD
# nonparametric bootstrapping for standard deviation
# OBSERVED is the column containing the original data
# T = sample size
# R =number of random bootstrap samples desired.
# (1 - alpha)*100 percent confidence for interval estimate
# SD is the column which will contain the SDs of resamples
MCONSTANT T R mhat shat I L H D alpha k.1-k.12
MCOLUMN OBSERVED boots SD C.1-C.T
# k.7 k.8 are quantiles of Chi-Square distribution with D degrees of freedom
Let D=T-1 # degrees of freedom
Let k.1 = alpha/2
Let k.2 = 1-k.1
InvCDF k.1 k.7;
ChiSquare D.
InvCDF k.2 k.8;
ChiSquare D.
# L and H are percentile ranks of the bootstrap distribution
Let L=Floor(alpha*R/2)
Let H = R +1 - Floor(alpha*R/2)
# k.9 to k.10 is the normal theory confidence interval of the standard deviation
Let shat = STDEV(OBSERVED)* SQRT((T-1)/T)
Let mhat = MEAN(OBSERVED)
Let k.9 = shat*SQRT((T-1)/k.8)
Let k.10 = shat*SQRT((T-1)/k.7)
Print k.9 k.10 #normal theory confidence interval
NOTE Normal Theory Confidence Interval
Do I = 1:T
sample R OBSERVED C.I;
Replace.
enddo
# SD = simulated population of sample standard deviations
RSTDEV C.1-C.T SD
# boots = simulated sample SD / original population SD
Let boots = SD/shat
Sort boots boots;
By boots.
# k.5 to k.6 is the bootstrap confidence interval of the standard deviation
Let k.3 = boots(L)
Let k.4 = boots(H)
Let k.5 = STDEV(OBSERVED)/ k.4
Let k.6 = STDEV(OBSERVED)/ k.3
PRINT k.5 k.6 #nonparametric bootstrap confidence interval
NOTE Nonparametric bootstrap confidence interval
Let k.11 = MEAN(SD)-shat
Print k.11
NOTE Approximate bias of sample SD
Let k.12 = SQRT(STDEV(SD)**2 + k.11**2)
Print k.12
NOTE Approximate root mean square error
ENDMACRO
```

Store this macro (as a text file with line breaks) in the Macro folder under the name bootsSDnp.MAC.

The Macro folder will be in you program files for Minitab (probably in a folder named Mtbwinst or Mtbwin if you are using the Windows operating system).

You can use this macro with the data in Example 17.2 in a Minitab session to find the following:

1. a normal theory 90% confidence interval for σ,

2. a nonparametric bootstrap 90% confidence interval for *sigma* based on 4000 resamples, and

3. the approximate bias and root mean square error.

Using the macro Enter the original sample of 20 scores in a column **C1**:

Choose **Edit → Command Line Editor** and fill in the single line:

%bootsSDnp C1 20 4000 0.10 C2

Then column C2 will contain the standard deviations of your 4000 resamples. You should see something like this in your Session Window.

```
MTB >   %bootsSDnp C1 20 4000 0.10 C2
  Executing from file:C:\Program Files\MTBWINST\MACROS\bootsSDnp.MAC

  Data Display
       k.9    10.3964
       k.10   17.9454
  Normal Theory Confidence Interval

  Data Display
       k.5    10.2524
       k.6    21.7874
  Nonparametric bootstrap confidence interval

  Data Display
       k.11   -0.295895
  Approximate bias of sample SD

  Data Display
       k.12   2.77001
  Approximate root mean square error
```

17.6 COMPUTER-ASSIST. NONPARAMETRIC RESAMPLING

EXCEL Make sure you have loaded the Analysis Tool Pak described earlier.

1. Create a cumulative distribution table for your data by placing the 20 data points into cells **A3..A22** and sort the data.

2. Place the cumulative probabilities 0, 1/20. ..., 20/20 into cells **B2..B22**.

3. Generate a random sample by putting the function
   =**LOOKUP(RAND(),$B$2:$B$21,$A$3:$A$22)**
   in cell **C2**.

   This is the key step—placing a randomly selected value from the data in column A into cell C2. It represents the first sample in the first resample.

4. Select **C2..V2** and fill right.

   This completes the first resample of size 20 from the original 20 data points.

5. Select **C2..V2** through **C1001..V1001** and fill down.

   This completes the 1000 resamples of size 20 from the original 20 data points.

6. Copy **C2..V2** through **C1001..V1001** and paste special-values only.

   This keeps the 1000 resamples of size 20 from being recalculated.

7. Place the function =**STDEV(C2:V2)** in cell **W2**.

   This computes the standard deviation of the first resample.

8. Select **W2..W1001** and fill down to compute the standard deviations of all 1000 resamples.

9. To compute the S^*/σ^* values in cells **X2..X1001**, first put the function =**STDEVP(A3:A22)** in cell **X1**.

10. Then put the function =**W2/$X$1** in cell **X2** then select **X2..X1001** and fill down.

11. Finally, copy **X2..X1001** and paste special-values only, and then sort **X2..X1001**.

Bootstrap confidence interval To compute a 90% nonparametric bootstrap confidence interval, compute the following functions:

=**STDEV(A3:A22)/X952**   and   =**STDEV(A3:A22)/X50**

Bias To approximate the bias, compute the difference between the mean of the 1000 SDs and the mean of the scores:

= **AVERAGE(W2:W1001) − AVERAGE(A3:A22)**

MAPLE A brief Maple session follows. We only generate 100 resamples. You will probably want to generate more. The key step is using the **rand( )** function to generate random samples from the original data.

```
> with(stats): with(stats[statplots]):   # Loads packages for statistics
> SampleData := [107.853, 109.640, 93.927, 104.021, 118.092, 67.698, 111.109, 101.384,
    124.980, 102.511, 118.291, 96.477, 117.134, 110.180, 110.413, 80.494, 109.089,
    98.480, 96.813, 113.998];
    SampleData := [107.853, 109.640, 93.927, 104.021, 118.092, 67.698,
    111.109, 101.384, 124.980, 102.511, 118.291, 96.477, 117.134,
    110.180, 110.413, 80.494, 109.089, 98.480, 96.813, 113.998]
> SampleSize := describe[count](SampleData);
    SampleSize := 20
> RandomSelect := rand(1 .. SampleSize):
> seq(SampleData[RandomSelect()], i = 1 .. SampleSize);  # one resample
    102.511, 67.698, 124.980, 109.640, 104.021, 111.109, 80.494,
    107.853, 111.109, 113.998, 113.998, 96.477, 107.853, 104.021,
    101.384, 96.477, 109.640, 113.998, 111.109, 93.927
> # 100 resamples of sample standard deviation
> ResampleSDs100 := seq(describe[standarddeviation[1]]([seq(SampleData[RandomSelect()],
    i = 1 .. SampleSize)]), i = 1..100);
    ResampleSDs100 := 14.14624804, 13.53941787, 10.07678529, 13.86275216, 13.02995620,
    8.992051838, 12.42627755, 16.13990828, 7.727533010, 14.08140651, 12.04361410,
    16.82160794, 13.79886611, 7.551474966, 12.75875186, 6.887027579, 11.24883794,
    13.06250852, 10.63465803, 10.11342719, 13.57139657, 8.418290135, 14.91172848,
    14.01581212, 10.09480131, 15.71270654, 12.67080187, 6.594723448, 15.25637639,
    14.38669405, 13.17235504, 14.45404387, 9.435920196, 6.837858315, 7.787194160,
    16.19415965, 15.52324871, 13.12922949, 15.30361759, 14.39096070, 11.59010451,
    13.92314703, 9.721835199, 16.71127176, 16.22337770, 17.65661766, 10.93064786,
    7.962090042, 9.402166671, 18.96640071, 10.69232666, 15.91284167, 12.36057229,
    10.34695518, 10.06978681, 10.32646140, 8.930564372, 11.48140846, 11.51698217,
    12.76077904, 15.71189769, 14.61826189, 16.43800818, 12.69488203, 7.089575104,
    17.12766506, 14.90126370, 16.52765657, 9.662454096, 12.87894910, 11.56434813,
    6.223353579, 12.40467982, 16.31532932, 13.97939608, 13.74192486, 10.72659663,
    10.11458000, 13.83400831, 7.970516153, 12.28662535, 14.84586266, 10.16362577,
    13.75512890, 10.21201210, 7.240585438, 12.31000435, 9.222148097, 16.20353129,
    12.26543823, 12.49800711, 14.64913411, 11.69518259, 15.60116082, 17.40385217,
    15.86220428, 11.58998830, 8.419143337, 12.45876002, 9.621447602
> sigmaStar := evalf(describe[standarddeviation[0]](SampleData));
> # bootstrap population SD
    sigmaStar := 13.09493054
> # Approximate mean relative error.
> describe[mean]([seq(ResampleSDs100[j]/sigmaStar , j = 1 .. 100)]) -1;
    -.0522129527
> # Approximate bias of the sample SD as an estimator of the population SD.
> describe[mean]([ResampleSDs100])- sigmaStar;
    -.68372499
```

Example 17.7 *Gopain dosage again.* We now show how to compute the P-value for the test described in Section 17.4 using D. J. Finney's Gopain example on page 488.

> Maple with(combinat, choose: with(combinat, numbcomb):

> # choosing subsets of size 2 and counting the number of subsets of size 5 of 12.

> choose([14,8,3,20],2);numbcomb(12,5);

  [[3, 8], [3, 14], [3, 20], [8, 14], [8, 20], [14, 20]]

  792

> # listing all the subsets of size 5 using the Gopain data adjusted to remove ties.

> cGoPain := choose([14,8,3,20,18,9,12,4,10,7,1,7.9999],5):

> # creating an indicator function small6:=x-¿piecewise(x_i=6,1,x_i6,0); small6 := x -¿ piecewise(x ¡= 6, 1, 6 ¡ x, 0)

> # finding the proportion of subsets of size 5 with mean no more than 6 with(stats): describe[mean]([seq(small6(describe[mean](cGoPain[j])), j=1..numbcomb(12,5))]); 1/24

REVIEW EXERCISES FOR CHAPTER 17

- **17.11** Use the data in Exercise 17.5 on page 478. Construct an approximate 98% nonparametric bootstrap-t confidence interval for the expected difference.

  17.12 Find the box plot for the Darwin data in Exercise 17.5. Discuss skewness and outlying values.

- **17.13** Use the Darwin data in Exercise 17.5 on page 478. Construct an approximate 90% nonparametric bootstrap confidence interval for the standard deviation.

  17.14 Use the iris data in Exercise 17.8 on page 487. Approximate the bias and standard error of the sample correlation coefficient r as an estimate of the population correlation coefficient ρ. Use nonparametric bootstrap methods.

- **17.15** Use the data in Exercise 17.5 on page 478. Construct an approximate 90% nonparametric bootstrap-t confidence interval for the population's median score. Base your estimate on your bootstrap approximation to the sampling distribution of the statistic

$$T = \frac{\text{med}_{\text{sample}} - \text{med}_{\text{population}}}{\text{MAD}_{\text{population}}}$$

Refer to Section 2.11 for a description of MAD, the mean absolute deviation from the median.

17.16 Two random variables X and Y are said to be **bivariate normal** if any linear combination $V = aX + bY$ is normal (provided a and b are not both 0). Use the data in Example 17.5 to construct an approximate 90% parametric bootstrap confidence interval for the correlation coefficient of the species' petal and sepal lengths. Assume the population is bivariate normal and the data consists of a random sample from this population. (Hint: See Exercise 10.27. If Z_1 and Z_2 are independent standard normal random variables, then Z_1 and $\rho Z_1 + \sqrt{1-\rho^2} Z_2$ are standard normal random variables with correlation ρ.)

17.17 Use the data of Exercise 17.5. Approximate the bias $\beta_S = E(S - \sigma)$ when the sample standard deviation s is used to estimate the population standard deviation σ. How many resamples are needed to reduce the standard error associated with resampling to no more than 0.01? Use parametric bootstrap assuming the parent population is normal. Hint: Refer to Section 11.9 and 13.5 and use the formula

$$\text{SE}_{\beta_S^*} = \frac{s}{\sqrt{n}} \leq 0.01$$

with n now referring to the number of resamples R and s referring to an estimate of the standard deviation of s^*. This estimate of the standard deviation s^* would itself be based on a preliminary resampling of the data.

STATISTICAL TABLES

TABLE 1. 2250 Random Digits
Each row has 50 random digits, grouped into fives for convenience.

Row	1–5	6–10	11–15	16–20	21–25	26–30	31–35	36–40	41–45	46–50
1	47918	57733	55245	48623	26967	85071	53931	29229	61730	72057
2	65930	85921	86794	62673	94454	70013	60190	79248	61998	21559
3	05595	09039	88683	78929	93306	48269	79536	88977	63884	70066
4	98701	76059	94858	44220	63966	34620	62617	32459	34113	71629
5	86769	94430	31975	91215	03534	02226	53037	69408	39424	81758
6	04282	74387	38818	36396	36882	89524	73853	51195	08321	87907
7	83059	03093	58029	31341	96562	45535	20433	25045	98385	94098
8	37411	57451	69440	87614	33968	09784	63070	38187	89194	01890
9	09632	18651	84757	60646	47935	63107	82426	09975	02049	46430
10	56665	42663	09718	66489	43884	72752	66803	82277	68337	12776
11	52374	54219	69939	45657	00493	61885	27711	59387	31924	09216
12	96624	67381	72597	02507	09035	87279	22032	71572	85271	74017
13	89494	81528	36497	48239	06981	82652	69256	82355	05739	35372
14	37882	41491	43145	86144	00696	18916	70221	01916	05954	04479
15	03540	85798	72246	15976	47619	72019	00358	99663	59466	58521
16	93174	99109	14827	24949	16210	95105	89186	86281	65888	15034
17	04506	11485	45016	55018	22362	71351	47022	08020	82048	89495
18	31733	99700	10919	28339	12681	65711	82459	49468	34970	38804
19	24129	34494	31614	91923	85547	12990	62679	66022	92029	17150
20	46741	29678	90242	32246	94521	43876	43572	54754	20405	51653
21	09630	31028	97174	12325	92727	33416	99378	97802	81167	36996
22	31039	98746	07587	70782	98336	72230	65198	23630	50693	19627
23	96614	23530	37342	12370	28426	62469	27581	18963	61079	08077
24	62749	79640	32803	87390	12614	97437	24107	64792	34341	32465
25	77694	90634	92169	69615	70729	41610	85509	14030	65261	50654
26	87135	55696	09169	08851	07565	53854	37132	98449	43392	56709
27	48476	75960	37295	95550	49820	64435	38361	25766	20608	13265
28	77377	89648	74563	75984	79397	55515	31557	12010	39508	89860
29	48695	59889	34486	38042	55481	99040	45297	51568	41868	95177
30	97930	95390	56787	53750	55284	70695	44054	62081	15986	03071
31	82256	97023	16626	34014	40375	68129	99012	33597	74464	17954
32	37709	08368	47847	88008	07793	04088	73370	80955	93051	06130
33	48649	26969	47986	99881	58308	82071	67103	66885	33592	73525
34	73703	78278	83809	98615	97006	89821	24610	29918	64731	89015
35	41365	12341	97005	67343	74850	02496	66677	23844	27174	80972
36	13219	56378	88897	61505	50059	20614	33569	21847	96682	95728
37	37243	61104	41333	59790	33878	77664	64749	31926	20670	44151
38	04470	77713	20113	72811	42421	43062	62535	94072	51364	50026
39	30289	27372	85581	95034	30635	97101	47653	30765	57149	75662
40	06115	71564	53228	75495	42429	92792	61959	54487	90050	81634
41	37615	67524	12225	40337	63097	57784	83963	64591	33707	59820
42	34515	72898	07746	57148	65304	27655	30625	62395	80328	01465
43	55959	30708	10908	66201	55243	98039	76326	81336	25544	54259
44	70380	55764	38828	30082	54542	81691	25065	32965	11271	70586
45	30082	36935	37949	46956	93497	37846	23290	44794	80548	38758

TABLE 2a. Cumulative Binomial Distribution

$$F_K(k) = Pr(K \leq k) = \sum_{t \leq k} \binom{n}{t} \pi^t (1-\pi)^{n-t}$$

n	k	π = 0.10	0.20	0.25	0.30	0.40	0.50	0.60	0.70	0.80	0.90	0.95
7	0	0.4783	0.2097	0.1335	0.0824	0.0280	0.0078	0.0016	0.0002	0.0000	0.0000	0.0000
	1	0.8503	0.5767	0.4449	0.3294	0.1586	0.0625	0.0188	0.0038	0.0004	0.0000	0.0000
	2	0.9743	0.8520	0.7564	0.6471	0.4199	0.2266	0.0963	0.0288	0.0047	0.0002	0.0000
	3	0.9973	0.9667	0.9294	0.8740	0.7102	0.5000	0.2898	0.1260	0.0333	0.0027	0.0002
	4	0.9998	0.9953	0.9871	0.9712	0.9037	0.7734	0.5801	0.3529	0.1480	0.0257	0.0038
	5	1.0000	0.9996	0.9987	0.9962	0.9812	0.9375	0.8414	0.6706	0.4233	0.1497	0.0444
	6	1.0000	1.0000	0.9999	0.9998	0.9984	0.9922	0.9720	0.9176	0.7903	0.5217	0.3017
8	0	0.4305	0.1678	0.1001	0.0576	0.0168	0.0039	0.0007	0.0001	0.0000	0.0000	0.0000
	1	0.8131	0.5033	0.3671	0.2553	0.1064	0.0352	0.0085	0.0013	0.0001	0.0000	0.0000
	2	0.9619	0.7969	0.6785	0.5518	0.3154	0.1445	0.0498	0.0113	0.0012	0.0000	0.0000
	3	0.9950	0.9437	0.8862	0.8059	0.5941	0.3633	0.1737	0.0580	0.0104	0.0004	0.0000
	4	0.9996	0.9896	0.9727	0.9420	0.8263	0.6367	0.4059	0.1941	0.0563	0.0050	0.0004
	5	1.0000	0.9988	0.9958	0.9887	0.9502	0.8555	0.6846	0.4482	0.2031	0.0381	0.0058
	6	1.0000	0.9999	0.9996	0.9987	0.9915	0.9648	0.8936	0.7447	0.4967	0.1869	0.0572
	7	1.0000	1.0000	1.0000	0.9999	0.9993	0.9961	0.9832	0.9424	0.8322	0.5695	0.3366
9	0	0.3874	0.1342	0.0751	0.0404	0.0101	0.0020	0.0003	0.0000	0.0000	0.0000	0.0000
	1	0.7748	0.4362	0.3003	0.1960	0.0705	0.0195	0.0038	0.0004	0.0000	0.0000	0.0000
	2	0.9470	0.7382	0.6007	0.4628	0.2318	0.0898	0.0250	0.0043	0.0003	0.0000	0.0000
	3	0.9917	0.9144	0.8343	0.7297	0.4826	0.2539	0.0994	0.0253	0.0031	0.0001	0.0000
	4	0.9991	0.9804	0.9511	0.9012	0.7334	0.5000	0.2666	0.0988	0.0196	0.0009	0.0000
	5	0.9999	0.9969	0.9900	0.9747	0.9006	0.7461	0.5174	0.2703	0.0856	0.0083	0.0006
	6	1.0000	0.9997	0.9987	0.9957	0.9750	0.9102	0.7682	0.5372	0.2618	0.0530	0.0084
	7	1.0000	1.0000	0.9999	0.9996	0.9962	0.9805	0.9295	0.8040	0.5638	0.2252	0.0712
	8	1.0000	1.0000	1.0000	1.0000	0.9997	0.9980	0.9899	0.9596	0.8658	0.6126	0.3698
10	0	0.3487	0.1074	0.0563	0.0282	0.0060	0.0010	0.0001	0.0000	0.0000	0.0000	0.0000
	1	0.7361	0.3758	0.2440	0.1493	0.0464	0.0107	0.0017	0.0001	0.0000	0.0000	0.0000
	2	0.9298	0.6778	0.5256	0.3828	0.1673	0.0547	0.0123	0.0016	0.0001	0.0000	0.0000
	3	0.9872	0.8791	0.7759	0.6496	0.3823	0.1719	0.0548	0.0106	0.0009	0.0000	0.0000
	4	0.9984	0.9672	0.9219	0.8497	0.6331	0.3770	0.1662	0.0473	0.0064	0.0001	0.0000
	5	0.9999	0.9936	0.9803	0.9527	0.8338	0.6230	0.3669	0.1503	0.0328	0.0016	0.0001
	6	1.0000	0.9991	0.9965	0.9894	0.9452	0.8281	0.6177	0.3504	0.1209	0.0128	0.0010
	7	1.0000	0.9999	0.9996	0.9984	0.9877	0.9453	0.8327	0.6172	0.3222	0.0702	0.0115
	8	1.0000	1.0000	1.0000	0.9999	0.9983	0.9893	0.9536	0.8507	0.6242	0.2639	0.0861
	9	1.0000	1.0000	1.0000	1.0000	0.9999	0.9990	0.9940	0.9718	0.8926	0.6513	0.4013
11	0	0.3138	0.0859	0.0422	0.0198	0.0036	0.0005	0.0000	0.0000	0.0000	0.0000	0.0000
	1	0.6974	0.3221	0.1971	0.1130	0.0302	0.0059	0.0007	0.0000	0.0000	0.0000	0.0000
	2	0.9104	0.6174	0.4552	0.3127	0.1189	0.0327	0.0059	0.0006	0.0000	0.0000	0.0000
	3	0.9815	0.8389	0.7133	0.5696	0.2963	0.1133	0.0293	0.0043	0.0002	0.0000	0.0000
	4	0.9972	0.9496	0.8854	0.7897	0.5328	0.2744	0.0994	0.0216	0.0020	0.0000	0.0000
	5	0.9997	0.9883	0.9657	0.9218	0.7535	0.5000	0.2465	0.0782	0.0117	0.0003	0.0000
	6	1.0000	0.9980	0.9924	0.9784	0.9006	0.7256	0.4672	0.2103	0.0504	0.0028	0.0001
	7	1.0000	0.9998	0.9988	0.9957	0.9707	0.8867	0.7037	0.4304	0.1611	0.0185	0.0016
	8	1.0000	1.0000	0.9999	0.9994	0.9941	0.9673	0.8811	0.6873	0.3826	0.0896	0.0152
	9	1.0000	1.0000	1.0000	1.0000	0.9993	0.9941	0.9698	0.8870	0.6779	0.3026	0.1019
	10	1.0000	1.0000	1.0000	1.0000	1.0000	0.9995	0.9964	0.9802	0.9141	0.6862	0.4312

TABLE 2b. Cumulative Binomial Distribution

n	k	0.10	0.20	0.25	0.30	0.40	π 0.50	0.60	0.70	0.80	0.90	0.95
12	0	0.2824	0.0687	0.0317	0.0138	0.0022	0.0002	0.0000	0.0000	0.0000	0.0000	0.0000
	1	0.6590	0.2749	0.1584	0.0850	0.0196	0.0032	0.0003	0.0000	0.0000	0.0000	0.0000
	2	0.8891	0.5583	0.3907	0.2528	0.0834	0.0193	0.0028	0.0002	0.0000	0.0000	0.0000
	3	0.9744	0.7946	0.6488	0.4925	0.2253	0.0730	0.0153	0.0017	0.0001	0.0000	0.0000
	4	0.9957	0.9274	0.8424	0.7237	0.4382	0.1938	0.0573	0.0095	0.0006	0.0000	0.0000
	5	0.9995	0.9806	0.9456	0.8822	0.6652	0.3872	0.1582	0.0386	0.0039	0.0001	0.0000
	6	0.9999	0.9961	0.9857	0.9614	0.8418	0.6128	0.3348	0.1178	0.0194	0.0005	0.0000
	7	1.0000	0.9994	0.9972	0.9905	0.9427	0.8062	0.5618	0.2763	0.0726	0.0043	0.0002
	8	1.0000	0.9999	0.9996	0.9983	0.9847	0.9270	0.7747	0.5075	0.2054	0.0256	0.0022
	9	1.0000	1.0000	1.0000	0.9998	0.9972	0.9807	0.9166	0.7472	0.4417	0.1109	0.0196
	10	1.0000	1.0000	1.0000	1.0000	0.9997	0.9968	0.9804	0.9150	0.7251	0.3410	0.1184
	11	1.0000	1.0000	1.0000	1.0000	1.0000	0.9998	0.9978	0.9862	0.9313	0.7176	0.4596
13	0	0.2542	0.0550	0.0238	0.0097	0.0013	0.0001	0.0000	0.0000	0.0000	0.0000	0.0000
	1	0.6213	0.2336	0.1267	0.0637	0.0126	0.0017	0.0001	0.0000	0.0000	0.0000	0.0000
	2	0.8661	0.5017	0.3326	0.2025	0.0579	0.0112	0.0013	0.0001	0.0000	0.0000	0.0000
	3	0.9658	0.7473	0.5843	0.4206	0.1686	0.0461	0.0078	0.0007	0.0000	0.0000	0.0000
	4	0.9935	0.9009	0.7940	0.6543	0.3530	0.1334	0.0321	0.0040	0.0002	0.0000	0.0000
	5	0.9991	0.9700	0.9198	0.8346	0.5744	0.2905	0.0977	0.0182	0.0012	0.0000	0.0000
	6	0.9999	0.9930	0.9757	0.9376	0.7712	0.5000	0.2288	0.0624	0.0070	0.0001	0.0000
	7	1.0000	0.9988	0.9944	0.9818	0.9023	0.7095	0.4256	0.1654	0.0300	0.0009	0.0000
	8	1.0000	0.9998	0.9990	0.9960	0.9679	0.8666	0.6470	0.3457	0.0991	0.0065	0.0003
	9	1.0000	1.0000	0.9999	0.9993	0.9922	0.9539	0.8314	0.5794	0.2527	0.0342	0.0031
	10	1.0000	1.0000	1.0000	0.9999	0.9987	0.9888	0.9421	0.7975	0.4983	0.1339	0.0245
	11	1.0000	1.0000	1.0000	1.0000	0.9999	0.9983	0.9874	0.9363	0.7664	0.3787	0.1354
	12	1.0000	1.0000	1.0000	1.0000	1.0000	0.9999	0.9987	0.9903	0.9450	0.7458	0.4867
14	0	0.2288	0.0440	0.0178	0.0068	0.0008	0.0001	0.0000	0.0000	0.0000	0.0000	0.0000
	1	0.5846	0.1979	0.1010	0.0475	0.0081	0.0009	0.0001	0.0000	0.0000	0.0000	0.0000
	2	0.8416	0.4481	0.2811	0.1608	0.0398	0.0065	0.0006	0.0000	0.0000	0.0000	0.0000
	3	0.9559	0.6982	0.5213	0.3552	0.1243	0.0287	0.0039	0.0002	0.0000	0.0000	0.0000
	4	0.9908	0.8702	0.7415	0.5842	0.2793	0.0898	0.0175	0.0017	0.0000	0.0000	0.0000
	5	0.9985	0.9561	0.8883	0.7805	0.4859	0.2120	0.0583	0.0083	0.0004	0.0000	0.0000
	6	0.9998	0.9884	0.9617	0.9067	0.6925	0.3953	0.1501	0.0315	0.0024	0.0000	0.0000
	7	1.0000	0.9976	0.9897	0.9685	0.8499	0.6047	0.3075	0.0933	0.0116	0.0002	0.0000
	8	1.0000	0.9996	0.9978	0.9917	0.9417	0.7880	0.5141	0.2195	0.0439	0.0015	0.0000
	9	1.0000	1.0000	0.9997	0.9983	0.9825	0.9102	0.7207	0.4158	0.1298	0.0092	0.0004
	10	1.0000	1.0000	1.0000	0.9998	0.9961	0.9713	0.8757	0.6448	0.3018	0.0441	0.0042
	11	1.0000	1.0000	1.0000	1.0000	0.9994	0.9935	0.9602	0.8392	0.5519	0.1584	0.0301
	12	1.0000	1.0000	1.0000	1.0000	0.9999	0.9991	0.9919	0.9525	0.8021	0.4154	0.1530
	13	1.0000	1.0000	1.0000	1.0000	1.0000	0.9999	0.9992	0.9932	0.9560	0.7712	0.5123
15	0	0.2059	0.0352	0.0134	0.0047	0.0005	0.0000	0.0000	0.0000	0.0000	0.0000	0.0000
	1	0.5490	0.1671	0.0802	0.0353	0.0052	0.0005	0.0000	0.0000	0.0000	0.0000	0.0000
	2	0.8159	0.3980	0.2361	0.1268	0.0271	0.0037	0.0003	0.0000	0.0000	0.0000	0.0000
	3	0.9444	0.6482	0.4613	0.2969	0.0905	0.0176	0.0019	0.0001	0.0000	0.0000	0.0000
	4	0.9873	0.8358	0.6865	0.5155	0.2173	0.0592	0.0093	0.0007	0.0000	0.0000	0.0000
	5	0.9978	0.9389	0.8516	0.7216	0.4032	0.1509	0.0338	0.0037	0.0001	0.0000	0.0000
	6	0.9997	0.9819	0.9434	0.8689	0.6098	0.3036	0.0950	0.0152	0.0008	0.0000	0.0000
	7	1.0000	0.9958	0.9827	0.9500	0.7869	0.5000	0.2131	0.0500	0.0042	0.0000	0.0000
	8	1.0000	0.9992	0.9958	0.9848	0.9050	0.6964	0.3902	0.1311	0.0181	0.0003	0.0000

TABLE 2c. Cumulative Binomial Distribution

n	k	0.10	0.20	0.25	0.30	0.40	π 0.50	0.60	0.70	0.80	0.90	0.95
	9	1.0000	0.9999	0.9992	0.9963	0.9662	0.8491	0.5968	0.2784	0.0611	0.0022	0.0001
	10	1.0000	1.0000	0.9999	0.9993	0.9907	0.9408	0.7827	0.4845	0.1642	0.0127	0.0006
	11	1.0000	1.0000	1.0000	0.9999	0.9981	0.9824	0.9095	0.7031	0.3518	0.0556	0.0055
	12	1.0000	1.0000	1.0000	1.0000	0.9997	0.9963	0.9729	0.8732	0.6020	0.1841	0.0362
	13	1.0000	1.0000	1.0000	1.0000	1.0000	0.9995	0.9948	0.9647	0.8329	0.4510	0.1710
	14	1.0000	1.0000	1.0000	1.0000	1.0000	1.0000	0.9995	0.9953	0.9648	0.7941	0.5367
16	0	0.1853	0.0281	0.0100	0.0033	0.0003	0.0000	0.0000	0.0000	0.0000	0.0000	0.0000
	1	0.5147	0.1407	0.0635	0.0261	0.0033	0.0003	0.0000	0.0000	0.0000	0.0000	0.0000
	2	0.7892	0.3518	0.1971	0.0994	0.0183	0.0021	0.0001	0.0000	0.0000	0.0000	0.0000
	3	0.9316	0.5981	0.4050	0.2459	0.0651	0.0106	0.0009	0.0000	0.0000	0.0000	0.0000
	4	0.9830	0.7982	0.6302	0.4499	0.1666	0.0384	0.0049	0.0003	0.0000	0.0000	0.0000
	5	0.9967	0.9183	0.8103	0.6598	0.3288	0.1051	0.0191	0.0016	0.0000	0.0000	0.0000
	6	0.9995	0.9733	0.9204	0.8247	0.5272	0.2272	0.0583	0.0071	0.0002	0.0000	0.0000
	7	0.9999	0.9930	0.9729	0.9256	0.7161	0.4018	0.1423	0.0257	0.0015	0.0000	0.0000
	8	1.0000	0.9985	0.9925	0.9743	0.8577	0.5982	0.2839	0.0744	0.0070	0.0001	0.0000
	9	1.0000	0.9998	0.9984	0.9929	0.9417	0.7728	0.4728	0.1753	0.0267	0.0005	0.0000
	10	1.0000	1.0000	0.9997	0.9984	0.9809	0.8949	0.6712	0.3402	0.0817	0.0033	0.0001
	11	1.0000	1.0000	1.0000	0.9997	0.9951	0.9616	0.8334	0.5501	0.2018	0.0170	0.0009
	12	1.0000	1.0000	1.0000	1.0000	0.9991	0.9894	0.9349	0.7541	0.4019	0.0684	0.0070
	13	1.0000	1.0000	1.0000	1.0000	0.9999	0.9979	0.9817	0.9006	0.6482	0.2108	0.0429
	14	1.0000	1.0000	1.0000	1.0000	1.0000	0.9997	0.9967	0.9739	0.8593	0.4853	0.1892
	15	1.0000	1.0000	1.0000	1.0000	1.0000	1.0000	0.9997	0.9967	0.9719	0.8147	0.5599
17	0	0.1668	0.0225	0.0075	0.0023	0.0002	0.0000	0.0000	0.0000	0.0000	0.0000	0.0000
	1	0.4818	0.1182	0.0501	0.0193	0.0021	0.0001	0.0000	0.0000	0.0000	0.0000	0.0000
	2	0.7618	0.3096	0.1637	0.0774	0.0123	0.0012	0.0001	0.0000	0.0000	0.0000	0.0000
	3	0.9174	0.5489	0.3530	0.2019	0.0464	0.0064	0.0005	0.0000	0.0000	0.0000	0.0000
	4	0.9779	0.7582	0.5739	0.3887	0.1260	0.0245	0.0025	0.0001	0.0000	0.0000	0.0000
	5	0.9953	0.8943	0.7653	0.5968	0.2639	0.0717	0.0106	0.0007	0.0000	0.0000	0.0000
	6	0.9992	0.9623	0.8929	0.7752	0.4478	0.1662	0.0348	0.0032	0.0001	0.0000	0.0000
	7	0.9999	0.9891	0.9598	0.8954	0.6405	0.3145	0.0919	0.0127	0.0005	0.0000	0.0000
	8	1.0000	0.9974	0.9876	0.9597	0.8011	0.5000	0.1989	0.0403	0.0026	0.0000	0.0000
	9	1.0000	0.9995	0.9969	0.9873	0.9081	0.6855	0.3595	0.1046	0.0109	0.0001	0.0000
	10	1.0000	0.9999	0.9994	0.9968	0.9652	0.8338	0.5522	0.2248	0.0377	0.0008	0.0000
	11	1.0000	1.0000	0.9999	0.9993	0.9894	0.9283	0.7361	0.4032	0.1057	0.0047	0.0001
	12	1.0000	1.0000	1.0000	0.9999	0.9975	0.9755	0.8740	0.6113	0.2418	0.0221	0.0012
	13	1.0000	1.0000	1.0000	1.0000	0.9995	0.9936	0.9536	0.7981	0.4511	0.0826	0.0088
	14	1.0000	1.0000	1.0000	1.0000	0.9999	0.9988	0.9877	0.9226	0.6904	0.2382	0.0503
	15	1.0000	1.0000	1.0000	1.0000	1.0000	0.9999	0.9979	0.9807	0.8818	0.5182	0.2078
	16	1.0000	1.0000	1.0000	1.0000	1.0000	1.0000	0.9998	0.9977	0.9775	0.8332	0.5819
18	0	0.1501	0.0180	0.0056	0.0016	0.0001	0.0000	0.0000	0.0000	0.0000	0.0000	0.0000
	1	0.4503	0.0991	0.0395	0.0142	0.0013	0.0001	0.0000	0.0000	0.0000	0.0000	0.0000
	2	0.7338	0.2713	0.1353	0.0600	0.0082	0.0007	0.0000	0.0000	0.0000	0.0000	0.0000
	3	0.9018	0.5010	0.3057	0.1646	0.0328	0.0038	0.0002	0.0000	0.0000	0.0000	0.0000
	4	0.9718	0.7164	0.5187	0.3327	0.0942	0.0154	0.0013	0.0000	0.0000	0.0000	0.0000
	5	0.9936	0.8671	0.7175	0.5344	0.2088	0.0481	0.0058	0.0003	0.0000	0.0000	0.0000
	6	0.9988	0.9487	0.8610	0.7217	0.3743	0.1189	0.0203	0.0014	0.0000	0.0000	0.0000
	7	0.9998	0.9837	0.9431	0.8593	0.5634	0.2403	0.0576	0.0061	0.0002	0.0000	0.0000
	8	1.0000	0.9957	0.9807	0.9404	0.7368	0.4073	0.1347	0.0210	0.0009	0.0000	0.0000

TABLE 2d. Cumulative Binomial Distribution

n	k	0.10	0.20	0.25	0.30	0.40	π 0.50	0.60	0.70	0.80	0.90	0.95
	9	1.0000	0.9991	0.9946	0.9790	0.8653	0.5927	0.2632	0.0596	0.0043	0.0000	0.0000
	10	1.0000	0.9998	0.9988	0.9939	0.9424	0.7597	0.4366	0.1407	0.0163	0.0002	0.0000
	11	1.0000	1.0000	0.9998	0.9986	0.9797	0.8811	0.6257	0.2783	0.0513	0.0012	0.0000
	12	1.0000	1.0000	1.0000	0.9997	0.9942	0.9519	0.7912	0.4656	0.1329	0.0064	0.0002
	13	1.0000	1.0000	1.0000	1.0000	0.9987	0.9846	0.9058	0.6673	0.2836	0.0282	0.0015
	14	1.0000	1.0000	1.0000	1.0000	0.9998	0.9962	0.9672	0.8354	0.4990	0.0982	0.0109
	15	1.0000	1.0000	1.0000	1.0000	1.0000	0.9993	0.9918	0.9400	0.7287	0.2662	0.0581
	16	1.0000	1.0000	1.0000	1.0000	1.0000	0.9999	0.9987	0.9858	0.9009	0.5497	0.2265
	17	1.0000	1.0000	1.0000	1.0000	1.0000	1.0000	0.9999	0.9984	0.9820	0.8499	0.6028
19	0	0.1351	0.0144	0.0042	0.0011	0.0001	0.0000	0.0000	0.0000	0.0000	0.0000	0.0000
	1	0.4203	0.0829	0.0310	0.0104	0.0008	0.0000	0.0000	0.0000	0.0000	0.0000	0.0000
	2	0.7054	0.2369	0.1113	0.0462	0.0055	0.0004	0.0000	0.0000	0.0000	0.0000	0.0000
	3	0.8850	0.4551	0.2631	0.1332	0.0230	0.0022	0.0001	0.0000	0.0000	0.0000	0.0000
	4	0.9648	0.6733	0.4654	0.2822	0.0696	0.0096	0.0006	0.0000	0.0000	0.0000	0.0000
	5	0.9914	0.8369	0.6678	0.4739	0.1629	0.0318	0.0031	0.0001	0.0000	0.0000	0.0000
	6	0.9983	0.9324	0.8251	0.6655	0.3081	0.0835	0.0116	0.0006	0.0000	0.0000	0.0000
	7	0.9997	0.9767	0.9225	0.8180	0.4878	0.1796	0.0352	0.0028	0.0000	0.0000	0.0000
	8	1.0000	0.9933	0.9713	0.9161	0.6675	0.3238	0.0885	0.0105	0.0003	0.0000	0.0000
	9	1.0000	0.9984	0.9911	0.9674	0.8139	0.5000	0.1861	0.0326	0.0016	0.0000	0.0000
	10	1.0000	0.9997	0.9977	0.9895	0.9115	0.6762	0.3325	0.0839	0.0067	0.0000	0.0000
	11	1.0000	1.0000	0.9995	0.9972	0.9648	0.8204	0.5122	0.1820	0.0233	0.0003	0.0000
	12	1.0000	1.0000	0.9999	0.9994	0.9884	0.9165	0.6919	0.3345	0.0676	0.0017	0.0000
	13	1.0000	1.0000	1.0000	0.9999	0.9969	0.9682	0.8371	0.5261	0.1631	0.0086	0.0002
	14	1.0000	1.0000	1.0000	1.0000	0.9994	0.9904	0.9304	0.7178	0.3267	0.0352	0.0020
	15	1.0000	1.0000	1.0000	1.0000	0.9999	0.9978	0.9770	0.8668	0.5449	0.1150	0.0132
	16	1.0000	1.0000	1.0000	1.0000	1.0000	0.9996	0.9945	0.9538	0.7631	0.2946	0.0665
	17	1.0000	1.0000	1.0000	1.0000	1.0000	1.0000	0.9992	0.9896	0.9171	0.5797	0.2453
	18	1.0000	1.0000	1.0000	1.0000	1.0000	1.0000	0.9999	0.9989	0.9856	0.8649	0.6226
20	0	0.1216	0.0115	0.0032	0.0008	0.0000	0.0000	0.0000	0.0000	0.0000	0.0000	0.0000
	1	0.3917	0.0692	0.0243	0.0076	0.0005	0.0000	0.0000	0.0000	0.0000	0.0000	0.0000
	2	0.6769	0.2061	0.0913	0.0355	0.0036	0.0002	0.0000	0.0000	0.0000	0.0000	0.0000
	3	0.8670	0.4114	0.2252	0.1071	0.0160	0.0013	0.0000	0.0000	0.0000	0.0000	0.0000
	4	0.9568	0.6296	0.4148	0.2375	0.0510	0.0059	0.0003	0.0000	0.0000	0.0000	0.0000
	5	0.9887	0.8042	0.6172	0.4164	0.1256	0.0207	0.0016	0.0000	0.0000	0.0000	0.0000
	6	0.9976	0.9133	0.7858	0.6080	0.2500	0.0577	0.0065	0.0003	0.0000	0.0000	0.0000
	7	0.9996	0.9679	0.8982	0.7723	0.4159	0.1316	0.0210	0.0013	0.0000	0.0000	0.0000
	8	0.9999	0.9900	0.9591	0.8867	0.5956	0.2517	0.0565	0.0051	0.0001	0.0000	0.0000
	9	1.0000	0.9974	0.9861	0.9520	0.7553	0.4119	0.1275	0.0171	0.0006	0.0000	0.0000
	10	1.0000	0.9994	0.9961	0.9829	0.8725	0.5881	0.2447	0.0480	0.0026	0.0000	0.0000
	11	1.0000	0.9999	0.9991	0.9949	0.9435	0.7483	0.4044	0.1133	0.0100	0.0001	0.0000
	12	1.0000	1.0000	0.9998	0.9987	0.9790	0.8684	0.5841	0.2277	0.0321	0.0004	0.0000
	13	1.0000	1.0000	1.0000	0.9997	0.9935	0.9423	0.7500	0.3920	0.0867	0.0024	0.0000
	14	1.0000	1.0000	1.0000	1.0000	0.9984	0.9793	0.8744	0.5836	0.1958	0.0113	0.0003
	15	1.0000	1.0000	1.0000	1.0000	0.9997	0.9941	0.9490	0.7625	0.3704	0.0432	0.0026
	16	1.0000	1.0000	1.0000	1.0000	1.0000	0.9987	0.9840	0.8929	0.5886	0.1330	0.0159
	17	1.0000	1.0000	1.0000	1.0000	1.0000	0.9998	0.9964	0.9645	0.7939	0.3231	0.0755
	18	1.0000	1.0000	1.0000	1.0000	1.0000	1.0000	0.9995	0.9924	0.9308	0.6083	0.2642
	19	1.0000	1.0000	1.0000	1.0000	1.0000	1.0000	1.0000	0.9992	0.9885	0.8784	0.6415

TABLE 3a. Cumulative Distribution of Standard Normal Curve

Cumulative probability (shaded area) $\Phi(z) = Pr(Z \leq z)$ for negative z

z	0.00	0.01	0.02	0.03	0.04	0.05	0.06	0.07	0.08	0.09
-4.1	0.00002	0.00002	0.00002	0.00002	0.00002	0.00002	0.00002	0.00002	0.00001	0.00001
-4.0	0.00003	0.00003	0.00003	0.00003	0.00003	0.00003	0.00002	0.00002	0.00002	0.00002
-3.9	0.00005	0.00005	0.00004	0.00004	0.00004	0.00004	0.00004	0.00004	0.00003	0.00003
-3.8	0.00007	0.00007	0.00007	0.00006	0.00006	0.00006	0.00006	0.00005	0.00005	0.00005
-3.7	0.00011	0.00010	0.00010	0.00010	0.00009	0.00009	0.00008	0.00008	0.00008	0.00008
-3.6	0.00016	0.00015	0.00015	0.00014	0.00014	0.00013	0.00013	0.00012	0.00012	0.00011
-3.5	0.00023	0.00022	0.00022	0.00021	0.00020	0.00019	0.00019	0.00018	0.00017	0.00017
-3.4	0.00034	0.00032	0.00031	0.00030	0.00029	0.00028	0.00027	0.00026	0.00025	0.00024
-3.3	0.00048	0.00047	0.00045	0.00043	0.00042	0.00040	0.00039	0.00038	0.00036	0.00035
-3.2	0.00069	0.00066	0.00064	0.00062	0.00060	0.00058	0.00056	0.00054	0.00052	0.00050
-3.1	0.00097	0.00094	0.00090	0.00087	0.00084	0.00082	0.00079	0.00076	0.00074	0.00071
-3.0	0.00135	0.00131	0.00126	0.00122	0.00118	0.00114	0.00111	0.00107	0.00104	0.00100
-2.9	0.00187	0.00181	0.00175	0.00169	0.00164	0.00159	0.00154	0.00149	0.00144	0.00139
-2.8	0.00256	0.00248	0.00240	0.00233	0.00226	0.00219	0.00212	0.00205	0.00199	0.00193
-2.7	0.00347	0.00336	0.00326	0.00317	0.00307	0.00298	0.00289	0.00280	0.00272	0.00264
-2.6	0.00466	0.00453	0.00440	0.00427	0.00415	0.00402	0.00391	0.00379	0.00368	0.00357
-2.5	0.00621	0.00604	0.00587	0.00570	0.00554	0.00539	0.00523	0.00508	0.00494	0.00480
-2.4	0.00820	0.00798	0.00776	0.00755	0.00734	0.00714	0.00695	0.00676	0.00657	0.00639
-2.3	0.01072	0.01044	0.01017	0.00990	0.00964	0.00939	0.00914	0.00889	0.00866	0.00842
-2.2	0.01390	0.01355	0.01321	0.01287	0.01255	0.01222	0.01191	0.01160	0.01130	0.01101
-2.1	0.01786	0.01743	0.01700	0.01659	0.01618	0.01578	0.01539	0.01500	0.01463	0.01426
-2.0	0.02275	0.02222	0.02169	0.02118	0.02068	0.02018	0.01970	0.01923	0.01876	0.01831
-1.9	0.02872	0.02807	0.02743	0.02680	0.02619	0.02559	0.02500	0.02442	0.02385	0.02330
-1.8	0.03593	0.03515	0.03438	0.03362	0.03288	0.03216	0.03144	0.03074	0.03005	0.02938
-1.7	0.04457	0.04363	0.04272	0.04182	0.04093	0.04006	0.03920	0.03836	0.03754	0.03673
-1.6	0.05480	0.05370	0.05262	0.05155	0.05050	0.04947	0.04846	0.04746	0.04648	0.04551
-1.5	0.06681	0.06552	0.06426	0.06301	0.06178	0.06057	0.05938	0.05821	0.05705	0.05592
-1.4	0.08076	0.07927	0.07780	0.07636	0.07493	0.07353	0.07215	0.07078	0.06944	0.06811
-1.3	0.09680	0.09510	0.09342	0.09176	0.09012	0.08851	0.08692	0.08534	0.08379	0.08226
-1.2	0.11507	0.11314	0.11123	0.10935	0.10749	0.10565	0.10383	0.10204	0.10027	0.09853
-1.1	0.13567	0.13350	0.13136	0.12924	0.12714	0.12507	0.12302	0.12100	0.11900	0.11702
-1.0	0.15866	0.15625	0.15386	0.15151	0.14917	0.14686	0.14457	0.14231	0.14007	0.13786
-0.9	0.18406	0.18141	0.17879	0.17619	0.17361	0.17106	0.16853	0.16602	0.16354	0.16109
-0.8	0.21186	0.20897	0.20611	0.20327	0.20045	0.19766	0.19489	0.19215	0.18943	0.18673
-0.7	0.24196	0.23885	0.23576	0.23270	0.22965	0.22663	0.22363	0.22065	0.21770	0.21476
-0.6	0.27425	0.27093	0.26763	0.26435	0.26109	0.25785	0.25463	0.25143	0.24825	0.24510
-0.5	0.30854	0.30503	0.30153	0.29806	0.29460	0.29116	0.28774	0.28434	0.28096	0.27760
-0.4	0.34458	0.34090	0.33724	0.33360	0.32997	0.32636	0.32276	0.31918	0.31561	0.31207
-0.3	0.38209	0.37828	0.37448	0.37070	0.36693	0.36317	0.35942	0.35569	0.35197	0.34827
-0.2	0.42074	0.41683	0.41294	0.40905	0.40517	0.40129	0.39743	0.39358	0.38974	0.38591
-0.1	0.46017	0.45620	0.45224	0.44828	0.44433	0.44038	0.43644	0.43251	0.42858	0.42465
-0.0	0.50000	0.49601	0.49202	0.48803	0.48405	0.48006	0.47608	0.47210	0.46812	0.46414

TABLE 3b. Cumulative Distribution of Standard Normal Curve

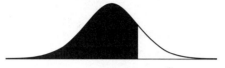

Cumulative probability (shaded area) $\Phi(z) = Pr(Z \leq z)$ for positive z

z	0.00	0.01	0.02	0.03	0.04	0.05	0.06	0.07	0.08	0.09
0.0	0.50000	0.50399	0.50798	0.51197	0.51595	0.51994	0.52392	0.52790	0.53188	0.53586
0.1	0.53983	0.54380	0.54776	0.55172	0.55567	0.55962	0.56356	0.56749	0.57142	0.57535
0.2	0.57926	0.58317	0.58706	0.59095	0.59483	0.59871	0.60257	0.60642	0.61026	0.61409
0.3	0.61791	0.62172	0.62552	0.62930	0.63307	0.63683	0.64058	0.64431	0.64803	0.65173
0.4	0.65542	0.65910	0.66276	0.66640	0.67003	0.67364	0.67724	0.68082	0.68439	0.68793
0.5	0.69146	0.69497	0.69847	0.70194	0.70540	0.70884	0.71226	0.71566	0.71904	0.72240
0.6	0.72575	0.72907	0.73237	0.73565	0.73891	0.74215	0.74537	0.74857	0.75175	0.75490
0.7	0.75804	0.76115	0.76424	0.76730	0.77035	0.77337	0.77637	0.77935	0.78230	0.78524
0.8	0.78814	0.79103	0.79389	0.79673	0.79955	0.80234	0.80511	0.80785	0.81057	0.81327
0.9	0.81594	0.81859	0.82121	0.82381	0.82639	0.82894	0.83147	0.83398	0.83646	0.83891
1.0	0.84134	0.84375	0.84614	0.84849	0.85083	0.85314	0.85543	0.85769	0.85993	0.86214
1.1	0.86433	0.86650	0.86864	0.87076	0.87286	0.87493	0.87698	0.87900	0.88100	0.88298
1.2	0.88493	0.88686	0.88877	0.89065	0.89251	0.89435	0.89617	0.89796	0.89973	0.90147
1.3	0.90320	0.90490	0.90658	0.90824	0.90988	0.91149	0.91308	0.91466	0.91621	0.91774
1.4	0.91924	0.92073	0.92220	0.92364	0.92507	0.92647	0.92785	0.92922	0.93056	0.93189
1.5	0.93319	0.93448	0.93574	0.93699	0.93822	0.93943	0.94062	0.94179	0.94295	0.94408
1.6	0.94520	0.94630	0.94738	0.94845	0.94950	0.95053	0.95154	0.95254	0.95352	0.95449
1.7	0.95543	0.95637	0.95728	0.95818	0.95907	0.95994	0.96080	0.96164	0.96246	0.96327
1.8	0.96407	0.96485	0.96562	0.96638	0.96712	0.96784	0.96856	0.96926	0.96995	0.97062
1.9	0.97128	0.97193	0.97257	0.97320	0.97381	0.97441	0.97500	0.97558	0.97615	0.97670
2.0	0.97725	0.97778	0.97831	0.97882	0.97932	0.97982	0.98030	0.98077	0.98124	0.98169
2.1	0.98214	0.98257	0.98300	0.98341	0.98382	0.98422	0.98461	0.98500	0.98537	0.98574
2.2	0.98610	0.98645	0.98679	0.98713	0.98745	0.98778	0.98809	0.98840	0.98870	0.98899
2.3	0.98928	0.98956	0.98983	0.99010	0.99036	0.99061	0.99086	0.99111	0.99134	0.99158
2.4	0.99180	0.99202	0.99224	0.99245	0.99266	0.99286	0.99305	0.99324	0.99343	0.99361
2.5	0.99379	0.99396	0.99413	0.99430	0.99446	0.99461	0.99477	0.99492	0.99506	0.99520
2.6	0.99534	0.99547	0.99560	0.99573	0.99585	0.99598	0.99609	0.99621	0.99632	0.99643
2.7	0.99653	0.99664	0.99674	0.99683	0.99693	0.99702	0.99711	0.99720	0.99728	0.99736
2.8	0.99744	0.99752	0.99760	0.99767	0.99774	0.99781	0.99788	0.99795	0.99801	0.99807
2.9	0.99813	0.99819	0.99825	0.99831	0.99836	0.99841	0.99846	0.99851	0.99856	0.99861
3.0	0.99865	0.99869	0.99874	0.99878	0.99882	0.99886	0.99889	0.99893	0.99896	0.99900
3.1	0.99903	0.99906	0.99910	0.99913	0.99916	0.99918	0.99921	0.99924	0.99926	0.99929
3.2	0.99931	0.99934	0.99936	0.99938	0.99940	0.99942	0.99944	0.99946	0.99948	0.99950
3.3	0.99952	0.99953	0.99955	0.99957	0.99958	0.99960	0.99961	0.99962	0.99964	0.99965
3.4	0.99966	0.99968	0.99969	0.99970	0.99971	0.99972	0.99973	0.99974	0.99975	0.99976
3.5	0.99977	0.99978	0.99978	0.99979	0.99980	0.99981	0.99981	0.99982	0.99983	0.99983
3.6	0.99984	0.99985	0.99985	0.99986	0.99986	0.99987	0.99987	0.99988	0.99988	0.99989
3.7	0.99989	0.99990	0.99990	0.99990	0.99991	0.99991	0.99992	0.99992	0.99992	0.99992
3.8	0.99993	0.99993	0.99993	0.99994	0.99994	0.99994	0.99994	0.99995	0.99995	0.99995
3.9	0.99995	0.99995	0.99996	0.99996	0.99996	0.99996	0.99996	0.99996	0.99997	0.99997
4.0	0.99997	0.99997	0.99997	0.99997	0.99997	0.99997	0.99998	0.99998	0.99998	0.99998

TABLE 4. Cumulative Distribution of Student's t Curves

	Cumulative probability (shaded region) $Pr(T \leq t)$ for positive t									
df	0.80	0.90	0.95	0.96	0.97	0.975	0.98	0.99	0.995	0.9995
1	1.3764	3.0777	6.3137	7.9158	10.5789	12.7062	15.8945	31.8210	63.6559	636.5776
2	1.0607	1.8856	2.9200	3.3198	3.8964	4.3027	4.8487	6.9645	9.9250	31.5998
3	0.9785	1.6377	2.3534	2.6054	2.9505	3.1824	3.4819	4.5407	5.8408	12.9244
4	0.9410	1.5332	2.1318	2.3329	2.6008	2.7765	2.9985	3.7469	4.6041	8.6101
5	0.9195	1.4759	2.0150	2.1910	2.4216	2.5706	2.7565	3.3649	4.0321	6.8685
6	0.9057	1.4398	1.9432	2.1043	2.3133	2.4469	2.6122	3.1427	3.7074	5.9587
7	0.8960	1.4149	1.8946	2.0460	2.2409	2.3646	2.5168	2.9979	3.4995	5.4081
8	0.8889	1.3968	1.8595	2.0042	2.1892	2.3060	2.4490	2.8965	3.3554	5.0414
9	0.8834	1.3830	1.8331	1.9727	2.1504	2.2622	2.3984	2.8214	3.2498	4.7809
10	0.8791	1.3722	1.8125	1.9481	2.1202	2.2281	2.3593	2.7638	3.1693	4.5868
11	0.8755	1.3634	1.7959	1.9284	2.0961	2.2010	2.3281	2.7181	3.1058	4.4369
12	0.8726	1.3562	1.7823	1.9123	2.0764	2.1788	2.3027	2.6810	3.0545	4.3178
13	0.8702	1.3502	1.7709	1.8989	2.0600	2.1604	2.2816	2.6503	3.0123	4.2209
14	0.8681	1.3450	1.7613	1.8875	2.0462	2.1448	2.2638	2.6245	2.9768	4.1403
15	0.8662	1.3406	1.7531	1.8777	2.0343	2.1315	2.2485	2.6025	2.9467	4.0728
16	0.8647	1.3368	1.7459	1.8693	2.0240	2.1199	2.2354	2.5835	2.9208	4.0149
17	0.8633	1.3334	1.7396	1.8619	2.0150	2.1098	2.2238	2.5669	2.8982	3.9651
18	0.8620	1.3304	1.7341	1.8553	2.0071	2.1009	2.2137	2.5524	2.8784	3.9217
19	0.8610	1.3277	1.7291	1.8495	2.0000	2.0930	2.2047	2.5395	2.8609	3.8833
20	0.8600	1.3253	1.7247	1.8443	1.9937	2.0860	2.1967	2.5280	2.8453	3.8496
21	0.8591	1.3232	1.7207	1.8397	1.9880	2.0796	2.1894	2.5176	2.8314	3.8193
22	0.8583	1.3212	1.7171	1.8354	1.9829	2.0739	2.1829	2.5083	2.8188	3.7922
23	0.8575	1.3195	1.7139	1.8316	1.9783	2.0687	2.1770	2.4999	2.8073	3.7676
24	0.8569	1.3178	1.7109	1.8281	1.9740	2.0639	2.1715	2.4922	2.7970	3.7454
25	0.8562	1.3163	1.7081	1.8248	1.9701	2.0595	2.1666	2.4851	2.7874	3.7251
26	0.8557	1.3150	1.7056	1.8219	1.9665	2.0555	2.1620	2.4786	2.7787	3.7067
27	0.8551	1.3137	1.7033	1.8191	1.9632	2.0518	2.1578	2.4727	2.7707	3.6895
28	0.8546	1.3125	1.7011	1.8166	1.9601	2.0484	2.1539	2.4671	2.7633	3.6739
29	0.8542	1.3114	1.6991	1.8142	1.9573	2.0452	2.1503	2.4620	2.7564	3.6595
30	0.8538	1.3104	1.6973	1.8120	1.9546	2.0423	2.1470	2.4573	2.7500	3.6460
31	0.8534	1.3095	1.6955	1.8100	1.9522	2.0395	2.1438	2.4528	2.7440	3.6335
32	0.8530	1.3086	1.6939	1.8081	1.9499	2.0369	2.1409	2.4487	2.7385	3.6218
33	0.8526	1.3077	1.6924	1.8063	1.9477	2.0345	2.1382	2.4448	2.7333	3.6109
34	0.8523	1.3070	1.6909	1.8046	1.9457	2.0322	2.1356	2.4411	2.7284	3.6007
35	0.8520	1.3062	1.6896	1.8030	1.9438	2.0301	2.1332	2.4377	2.7238	3.5911
36	0.8517	1.3055	1.6883	1.8015	1.9419	2.0281	2.1309	2.4345	2.7195	3.5821
37	0.8514	1.3049	1.6871	1.8001	1.9402	2.0262	2.1287	2.4314	2.7154	3.5737
38	0.8512	1.3042	1.6860	1.7988	1.9386	2.0244	2.1267	2.4286	2.7116	3.5657
39	0.8509	1.3036	1.6849	1.7975	1.9371	2.0227	2.1247	2.4258	2.7079	3.5581
40	0.8507	1.3031	1.6839	1.7963	1.9357	2.0211	2.1229	2.4233	2.7045	3.5510
50	0.8489	1.2987	1.6759	1.7870	1.9244	2.0086	2.1087	2.4033	2.6778	3.4960
60	0.8477	1.2958	1.6706	1.7808	1.9170	2.0003	2.0994	2.3901	2.6603	3.4602
80	0.8461	1.2922	1.6641	1.7732	1.9078	1.9901	2.0878	2.3739	2.6387	3.4164
100	0.8452	1.2901	1.6602	1.7687	1.9024	1.9840	2.0809	2.3642	2.6259	3.3905
∞	0.8416	1.2815	1.6448	1.7507	1.8808	1.9600	2.0537	2.3264	2.5758	3.2905
	60%	80%	90%	92%	94%	95%	96%	98%	99%	99.9%

Two-sided confidence level

TABLE 5a. Exact two-sided 95% confidence intervals for binomial π

The numbers printed along the curves indicate the sample size n. Given a statistic p, draw a vertical line at p. Observe where the vertical lines intersect the two curves marked n. Read their heights π_1 and π_2 off the vertical scale. These are the end points of the exact 95% confidence interval.

95% CI: $\pi_1 \leq \pi \leq \pi_2$

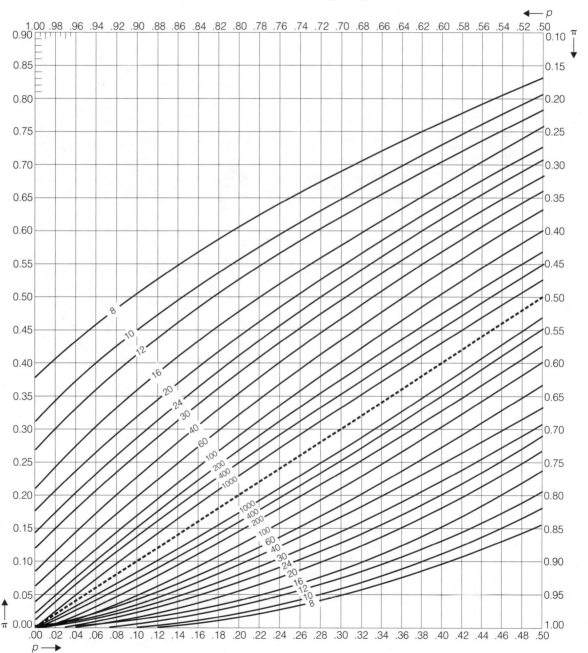

TABLE 5b. Exact two-sided 99% confidence intervals for binomial π

The numbers printed along the curves indicate the sample size n. Given a statistic p, draw a vertical line at p. Observe where the vertical lines intersect the two curves marked n. Read their heights π_1 and π_2 off the vertical scale. These are the end points of the exact 99% confidence interval.

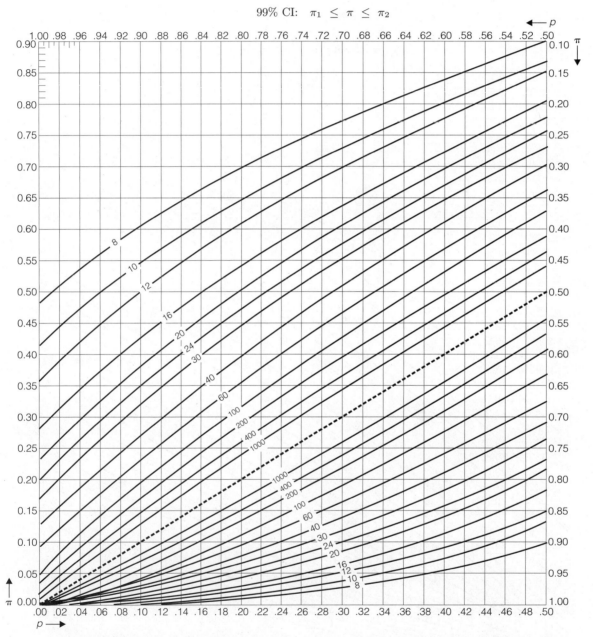

Tables 5a and 5b have been reproduced with permission of the Biometrika Trustees, from E. S. Pearson and H. O. Hartley, (editors), *Biometrika Tables for Statisticians*, third edition, volume 1, Cambridge University Press, Cambridge, 1966.

TABLE 6. Cumulative Chi-Square Distribution

	Cumulative probability (represented by area of unshaded region in graph below)									
df	0.010	0.025	0.050	0.100	0.750	0.900	0.950	0.975	0.990	0.999
1	0.0002	0.0010	0.0039	0.0158	1.3233	2.7055	3.8415	5.0239	6.6349	10.8274
2	0.0201	0.0506	0.1026	0.2107	2.7726	4.6052	5.9915	7.3778	9.2104	13.8150
3	0.1148	0.2158	0.3518	0.5844	4.1083	6.2514	7.8147	9.3484	11.3449	16.2660
4	0.2971	0.4844	0.7107	1.0636	5.3853	7.7794	9.4877	11.1433	13.2767	18.4662
5	0.5543	0.8312	1.1455	1.6103	6.6257	9.2363	11.0705	12.8325	15.0863	20.5147
6	0.8721	1.2373	1.6354	2.2041	7.8408	10.6446	12.5916	14.4494	16.8119	22.4575
7	1.2390	1.6899	2.1673	2.8331	9.0371	12.0170	14.0671	16.0128	18.4753	24.3213
8	1.6465	2.1797	2.7326	3.4895	10.2189	13.3616	15.5073	17.5345	20.0902	26.1239
9	2.0879	2.7004	3.3251	4.1682	11.3887	14.6837	16.9190	19.0228	21.6660	27.8767
10	2.5582	3.2470	3.9403	4.8652	12.5489	15.9872	18.3070	20.4832	23.2093	29.5879
11	3.0535	3.8157	4.5748	5.5778	13.7007	17.2750	19.6752	21.9200	24.7250	31.2635
12	3.5706	4.4038	5.2260	6.3038	14.8454	18.5493	21.0261	23.3367	26.2170	32.9092
13	4.1069	5.0087	5.8919	7.0415	15.9839	19.8119	22.3620	24.7356	27.6882	34.5274
14	4.6604	5.6287	6.5706	7.7895	17.1169	21.0641	23.6848	26.1189	29.1412	36.1239
15	5.2294	6.2621	7.2609	8.5468	18.2451	22.3071	24.9958	27.4884	30.5780	37.6978
16	5.8122	6.9077	7.9616	9.3122	19.3689	23.5418	26.2962	28.8453	31.9999	39.2518
17	6.4077	7.5642	8.6718	10.0852	20.4887	24.7690	27.5871	30.1910	33.4087	40.7911
18	7.0149	8.2307	9.3904	10.8649	21.6049	25.9894	28.8693	31.5264	34.8052	42.3119
19	7.6327	8.9065	10.1170	11.6509	22.7178	27.2036	30.1435	32.8523	36.1908	43.8194
20	8.2604	9.5908	10.8508	12.4426	23.8277	28.4120	31.4104	34.1696	37.5663	45.3142
21	8.8972	10.2829	11.5913	13.2396	24.9348	29.6151	32.6706	35.4789	38.9322	46.7963
22	9.5425	10.9823	12.3380	14.0415	26.0393	30.8133	33.9245	36.7807	40.2894	48.2676
23	10.1957	11.6885	13.0905	14.8480	27.1413	32.0069	35.1725	38.0756	41.6383	49.7276
24	10.8563	12.4011	13.8484	15.6587	28.2412	33.1962	36.4150	39.3641	42.9798	51.1790
25	11.5240	13.1197	14.6114	16.4734	29.3388	34.3816	37.6525	40.6465	44.3140	52.6187
26	12.1982	13.8439	15.3792	17.2919	30.4346	35.5632	38.8851	41.9231	45.6416	54.0511
27	12.8785	14.5734	16.1514	18.1139	31.5284	36.7412	40.1133	43.1945	46.9628	55.4751
28	13.5647	15.3079	16.9279	18.9392	32.6205	37.9159	41.3372	44.4608	48.2782	56.8918
29	14.2564	16.0471	17.7084	19.7677	33.7109	39.0875	42.5569	45.7223	49.5878	58.3006
30	14.9535	16.7908	18.4927	20.5992	34.7997	40.2560	43.7730	46.9792	50.8922	59.7022
31	15.6555	17.5387	19.2806	21.4336	35.8871	41.4217	44.9853	48.2319	52.1914	61.0980
32	16.3622	18.2908	20.0719	22.2706	36.9730	42.5847	46.1942	49.4804	53.4857	62.4873
33	17.0735	19.0467	20.8665	23.1102	38.0575	43.7452	47.3999	50.7251	54.7754	63.8694
34	17.7891	19.8062	21.6643	23.9522	39.1408	44.9032	48.6024	51.9660	56.0609	65.2471
35	18.5089	20.5694	22.4650	24.7966	40.2228	46.0588	49.8018	53.2033	57.3420	66.6192
36	19.2326	21.3359	23.2686	25.6433	41.3036	47.2122	50.9985	54.4373	58.6192	67.9850
37	19.9603	22.1056	24.0749	26.4921	42.3833	48.3634	52.1923	55.6680	59.8926	69.3476
38	20.6914	22.8785	24.8839	27.3430	43.4619	49.5126	53.3835	56.8955	61.1620	70.7039
39	21.4261	23.6543	25.6954	28.1958	44.5395	50.6598	54.5722	58.1201	62.4281	72.0550
40	22.1642	24.4331	26.5093	29.0505	45.6160	51.8050	55.7585	59.3417	63.6908	73.4029
	0.990	0.975	0.950	0.900	0.250	0.100	0.050	0.025	0.010	0.001

P-value for χ^2 test (represented by area of shaded region)

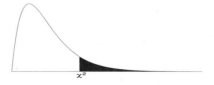

TABLE 7. Means and standard deviations of special random variables

NAME	DENSITY	MEAN	SD
Discrete Uniform	$f(k) = \dfrac{1}{n}$ for $k = 1, 2, \ldots, n$	$\dfrac{n+1}{2}$	$\sqrt{\dfrac{n^2-1}{12}}$
Bernoulli	$f(k) = \pi^k(1-\pi)^{1-k}$ for $k = 0, 1$	π	$\sqrt{\pi(1-\pi)}$
Binomial	$f(k) = \binom{n}{k}\pi^k(1-\pi)^{n-k}$ for $k = 0, 1, \ldots, n$	$n\pi$	$\sqrt{n\pi(1-\pi)}$
Hypergeometric	$f(k) = \dfrac{\binom{N_1}{k}\binom{N-N_1}{n-k}}{\binom{N}{n}}$ for $k = 1, 2, \ldots, n$	$n\pi = n\dfrac{N_1}{N}$	$\sqrt{\dfrac{N-n}{N-1}}\sqrt{n\pi(1-\pi)}$
Continuous Uniform	$f(u) = \dfrac{1}{\beta-\alpha}$ for $\alpha \leq u \leq \beta$	$\dfrac{\alpha+\beta}{2}$	$\dfrac{\beta-\alpha}{\sqrt{12}}$
Normal	$f(x) = \dfrac{1}{\sqrt{2\pi}\sigma}e^{-\frac{1}{2}(\frac{x-\mu}{\sigma})^2}$ for $-\infty < x < \infty$	μ	σ
Geometric	$f(k) = (1-\pi)^{k-1}\pi$ for $k = 1, 2, \ldots,$	$\dfrac{1}{\pi}$	$\dfrac{\sqrt{1-\pi}}{\pi}$
Exponential	$f(t) = \frac{1}{\theta}e^{-\frac{t}{\theta}}$ for $t \geq 0$	θ	θ
Poisson	$f(y) = e^{-\lambda}\dfrac{\lambda^y}{y!}$ for $y = 0, 1, \ldots$	λ	$\sqrt{\lambda}$
Sample Count K	Binomial if n is small Normal if n is large $\{n\pi \geq 5 \,\&\, n(1-\pi) \geq 5\}$ Poisson if n large, π small $\{n \geq 100 \,\&\, n\pi \leq 7\}$	$n\pi$	$f\sqrt{n\pi(1-\pi)}$
Sample Proportion P	Normal if n is large $\{n\pi \geq 5 \,\&\, n(1-\pi) \geq 5\}$ Poisson if n large, π small $\{n \geq 100 \,\&\, n\pi \leq 7\}$	π	$f\sqrt{\dfrac{\pi(1-\pi)}{n}}$
Sample SUM	Normal if n large $\{n \geq 40\}$	$n\mu$	$f\sqrt{n}\sigma$
Sample mean $\overline{X}$	Normal if n large $\{n \geq 40\}$	μ	$f\dfrac{\sigma}{\sqrt{n}}$

Reduction factor: $f = \begin{cases} \sqrt{\dfrac{N-n}{N-1}} & \text{if sampling without replacement and } N \text{ small } \{n \geq N/20\}; \\ 1 & \text{if sampling with replacement or if } n \text{ is small } \{n \leq N/20\}. \end{cases}$

ANSWERS TO SELECTED EXERCISES

Chapter 1. WHAT IS STATISTICS

1.1 **a.** $480 \pm \sqrt{1700}$ **b.** $700 \pm \sqrt{1925}$

1.7 **a.** $n > 100$ **b.** 400 **c.** 900.

1.9 $p = 57/939$; $\text{SE}_p \leq \frac{1}{2\sqrt{939}}$

1.11 statistic; parameter.

1.15 **a.** 40 seconds give or take $\sqrt{50}$ seconds
b. 80 seconds give or take 10 seconds

1.17 2 pounds $\pm\sqrt{2}$ ounces, 16 lb ± 4 oz

1.19 Because polio was not a common disease, many children would have to be in the placebo group just to observe a few cases. (If no cases were observed in the placebo group, it would be impossible to demonstrate the effectiveness of the vaccine!)

Chapter 2. HOW TO DESCRIBE AND SUMMARIZE DATA

2.1

Blood type	Freq.	%	US %
O	17	42.5	45.4
A	11	27.5	39.5
B	7	17.5	11.1
AB	5	12.5	4.0

Comparing blood type distributions

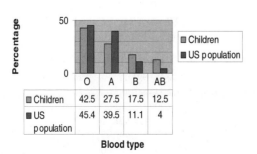

2.3

Family size	f	$d\%$
2	1	0.05
3	4	0.20
4	6	0.30
5	4	0.20
6	3	0.15
7	1	0.05
8	1	0.05

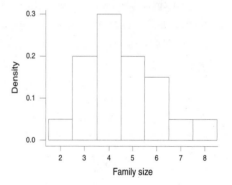

2.5 Stem-and-leaf of family size $n = 20$

```
2 | 0
3 | 0000
4 | 000000
5 | 0000
6 | 000
7 | 0
8 | 0
```

Five-number summary: 2 3.5 4 5.5 8

2.7

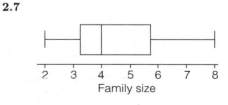

2.9 $\bar{x} = 3$

2.11

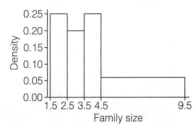

The mean family size of the 60 families is 4.20. Polling the 252 people will give you a larger mean of 5.12 because families with more people are polled more often than small families. This gives the perception that there are many large families.

2.13 $\bar{x} = 2.3$, $s = 0.7118$

2.15 $s = \sqrt{3}$, $SE_{\bar{x}} = \sqrt{6}/4$

2.17 $\bar{x} = 4.55$, $s = 1.504$

2.19 a. $\bar{x} = 6.99$, $s = 2.139$, $SE_{\bar{x}} = 2.139/\sqrt{16} = 0.535$ **b.** Another baby will weigh about 6.99 lb give or take 2.139 lb. **c.** The population mean is about 6.99 lb give or take 0.535 lb or so.

2.21 For the grouped data, $\bar{x} = 69.182$, $s = 2.567$. The grouped mean is slightly larger than the raw mean; and the grouped standard deviation is slightly smaller than raw one.

2.23 $x_{.80} = 4.5 + \frac{0.8-0.5}{0.875-0.5}(6.5-4.5) = 6.1$

$x_{.30} = 2.5 + \frac{0.3-0.125}{0.5-0.125}(4.5-2.5) = 3.4\overline{3}$

2.25

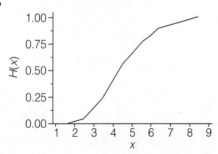

$x_{.25} = 3.5$

2.27 $\bar{x} = \sum_j m_j \frac{f_j}{n} = 118{,}192$ dollars.

For some other upper bound b, the mean becomes $118192 + \frac{b-300000}{2} \cdot \frac{1468}{4555}$
$= 69849.52 + 0.1611416b$ dollars.

For $b = \$342{,}248.56$, the mean becomes $\$125{,}000$. For $b = \$497{,}391.61$, the mean becomes $\$150{,}000$.

2.29 The cumulative frequency at $\$75{,}000$ is $F = (79 + 103 + 20 + \cdots + 366) = 1910$. The cumulative frequency at $\$100{,}000$ is $F = 1910 + 483 = 2393$. The median occurs at the cumulative frequency of $4555/2 = 2277.5$. So the median is $\$94{,}022$ because
$x_{.50} =$
$75000 + \frac{2277.5-1910}{2392-1910}(100000 - 75000)$
$= 94022$

2.33 Note: $\int_a^b I_j(x)\,dx = a_j - a_{j-1} = w_j$.
So $\int_a^b h(x)\,dx = \int_a^b \sum_{j=1}^k I_j(x) \frac{f_j}{nw_j}\,dx$
$= \sum_{j=1}^k \frac{f_j}{nw_j} \int_a^b I_j(x)\,dx = \sum_{j=1}^k \frac{f_j}{n} = 1$

2.35 The cumulative distribution function is constant on intervals between observed values. So $F'(x) = 0$ on these open intervals. The derivative does not exist at the n observed values.

2.39 a.

Class	x	f
7.5–12.5	10	2
12.5–17.5	15	4
17.5–22.5	20	7
22.5–27.5	25	2

Each class contains its left endpoint but not its right one.

b. 20 **c.** 18 **d.** 4.55
e. 9.9 13.55 17.7 19.55 24.5 **f.** 6.0

2.41 a.

Class	x	f
64–66	65	12
66–68	67	13
68–70	69	5
70–72	71	2
72–74	73	1

Each class contains its left endpoint but not its right one.

b. 67 **c.** 67.0 **d.** 2.06
e. 64.3 65.5 66.8 67.8 72.2 **f.** 2.3

2.43 a.

Class	x	f
0.00–50.00	25	0
50.00–100.00	75	1
100.00–150.00	125	2
150.00–200.00	175	4
200.00–250.00	225	2
250.00–300.00	275	1

b.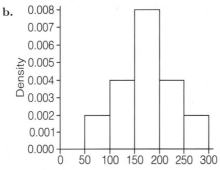

c. $\bar{x} = 175$, $s = 57.7$
d. $\bar{x} = 173.1$, $s = 53.2$ **e.** 175.00

2.45 Hint: Use the 68% and 95% empirical rules as given on page 49.

2.47 The mean and standard deviation increase to 17.27 and 33.45, respectively. The median and IQR remain the same.

2.49 Draw a horizontal line through the cumulative frequency 19.8, which is 60% of $n = 33$. Then draw a vertical line where this horizontal line intersects the ogive. The 60th percentile corresponds to a height of about 70 inches.

Chapter 3. PROBABILITY

3.1 The program depends on the technology you use. Here is one that works on the TI-83 Plus graphing calculator. It takes a minute or so to run on the TI-83 Plus.
PRGM→NEW
Name=**LAB1**
0 STO→ C
0 STO→ I
FOR(I,1,100,1)
RAND STO→ X
RAND STO→ Y
IF $(X^2 + Y^2) < 1$
C+1 STO→ C
END
DISP $4 \times C \div 100$
QUIT

Notes: Variable **C** is for the number of successes; **I** is for the number of trials; **FOR** is under **PRGM→CTL**; **RAND** is under **MATH→PRB**; **IF** is under **PRGM→CTL**; $<$ is under **TEST→TEST**; **END** is under **PRGM→CTL**; **DISP** is under **PRGM→I/O**.

3.3 **a.** 1/4 **b.** 1/4 **c.** 1/2 **d.** 1/8

3.5 $Pr(E \cap F) = 0.15, Pr(E \cap F') = 0.20$, $Pr(F \cap E') = 0.40, Pr((E \cup F)') = 0.25$

3.7 $Pr(F) = 99589/191596$, $Pr(MSA) = 6822/191596$, $Pr(F|MSA) = 4026/6822$, $Pr(F \cap MSA) = 4026/191596$. Different denominators reflect different populations.

3.9 **a.** 0.15 **b.** 0.55 **c.** 0.45 **d.** 0.75

3.11 $1 - (\frac{5}{6})^4$. Addition rule cannot be used because events are not disjoint.

3.13 $1 - (\frac{35}{36})^{24}$

3.15 $Pr(F \mid E) = \frac{Pr(F \cap E)}{Pr(E)} = \frac{Pr(F)Pr(E \mid F)}{Pr(E)} = \frac{Pr(F)Pr(E)}{Pr(E)}$

3.17 Branches: $Pr(E') = 0.70$, $Pr(F' \mid E) = 0.20$, $Pr(F' \mid E') = 0.40$. Terminal Probabilities: $Pr(E \cap F) = 0.24$, $Pr(E \cap F') = 0.06$, $Pr(E' \cap F) = 0.42$, $Pr(E' \cap F') = 0.28$

3.19 0.88305

3.23 **a.** $PV^+ = \frac{8}{8+62}$
b. 0.0159, 0.1143 **c.** 0.0159, 0.0036

3.25 0.4706, 0.5294

3.27 **a.** 0.54 **b.** 0.2963 **c.** 0.3333 **d.** 0.3704

3.29 **a.** $x \cdot (1 - \text{sensitivity})$, $(1 - x) \cdot (1 - \text{specificity})$
b. $x \cdot \text{sensitivity} + (1 - x) \cdot (1 - \text{specificity})$
c. $\frac{x \cdot \text{sensitivity}}{x \cdot \text{sensitivity} + (1 - x) \cdot (1 - \text{specificity})}$

3.33 (a) and (b)

3.35 **a.** 0.10 **b.** 0.60 **c.** 0.40 **d.** 0.40

3.37 **a.** 0.0625 **b.** 0.0625 **c.** 0.9375 **d.** 0.0625

3.39 **a.** 1/9 **b.** 4/9 **c.** 5/9 **d.** 4/9
e. 1/15, 2/5, 7/15, 8/15

3.41 **a.** 1/55 **b.** 6/55 **c.** 1/6

3.43 5/9

3.45 **a.** disjoint **b.** exactly one **c.** add **d.** independent **e.** both **f.** multiply

3.47 0.8926

3.49 1/20

3.51 a. 10 **b.** 4 **c.** 0.5 **d.** 0.5 **e.** 0.3438

3.53 a. 0.3333

3.55 a. 0.6200 **b.** 0.3226

3.57 0.0313, 0.4688, 0.5000

3.59 0.3324, 0.999995

3.61 0.5761

3.63 1/7

3.65 0.02814

3.67 $1 - (0.99)^5 = 0.0490$

3.69 0.15/0.50

3.71 $1024 - (1 + 10 + 45 + 120) = 848$

3.73 a. 32.1% **b.** 76.9%

3.75 a. Prevalence $= .30$ **b.** Sensitivity $= .80$
c. Specificity $= 22/70$
d. Predictive value positive $= 1/3$
e. Predictive value negative $= 22/28$

Chapter 4. DISCRETE RANDOM VARIABLES

4.1 U and X are discrete.

4.3
$$F(x) = \begin{cases} 0 & : \text{for } x < 0 \\ 27/64 & : \text{for } 0 \leq x < 1 \\ 54/64 & : \text{for } 1 \leq x < 2 \\ 63/64 & : \text{for } 2 \leq x < 3 \\ 1 & : \text{for } x \geq 3 \end{cases}$$

4.5 $f(0) = 0.75, f(1) = 0.25$, and $f(x) = 0$ elsewhere.

4.9 $\mu = 3/4, \sigma = 3/4$

4.11 $E(X) = 1.5, \text{SD}(X) = \sqrt{0.75}$.
$E(Y) = 1.5, \text{SD}(Y) = \sqrt{0.75}$.
Note $X + Y = 3$ for each outcome, so $E(X+Y) = 3, \text{SD}(X+Y) = 0$.

4.13 $E(X) = 1.56, \text{SD}(X) = \sqrt{0.8064}$.
$E(Y) = 1.44, \text{SD}(Y) = \sqrt{0.8064}$.
Note $X + Y = 3$ for each outcome, so $E(X+Y) = 3, \text{SD}(X+Y) = 0$.

4.15 $\mu = 3/2, \sigma = \sqrt{3}/2$

4.17 a. 5.5000 **b.** 2.8723

4.19 $\mu = 1/2, \sigma = 1/2$

4.21 a. $f(s) = (6-|s-7|)/36$ if $s = 2, 3, \ldots 12$, or a table of probabilities can be made.

b.

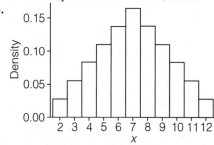

c. $\mu = 7$ **d.** $\sigma = 2.4152$

4.23 a. $f(0) = 30/56$, $f(1) = 24/56$,
$f(2) = 2/56$, $f(x) = 0$ otherwise
c. $\mu = 1/2$ **d.** $\sigma = \sqrt{9/28} = .5669$

4.25 $f(0) = 1/8, f(1) = 3/8, f(2) = 3/8$,
$f(3) = 1/8$. Note that $\sum f(x) = 1$.

4.27 a. $f(0) = 0.8145, f(1) = 0.1715$,
$f(2) = 0.0135$, $f(3) = 0.0005$,
$f(4) = 0.0000$, $f(x) = 0$ otherwise
b. 0.2000 **c.** 0.4359

4.29 a. $f(0) = (35/40)(34/39)$,
$f(1) = 2(5/40)(35/39)$,
$f(2) = (5/40)(4/39)$,
$f(x) = 0$ otherwise
b. $f(0) = (0.96)^2$, $f(1) = 2(0.96)(0.04)$,
$f(2) = (0.04)^2$, $f(x) = 0$ otherwise

Chapter 5. RANDOM VARIABLES FOR SUCCESS/FAILURE EXPERIMENTS

5.1 $\mu = 0.0769, \sigma = 0.2665$

5.3 $mu = 0.7506, \sigma = 0.4327$

5.9 a. $100/3, \sqrt{200}/3$ **b.** $300, \sqrt{200}$
c. $200/3, \sqrt{200}/3$ **d.** $600, \sqrt{200}$.

5.11 a. $(0.90)^{10} = 0.3487$
b. $10(0.10)(0.90)^9 = 0.3874$ **c.** 0.1937
d. $0.3487 + 0.3874 = 0.7361$
e. $1 - 0.7361 = 0.2639$ **f.** 0.0702

5.13 a. $Pr(K = 1) = \binom{100}{1}(0.01)^1(0.99)^{99} = 0.3697$; so about 1/3 of the time you get the expected count. **b.** 0.366
c. $1 - 0.366 = 0.634$ Use rule of opposites.
d. 16 boxes contains 1600 widgets, so you expect 16 faulty ones.
$f_K(16) = Pr(K = 16)$
$= \binom{1600}{16}(0.01)^{16}(0.99)^{1584} = 0.0997$

5.15 5/16

5.17 a. $\mu = 12.5$, $\sigma = 2.5$
b. $\mu = 12.5$, $\sigma = 2.4391$

5.19 a. $\binom{13}{7}\binom{39}{6}/\binom{52}{13} = 0.00882$
b. $13/4$, $\sqrt{39/51}\sqrt{39/16} = 1.3653$
c. 4×0.00882

5.21 a. 0.4745 **b.** 0.1279 **c.** 0.9697

5.23 a.

x	$f(x)$
0	0.1296
1	0.3456
2	0.3456
3	0.1536
4	0.0256

b.

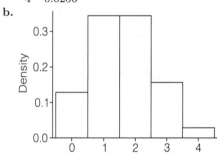

5.25 0.6250

5.27 a. 0.2817 **b.** 0.00103 **c.** 0.4696 **d.** 0.7183
e. $\mu_K = 1$ **f.** $\sigma_K = 0.7947$

5.29 0.7054

5.31 $1 - (47/48)^{12} = 0.2232$

5.33 a. 0.1638 **b.** 0.1648
c. Yes, because $N > 20n$.

Chapter 6. INTRODUCTION TO HYPOTHESES TESTING

6.1 For each n, $\alpha = 2^{-n}$, $\beta = 0$.

6.3 a. $\alpha = 2^{-12}$ **b.** $\beta = 1 - (0.75)^{12}$

6.5 a. 0.0032 **b.** 0.7251 **c.** 0.1184

6.7 $H_0: \pi = 0.5$; $H_a: \pi > 0.5$.
Decision rule: Retain H_0 if $K \leq 5$ and reject H_0 if $K \geq 6$. $\alpha = 0.3770$.
$\beta = 0.3669$ for $\pi = 0.60$ and
$\beta = 0.1503$ for $\pi = 0.70$.

6.9 There signs are 4 (+) and one (0).
The P-value $= 2^{-4} = 0.0625$.
Decide that pulse goes down.

6.11 The signs are 10 (+), one (−) and one (0).
P-value $= 0.0059$.
Decide that men exaggerate their height.

6.13 $n = 95$, $\pi = 0.50$,
P-value $= Pr(K \geq 57) = 0.0321$

6.15 For $\pi = 0.5$ and $n = 54$,
P-value $= Pr(K \leq 4) = 1.901 \times 10^{-11}$

6.17 $n = 83$, $\pi = 1/3$, P-value $=$
$Pr(K \geq 38) = 0.0123$. There is enough evidence to conclude that skipping breakfast increases late morning fatigue.

6.19 $n = 12$, P-value $= 0.0730$.
Insufficient evidence to conclude that drug increases incidence of headaches.

6.21

	Fatal	Other	Total
Vaccine	0	200745	200745
Placebo	4	201225	201229
Total	4	401960	401964

P-value $= \dfrac{\binom{200745}{0}\binom{201229}{4}}{\binom{401964}{4}} =$

$\dfrac{201229 \cdot 201228 \cdot 201227 \cdot 201226}{401964 \cdot 401963 \cdot 401962 \cdot 401961} \approx .0628$

6.23 a. P-value $= binomcdf(4, 10/14, 1) = 0.0733$. Retain at the $\alpha = 0.05$ level.
b. P-value $= 0.0410$. Reject at $\alpha = 0.05$. Conclusion: Under Fisher's exact test, there is sufficient evidence to conclude that the rate at which women are assigned as loan officers is less than corresponding rate for men. The binomial exact test is not recommended because $n = 4 > N/20 = 14/20$.

6.25 Fisher's exact P-value $= F_K(1) = 0.0570$.

6.27 In the sample, $k = 12$ of $n = 20$ earn more than \$50,000 and $k_s = 8$ earn less.
P-value $= Pr(K \leq 8 \mid n = 20, \pi = 0.5)$
$= 0.2517$. Retain H_0; median is \$50,000.

6.29 The null hypothesis is that the median percentile rank is 50. The alternative hypothesis is that the median percentile rank is above 50. Note that only 14 of the 34 percentile ranks are above 50%; the rest are below 50%. If the null hypothesis is true, then the number of percentile ranks below 50% follows the binomial distribution with $n = 34$ and $\pi = 0.50$. The P-value is $Pr(K \leq 14 \mid \pi = 0.50) = F_K(14) =$

0.1958. Retain the null hypothesis. There is insufficient evidence (P-value $= 0.1958$) to decide that the median percentile rank of white non-Hispanic girls is above 50 on the CDC Growth Chart.

6.31 $H_0 : \pi = 0.08, H_a : \pi > 0.08$.
Binomial exact test: $n = 100, \pi = 0.08$, with $\alpha = 0.05$. P-value $= Pr(K \geq 15) = 0.0133$. Reject H_0.

6.33 P-value $= 9.08 \times 10^{-7}$;
here $n = 100, \pi = 102/350 = 0.291$.

6.35 Fisher's exact P-value $= 0.0254$.
Retain H_0.

6.37 For binomial exact test:
P-value $= binomcdf(21, 15/48, 3) = 0.068$

6.39 P-value $= Pr(K \geq 7) = 1 - binomcdf(11, .5, 6) = 0.2744$

6.41 P-value $= Pr(K \geq 7) =$
$$\sum_{k=7}^{11} \frac{\binom{13}{k}\binom{17}{11-k}}{\binom{30}{11}} = 0.0927$$

Chapter 7. CONTINUOUS RANDOM VARIABLES

7.1 $$f(x) = \begin{cases} 0 & : \text{for } x < 0 \\ \frac{x}{2} & : \text{for } 0 \leq x < 2 \\ 0 & : \text{for } x \geq 2 \end{cases}$$

7.3 **a.** $F(x) = 0$ for $x < 0$,
$F(x) = (1/2)(3x^2 - x^3)$ for $0 < x < 1$,
$F(x) = 1$ for $x > 1$
b. $11/128$ **c.** $1 - 11/128$ **d.** $81/128$
e. $81/128 - 11/128$

7.5 $F(x) = 0$ for $x < -1$,
$F(x) = x^2/2 + x + 1/2$ for $-1 < x < 0$,
$F(x) = -x^2/2 + x + 1/2$ for $0 < x < 1$,
$F(x) = 1$ for $x > 1$
b. $1/8$ **c.** 0.5 **d.** $7/8$ **e.** $7/8 - 1/8$

7.7 **a.**
$$F(x) = \begin{cases} 0 & \text{for } x < 0 \\ x^2 & \text{for } 0 \leq x < 1 \\ 1 & \text{for } 1 \leq x \end{cases}$$
b. $F(0.25) = 1/16$
c. $Pr(X \geq 0.25) = 15/16$
d. $F(0.75) = 9/16$
e. $Pr(0.25 < X \leq 0.75) = \frac{9-1}{16}$

7.9 $x_{.25} = 1, med = \sqrt{2}$.

7.11 $1 - F(t) = e^{t/80}$.
So $t_q = -80\ln(1-q)$, and
$IQR = t_{.75} - t_{.25} = -80\ln(\frac{1}{4}) + 80\ln(\frac{3}{4})$
$= 80\ln 3$.

7.13 median $= 3.879$

7.15 mode $= 0$, median $= 0$,
$IQR = 2 - \sqrt{2} = 0.5858$

7.17 $\mu = 5/8, E(X^2) = 9/20$,
$\sigma = \sqrt{9/20 - (5/8)^2}$

7.19 $\mu = 1, \sigma = \sqrt{1/3}$

7.21 $\mu = E(X) = \int_3^6 x/3\,dx = 9/2$,
$E(X^2) = \int_3^6 x^2/3\,dx = 21$,
$Var(X) = 21 - (9/2)^2 = 3/4$,
$\sigma = SD(X) = \sqrt{3}/2$

7.23 $\mu = E(X) = \int_0^1 2x^2\,dx = 2/3$,
$E(X^2) = \int_0^1 2x^3\,dx = 1/2$,
$Var(X) = 1/2 - (2/3)^2 = 1/18$,
$\sigma = SD(X) = \sqrt{2}/6$

7.25 $\dfrac{\int_0^1 (x - 2/3)^3 (2x)\,dx}{\sqrt{8}} = \dfrac{1}{270\sqrt{2}}$

7.27 **a.** 68%, more than 0%
b. 95%, more than 75%
c. 99.7%, more than 88.89%

7.29 a.

b. 0.5000 **c.** 4.7500 **d.** 2.4375 **e.** 5.0000
f. 7.0000

ANSWERS TO SELECTED EXERCISES

7.31 a.
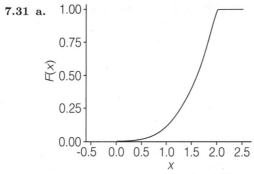
b. 0.8750 **c.** 1.5000 **d.** 0.3873 **e.** 0.9523

7.35 $\mu = 0$, $E(X^2) = 1/6$, $\sigma = \sqrt{1/6 - 0^2}$

7.37 b. $f(x) = 2x$ for $0 < x < 1$ **c.** 0.5000
d. 0.6667 **e.** $1/2 - 4/9 = 1/18$

7.39 a. 5.0000 **b.** 5.0000 **c.** 0.00 **d.** 3.4657

7.41 a. 0.3679 **b.** $f(t) = 0$ if $t < 0$,
$f(t) = e^{-t/1000}/1000$ if $0 < t$

7.43 a. 0.3762 **b.** 0.2199 **c.** 0
d. $\mu = 0$, $\sigma = \sqrt{7.2}$

7.45 $\mu = 57.2$, $\sigma^2 = 162$.

Chapter 8. NORMAL RANDOM VARIABLES

8.1 a. 1.4 **b.** 0 **c.** -6.6 **d.** 1 **e.** -1

8.3 a. -56 **b.** -50 **c.** -42.8 **d.** -65
e. -23

8.5 a. 0.39435 **c.** 0.29564

8.7 a. 0.8413 **b.** 0.3413 **c.** 0.3413
d. 0.6147 **e.** 0.0530 **f.** 0.1056

8.9 0, 1, 0

8.11 $Q_1 = 124.765$, $med = 145$, $Q_3 = 165.235$, $IQR = 40.47$

8.13 36.94%, 63.06%, 63.06%, 7.12%, 0.26%

8.15 156 mg/dl, 214 mg/dl, 225 mg/dl

8.17 2.29%

8.19 Aras. He is in the 97th percentile, whereas Rimas is in the 84th percentile.

8.21 a. $Pr(IQ = 101) = \Phi(0.1) - \Phi(0.03333) = 0.0265$ **b.** $Pr(IQ > 130) = Pr(IQ > 130.5) = 1 - \Phi(2.0333) = 0.0210$
c. $Pr(IQ \geq 130) = Pr(IQ > 129.5) = 0.0246$
d. $Pr(97.5 \leq IQ \leq 102.5) = 0.1324$

8.23 $Pr(K = 50) = Pr(49.5 < K < 50.5)$
$\approx \Phi(0.1) - \Phi(-0.1) = 0.07966$

8.25 Normal approx. 0.4393; exact value 0.4408

8.27 a. 0.24196 **b.** 0.18406

8.29 a. 0.0821 **b.** 0.0800 **c.** 15.8 successes and 24.2 failures expected. Both are at least 5.

8.31 $IQ_{.05} = 75.3$, $IQ_{.75} = 110.1$

8.33 a. 0.0668 **b.** 0.3829 **c.** 2.5347 minutes

8.35 a. 135
Answers without continuity correction:
b. 0.0912 **c.** 0.00007 **d.** 0.3179
e. 0.0028
Answers with continuity correction:
b. 0.0859 **c.** 0.00005 **d.** 0.3017
e. 0.0023

8.37 a. 0.1587 **b.** 0.1587 **c.** 16.0822

8.39 6.68%

8.41 a. 0.0035 **b.** 164.5 claims

8.43 0.0166

8.45 0.0605

8.47 a. 100 feet **b.** 2.5 inches **c.** 0.0228
d. No. By the Central Limit Theorem, for large samples, the sum of the lengths of the bricks follow the normal curve, even if the individual lengths do not.

8.49 $Pr(K > 572.5) = 0.00000212$. A lot of the people were not telling the truth.

Chapter 9. WAITING TIME RANDOM VARIABLES

9.3 $f(t) = 0.2174e^{-0.2174t}$ for $t \geq 0$,
$f(t) = 0$ for $t < 0$.
$F(t) = 1 - e^{-0.2174t}$ for $t \geq 0$,
$F(t) = 0$ for $t < 0$.

9.9 a. 0.7334 **b.** 0.2274 **c.** 0.0352 **d.** 0.0039

9.11 a. 0.4060 **b.** 0.3660 **c.** 0.5940

9.13 $\lambda = n\pi = 4000(0.003) = 12$,
$Pr(K = 12) \approx e^{-\lambda}\frac{\lambda^{12}}{12!} = 0.1144$

9.15 a. 0.3662 **b.** 0.3159 **c.** 0.3625

9.17 0.1146

9.19 0.3000

9.23 a. 0.0183 **b.** 1/9

9.25 **a.** 0.1353 **b.** 0.3834 **c.** 0.5167

9.27 **a.** 0.0498 **b.** 0.2240 **c.** 0.8009

9.29 **a.** 0.2707 **b.** 0.2707 **c.** 0.5940

9.31 Roll 2 dice 36 times.

Ace pairs	Exact probability	Poisson approx.
0	0.36271	0.36788
1	0.37307	0.36788
2	0.18654	0.18394
3	0.06040	0.06131
4	0.01424	0.01533

9.33 0.1200

9.35 $1 - 0.9319$

9.37 0.0293

9.39 0.0131

9.41 **a.** $\mu = 9.49$ years **b.** 0.09

9.43 **a.** 3.8673

b. Rutherford & Geiger data

Emissions	Observed	Expected
0	57	54.5
1	203	210.9
2	383	407.9
3	525	525.8
4	532	5.8.4
5	408	393.2
6	273	253.4
7	139	140.0
8	45	67.7
9	27	29.1
10	16	11.2

9.45 $\left(e^{-3\lambda}\frac{(3\lambda)^0}{0!}\right)\left(e^{-5\lambda}\frac{(5\lambda)^6}{6!}\right)/\left(e^{-8\lambda}\frac{(8\lambda)^6}{6!}\right) = (5/8)^6$, where λ is the mean number of items confiscated per hour.

9.47 Q has the density $f_Q(q) = 1$ for $0 < q < 1$ and $f_Q(q) = 0$ otherwise.

Chapter 10. TWO OR MORE RANDOM VARIABLES

10.1 **a.** 1/12 **b.** 11/24

10.3 **a.** 1/3 **b.** They are not independent because the joint probability table is not a multiplication table of the marginal probabilities. For example, $f_{XY}(0,0) = \frac{1}{24}$ but $f_X(0)f_Y(1) = \frac{6}{24}\frac{5}{24} \neq \frac{1}{24}$.

10.5 They are not independent because the joint probability table is not a multiplication table of the marginal probabilities. For example, $f_{XY}(0,1) = 0$ but $f_X(0)f_Y(1) = \frac{1}{32}\frac{8}{32} \neq 0$.
They are uncorrelated because $\text{Cov}(X,Y) = E(XY) - E(X)E(Y) = 5 - (2.5)(2) = 0$

10.9 Hint: $\text{Var}(X+Y) = 0$

10.11 **a.** $E(X+Y) = 4.5, \text{Var}(X+Y) = 2.25$
b. Yes.

10.15 $E(X_1) = E(X_2) = 3/5, E(X_1X_2) = 1/34$.
So $\text{Cov}(X_1, X_2) = 1/3 - (3/5)^2 = -2/75$.
$\text{Var}(X) = 6/25$. By Theorem 10.7, $\text{Var}(X_1 + X_2) = 6/25 - 2(2/75) + 6/25 = 32/75$.

10.17 **a.** 24.3500, **b.** 4.9500 **c.** 0.3400

10.19 **a.** 259.5, 86.25, 141.5 **b.** 5, 0.3999
c. No. Correlation is not zero.

10.21

			y		
		0	1	2	3
	0	0.00	0.00	0.15	0.05
z	1	0.00	0.30	0.30	0.00
	2	0.05	0.15	0.00	0.00

a. 2.50 **b.** 1.20 **c.** -0.30

10.23 **a.** $\mu_C = 350, \sigma_C = 70$ **b.** 0.3605

10.25 $f(y_1, y_2, y_3) = f(y_1)f(y_2)f(y_3)$

10.27 **a.** 0 **b.** $a^2 + b^2$ **c.** $a/\sqrt{a^2+b^2}$, assuming a and b are not both 0.

10.29 **b.** $f_X(x) = 1$ if $0 < x < 1$, $f_X(x) = 0$ otherwise.
c. Yes, because $f_{XY}(x,y) = f_X(x)f_Y(y)$.
d. 0.5000, 0.2887 **e.** 0.2500

Chapter 11. SAMPLING AND THE LAW OF AVERAGES

11.1 Put 2 cards numbered 0 and 1 into the box. Expect to get 0.5 heads, give or take 0.5 heads or so.

11.3 **a.** 200 head, give or take 10 or so.
b. 190, 210

11.5 **a.** 32, 4.38 **b.** Let K be the number of Democrats in the sample. $Pr(K < 40) = Pr(K < 39.5)$ (continuity correction) $= Pr(z < 1.712) = 0.9565$

11.7 0.0146

11.9 0.0127

11.11 a. 0.1587 **b.** 0.02275, **c.** 0.0006

11.13 a. $\mu = 140$, $\sigma = 13.23$ **b.** 0.7752
c. $140 + 2.326(13.23) = 171$

11.15 (b) Twice as accurate.

11.17 $K - \mu_K$ is more likely to be smaller for Hospital A; $P - \mu_P$ is more likely to be smaller for Hospital B.

11.19 a. 1 **b.** 4/9
d. $1 - (1.96)(2)$ and $1 + (1.96)(2)$
e. $1 - (1.96)(2/3)$ and $1 + (1.96)(2/3)$
f. $1 - (2.576)(2/3)$ and $1 + (2.576)(2/3)$

11.21 a. 3.5 **b.** $\bar{x}$ **c.** 0.1708
d. the sampling error

11.23 a. 0.9876 **b.** 0.0062,

11.25 a. 43,000, 750 **b.** P-value $= 0.0038$
c. Agree, because the P-value is small.

11.27 a. 0.2119 **b.** 0.0057

11.29 a. Estimated bias: $63.56\% - 72\% = -8.44\%$
b. Estimated precision: $s = 1.2\%$
c. Estimated accuracy: 8.52%

11.31 Use Theorem 10.6.

11.35 The largest number of aces is 100, the smallest is 0.
You expect 16.7, give or take 3.7 or so.

11.37 a. Expect 12% to be left-handed, give or take 1.028% or so.
b. $Pr(P < 0.10) = 0.0258$

11.39 a. $\pi = 0.60$ **b.** $Pr(K \geq 2) = 3(0.40)(0.60)^2 + (0.60)^3 = 0.648$
c. $Pr(K \geq 51) \approx 1 - \Phi(\frac{50.5 - 60}{\sqrt{100 \cdot .60 \cdot .40}})$
$= 1 - \Phi(-1.939) = 0.9738$

11.41 0.0062

11.43 a. $5000, $250 **b.** 0.0228

11.45 a. 0.2636 **b.** 0.0092

11.47 More than 64

11.49 a. 278 **b.** 625 **c.** 2500

11.51 2500

11.53 $\mu = 16 \cdot 2$, $\sigma = \sqrt{16}(2)$

11.55 a. 0.3000, 0.4583 **b.** 0.6000, 0.4899
c. 0.1800, 0.3842 **d.** All are Bernoulli
e. -0.3000, 0.6708

11.57 Let S denote the sample standard deviation based on a random sample of size n from a population with mean μ and standard deviation σ. We know that variances of nonconstant random variables are positive, and from Theorem 7.2, that $\text{Var}(S) = E(S^2) - E^2(S)$. We know from Theorem 11.1, that sample variances are unbiased estimators of population variances; that is $E(S^2) = \sigma^2$. Putting it all together, $\text{Var}(S) = E(S^2) - E^2(S) = \sigma^2 - E^2(S) > 0$. So $E^2(S) < \sigma^2$ and $E(S) < \sigma$. Thus, on average, the sample standard deviation will underestimate the population standard deviation.

Chapter 12. THE z AND t TESTS OF HYPOTHESES

12.3 a. $H_0: \mu = \$24,000$; $H_a: \mu < \$24,000$.
Test statistic $z_s = \frac{\bar{x} - 24000}{s/\sqrt{100}} = -1.25$.
Rejection region: $z < z_c = -1.6448$ or $\bar{x} < \bar{x}_c = 23342.06$. Conclusion: Retain the null hypothesis; insufficient evidence to conclude that average income of 35-year-olds in North Dakota is below the national average.
b. $\beta = Pr(\text{Retain } H_0 | \mu = 23600) = Pr(\bar{x} > 23342.06 | \mu = 23600) = 0.7405$.

12.5 P-value $= 0.01$, the observed difference is not well-explained by chance.

12.7 $H_0: \pi = 0.50$; $H_a: \pi \neq 0.50$.
Under the assumption of the null hypothesis, $\sigma = \sqrt{\pi(1-\pi)} = 0.50$.
Test statistic $z_s = \frac{p - 0.50}{\sigma/\sqrt{121}} = 3.08$.
Rejection region: $|z| > 1.96$. P-value $= 2Pr(z > 3.08) = 0.00208$. Conclusion: Reject the null hypothesis; there is strong evidence to conclude that $\pi \neq 0.50$.

12.9 a. $H_0: \pi = 0.55$; $H_a: \pi > 0.55$. Under the assumption of the null hypothesis, $\sigma = \sqrt{0.55(1-0.55)} = 0.4975$. Test statistic $z_s = \frac{p - 0.55}{\sigma/\sqrt{60}} = 1.8165$, without continuity correction, and $z_s = 1.6867$ with continuity correction. P-value $= Pr(z >$

1.8165) = 0.0346 without continuity correction, and P-value = $Pr(z > 1.6867)$ = 0.0458 with continuity correction. In either case, there is sufficient evidence to conclude that player has improved.
b. Since P-value > 0.01, we retain the null hypothesis; there is insufficient evidence at the $\alpha = 0.01$ level to conclude that the player has improved.
c. Rejection region for $\alpha = 0.01$:
$z > 2.364$ or $p > 41.97/60 = 0.6994$.
$\beta = Pr(\text{Retain } H_0 | \pi = 0.70) =$
$Pr(p < 0.6994 | \pi = 0.70) = 0.4961$

12.11 $H_0: \mu = 140, H_a: \mu \neq 140$; test statistic $z_s = \frac{\bar{x}-140}{s/\sqrt{81}} = -2.25$. Rejection region: $|z| > 1.96$; P-value = 0.0244. Conclusion: Reject the null hypothesis; conclude that the population mean is not 140.

12.13 Does not need adjustment.

12.15 $H_0: \mu = 15, H_a: \mu > 15$; test statistic $t_s = \frac{\bar{x}-15}{s/\sqrt{18}} = 1.6613$ with $df = 17$. Rejection region: $t > 1.7396$; P-value = 0.0575. Retain the null hypothesis. Conclusion: There is insufficient evidence to conclude that the population mean is greater than 15.

12.17 $H_0: \mu = \$45,000, H_a: \mu < \$45,000$; test statistic $\bar{x}_s = 43,000$ $t_s = \frac{\bar{x}_s - 45000}{s/\sqrt{49}}$; rejection region: $T < -1.68$; Retain the null hypothesis. Conclusion: there is insufficient evidence to conclude that that average starting salary is less than \$45,000

12.19 P-value = $2Pr(Z < -2.5) = 2(0.0062) = .0124$, so reject the manufacturer's claim at the 5% level of significance.

12.21 $P-\text{value} = Pr\left(Z > \frac{3.5-2.85}{.45/\sqrt{160}}\right) =$
$Pr(Z > 18.2708)$. The difference cannot be explained by sampling variability. Most likely people were not being truthful.

12.23 Find k_s for which $Pr(K < k_s) < .07$. Using the normal approximation we obtain $invNorm(210, 210*.9, \sqrt{210*.9*.1}) = 182.5841$. So if the sample count is 182 or less, then the null hypothesis is rejected.

12.25 P-value = $Pr(P \leq 0.09)$
$= \Phi(\frac{0.095 - 102/350}{\sqrt{(102/350)(248/350)/100}})$
$= \Phi(-4.3226) = 0.000008$

12.27 a. $H_0: \mu = 750, \sigma = 50$,
$H_a: \mu > 750, \sigma = 50$
b.
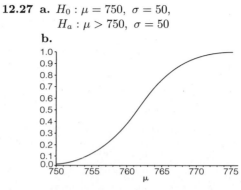
c. $Pr(\text{Retain } H_0 | \mu = 750) = 0.99$
d. $Pr(\text{Reject } H_0 | \mu = 760) = 0.3722$,
$Pr(\text{Reject } H_0 | \mu = 770) = 0.9529$

Chapter 13. ESTIMATION WITH CONFIDENCE

13.1 a. 95% CI: $\mu = 155.4 \pm 1.960 \cdot (8/\sqrt{9})$
b. 99% CI: $\mu = 155.4 \pm 2.576 \cdot (8/\sqrt{9})$

13.3 a. 120 lb, 280. lb
b. 95% CI: $192 \text{ lb} \leq \mu \leq 208 \text{ lb}$
c. 0.0000

13.5 a. 95% CI: $\mu \geq 18.3 - 1.645 \cdot (1.5/\sqrt{16})$
b. 99% CI: $\mu \geq 18.3 - 2.326 \cdot (1.5/\sqrt{16})$

13.7 a. 95% CI: $\mu = 28.2 \pm 1.8$ years
b. 95% CI: $\mu \leq 28.2 + 1.5$ years

13.9 a. $\mu = 87 \pm 1.99 \cdot (0.5/\sqrt{70})$
b. $\mu = 87 \pm 2.37 \cdot (0.5/\sqrt{70})$
c. Considering the size of the sample, the shape of the population is not important.

13.11 a. 95% CI: $\mu = 85 \pm 3.1824 \cdot (7.7/\sqrt{4})$
b. 99% CI: $\mu = 85 \pm 5.8408 \cdot (7.7/\sqrt{4})$

13.13 90% CI: $\mu = 150 \pm 13$ mg/dl

13.15 a. $\bar{x} = 924.8, s = 136.6$
b. 95% CI: $\mu \leq 924.8 + 1.8331 \times 43.2 = 1003.99$ **c.** Yes, 1000 hours is consistent with the 95% CI.

13.19 $n \geq 1037$

13.21 $n \geq 374$

13.23 a. 95% CI:
$\pi \geq 0.44 - 1.645\left(\frac{1}{2\sqrt{150}}\right) = 0.3728$
b. 95% CI:
$\pi \geq 0.44 - 1.645\left(\sqrt{\frac{(0.44)(0.56)}{150}}\right) = 0.3733$

13.25 $\pi = 50.05\% \pm 0.83\%$

13.31 99% CI: $2.08\% \leq \pi \leq 16.28\%$

13.33 a. [0.0279, 0.2366] b. [0.0070, 0.1930]

13.35 [0.2465, 0.5930]

13.39 a. 12.37 g/l, 0.21 g/l
b. 12.37 g/l; 0.03 g/l
c. 12.37 g/l, 0.21 g/l
d. $12.37 \pm (1.96)(0.03)$ g/l

13.41 98% CI: $34.3 \leq \mu \leq 36.7$ cm

13.43 a. 16.67 micrograms
b. [0.099983 gm, 0.100017 gm]
c. [0.099923 gm, 0.100077 gm] d. 0.7995
f. yes. 9, 0.9 g. 0.0702, 0.6778

13.47 $n = 34$, $\bar{x} = 0.8176$, $s = 0.1931$,
90% CI: [0.4900, 1.1435]

13.49 $n = 59, k = 11$, 99% CI: $\pi \geq 7.04\%$

13.51 95% CI: $\mu = 35 \pm 0.98$ minutes

13.53 a. 978.721 cm/sec^2, SE$_{\bar{x}} = 0.003$ cm/sec^2
c. True
e. False; it should be 979.721 ± 0.0882.

Chapter 14. TWO-SAMPLE INFERENCE

14.1 95% CI: $\mu_d = 3 \pm 2.576(\sqrt{6}/\sqrt{6})$

14.3 $\mu_d = 1.58 \pm 0.906$ or $0.67 \leq \mu_d \leq 2.49$

14.5 SE $= \sqrt{2.75}$, $df = 7$,
95% CI: $\mu_X - \mu_Y = 15 \pm 3.92$

14.7 SE $= \sqrt{5.096}$, $df = 9$, $t = 1.7719$,
P-value $= 0.05508$. Retain $\mu_X - \mu_Y = 0$.

14.9 a. $H_0 : \mu_X - \mu_Y = 0$
b. $H_a : \mu_X - \mu_Y \neq 0$
c. $t = \dfrac{26240 - 25930}{\sqrt{\dfrac{1000^2}{100} + \dfrac{1200^2}{81}}} = 1.86$.
Retain H_0 if $|t| \leq 1.96$.
Reject H_0 if $|t| \geq 1.96$.
d. There is no significant difference between the average salary of union and non-union workers performing these tasks.

14.11 Let X be candy and Y be no candy.
$H_0 : \mu_X - \mu_Y = 0$; $H_a : \mu_X - \mu_Y > 0$.
$df = 45$, $t = \dfrac{17.8 - 15.1}{\sqrt{\dfrac{3.06^2}{46} + \dfrac{1.89^2}{46}}} = 5.0915$,
P-value $< .0005$. So reject H_0.

14.13 a. Welch's formula yields 30.52; $df = 30$.
b. 95% CI: $\mu_X - \mu_Y = -4 \pm 2.0423 \cdot 1.0006$

14.15 a. Welch's formula yields 13.63, $df = 13$,
SE $= \sqrt{5.096}$ b. P-value $= 0.0499$.
Accept $\mu_X - \mu_Y > 0$

14.17 a. $df = 49$, 95% CI: $\mu_{Veg} - \mu_{Nonveg}$
$= 29 \pm 2.0096(27.8581)$
b. $df = 27$, 95% CI: $\mu_{Veg} - \mu_{Nonveg}$
$= 29 \pm 2.0518(27.8581)$

14.19 a. $C =$ Control, $T =$ Treatment.
$\bar{x}_C = 15$, $s_C^2 = 13$, $n_C = 3$, $\bar{x}_T = 20$,
$s_T^2 = 9.5$, $n_T = 4$, $df = 3$,
90% CI: $\mu_T - \mu_C = 5 \pm 2.3534(2.4967)$
b. Retain the null hypothesis of no difference in mean weight gains for these two groups.

14.21 $s_p = 2.4745$, $df = 12$,
95% CI: $\mu_X - \mu_Y = -4 \pm 2.1788 \cdot 1.3364$

14.23 $s_p = 5.1122$, $df = 33$, SE $= 1.9128$,
P-value $= 0.0221$. Reject $\mu_X - \mu_Y = 0$

14.25 The t statistic is $t_s = 2.21$, using $s_p = 16.974$, so the difference is not significant at the 1% level.

14.27 $s_p = 2.9205$, SE $= 1.3419$,
95% CI: $\mu_1 - \mu_2 = 3.6 \pm 2.1098 \cdot 1.3419$

14.29 The z statistic is $z_s = -0.375$ and P-value $= 0.35$. Retain the hypothesis of no difference in the proportion of games won by third quarter leader in these two sports.

14.31 $z = 2.347$, P-value $= 0.009$. Conclude the new treatment is significantly better at the 1% significance level.

14.33 a. $df = 4$, 90% CI: 10 ± 10.73
b. $df = 6$, 90% CI: 10 ± 9.78

14.35 $z = 2.236$, P-value $= 0.013$. Conclude flu shots reduce the probability of getting flu (at the 5% significance level).

14.37 $z = 2.154$, two-sided P-value $= 0.0312$

14.39 Let X be the first attempt, Y the second.
$\bar{x} = 550$, $\bar{y} = 530$, $s_X = 52.4$, $s_Y = 70.0$.
95% CI: $\mu_Y - \mu_X = 20 \pm 92.47$

14.43 a. Use the result from Exercise 9.46. Let $f_Z(z)$ denote the standard normal density. Then $f_Q(q) = -f_Z(z_q)/Q'(x_q) = f_Z(z_q)/f_Z(z_q) = 1$
b. Let $f_Z(z)$ denote the standard normal

density and let $f_Z(z|1)$ denote the normal density with mean 1 and standard deviation 1. Then $f_Q(q) = -f_Z(z_q)/Q'(x_q) = f_Z(z_q)/f_Z(z_q|1) = e^{(z_q-0.5)}$ Here z_q denotes the qth quantile of a standard normal random variable.

Chapter 15. CORRELATION AND REGRESSION

15.1 b. $y = 45 + 0.5x$ c. $r = 0.4583$

15.3 $\sum z_x z_y = (-1)(0) + (0)(-1) + (+1)(+1) = 1$, $r = 1/2$

15.5 $\bar{x} = 25$, $\bar{y} = 4$, $s_x = 10$, $s_y = 2$, $r = -0.95$

15.9 $\hat{y} = 63.9583 + 0.2292x$

15.11 $\hat{y} = 1.50 + 0.250x$, SSE $= 1.5$.

15.13 a. $\hat{y} = 8.75 - 0.19x$ b. $s_e = 0.6982$

15.15 a. 18 b. 18 c. 19.3 d. 16.4

15.17 a. 15.9% b. 68%

15.19 $\hat{x} = 69.5$, prediction interval is from 64.5 to 74.5.

15.21 Average $= 107.5$

15.25 95% CI: $-0.1 \leq \beta \leq 1.9$, P-value $= 0.26$

15.27 90% CI: $-0.29 \leq \beta \leq 0.69$, P-value $= 0.22$

15.33 a. $\hat{y} = 40 + 0.6x$
b. $\hat{y} = 77.8$
c. 95% CI: $-0.59 \leq \beta \leq 1.79$

15.35 a. $r = 0.97506$, b. $y = 93.375 + 1.0904x$
c. $\hat{y} = 101.8052 + 1.0632x$
d. 95% CI: $0.6 \leq \beta \leq 1.5$
e. 399.5 (when humerus is 280)

Chapter 16. INFERENCE WITH CATEGORICAL DATA

16.1 Use integration by parts $\int u\, dv = uv - \int v\, du$ with $u = x^{a/2}$ and $dv = e^{-x/2}$ to show
$$\int_0^\infty x^{a/2} e^{-x/2}\, dx = a \int_0^\infty x^{(a/2)-1} e^{-\frac{x}{2}}\, dx$$
for $a \geq 1$. Use this to show
$$E(T) = \frac{\int_0^\infty x^{(\nu+2)/2} e^{-x/2}\, dx}{\int_0^\infty x^{\nu/2} e^{-x/2}\, dx} = \nu$$

16.3 Refer to Section 10.1. To show that the marginal density function of K_1 is binomial, factor the joint density of $K_1, \ldots, K_c$ into two parts, and then sum over all possible values of $k_2, \ldots, k_c$.
$$f_{K_1}(k_1) = \sum_{k_2, \ldots k_c} f(k_1, k_2, \ldots k_c) = A \cdot B$$
where the first factor A is
$$A = \frac{n!}{k_1!(n-k_1)!} \pi_1^{k_1} (1 - \pi_1)^{n-k_1}$$
and the second factor is $B =$
$$\sum_{k_2, \ldots k_c} \frac{(n-k_1)!}{k_2! \ldots k_c!} \left(\frac{\pi_2}{1-\pi_1}\right)^{k_2} \cdots \left(\frac{\pi_c}{1-\pi_1}\right)^{k_c}$$
Notice that the sum B is that of a multinomial probability mass function of with $c - 1$ categories and sample size $n - k_1$. Since the sum is over all possible values, it must sum to one.

16.5 Many mutually exclusive different values of K_{c-1} and K_c lead to the new $K = K_{c-1} + K_c$. Sum the probability mass function for $K_1, \ldots, K_c$ over all these values to find the probability mass function of K_j, $j = 1, \ldots, c-2$, and K.
$$f(k_1, k_2, \ldots, k_{c-2}, k) =$$
$$\sum_{k_{c-1}+k_c=k} \frac{n!}{k_1! \cdots k_c!} \pi_1^{k_1} \cdots \pi_{c-1}^{k_{c-1}} \pi_c^{k_c}$$
$$= \left(\frac{n!}{k_1! \cdots k_{c-2}! k!} \pi_1^{k_1} \cdots \pi_{c-2}^{k_{c-2}}\right) \cdot$$
$$\left(\sum_{k_{c-1}+k_c=k} \frac{k!}{k_{c-1}! k_c!} \pi_{c-1}^{k_{c-1}} \pi_c^{k_c}\right)$$
$$= \frac{n!}{k_1! \cdots k_{c-2}! k!} \pi_1^{k_1} \pi_2^{k_2} \cdots \pi_{c-2}^{k_{c-2}} \pi^k$$
Use the binomial theorem to see that the last sum above equals $(\pi_{c-1} + \pi_c)^k$.

16.7 Let χ_n^2 denote the chi-square statistic based on a sample size of n. You can use induction on n. For $n = 1$ we have $\chi_1^2 = \sum_{j=1}^c \frac{(X_j - \pi_j)^2}{\pi_j}$, where exactly one of the X_j is 1 and the other $c - 1$ are all 0. Also $Pr(X_j = 1) = \pi_j$, which is a special case of Formula 16.1. The case $n = 1$ can then be proved using the fact $\chi_1^2 = \frac{(1-\pi_j)^2}{\pi_j} + (1 - \pi_j)$. The proof can be finished by the use of the induction principle, which permits you to assume the formula for n in order to demonstrate it correct for $n + 1$.

16.9 $\chi^2 = 2.0889$, $df = 3$, P-value $= 0.5542$
There is little evidence (P-value > 0.25)

to reject the 9:3:3:1 model for the population.

16.11 $\chi^2 = 6.658$, $df = 1$, P-value $= 0.0099$
This is not within historical limits.

16.13 $\chi^2 = 3.4$, P-value $= 7056/6^5 = 0.9074$
Retain the hypothesis that the die is fair.

16.17 $\chi^2 = 6.91$, $df = 4$, P-value $= 0.14$
Retain the null hypothesis that the number of overcharges is independent of the type of retail store.

16.19 $\chi^2 = 2.55$, $df = 2$, P-value $= 0.28$
Retain the null hypothesis that the overcharge rates remained the same.

16.21 $\chi^2 = 8.579$, $df = 2$, P-value $= 0.013$
Two of the cells have expected counts below 5.

16.23 Expected counts

	Cancer	No cancer	Total
Raloxifen	40.61	5088.39	5129
Placebo	20.39	2555.61	2576
Total	61	7644	7705

$\chi^2 = 8.525 + 0.068 + 16.975 + 0.135$
$= 25.704$, $df = 1$, P-value $= 0.000$

16.25 $\chi^2 = 2.916$, $df = 1$, P-value $= 0.088$

16.27 Yes. $\chi^2 = 6.64$, $df = 1$, P-value $= 0.005$

16.29 Expect 5.6667 scores in each interval. The null hypothesis is that the scores are uniformly distributed (discrete uniform over the 6 values). The alternative is that the scores are not uniformly distributed. The P-value is 0.8955, so the null hypothesis would be retained (at any of the usual levels of significance).

16.31 H_0: Facial wrinkles and smoking are independent, H_a: Severity of facial wrinkles are associated with smoking,
$\chi^2 = 9.42$, $df = 4$, P-value $= 0.05133$
Severity of facial wrinkles are associated with smoking.

16.33 $\chi^2 = 5.61$, $df = 1$, P-value $= 0.018/2 = 0.009$ for one-sided alternative. Men are more likely to be ambidextrous.

16.35 $\chi^2 = 10.28$, $df = 2$, P-value $= 0.006$

Chapter 17. RESAMPLING METHODS

17.3 A box plot shows that the scores are skewed toward lower values. The smallest score 67.698 is an outlying value.

17.5 $n = 15$, $\bar{x} = 20.93$, $s = 37.74$
98% CI: $-4.6 \leq \mu_d \leq 46.5$

17.7 Four approximate 98% nonparametric bootstrap-t confidence interval for the population's median score are shown below. Each is based on 5000 resamples. Your answers may differ somewhat from these.

Int.	L. bound	U. bound
1	104.077	118.736
2	104.154	118.551
3	104.277	118.347
4	104.166	118.470

17.9 P-value $= 3/\binom{8}{3} = 3/56 \approx 0.0536$

17.11 Four approximate 98% nonparametric bootstrap-t confidence intervals for the expected difference are shown below. Each is based on 5000 resamples. Your answer may differ somewhat from these.

Int.	L. bound	U. bound
1	-12.8985	41.9948
2	-14.3725	41.8781
3	-14.4558	41.8608
4	-13.0420	41.9445

17.13 Four approximate 90% nonparametric bootstrap confidence intervals for the standard deviation are shown below. Each is based on 5000 resamples. Your answers may differ somewhat from these.

Int.	L. bound	U. bound
1	28.7365	68.5304
2	28.5987	69.6579
3	28.7847	68.8555
4	28.7481	68.9575

17.15 Four approximate 90% nonparametric bootstrap-t confidence interval for the population's median score are shown below. Each is based on 5000 resamples. Your answers may differ somewhat from these.

Int.	L. bound	U. bound
1	5.278	35.39
2	5.148	35.25
1	4.869	35.32
4	5.411	35.49

Index

SYMBOLS

•, xiv
C, xiv
⋆, xiv
AV, 5
IQR, 39
K, 147
MAD, 46
MA, 5
MSE, 323
N, 155, 182, 186
N_1, 155, 182, 186
P-value, 172
$PV+$, 108
$PV-$, 109
RMS, 5, 46, 446
SD, 46
SE, 5
SSE, 423
SUM, 234, 305
α, 165–168, 335, 336, 435
β, 167, 337, 435, 437
χ^2, 446, 447
μ, 132, 205, 206
π, 6, 144, 147
ρ, 282
σ, 47, 135, 208
θ, 248
df, 390
r, 413
s, 46
s^2, 47
s_e, 423
68% rule, 49, 212, 221
95% rule, 49, 212, 222
99.7% rule, 49, 213, 222

A

absolute density, 29, 33
accept hypothesis, 164
accuracy, 9, 323
    of mean, 49
    of regression, 428
addition rule, 93, 95, 96, 102
alleles, 152
almost certain, 91
almost impossible, 91
α, 165–168, 335, 336, 435
alternative hypothesis, 164
approximation
    normal, 227, 300
    Poisson, 255, 300
arithmetic mean, 42
association, 439
average. *See* mean.

B

bar graph, 22, 23
Barron, E. N., xvi
Bayes' formula, 109–111
Bayes' method, 107
Bernoulli box, 145
Bernoulli random variable, 143, 144
β, 167, 337, 435, 437
bias, 10, 294, 323
binary experiment, 143
binomial
    coefficient, 149, 255
    exact interval, 374
    exact test, 176
    probabilities, 149, 154
    random variable, 147, 150, 227, 255
birthday paradox, 106

bivariate normal, 502
bivariate random variable. *See* Chapter 10.
blind, 12
bootstrap, 471, 472
    distribution, 472
    nonparametric, 472, 479
    parametric, 472, 473
    permutation, 488
bootstrapping, 345, 358, 370
Bortkiewicz, 70
boundaries, 29
box model, 143, 145, 147, 155, 241, 291
box plot, 40
box-and-whisker plot, 40

C

case, 20
categorical variable, 22
causation, 439
census, 2, 10
central limit theorem, 229, 230, 233, 234, 299, 300, 333
centroid, 411
certain event, 81, 91
chance. *See* probability.
Chebyshev's Inequality, 49, 213
chi-square
    statistic, 443, 446
    test, 462
χ^2, 447
χ^2, 446
class, 29
class mark, 30
coefficient of skewness, 147, 162, 212, 218
common response, 439
common rules, 49, 212
complement event, 91
compound hypothesis, 167, 334, 467
computer simulation, 358, 473, 491, 495
conditional probability, 99
confidence, 9, 351
confidence interval, 352, 356
    binomial exact, 374
    bootstrap-t, 480
    coverage, 377
    for difference, 402

    for independent samples, 389
    for mean, 354, 364, 474, 479
    for paired samples, 385, 388
    parametric bootstrap, 474
    for proportion, 369, 374, 377
    score, 384
    for small samples, 374
    for standard deviation, 474, 482
confidence level, 172
confounding factors, 439
contingency table, 444, 457, 464
continuity correction, 230, 231
continuous variable, 26, 122
control group, 179
correlation, 15, 278, 281, 407, 408, 439, 484
correlation coefficient, 282, 411, 413, 414, 416
count, 147
covariance, 278, 283
coverage functions, 377
critical region, 335
critical value, 172, 335
cumulative distribution function, 52, 125, 272
    binomial, 154
    hypergeometric, 159
    normal, 224
    Poisson, 254

D

data
    reported, 294
    set, 9, 20
data snooping, 172, 336, 446
De Morgan's Laws, 103
decision rule, 165
degrees of freedom, 46, 390, 393, 437
density, 29
    bivariate, 272
    function, 124
    joint, 267
    marginal, 271
dependent draws, 101
dependent events, 98
dependent variable, 408
depth, 57

descriptive statistics, 1, 14, 19
design of experiments, 13
deterministic relationship, 282
deviation
    mean absolute, 46
    standard, 46
df, 390
dichotomous experiment, 78, 143
dichotomous variable, 2
difference of means, 385
discrete variable, 26, 82, 122
disjoint events, 81, 101
distribution, 126
    bootstrap, 472
dominant, 152
double-blind, 12
draws with replacement, 3, 101
draws without replacement, 101

E

effective, 186
efficiency, 323
empirical
    distribution, 52, 140
    frequency table, 27, 223, 273, 458
    histogram, 129, 131, 140
    mean, 132
    probability, 86, 88
empirical rules, 49
error
    margin of, 367
    Type I, 164
    Type II, 164
error analysis, 5, 151
estimate
    difference of means, 385
    interval, 351
    of mean, 352, 356, 358
    point, 323
    of proportion, 369
    of standard deviation, 474
estimate of parameter, 14
estimated standard error, 50
estimation, 1
    for categorical data, 443
    one-sided, 356

    with regression, 428
    two-sided, 352
estimation error, 10
estimator, 47, 323
event, 80
Excel. *See* Notes on Technology.
execution, 5
expected value, 132, 205, 279
experiment, 5, 78
exploratory data analysis, 56
exponential random variable, 246
extrapolation, 440
eyeballing scatter plot, 418

F

failure, 143
fair game, 309
false negative, 108
false positive, 108
far out points, 40
favorable outcomes, 80
Fisher's exact test, 186
fitting scatter plot, 418
five-number summary, 37, 38, 412
forensics, 434
frequency polygon, 31
frequency table, 27, 458

G

gambler's ruin, 309
Geiger counter, 70
genetics, 76, 146, 152, 456
geometric random variable, 241
Giaquinto, Anthony, xvi
goodness of fit, 443, 450
Gosset, W. S., 322
grouped data, 31

H

highly significant, 172, 336
hinges, 66
histogram, 28
    empirical, 129, 131, 140
    probability, 128, 131

theoretical, 128, 131
histogram function, 65
historical standard deviation, 335
homoscedasticity, 428
hypergeometric
    probabilities, 156, 159
    random variable, 155
hypothesis
    alternative, 164
    compound, 167, 334
    null, 347
    simple, 167, 334
    test, 12, 14, 163
        binomial exact, 176
        chi-square, 450, 457, 462, 464
        compound, 467
        contingency table, 457, 462, 464
        correlation, 434
        for difference, 388, 389, 402
        Fisher's exact, 186
        goodness of fit, 450
        for mean, 348
        for mean. *See also* Chapter 12.
        sign, 170, 173
        t, 333
        two-sided, 339
        z, 333

I

impossible event, 81, 90
inclusion rule, 92
incompatible, 81
independent, 276
    draws, 101
    events, 97–102, 276
    random variables, 275, 276, 408
    samples, 389
inference
    with regression, 428
inference. *See* estimation *or* hypothesis.
inference with categorical data, 443
inferential statistics, 9
interpolate, 440
interquartile range, 39
interval, 29
inverse exponential random variable, 259

IQR, 39

J

joint density, 267
joint mean, 411
joint random variables. *See* Chapter 10.
Jordan, Steven L., xvi

K

K, 147

L

large
    sample, 157, 231, 255, 300, 317, 333, 370, 450, 451, 461
    study, 188
law of averages
    for count, 299
    for mean, 316
    for proportion, 312
    for sum, 305
    for T statistic, 321
    for $Z_{\overline{X}}$, 320
law of large numbers, 131, 140
leaf number, 36
least squares regression, 423
level of significance, 335
linear regression, 408
linear relationship, 282, 415

M

MA, 5
MAD, 46
Maher, Richard, xvi
major axis, 414, 419
margin of error, 2, 9, 367
marginal density, 271
marginal probabilities, 271
Markov's Lemma, 213
mass points, 124
matched pair samples, 385
maximum, 37
mean, 42, 65, 132, 279

absolute deviation, 46
grouped, 43
sample, 42
mean squared error, 323
measure, 195
measure zero, 91
measurement, 20, 27
measurement error, 10
median, 26, 38, 131, 202
medical testing, 108
memoryless process, 242, 246
Mendel's genetic model, 76, 146, 152, 456
minimum, 37
Minitab. *See* Notes on Technology.
mode, 26, 131, 202, 204
monoclonal, 76
Monte Carlo method, 88
MSE, 323
μ, 132, 205, 206
multinomial
coefficient, 447
random variable, 447
multiple regression, 408
multiplication rule, 99, 102, 276
multivariate random variable. *See* Chapter 10.
mutually exclusive, 81

N

N, 155, 182, 186
N_1, 182, 186
negative correlation, 409
nominal data, 25
nonparametric bootstrap, 472, 479
nonresponse bias, 10
normal
bivariate, 502
distribution function, 224
probabilities, 224
normal approximation, 227, 300
normal density functions, 203
Notes on Technology
binomial probabilities, 154
box plot, 42
chi-square test, 462
confidence interval for mean, 364
confidence interval for proportions, 377
contingency table test, 462
correlation coefficient, 416
difference of means, 402
getting started, 41
histogram, 42
hypergeometric probabilities, 159
inverse normal, 224
normal cumulative distribution, 224
paired samples, 388
Poisson probabilities, 254
random numbers, 88
regression line, 426
resampling, 491
scatter plot, 416
sign test, 173
stem plot, 42
summary statistics, 66
test of hypothesis for mean, 348
null hypothesis, 12, 164, 347

O

observation, 20
ogive, 52, 53, 65
one-sided confidence interval, 356
opposite event, 91
ordinal data, 25
outcome, 78, 408
outcome set, 78
outliers, 40

P

P-value, 171, 172
paired samples, 385
pairwise disjoint, 95
parameter, 3, 292
parametric bootstrap, 472, 473
Park, Henry, xvi
partition, 110
Pascal's Triangle, 150
percentile, 55, 131, 202, 222
permutation bootstrap, 472
permutation tests, 488

Petersen, Hans-Juergen, xvi
π, 6, 144, 147
placebo effect, 433
placebo group, 12
point estimator, 323
point of averages, 411
Poisson
    approximation, 255, 300
    limit theorem, 257
    probabilities, 254
    process, 246
    random variable, 251
polyclonal, 76
pooled variance, 396
population, 2, 292
    box model, 3, 101, 143, 145
    mean, 3, 47, 132
    percentage, 3, 6
    proportion, 144
    small, 157
positive correlation, 409
power of a test, 167, 337
precision, 9, 323
predictive value negative, 109
predictive value positive, 108
predictor, 408
prevalence, 108
probability, 83, 351
    bivariate, 268, 271
    conditional, 97, 99
    density, 124, 197
    empirical, 86
    function, 124
    histogram, 128
    independence, 97
    joint, 270, 271
    marginal, 271, 276
    mass, 124
    model, 14, 143
    rules of, 90
    subjective, 89
    table, 125
    theoretical, 83
    theory, 13, 14, 75, 76
$PV-$, 109
$PV+$, 108

Pythagorean property of error, 6

Q

quantile, 52, 55, 65, 222
quantitative variable, 2, 22
quartile, 38

R

r, 413
random error, 4, 292, 293, 323
random error rule, 49
random experiment, 5, 78
random numbers, 4, 88
random sampling error, 4, 292, 293, 323
random variable, 121
    Bernoulli, 143
    binomial, 147, 227, 255
    categorical, 444
    continuous, 27, 122, 195
    discrete, 27, 82, 121, 122, 143
    exponential, 246, 247
    geometric, 241, 242
    hypergeometric, 155
    independent, 151, 275, 276
    inverse exponential, 259
    joint, 267
    multinomial, 447
    multivariate, 267
    normal, 219
    Poisson, 251
    quantitative, 444
    uniform continuous, 216
    uniform discrete, 138
    waiting time, 241
randomized controlled experiment, 179
range, 39, 49
ratio data, 26
recessive, 152
reduction factor, 157, 158, 300, 306, 308, 312, 317, 320
regression
    analysis, 15, 407
    curve, 421
    effect, 431, 433
    equation, 440

fallacy, 433
line, 421, 423, 426
paradox, 431
procedure, 428
theory, 408
toward the mean, 431
reject hypothesis, 164
rejection region, 335
relative density, 29, 33
resamples, 472
resampling, 481, 488
nonparametric, 472
parametric, 472
permutation, 472, 488
residual, 423
residual standard deviation, 423, 430
response bias, 10
response variable, 408
retain hypothesis, 164
ρ, 282
RMS, 5, 46, 446
root mean square, 5
Rutherford, 70

S

s, 46
Saleski, Alan, xvi
Salk vaccine trials, 12
sample, 2, 296
large, 157, 231, 255, 317, 333, 370, 450, 451, 461
matched pair, 385
mean, 42
percentage, 3
simple random, 3, 296, 457
sample size
for estimating a mean, 367
for estimating a mean, 367
for estimating a proportion, 372
for opinion polls, 372
sample space, 78
sampling error, 10
versus bias, 10
sampling experiment, 291
sampling theory, 1, 9, 14, 77, 291
scatter plot, 408, 409, 416

score confidence interval, 384
screening, 108
SD. *See* standard deviation.
SE. *See* standard error.
s_e, 423
sensitivity, 108
σ, 47, 135, 208
σ^2, 47, 135, 208
s^2, 47
sign test, 170, 173
significance, 172, 336
significant, 172, 185, 336
simple hypothesis, 167, 334
simple random sample, 3, 296
simple regression, 408
simulation, 354
skew, 29, 44, 58
skewness, 147, 162, 212, 218
small population, 157, 158
specificity, 108
SSE, 423
standard deviation, 45, 135, 208
grouped, 48
historical, 335
pooled, 396
population, 47
residual, 423, 430
sample, 46
standard error, 5, 49, 293, 300
estimated, 50
of mean, 50
for P, 371
standard normal curve, 220
standardized random variable, 211
standardized scale, 210
statistic, 3, 296, 297
statistical relationship, 14
stem number, 36
stem plot, 35
stem-and-leaf plot, 35
stimulus, 408
stochastic relationship, 282
strength of relationship. *See* correlation.
Student t curve, 321
Sturge's Rule, 31, 32
subject, 2, 20

subjective probability, 89
success, 143
success count, 147
SUM, 234, 305
summary statistics, 66
sure, 91
symmetry, 58, 83

T

t curves, 322, 358
T statistic, 321
t test, 333
taxonomies, 10
test
    binomial exact, 176
    for categorical data, 443
    chi-square, 450, 457, 462, 464
    compound hypothesis, 467
    contigency table, 457, 464
    contingency table, 462
    correlation, 434
    for difference, 385, 388, 389, 402
    Fisher's exact, 186
    goodness of fit, 450
    for mean, 348
    nonparametric bootstrap, 479
    parametric bootstrap, 473
    permuation, 488
    power, 167
    sign, 170, 173
    t, 358
    two-sided, 339
    z, 333
test statistic, 171, 173, 177, 180, 189, 336, 342, 343, 450
theoretical histogram, 128
theoretical probability, 83
TI-83 Plus. *See* Notes on Technology.
treatment group, 12, 179
tree diagram, 105
trial, 5, 78, 147
trivial effect, 185
true negative, 109
true positive, 108
two-sided z test, 339
two-sided confidence interval, 352

Type I error, 164, 165
Type II error, 164, 165

U

unbiased estimator, 47
uncorrelated, 282
uniform random variable, 138, 216
unit, 2
urn models, 143

V

vaccine group, 12
variable, 2, 20
    categorical, 22
    continuous, 26
    dichotomous, 2
    discrete, 26
    nominal, 25
    ordinal, 25
    outcome, 408
    quantitative, 2, 22
    random, 121
    ratio, 26
    response, 408
variance, 45, 47, 135, 151, 208
    grouped, 48
    pooled, 396
Venn diagram, 93, 94

W

Welch's formula, 393
whiskers, 40

X

χ^2, 446

Z

z-scale, 210
z-score, 211
Z statistic, 320
z test, 333
zero correlation, 410